Acoustical Imaging

Volume 9

Visualization and Characterization

Acoustical Imaging

A Continuation Order Plan is available for this series. A continuation order will bring
delivery of each new volume immediately upon publication. Volumes are billed only upon
actual shipment. For further information please contact the publisher.

Acoustical Imaging
Volume 9
Visualization and Characterization

Edited by
Keith Y. Wang
University of Houston
Houston, Texas

PLENUM PRESS · NEW YORK AND LONDON

The Library of Congress cataloged the first volume of this series as follows:

International Symposium on Acoustical Holography.

   Acoustical holography; proceedings. v. 1-
New York, Plenum Press, 1967-

   v. illus. (part col.), ports. 24 cm.

Editors: 1967-   A. F. Metherell and L. Larmore (1967 with H. M. A. el-Sum)
Symposiums for 1967-   held at the Douglas Advanced Research Laboratories,
Huntington Beach, Calif.

   1. Acoustic holography—Congresses—Collected works. I. Metherell. Alexander A.,
ed. II. Larmore, Lewis, ed. III. el-Sum, Hussein Mohammed Amin, ed. IV. Douglas
Advanced Research Laboratories. v. Title.
QC244.5.I 5   69-12533

Library of Congress Catalog Card Number 69-12533
ISBN 0-306-40477-X

Proceedings of the Ninth International Symposium on
Acoustical Imaging, held in
Houston, Texas, December 3–6, 1979

© 1980 Plenum Press, New York
A Division of Plenum Publishing Corporation
227 West 17th Street, New York, N.Y. 10011

PREFACE

     This volume contains forty-six of the papers presented at
the Ninth International Symposium on Acoustical Imaging held
December 3-6, 1979, in Houston, Texas. The theme of the conference was the integration of applications technology. The major
objective of the conference was to promote interaction among
researchers working on different applications of acoustical
imaging.

     In addition to serving as a state-of-the-art research ref-
erence, this volume includes six tutorial review papers. For
convenience, all the papers are grouped under the following
headings: methods, transducers, processing and display, phased
array considerations, acoustic microscopy/non-destructive evaluation, reconstructive tomography and inversion techniques, tissue
characterization and impediography, medical applications, underwater applications, and seismic applications.

     The editor would like to thank the authors and the confer-
ence participants. The editor would also like to express his
appreciation for the assistance in evaluating the abstracts from
the following members of the Program Committee: Mahfuz Ahmed,
University of California Irvine Medical Center; Pierre Alais,
Paris University, France; C. B. Burckhardt, Hoffman-Laroche,
Basel,Switzerland; B. P. Hildebrand, Spectron Development Labora-
tories, Inc., Costa Mesa, California; Larry W. Kessler, Sonoscan,
Inc., Bensenville, Illinois; Rolf Mueller, University of Minnesota,
Minneapolis, Minnesota; John P. Powers, Naval Post Graduate School,
Monterey, California; Jerry L. Sutton, Naval Ocean Systems Center,
San Diego, California; F. L. Thurstone, Duke University, Durham,
North Carolina; Robert C. Waag, University of Rochester, Rochester,
New York; and Glen Wade, University of California, Santa Barbara,
California. The session chairmen, Mahfuz Ahmed, Pierre Alais,
Robert S. Andrews, C. B. Burckhardt, Gerald H. F. Gardner, B. P.
Hildebrand, Larry W. Kessler, Charles R. Meyer, Rolf Mueller,
John P. Powers, Gary Ruckgaber, Jerry Sutton, F. L. Thurstone,
and Glen Wade, were also instrumental in the success of the

vi   PREFACE

conference.   Special thanks are also due to Fay Wang for her
assistance in the compilation and preparation of the proceedings.

The Symposium was cosponsored by the University of Houston's
College of Engineering, College of Natural Sciences and Mathematics and Seismic Acoustics Laboratory and the Physics Program of
the Office of Naval Research under contract N00014-79-G-0038 in
cooperation with the IEEE Group on Sonics and Ultrasonics.

The Tenth International Symposium on Acoustical Imaging is
scheduled for October 13-16, 1980, in Cannes, France, under the
chairmanship of Dr. Pierre Alais and Dr. Alex F. Metherell.   The
Eleventh International Symposium on Acoustical Imaging is scheduled for April 6-8, 1981, in Monterey, California, under the chairmanship of Dr. John P. Powers.   The Twelfth International Symposium on Acoustical Imaging is being planned for 1982 in London or
Oxford, England, and will be organized by Dr. Eric A. Ash and
Dr. C. R. Hill.

Keith Wang

CONTENTS

RECONSTRUCTIVE TOMOGRAPHY AND INVERSION TECHNIQUES

TISSUE CHARACTERIZATION AND IMPEDIOGRAPHY

MEDICAL APPLICATIONS

ALTERNATIVE SCANNING GEOMETRIES FOR HIGH-SPEED

IMAGING WITH LINEAR ULTRASONIC ARRAYS *

B. P. Hildebrand

Spectron Development Laboratories, Inc.
3303 Harbor Blvd., Suite G-3
Costa Mesa, California 92626

S. R. Doctor

Battelle Northwest
Battelle Blvd.
Richland, Washington 99352

INTRODUCTION

A program entitled, "Development of an Ultrasonic Imaging System for the Inspection of Nuclear Reactor Pressure Vessels", was initiated in 1976 under the sponsorship of the Electric Power Research Institute (EPRI). One objective was the development of a high-speed system capable of inspecting two lineal feet of weld per minute according to ASME Section XI code requirements. A further objective was to provide high resolution images of significant flaws.

Battelle Northwest, the contractor, determined that the only way these objectives could be met, was to use a long linear array of piezoelectric elements to take advantage of high-speed electronic scanning. The array is mechanically scanned parallel to itself to perform the required inspection.

The code requires that the weld be inspected at five different angles, typically ±60°, ±45° and 0°. That is, the weld must be examined from both sides as well as the top. This is required be- cause flaws tend to lie along the weld line and thus may be oriented nearly vertically to the surface. Consequently, the examination of

* This work was sponsored by the Electric Power Research Institute
  under Contract RP606-1.

two lineal feet of weld requires that eight square feet of surface area be inspected (for a twelve inch thick pressure vessel).

The imaging system is designed to operate in two modes, 1) high-speed search and locate and 2) imaging of significant flaws by holography. Both were to be implemented with the linear array.

In this paper, we describe the evolution of the holographic mode, beginning with the simple substitution of electronic for mechanical scanning, to the final configuration of a line source in conjunction with a linear receiver array.

ANÁLYSIS

Pulse-Echo Holography

Figure 1 is the simplified geometric arrangement as first proposed and implemented. Two linear arrays were used in this configuration, one to serve as source and the other as receiver. The elements of the source array are energized in sequence and the return echos received on an adjacent receiver. The round-trip distance from source to object to receiver is:

$$D(x,y) = 2[(x-x_o)^2 + (y-y_o)^2 + z_o^2]^{1/2} \qquad (1)$$

If we set $D(x,y)$ equal to a constant, K, and square both sides of Equation 1, we obtain:

$$(x-x_o)^2 + (y-y_o)^2 = (\frac{K}{2})^2 - z_o^2 \quad . \qquad (2)$$

This is the equation of a circle centered on (x_o, y_o) of radius $[(\frac{K}{2})^2 - z_o^2]^{1/2}$. If holography is performed by interfering the returning signals with a reference signal that is equivalent to a plane wave normal to the x-y plane, then the interference fringes record loci of equal distance $K = (2z_o + n\lambda)$. Thus, the pattern on the x-y plane becomes a family of concentric circles of radius $[n\lambda(z_o + \frac{n\lambda}{4})]^{1/2}$. Such a family of circles constitutes a Fresnel ring system capable of focusing light and hence creating an image of the point object and is the fundamental basis of holography. Although other complicating factors arise, such as aberration and twin image overlap, these problems have been studied and adequately solved in various publications. Most of the problems, arising from the desire to use light in the image forming process, derive from the large differential in the wavelengths of sound and light. When a computer is used to form the image, it assumes the same wavelength, so that these problems do not arise.

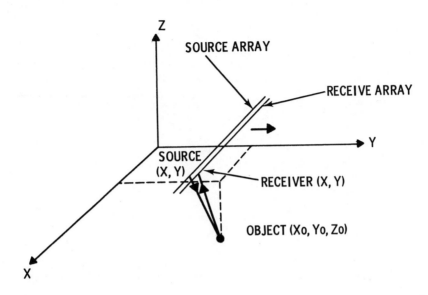

Figure 1.   Sketch Showing the Geometry of the Source Array/
            Receiver Array System.

When the system shown in Figure 1 was implemented it was found
that good results were obtained when the object was directly below
the aperture (0° longitudinal). However, when attempts were made
with shear waves to image an object at large angles from the perpendicular, the longitudinal reverberation between the metal surfaces
completely obscured the echo from the flaw. Therefore, the system
shown in Figure 2 was developed.

Point Source/Line Array

The source array was replaced by a single transducer focused on
the surface of the metal adjacent to the center of the receiver
array. This transducer can be mechanically tilted to provide illumination at large angles without also illuminating the back surface.
This prevents the reverberation problem referred to in the preceding
paragraph. The question was whether this arrangement would produce
an adequate image. Consider the simplified diagram shown in
Figure 3.

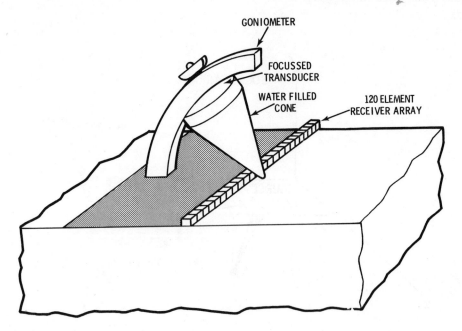

Figure 2.   Illuminating Transducer and Receiver Array
             Configuration.

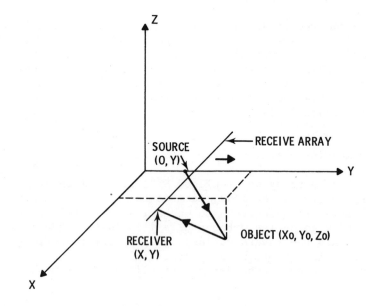

Figure 3.   Sketch Showing the Geometry of the Point
             Source/Receiver Array System.

The round-trip distance is:

$$D(x,y) = [x_o^2 + (y-y_o)^2 + z_o^2]^{1/2} + [(x-x_o)^2$$

$$+ (y-y_o)^2 + z_o^2]^{1/2} . \qquad (3)$$

Rearrangement of Equation 3 yields:

$$(x-x_o)^2 \left(\frac{D^2 + x_o^2}{2D^2} \right) - \frac{(x-x_o)^4}{4D^2} + (y-y_o)^2 =$$

$$\frac{D^4 + x_o^4 - 2D^2 x_o^2 - 4D^2 z_o^2}{4D^2} . \qquad (4)$$

By inspection of Figure 3, $D \geq (x_o^2 + y_o^2)^{1/2} + z_o$. Using the Fresnel approximation, $z_o \gg (x-x_o)$, the term in $(x-x_o)^4$ can be neglected yielding the result:

$$\frac{(x-x_o)^2}{a^2} + \frac{(y-y_o)^2}{b^2} = 1 \qquad (5a)$$

where

$$a^2 = \frac{D^4 - 4D^2 z_o^2}{2(D^2 + x_o^2)} \quad \text{and} \quad b^2 = \frac{D^2}{4} - z_o^2 . \qquad (5b)$$

This is the equation of an ellipse with major axis, a, and minor axis, b. Note that the ratio of major to minor axes is:

$$\frac{a}{b} = \sqrt{2} \left(\frac{D^2}{D^2 + x_o^2} \right)^{1/2} \qquad (6)$$

If the object lies under the center of the array ($x_o = 0$), then $a/b = \sqrt{2}$. For this case, a simple scale change $x' = x/\sqrt{2}$ results in the circularization of the ellipse. That is:

$$\left(x' - \frac{x_o}{\sqrt{2}} \right)^2 + (y-y_o)^2 = \frac{D^2}{4} - z_o^2 \qquad (7)$$

Setting $D = K = (x_o^2 + z_o^2)^{1/2} + z_o + n\lambda$, yields a family of concentric circles constituting a Fresnel zone pattern.

There are several differences between this method of gathering data and the holographic method. The first, of course, is the extraneous term in $(x-x_o)^4$. If the Fresnel approximation cannot be used, the ellipses will be distorted. Computer simulations have shown that this distortion is negligible for all practical situations.

The second major difference between this method of gathering data and the holographic method resides in Equation 6. The eccentricity of the ellipse is a function of the x-position of the object. Therefore, a simple $\sqrt{2}$ scale change will not circularize all ellipses in the aperture, but only those that are centered ($x_o = 0$). Although this effect is not likely to be large, it does introduce an unknown error and image distortion. Sometimes the flaw will not act like a diffuse scatterer, but rather like a specular reflector. This is particularly true of calibration blocks where flat-bottomed and side-drilled holes are used as reference standards. This type of flaw illustrates a serious deficiency with the arrangement described in this section. Namely, a flat-bottomed hole placed at $x_o > L/2$, where L is the length of the receiving array, will reflect all incident energy beyond the last array element. Thus, the effective aperture of the array in this case is only one-half the length of the array.

Experiments with this technique revealed the surprising result that the hologram of a side-drilled hole will look entirely different if the hole is aligned with the array than if it is perpendicular to it. Furthermore, the $\sqrt{2}$ correction will not make them similar. Thus, for such a cylindrical flaw, the system performance is not satisfactory.

Consider Figure 4a, where a line flaw is positioned parallel to the array. In this case the point of reflection is always halfway between the receiver element and the source. Hence, the pathlength becomes

$$D(x,y) = 2[(\tfrac{x}{2})^2 + (y-y_o)^2 + z_o^2]^{1/2} \tag{8}$$

Rearranging this equation yields

$$\frac{x^2}{D^2 - 4z_o^2} + \frac{(y-y_o)^2}{\dfrac{D^2 - 4z_o^2}{4}} = 1 \quad , \tag{9}$$

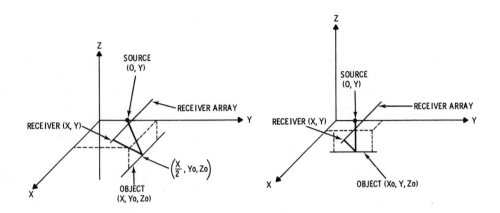

Figure 4. Geometry for Calculation of the Responses of the Point Source/Receiver Array System to (a) a Line Object Parallel to the Array and (b) Normal to the Array.

which is the equation of an ellipse with $a/b = 2$. The resulting family of ellipses obtained by setting $D = 2z_0 + n\lambda$ becomes a Fresnel pattern of high eccentricity.

If the flaw is aligned perpendicular to the array, the situation shown in Figure 4b pertains. In this case we obtain the equation

$$\frac{(x-x_o)^2}{(D-R_o)^2 - z_o^2} = 1 \tag{10}$$

where $R_o = (x_o^2 + z_o^2)^{1/2}$

This equation represents a one-dimensional Fresnel pattern when $D = R_o + z_o + n\lambda$, which is correct for a line object. Equation 9, however does not represent the proper pattern. Hence, for this type of flaw, the hologram is ambiguous.

For the reasons outlined above, the point source approach was re-evaluated. A new approach was considered and is described in the following section.

Line Source/Line Array

To overcome the limitations outlined in the preceding paragraphs, we proposed that the source consist of a focused line parallel and adjacent to the array as shown in Figure 5. This configuration preserves the capability for illuminating off-axis flaws without inducing reverberation. Obviously, it also rectifies the aperture limitation of the previous system. As shall be shown, it also eliminates the field distortion problem.

Consider Figure 6. Since the source is a line, the distance from the source to the object point will always be the shortest path, as shown. The round-trip distance from source to object to receiver array will, therefore, be:

$$D(x,y) = [(y-y_o)^2 + z_o^2]^{1/2} + [(x-x_o)^2 + (y-y_o)^2 + z_o^2]^{1/2} \tag{11}$$

When this equation is reduced, it becomes:

$$\frac{(x-x_o)^2}{2} - \frac{(x-x_o)^4}{4D^2} + (y-y_o)^2 = \frac{D^2}{4} - z_o^2 \ . \tag{12}$$

When the Fresnel approximation is applied, the result is the equation:

$$\frac{(x-x_o)^2}{\left(\dfrac{D^2 - 4z_o^2}{2}\right)} + \frac{(y-y_o)^2}{\left(\dfrac{D^2 - 4z_o^2}{4}\right)} = 1 \tag{13}$$

By inspection of Figure 6 we find $D \geq 2z_o$. Note that, setting $x' = x/\sqrt{2}$ gives the result:

$$\left(x' - \frac{x_o}{\sqrt{2}}\right)^2 + (y-y_o)^2 = \frac{K^2}{4} - z_o^2 \tag{14}$$

where

$$K = 2z_o + n\lambda \ .$$

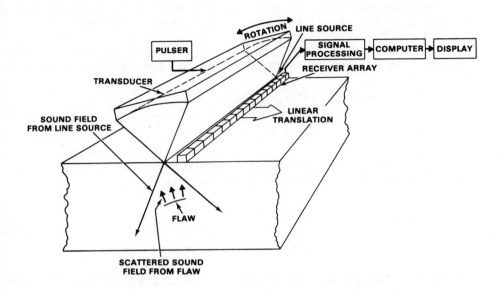

Figure 5.   Sketch of Line Source and Line Array.

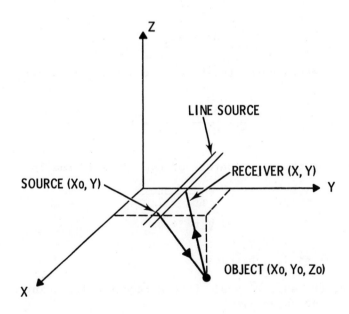

Figure 6.   Sketch Showing the Geometry of the Line Source/
            Receiver Array System.

Comparing Equations 14 and 2 we find exact equivalence. Again, the effect of violating the Fresnel approximation is negligible.

Figure 7a and 7b repeat the case of a line flaw. With the line parallel to the array as shown in Figure 7a, the resulting expression is

$$\frac{(y-y_o)^2}{(\frac{D}{2})^2 - z_o^2} = 1 \quad .$$ (15)

When the flaw is perpendicular to the array, as shown in Figure 7b the equation describing the pattern is

$$\frac{(x-x_o)^2}{(D-z_o)^2 - z_o^2} = 1 \quad .$$ (16)

Hence, both expressions represent one-dimensional Fresnel patterns. We must, however, examine the fringe spacing to get an idea of scale. In the first case (Equation 15), we set $D = 2z_o + n\lambda$, with the result

$$\frac{(y-y_o)^2}{\frac{n\lambda}{4} (n\lambda + 4z_o)} = 1 \quad .$$ (17)

In the second case (Equation 16), $D = 2z_o + n\lambda$, yielding

$$\frac{(x-x_o)^2}{n\lambda (n\lambda + 2z_o)} = 1 \quad .$$ (18)

Comparing the denominators of Equations 17 and 18 we have the ratio

$$2\sqrt{\frac{n\lambda + 2z_o}{n\lambda + 4z_o}} \quad ,$$

which for, $n\lambda \ll 2z_o$, becomes $\sqrt{2}$.

Hence, the same $\sqrt{2}$ scale change required for point objects also serves for line objects.

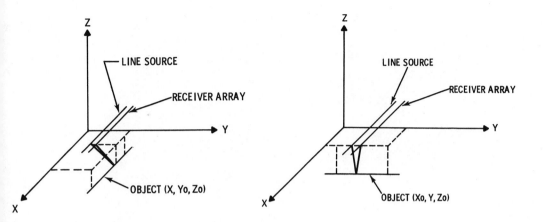

Figure 7. Geometry for Calculation of the Response of the Line
           Source/Receiver Array System to (a) a Line Object
           Parallel to the Array and (b) Normal to the Array.

SIMULATION

     The various data gathering configurations previously discussed
were simulated on the computer for verification. The exact equations were programmed without the Fresnel approximations used to
illuminate the discussion. Figure 8a shows the Fresnel ring pattern
to be expected from the holographic system with wavelength $\lambda = 0.1$,
and two object points at $(0,0,2)$ and $(-2,0,2)$. Figure 8b shows a
similar plot with object points at $(0,0,2)$ and $(-2,0,10)$. Note that
perfectly circular fringes occur, no matter where the object point
is situated.

     Figures 9a and 9b show the resulting fringes when a point source
and the same object points are used. Note, the slight variation in
fringe pattern as a function of object position. To emphasize this,
Figures 10a and 10b are the circularized versions of Figures 9a and
9b. The scale on x was changed by $\sqrt{2}$ as discussed. These figures
clearly indicate the non-circularity of the fringes for points not
at the center of the aperture. Actually, a careful measurement shows
that neither of the patterns is perfectly circular. The pattern at
the origin is slightly flattened and the offset pattern is slightly
lengthened in the y-direction. This is due to the fact that we are
working in the near field where the $(x-x_o)^4$ term has some influence.

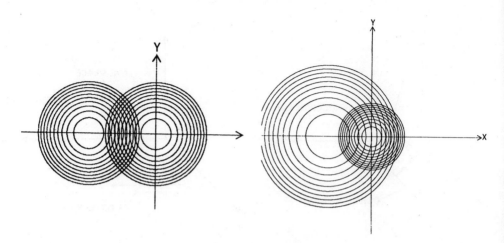

Figure 8.   Simulated Fringe Pattern from Two Point Objects at
            (a) (0,0,2) and (-2,0,2) and (b) (0,0,2) and (-2,0,10)
            Using the Source Array/Receiver Array System.

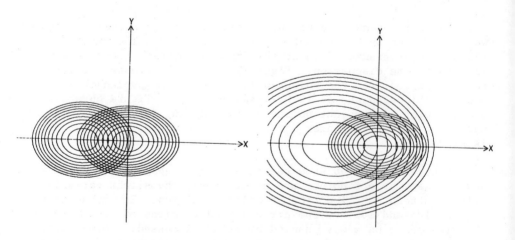

Figure 9.   Simulated Fringe Pattern from Two Point Objects at
            (a) (0,0,2) and (-2,0,2) and (b) (0,0,2) and (-2,0,10)
            Using the Point Source/Receiver Array System.

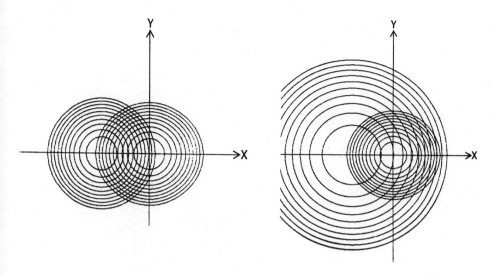

Figure 10.   Fringe Patterns of Figure 9 Circularized by a
             $\sqrt{2}$ Scale Change in x.

Finally, Figures 11a and 11b show the fringe patterns obtained
with the line source system. The scaled versions are shown in
Figures 12a and 12b, showing excellent agreement with the holographic
mode. We conclude that the line source system is an excellent compromise between practicality and optimum performance. Again, careful
measurement shows the lack of perfect circularity. This is due to
the fact that we are not in the Fresnel region. Now, however, both
patterns are flattened equally and presumably a scale factor somewhat different than $\sqrt{2}$ could be used to achieve perfect fringe circularity for objects at the same depth. Their scale factor would
change for different depths, so the pulse echo data could be used to
establish it. However, it is doubtful that the image would exhibit
any effect due to this slight deviation from fringe circularity.

For completeness, we present simulations of the response to the
line reflector or object in Figure 13. Figures 13a and 13b show
the Fresnel patterns obtained from a line object at (x,0,2) and
(0,y,2) respectively, using the point source. Figure 14 shows the
result when the line source is used.

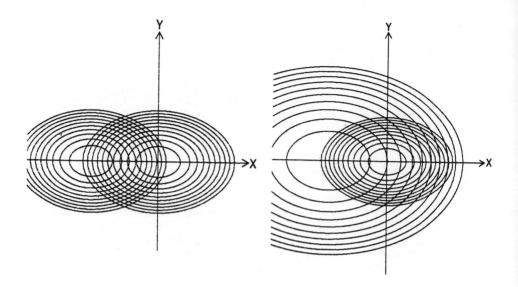

Figure 11.   Simulated Fringe Pattern from Two Point Objects at
              (a) (0,0,2) and (-2,0,2) and (b) (0,0,2) and (-2,0,10)
              Using the Line Source/Receiver Array System.

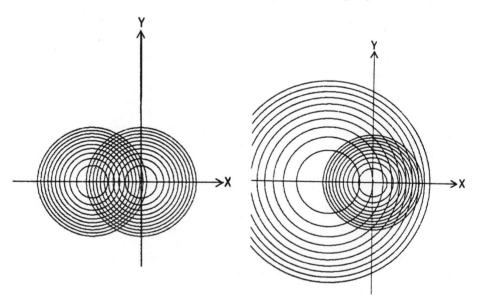

Figure 12.   Fringe Patterns of Figure 11 Circularized by a
              $\sqrt{2}$ Scale Change in x.

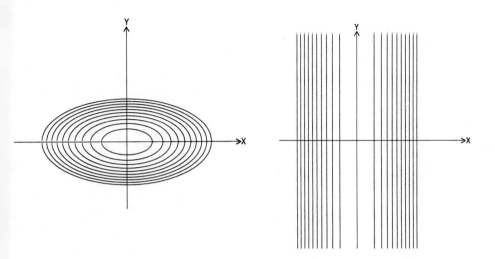

Figure 13.   Simulated Fringe Patterns Obtained with the Point
             Source/Receiver Array System for the Line Object
             (a) Parallel to the Array and (b) Normal to the Array.

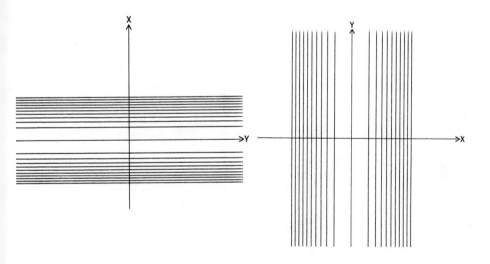

Figure 14.   Simulated Fringe Patterns Obtained with the Line
             Source/Receiver Array System for the Line Object
             (a) Parallel to the Array and (b) Normal to the Array.

EXPERIMENTAL VERIFICATION

The theoretical considerations discussed in the preceding pages were verified experimentally. Figure 15 shows a photograph of the point source/line array arrangement. Figure 16 shows an image of a row of flat-bottomed holes obtained by making a hologram with the data and reconstructing optically. This image exhibits good resolution, and since the total spread of the holes is only 50 mm, (2.0 inches) the aperture was adequate. The array had to be centered quite carefully in order to get data from all the holes. This simply illustrates the reduced effective aperture of this system.

In order to illustrate the problem of dealing with side-drilled holes, data was taken by scanning across the hole in orthogonal directions. Figure 17a shows the fringes obtained when the hole was parallel to the array and Figure 17b shows them when the hole was perpendicular to the array. Note the completely different shape of the fringe system in the two cases. The reason for not having a completely one-dimensional fringe system in Figure 17b is that the hole ended near the center of the array. This demonstrates the orientation sensitivity, at least for cylindrical flaws, of the point source method of gathering data.

Figure 15. Point Source/Line Array Arrangement.

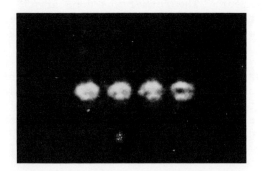

Figure 16. Photograph of the Image of Four Flat-Bottomed Holes
            Made with the Point Source/Receiver Array System
            and Reconstructed Optically.

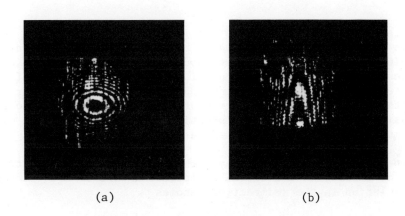

(a)   (b)

Figure 17. Fringe Pattern Obtained with the Point Source/
            Receiver Array System. The Object was a Side-
            Drilled Hole with Scan Direction (a) Perpendicular
            and (b) Parallel to the Hole.

Figure 18 is a photograph of the line source/linear array head
that was used to verify the concept. The line source was not as
long as the array, but it was adequate to demonstrate the method.
Figures 19a and 19b demonstrate the quality of the images obtained
with this arrangement. The former was done at 45° shear by tilting
the line source transducer by the appropriate angle. The latter is
the image of Flaw #1 in the Babcock and Wilcox test block with 0°
longitudinal illumination.

The superiority of the line source for cylindrical flaws is
clearly demonstrated by the holograms shown in Figure 20. These
fringe patterns result from scanning a side-drilled hole in both
directions. Note, that with this system, the pattern remains invar-
iant to the orientation of the hole.

RESOLUTION

Up to this point, the resolving capability of the various
imaging modalities has not been discussed. Since analyses of imaging
systems are well documented, we will only discuss them in terms of
the several scanning systems described in previous paragraphs. The
standard pulse-echo holographic arrangement is capable of resolving
point targets separated by

$$\delta = \frac{\lambda z_o}{2L}$$

where the length of scan equals the length of the array, L. This
formula can be derived in two ways; by calculating the response of
two point targets separated just enough that two distinct points are
visible in the image or by calculating half the separation of the
two outermost rings of the Fresnel pattern recorded by the array.
The latter method of deducing resolution is most pertinent to this
discussion since it allows an intuitive understanding of the differences in the modalities discussed earlier.

The point source/line array method will therefore have two
values of resolution due to the elliptical nature of the fringes.
The resolution in the x-direction becomes

$$\delta_x = \frac{\lambda z_o}{L}$$

and in the y-direction

$$\delta_y = \frac{\lambda z_o}{2L}$$

Figure 18.   Photograph of the Line Source/Receiver
             Array System.

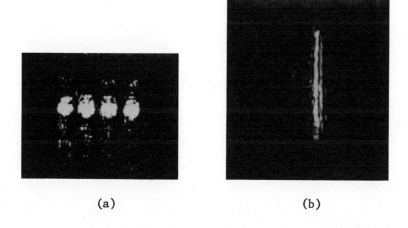

(a)   (b)

Figure 19.   (a) Photograph of the Image of Four Flat-Bottomed
                 Holes Made with the Line Source/Receiver Array
                 System and then Reconstructed by Computer.
             (b) Photograph of the Image of Flaw #1 in the Babcock
                 and Wilcox Test Block Reconstructed Optically.

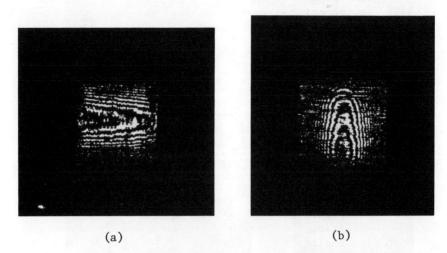

(a)   (b)

Figure 20.   Fringe Pattern Obtained with the Line Source/Receiver
             Array System.   The Object was a Side-Drilled Hole
             with Scan Direction (a) Perpendicular and (b) Parallel
             to the Hole.

Similarly, the line source/line array system has the same
unequal resolution in the two dimensions.   Thus, in the direction
of scan, the resolution is the same as for the standard pulse-echo
method, but in the array direction only half as good.   However, for
most applications, this is not a problem since resolution is still
very good.   As an example, suppose λ = 2.5 mm (0.1 in.), z_0 = 250 mm
(10 in.), and L = 150 mm (6 in.), then δ_x = 4 mm (0.17 in.) and
δ_y = 2 mm (0.085 in.).   This is certainly adequate for most applications.

CONCLUSION

During the course of this program the system evolved from one
using standard holographic techniques to one specifically tailored
to the linear array.   The evolution was forced, and guided by practical problems encountered in trying to use the array in ways not
wholly suitable.

The first impediment to using the array as a sequencing source
and receiver was the marginal energy available from the small array
elements.   This was alleviated by combining three elements in
parallel.

The next problem arose when attempting to image objects at shear angles. Since the small elements radiate in all directions, reverberation between the array and the back surface of the metal precluded the detection of small signals from the off-axis flaw. This was solved by using a focused source placed at the center of the receiving array.

After experimentation with the point source configuration, several problems of a practical nature were discovered. These were reduced field of view and flaw orientation sensitivity. Both these problems were solved by using a rectangular transducer, as long as the array, focused to a line adjacent and parallel to the receiver array. By mechanically rotating the transducer about its focus, a wedge of sound can be steered to obtain shear waves in the desired direction. This configuration also expands the field of view to the length of the array and solves the orientation sensitivity problem.

The final result is a practical, high-speed imaging system capable of utilizing the major advantage of arrays, namely, fast electronic sequencing. The line source provides a mechanically controllable, high amplitude source of sound. The F-number of the source, and its tilt with respect to the surface, control the region of insonification to illuminate only the flaw and thus prevent reverberation. The speed of data acquisition is thus dependent entirely on the velocity of sound in the material and the distance to the flaw.

ACKNOWLEDGEMENTS

The research program described in this paper was a joint effort by numerous colleagues whom we wish to acknowledge. They are F. L. Becker, V. L. Crow, T. J. Davis, D. K. Lemon, G. J. Posakony and T. P. Harrington of Battelle, and G. J. Dau and K. Stahlkopf of EPRI.

ACOUSTICAL IMAGING BY MEANS OF MULTI-FREQUENCY

HOLOGRAM MATRIX

T.Miyashita, J.Nakayama, and H.Ogura

Departments of Electrical Engineering and Electronics,
Kyoto Institute of Technology
Matsugasaki, Sakyo-ku, Kyoto 606, Japan

ABSTRACT

A multi-frequency holographic imaging method has been developed, which is an advanced application of the hologram-matrix imaging method suggested and developed by the authors. This new method is based on the idea that axial resolution as well as transverse one can be obtained by properly synthesizing transversely focused beams of different frequencies. This idea has been included into the hologram-matrix imaging method. An image is reconstructed by a linear transformation from hologram matrices of different frequencies. Grating-lobe artifacts usually encountered in the multi-frequency imaging methods are eliminated using nonuniformly spaced frequencies. Thus an imaging of objects placed in the near field by a holographic synthetic aperture method has become practical with a small transducer array and simple electronics.

A design way of transducer arrays and frequency series for this imaging method is introduced in this paper. Three examples of the design are evaluated in their point-spread functions and imaging abilities from simulated hologram matrices of a multi-point object. The transmitter is of one element, the receiver is composed of 31 or 37 elements, and the number of frequencies is 40.

Experimental results also shows that this method is promising. Measurement of the multi-frequency hologram matrices can be done within 30 ms. It may take only few minutes to reconstruct an image of 80×80 pixels by a high-speed mini-computer.

INTRODUCTION

Since the invention of the HISS radar[1], various types of imaging
methods by means of hologram matrix have been introduced by the
authors. Especially two-dimensional acoustical imaging has been
investigated theoretically and experimentally. The hologram-matrix
imaging method first measures a complex "hologram matrix", each
element of which is the complex amplitude of the field detected by a
receiver element scattered from an object illuminated by a transmitter element, and then reconstructs images numerically transforming
the matrix. In the first type, both transmitter and receiver arrays
were placed on a line. The axial resolution was poor.

In the second type, we introduced a method[2] in which a transmitter array and a receiver array were placed in the imaging plane
so that their synthesized beams may intersect with each other at
right angles at the center of the image region. Experiments[3] showed
that this method gives good two-dimensional resolution and a high S/N
ratio. It has, however, one demerit that the direct acoustical
coupling between the transmitter and the receiver must be excluded
by an acoustical absorber placed between them, although the electrical coupling is easily kept away with the electrical shield. This
may limit somewhat practical utility of this method.

The third type, multi-frequency method, removes the above
difficulties, *i.e.*, poor axial resolution in the first type and
necessity of placing an acoustical absorber in the second type.
First, we made imaging placing the transmitter and receiver arrays,
both of which are composed of 37 elements, on a line as in the first
type, but using five different frequencies. The experiment[4] gave
images of good resolution and a high S/N ratio. Second, we got an
idea that one of the transmitter and receiver arrays in the second
type can be replaced by a frequency series. The best way will be a
combination of a receiver array, one transmitter element and a
frequency series, which makes high-speed measurement of the hologram
matrix possible without high voltage- and high speed-switches otherwise necessary for a transmitter array. Experiments of imaging in
microwave and acoustical wave based on a similar idea were made
independently by Karg *et al.*[5] One should take caution, however, in
the design of the frequency series, because any uniformly spaced
frequency series produces periodic ambiguity, just like the grating
lobes of antenna array, in the half-infinite axial direction. In
this paper, nonuniformly spaced series are suggested and applied to
avoid the ambiguity. They give also the minimum point-spread for a
given side lobe level with a uniform weight.

Synthetic aperture techniques developed in the microwave and
seismology have been applied to the acoustical imaging by Johnson *et
al.*[6] named "synthetic focusing", which uses acoustical pulse. Our

method uses essentially continuous wave, although only a few hundred cycles of sinusoidal wave is transmitted in practice. This makes the hardware of the imaging system very simple and stable. Furthermore the software for the image reconstruction process becomes simple and has much versatility.

This multi-frequency method will be easily developed into a three-dimensional imaging with a planar transducer array as is discussed in the last section. This paper is, however, limited to a two-dimensional imaging, *i.e.*, tomography, to begin with.

FREQUENCY-SPREAD HOLOGRAPHIC IMAGING

As shown in the next section, frequency series, receiver-element positions and amplitude weights applied to the receiver elements and the frequency components are designed from continuous aperture and frequency distributions. Therefore, at first, the continuous distributions should be determined to satisfy the required resolution and S/N ratio from an investigation of the point-spread function. A hologram of a point target placed at α $(x_\alpha, 0, z_\alpha)$, as shown in Fig.1, is

$$H_\alpha(r,f) = G_R(R_{\alpha rf})G_T(T_{\alpha tf})\exp[-2\pi i(R_{\alpha rf}+T_{\alpha tf})] \ , \tag{1}$$

where r and t are x-coordinates of the receiver aperture and the transmitter, respectively, f frequency at which the hologram is made, $R_{\alpha rf}$ and $T_{\alpha tf}$ distances in terms of wavelength from the point target to the receiver and the transmitter, respectively, $G_R(R)$ and $G_T(T)$ decay factors due to wave divergence and absorption by the medium. An image reconstruction formula from the hologram $H(r,f)$ is

$$I(\mu) = |\int\int A_R(r)A_F(f)F_R(R_{\mu rf})F_T(T_{\mu tf})H(r,f)$$
$$\times \exp[i\{2\pi(R_{\mu rf}+T_{\mu tf})+\gamma(r)+\eta(f)\}]drdf|^2 \ , \tag{2}$$

where μ is a pixel of the reconstructed image, $A_R(r)$ and $A_F(f)$ aperture and frequency distribution functions, $\gamma(r)$ and $\eta(f)$ their phase distributions. $F_R(R)$ and $F_T(T)$ compensate the decay in the wave propagation. Substituting Eq.(1) into Eq.(2), we obtain

$$I(\mu,\alpha) = |\int \mathring{A}_F(f)R(\mu,\alpha|f)T(\mu,\alpha|f)df|^2 \ , \tag{3}$$

where

$$R(\mu,\alpha|f) = \int \mathring{A}_R(r)G_R(R_{\alpha rf})F_R(R_{\mu rf})\exp[i2\pi(R_{\mu rf}-R_{\alpha rf})]dr \ , \tag{4}$$

$$\mathring{A}_F(f) = A_F(f)\exp[i\eta(f)] \ , \qquad \mathring{A}_R(r) = A_R(r)\exp[i\gamma(r)] \ ,$$

$$T(\mu,\alpha|f) = G_T(T_{\alpha tf})F_T(T_{\mu tf})\exp[i2\pi(T_{\mu tf}-T_{\alpha tf})] \ .$$

General investigation of this near field point-spread function is not easy. We consider here the behavior of this function in the vicinity of the target. The outer surrounding region is taken into consideration in the next section. This function is the same as the field distribution when the synthesized beam is controlled to make a focus at α. Therefore, it will be also referred to as focus pattern hereafter. In the vicinity of the focus, $R(\mu,\alpha|f)$ can be simply described as

$$R(\mu,\alpha|f) = B(\bar{f}\bar{x}_\mu, 0, \bar{f}\bar{z}_\mu)\exp[i\Theta(\bar{f}\bar{x}_\mu, 0, \bar{f}\bar{z}_\mu)] \ , \tag{5}$$

where $\bar{f} = f/f_c$, $\bar{x}_\mu = x_\mu/\lambda_c$, $\bar{z}_\mu = z_\mu/\lambda_c$, f_c and λ_c are the center frequency and wavelength in the medium. If the focus point α is placed at $(0,0,z_\alpha)$ on the axis of the receiver aperture, the phase factor $\Theta(0,0,\bar{f}\bar{z}_\mu)$ can be approximated by a linear function of the position

$$\Theta(0,0,\bar{f}\bar{z}_\mu) = 2\pi\gamma_R(\bar{z}_\mu - \bar{z}_\alpha)\bar{f} \ , \tag{6}$$

where $\gamma_R = \lambda_c/\lambda_c' < 1$, λ_c' is the effective wavelength, and γ_R is a reduction factor. Furthermore, the amplitude factor $B(0,0,\bar{f}\bar{z}_\mu)$ can be assumed to be independent of frequency. Then the focus pattern is composed of two factors:

$$I(\mu,\alpha) = |R_c(\mu,\alpha)|^2|F(\mu,\alpha)|^2 \ , \tag{7}$$

where

$$F(\mu,\alpha) = \int \dot{A}_F(f_c\bar{f})\exp[2\pi i(\gamma_R + \gamma_T)(\bar{z}_\mu - \bar{z}_\alpha)\bar{f}]d\bar{f} \ , \tag{8}$$

and further, $R_c(\mu,\alpha)$ can be approximated by

$$R_c(\mu,\alpha) = \int \dot{A}_R'(\lambda_c\bar{u})\exp[-2\pi i\bar{x}_\mu\bar{u}]d\bar{u} \ , \tag{9}$$

where
$$\bar{u} = \bar{r}/\sqrt{\bar{r}^2 + \bar{z}_\alpha^2} \ ,$$

$$\dot{A}_R'(\lambda_c\bar{u}) = A_R(\lambda_c\bar{r})\{(\lambda_c\bar{r})^2 + (\lambda_c\bar{z}_\mu)^2\}^{3/2}/(\lambda_c\bar{z}_\mu)^2 \ . \tag{10}$$

Both factors are represented as Fourier transforms of $A_F(f_c\bar{f})$ and $A_R(\lambda_c\bar{u})$, respectively.

Various antenna illumination functions developed to obtain the specified far field patterns can be adopted in this case. We choose Taylor illumination functions. Let β_R and β_F be the full width at half maximum intensity of the main beams of the illumination functions adopted to the receiver aperture and the frequency distribution, respectively, then the aperture length of the receiver is given by

$$L_R = 2z_\alpha\tan(\sin^{-1}\beta_R/2w) \ , \tag{11}$$

and the frequency bandwidth is given by

$$L_F = 2f_c\delta = f_c\beta_F/2w \ ,\tag{12}$$

where w is the width of the two-dimensional focus in terms of wavelength. Simply, it is assumed that $\gamma_R = \gamma_T = 1$.

MULTI-FREQUENCY HOLOGRAM MATRIX IMAGING

In practice, the receiver-aperture and frequency distributions discussed in the previous section are realized by a finite number of receiver elements and frequency components as shown in Fig.1(b). The continuous distributions are transformed into amplitude-weighted and phase-controlled receiver element array and frequency series. In the transformation, the synthesized field distribution near the focus point is kept unchanged. Our attention must be turned to the surrounding noise due to the remainder of the grating lobes in the

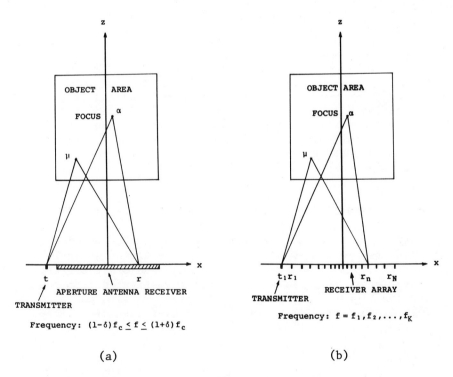

(a)   (b)

Fig. 1. (a) Frequency-spread holographic imaging.
(b) Multi-frequency hologram matrix imaging.
The transmitter may be placed at the center of the
receiver aperture or array.

near field. In the experiments, sinusoidal wave of short duration
is transmitted. Therefore, the object region relevant to the measured hologram is limited near the receiver array. However, the object
region is usually so large that some means of suppressing the grating
lobe noise is necessary. We adopt nonuniformly spaced frequency
series.

Frequencies f_k $(k = 1, 2, \ldots, K)$ and their amplitude weights A_{Fk}
are related with the continuous distribution $A_F(f)$ by

$$A_{Fk} = \int_{\nu_{k-1}}^{\nu_k} A_F(f) df \, , \qquad k = 1, 2, \ldots, K \, , \qquad (13)$$

where $\nu_k = (f_k + f_{k+1})/2$, $\nu_0 = (1 - \delta) f_c$, $\nu_K = (1 + \delta) f_c$. From these
equations, either amplitude-weighted and uniformly-spaced frequency
series or uniformly-weighted but nonuniformly-spaced series can be
designed. Similarly A_{Rn} and u_n $(n = 1, 2, \ldots, N)$ are determined for the
receiver array.

Then the image reconstruction formula becomes

$$I(\mu) = \left| \sum_{n=1}^{N} \sum_{k=1}^{K} A_{Rn} A_{Fk} F_R(R_{\mu nk}) F_T(T_{\mu tk}) H_{nk} \right.$$
$$\left. \times \exp[i\{2\pi(R_{\mu nk} + T_{\mu tk}) + \alpha_{nk} + \gamma_n + \eta_k\}] \right|^2 , \qquad (14)$$

where "hologram matrix" H_{nk} is the complex amplitude detected by the
n-th receiver element when an object is illuminated by the transmitter at the k-th frequency, and α_{nk}'s are introduced to compensate
nonuniform phase characteristics of the receiver elements and transmitter. Amplitude compensation may be included into A_{Rn} and A_{Fk},
if effective.

THREE EXAMPLES OF RECEIVER ARRAY AND FREQUENCY SERIES

We designed three representative examples of receiver array
and frequency series, whose specifications are shown in Table 1.
Their fundamental imaging characteristics and S/N ratios of their
focus patterns over a wide region are investigated from their point-
spread functions. Their imaging abilities are also estimated by the
images reconstructed from simulated multi-frequency hologram matrices.

Example I

The average spacing of the receiver elements is larger than the
wavelength, but their positions are distributed nonuniformly. The
point-spread function is calculated for a point-target placed at α
$(0, 0, 87\lambda_c)$. Its transverse and axial cross sections through α are

Table 1. Specification of the design examples.

$A_R(u)$: Amplitude illumination, L_R: Aperture length,
N: Number of receiver elements, w: Full width at half maximum intensity,
$A_F(f)$: Frequency distribution, L_F: Bandwidth, K: Number of frequencies.

Example	$A_R(u)$	L_R (λ_c)	N	Spacing (λ_c)	Type of Spacing	Type of Weight	w (λ_c)
I	Taylor -30dB ($\bar{n}$=5)	59.2	31	1.91 (average)	non-uniform	uniform	1.75
II	Taylor -30dB ($\bar{n}$=5)	59.2	37	1.6	uniform	non-uniform	3.10
III	Taylor -30dB ($\bar{n}$=5)	34.6875	37	0.9375	uniform	non-uniform	3.10

Example	$A_F(f)$	L_F (f_c)	K	Average Spacing (f_c)	Type of Spacing	Type of Weight
I	Taylor -25dB ($\bar{n}$=5)	0.3012	40	7.53×10^{-3}	non-uniform	uniform
II	Taylor -30dB ($\bar{n}$=5)	0.182	40	4.55×10^{-3}	non-uniform	uniform
III	Taylor -30dB ($\bar{n}$=5)	0.182	40	4.55×10^{-3}	non-uniform	uniform

shown in Fig.2. Two first-order grating lobe positions determined
from the average frequency spacing are $(0,0,21\lambda_c)$ and $(0,0,153\lambda_c)$.
As shown in Fig.2(b), those grating lobes are suppressed below -25
dB in front of the receiver and -20 dB far from it by nonuniformity
of the frequency spacing. A wide perspective view is also shown in
Fig.3 over an area of $158 \times 158 \ \lambda_c^2$. Remainder of the grating lobes
makes arc-shaped noise across a wide region, but it is relatively
low and not localized. A reconstructed image from a simulated
hologram matrix of seventeen point-targets placed on a circle with

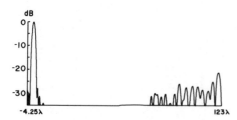
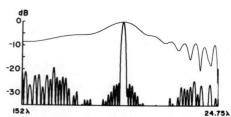

(a) Transverse cross section.   (b) Axial cross section.

Fig. 2. Point-spread function of Example I.
        Cross sections through the target point α $(0,0,87\lambda_c)$.
        A smooth curve superimposed in Fig.(b) is the point-spread
        function of a single frequency f_c.

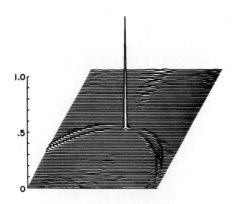

Fig. 3. Wide perspective view   Fig. 4. Perspective view of a
        of the point-spread   reconstructed image from
        function of Example I.   a simulated hologram
        (Area of view: $158 \times$ matrix. (Example I)
        $158 \ \lambda_c^2$)   Number of the point-
                                               targets is seventeen.

a diameter of 46.6 λ_c and at its center is shown in Fig.4. It is of
good quality without any remarkable artifact, although it is accompanied by relatively small background noise.

Example II

The receiver elements are placed with a uniform spacing of 1.6
λ_c which is longer than the wavelength. Remainders of the grating
lobes of the receiver array are somewhat localized in the point-
spread function, as shown in Fig.5. The point-target is placed at
$(0,0,100\lambda_c)$. However, the function is very clear within the object
region $(80 \times 80 \ \lambda_c^2)$. Consequently, an image of a circle with a
diameter of 46.6 λ_c reconstructed from a simulated hologram matrix
is very clear with   no noise inside the circle as shown in Fig.6,
although artifacts due to grating lobes appear outside the circle.
Therefore, this example will be usefull when no objects are located
outside the object region.

Example III

The spacing of the receiver elements is a little less than the
wavelength. In the point-spread function shown in Fig.7, where the
point target is placed at $(0,0,120\lambda_c)$, remainders of the grating
lobes of the frequency series appears far from the receiver array
along the axis. However, it should be noticed that the area of Fig.7

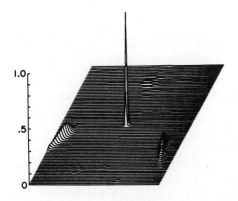

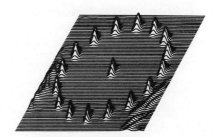

Fig. 5. Wide perspective view
     of the point-spread
     function of Example II.
     (Area of view: 158
     $\times 158 \ \lambda_c^2$)

Fig. 6. Perspective view of an
     image reconstructed from
     a simulated hologram
     matrix of seventeen
     point-targets.
     (Example II)

 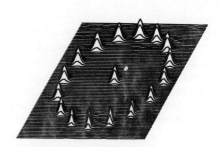

Fig. 7. Wide perspective view   Fig. 8. Perspective view of an
        of the point-spread   image reconstructed from
        function of Example III.   a simulated hologram matrix
        (Area of view: 237   of seventeen point-targets.
        × 237 λ_c^2)   (Example III)

is $240 \times 240 \ \lambda_c^2$. The first-order grating-lobe positions determined
from the average frequency spacing are $(0,0,10\lambda_c)$ and $(0,0,230\lambda_c)$.
Therefore, an image shown in Fig.8 reconstructed from a calculated
hologram matrix of a circle with a diameter of 70 λ_c is very clear
over all. These designed parameters will be usefull for higher
frequencies than those for the previous two examples.

EXPERIMENTAL RESULTS

Experimental investigations were made with acoustical wave.
Several objects composed of thin strings were imaged in a water tank
using the parameters of Example II of the previous section.

Specification of the Frequency Series and Transducer Array

We chose the center frequency f_c = 1 MHz, then λ_c = 1.5 mm in
water. According to Example II, the frequency series was determined
with nonuniform spacing. f_1 = 0.9146 MHz,..., f_{20} = 0.9985 MHz, f_{21}
= 1.0015 MHz,..., f_{40} = 1.0854 MHz. The maximum frequency spacing is
0.0104 MHz, the minimum is 0.0030 MHz, and the average is 0.00455
MHz.

The effective aperture length of the transducer array is 88.8
mm and the element spacing is 2.4 mm. Each element has a rectangular

aperture of 0.6 mm × 25 mm in order to have a relatively uniform element factor in the image plane, but sharply collimate in the direction perpendicular to the image plane. We made the transducer array cutting a $100 \times 25 \times 1.3$ mm^3 PZT plate of uniform characteristics attached on a flat surface of a backing material by a precise cutting machine. The average of resonance frequencies of the 37 elements is 1.230 MHz, their standard deviation is 0.014 MHz, and the maximum deviation is 0.037 MHz. These results show how this transducer array has uniform characteristics. An equivalent circuit of each transducer at its resonance frequency is parallel connection of 1 kΩ and 140 pF. Besides, we operated it below the resonance frequencies in order to get more additional uniformity in its frequency dependence. Furthermore, mutual coupling between adjacent elements is -40 dB even at the worst in the relevant frequency range.

We tried to measure overall electrical and acoustical characteristics of each transducer detecting uniform acoustical field in the water tank. Standard deviations of the sensitivities and phase shifts of the 37 elements are 1.1 dB and 10°, respectively, at 1 MHz. These data will not be so reliable, but we may expect that compensation for the nonuniformity of the phase characteristics will improve the reconstructed image. It is very easy to include the phase compensation because it is done in the image reconstruction process by a software.

Operation Mode and Characteristics of the Electrical Circuit

Measurements of the hologram matrices are made automatically controlled by a mini-computer. Various measurement modes and diagnostic modes are possible. Here we restrict ourselves to the measurement mode for the experimental results shown below. Sinusoidal electric signal of a duration of 330 cycles is applied to acoustic transmitter element, then the electrical signal obtained by a receiver element is decomposed into in-phase and quadrature components with respect to the electrical signal applied to the transmitter and integrated only for 40 cycles from 70 cycles after the end of the transmit. The results are digitized and transferred into the computer memory. These timing parameters depend on the region of the interest. Usually overall time from the transmission of the acoustic wave to the transfer of the results into the memory is fairly shorter than 500 μs. Magnitude of the electrical signal applied to the acoustic transmitter is as low as 4 V p-p in the experiments.

Electric characteristics of the measurement circuit of the in-phase and quadrature components, composed of a preamplifier, double-balanced mixers, integrators, sample and hold amplifiers, and analog to digital converters, were measured by changing the phase of the input signal over 720°. Calculated in amplitude and phase, the

standard and maximum deviations of the amplitude are 0.1 dB and 0.3
dB, respectively, and those of the phase are 1.8° and 4.0°. These
results are fairly satisfactory.

Imaging Experiments

"Point target" in the preceding sections means in a two-dimensional space, xz plane. All the test targets are uniform, $i.e.$, long
enough, in the y direction. Copper strings with a diameter of 0.4 mm
were strung up in parallel with each other making various cross
sections, line, circle, spiral, etc.

Reconstructed images of nine point-targets with a spacing of
10 mm aligned in the axial direction and in the transverse direction
are shown in Fig.9 in gray scale display and perspective view.

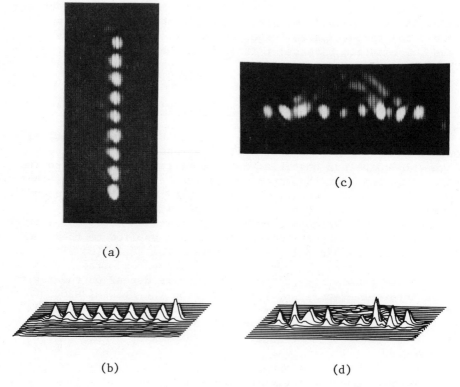

(a)

(c)

(b)   (d)

Fig. 9. Reconstructed images of line-shaped nine point-targets.
(Experiment) (a) Gray scale image when axially aligned,
(b) its side perspective view. (c) Gray scale image when
transversely aligned, (d) its front perspective view.

It is shown that the axial resolution by the multi-frequency synthesis is very satisfactory. On the contrary, the image of the transversely aligned point-targets is of poor quality, although the nine points are resolved. A plausible reason will be frequency dependence and spatial nonuniformity of the element factor of the transducer array.

Circle-shaped point-targets were imaged. There existed, however, no target at the center of the circle unlike the simulation in the previous section. Diameter of the circle is 70 mm. A reconstructed image is shown in Fig.10. It will be very satisfactory for the first experimental result of this method. Noticeable negative features of the image are distributed small noise which will have the same origin as Fig.9(c) and nonuniformity of the reconstructed targets in intensity.

A hologram matrix was measured also without any target in the water tank under the same condition, and a noise image was reconstructed. Its maximum intensity is 1.5×10^{-4} (-38 dB) time that of Fig.10. We hope that the above somewhat unsatisfactory features are not due to noise but due to nonuniformity in overall phase characteristic not only between the transverse elements but also between the different frequencies.

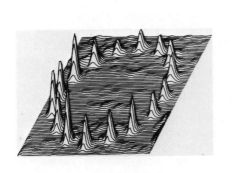 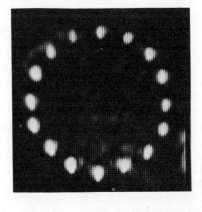

(a)   (b)

Fig. 10. Reconstructed image of circle-shaped sixteen
         point-targets. (Experiment)
         (a) Perspective view,
         (b) Gray scale view.

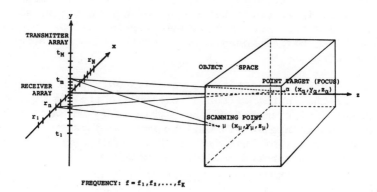

Fig. 11. A proposal of three-dimensional imaging by means of
multi-frequency hologram matrix.

DISCUSSION

Simultaneous detection of the signals of all the receiver
elements will make measurement of a 37 × 40 hologram matrix possible
within 30 ms. Image reconstruction of 80 × 80 pixels will be done in
one minute by a high-speed mini-computer with a multiply and divide
hardware for 32 bit integers.

The axial resolution does not depend on the distance from the
transducer array, but it is not the case with the transverse
resolution, which is understood from Eqs.8 and 9. The latter spreads
in proportic ı to the range distance. This, however, can be simply
kept constaى.c using different weight functions according to the
distance.

Reliable measurement of the phase compensation factors α_{nk}
will not easy but important. Once the method or technique is
developed, the multi-frequency hologram matrix imaging will give
better images than the conventional pulse echo method, with simple
electronics and much versatility.

This imaging method is easily developed to a three-dimensional
imaging using any planar array. One proposal is shown in Fig.11.

REFERENCES

1) H.Ogura, and K.Iizuka, "Hologram matrix and its application to
    a novel radar," *Proc. IEEE*, vol. 61, p. 1040 (1973).
    K.Iizuka, H.Ogura, J.L.Yen, V.K.Nguyen, and J.Weedmark, " A
    hologram matrix radar," *Proc. IEEE*, vol. 64, p. 1493 (1976).
2) H.Ogura, and S.Fukuoka, "Imaging of a two-dimensional target by
    means of hologram matrix," *Proc. IEEE*, vol. 64, p. 364 (1976).
3) T.Miyashita, J.Nakayama, H.Ogura, Y.Yoshida, and T.Soma, " A
    new acoustical imaging method by means of hologram matrix,"
    Proc. IEEE Ultrasonics Symposium, p. 268 (1977).
    J.Nakayama, T.Miyashita, N.Akagi, H.Ogura, Y.Yoshida, and T.Soma,
    "Imaging of a two dimensional target by means of hologram
    matrix-An ultrasound experiment," *Proc. IEEE*, vol. 66, p. 1287
    (1978).
4) J.Nakayama, H.Ogura, T.Miyashita, and T.Shibayama, "Two-
    dimensional imaging by means of multi-frequency hologram matrix
    -An ultrasound experiment," *Proc. IEEE*, to be published.
5) R.Karg, "Multifrequency microwave holography," *AEÜ*, Band 31,
    p. 150 (1977).
    H.Ermert and R.Karg, "Multifrequency acoustical holography,"
    IEEE Trans. Sonics and Ultrasonics, vol. SU-26, p. 279 (1979).
6) S.A.Johnson, J.F.Greenleaf, F.A.Duck, A.Chu, W.R.Samayoa, and
    B.K.Gilbert, "Digital computer simulation study of a real-time
    collection, post-processing synthetic focusing ultrasound
    cardiac camera," *Acoustical Holography*, vol. 6, p. 193 (Plenum
    Press, New York, 1975).

FOCUSED LINE SOURCE LINEAR ARRAY HOLOGRAPHY USING A

THREE DIMENSIONAL COMPUTER RECONSTRUCTION SYSTEM

H. Dale Collins, Tom E. Hall and Richard L. Wilson*

Battelle, Pacific Northwest Laboratories

P.O. Box 999, Richland, Washington 99352

ABSTRACT

This paper describes a unique holographic scanning configuration and computer reconstruction technique for the inspection of thick-walled nuclear pressure vessels.

The holographic imaging system consists of a focused line source and a 120 element receiver array operating at 2.31 MHz. The source and receive arrays are aligned parallel with each other, simulating simultaneous source-receiver scanning in the mechanical scanned direction. The holographic information is obtained by mechanically scanning perpendicular to the arrays and electronically multiplexing the receiver array. This hybrid scanning configuration produces directional resolution and astigmatism in the holographic reconstruction.

The computer reconstruction system is capable of generating near-real time holographic images via an array processor and FFT operation. The system has the unique ability of integrating up to ten images into one three dimensional composite image with the options of tilt and rotation.[1]

A computer spatial frequency filtering program provides the unique capabilities of image enhancement and object characterization.

Experimental imaging results are presented of various objects and defects in thick-walled metals graphically illustrating the capabilities of the holographic imaging system.
----
*Work was completed while employed at Holosonics, Inc., Richland, WA.

INTRODUCTION

This paper discusses the theory and application of a rapid holographic imaging system using a fast digital reconstruction technique with the options of three dimensional multi-image integration, image rotation/tilt and spatial frequency defect characterization. With the application of digital spatial frequency techniques the system can provide unique computer generated acoustic signatures of internal defects such as cracks, lack of fusion porosity, etc. This information can then be stored in the computer and used for defect correlation (i.e., adaptive learning).

This unique imaging configuration produces directional lateral resolution and correctable astigmatism in the holographic reconstruction.[2] The image characteristics associated with the variable or direction resolution are graphically illustrated in the experimental results of this paper.

DESCRIPTION OF THE HOLOGRAPHIC IMAGING SYSTEM

Figure 1 is a simplified block diagram of the composite holographic scanning configuration and the digital computer reconstruction technique. The rapid inspection scanning configuration consists of a focused line source and a 120 element

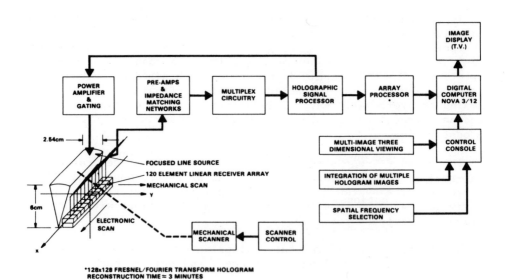

FIGURE 1.   SIMPLIFIED BLOCK DIAGRAM OF THE
HOLOGRAPHIC IMAGING SYSTEM

receiver array. The system is essentially a pitch-catch or send receiver imaging technique. The dual array geometry is shown in Figure 1. The holographic information is obtained by mechanically scanning in the "y" direction and electronically scanning the receiver array in the "x" direction.

The receiver array consists of 120 elements 1.0 mm in width and 150mm in length. The element center-to-center spacing is 1.25 mm. Both source and receiver arrays operate at 2.31 MHz. The first order grating lobes occur at approximately $\pm$ 30° in water. In steel the grating lobe angle exceeds 90°, and thus the lobes are eliminated. The system was designed to inspect large nuclear pressure vessels.

Hologram construction consists of gating the sinusoidal oscillator and driving the focused line source with the power amplifier. The object is illuminated with coherent sound and the reflected energy sampled by the 120 element receiver array. The object signals are multiplied with the electronic reference, time averaged and phase shifted to produce a digital complex hologram without the conjugate image.[3,4]

The computer then transforms the complex Fresnel hologram into a Fourier transform hologram by altering the phase of each data point as if a point source (i.e., reference was positioned on the object plane.

The hologram is then processed by operating on it with a two dimensional Fast Fourier Transform (FFT). The output is in the form of a complex value which is converted into an image by calcu- lating the intensity for each point. The system can process and integrate up to ten holographic images into one composite three dimensional image with the option of rotation and tilt.

Figures 2b, c and d illustrate the integration of three discrete holographic views of the complex three dimensional object (see Figure 2a). One on-axis and two sixty degree holograms were obtained and the single composite three dimensional image construc- ted. The computer system can generate composite images in real time with the options of rotation and tilt. A spatial frequency filtering program is also available to provide defect character- ization signatures and adaptive learning techniques.

ABERRATION CORRECTION ANALYSIS

One of the major problems inherent with the configuration is the astigmatism as a result of point source-receiver scanning in the "y" direction and stationary source point receiver scanning in the "x" direction.

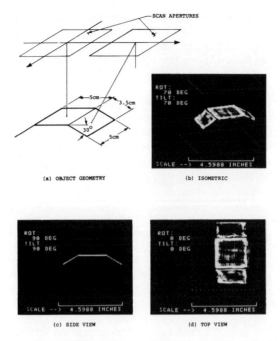

(a) OBJECT GEOMETRY   (b) ISOMETRIC

(c) SIDE VIEW   (d) TOP VIEW

FIGURE 2.   VARIOUS COMPOSITE VIEWS OF A THREE SIDED OBJECT

The hologram construction and reconstruction geometry used in the analysis is illustrated in Figure 3.   The phase at the receiver point (x, y, z) during the hologram construction is

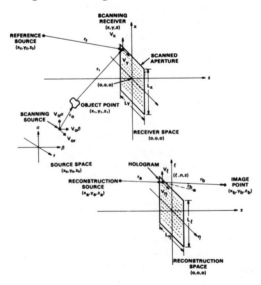

FIGURE 3.   GEOMETRY FOR SCANNED ACOUSTIC HOLOGRAPHY

$$\phi(x,y,z) = \phi_0(x,y,z) - \phi_r(x,y,z) \tag{1}$$

and

$$\phi(x,y,z) = \frac{2\pi}{\lambda_S} [r_0 + r_1 - r_2] \tag{2}$$

The phase at the receiver point (x,y,z) after illumination of the hologram by the reconstruction source is

$$\phi_1(x,y,z) = \pm \frac{2\pi}{\lambda_S} [r_0 + r_1 - r_2] - \frac{2\pi}{\lambda_L} r_a \tag{3}$$

where

λ_S = acoustic wavelength in construction medium

λ_L = reconstruction (light) wavelength

$+$ refers to the conjugate image

$-$ refers to the true image.

If the phase front Eq. (2) is to focus at the image point (x_b, y_b, z_b), then

$$\phi_1(x,y,z) = \frac{2\pi}{\lambda_L} r_b \tag{4}$$

which is termed the Gaussian-image sphere. The usual procedure is to expand the distance terms $(r_a, r_b, r_1, r_2$ and $r_0)$ in a binomial series and equate coefficients of x, y and z. We expand the distance terms about the origin of the (x,y,z) system and the distance r_0 is expanded about the α, β, γ system. The area in which the receiver scans is assumed small with respect to the distances and is centered at the (x,y,z) origin. A similar restriction holds for the source motion. Then we have

$$\phi_1(x,y,z) = \pm \frac{2\pi}{\lambda_S} \left\{ r_1 - \frac{xx_1}{r_1} - \frac{yy_1}{r_1} + \frac{x^2}{2r_1} + \frac{y^2}{2r_1} + r_0 - \frac{x_1\alpha}{r_0} - \frac{y_1\beta}{r_0} \right.$$

$$+ \frac{\alpha^2}{2r_0} + \frac{\beta^2}{2r_0} - r_2 + \frac{x_2 x}{r_2} + \frac{y_2 y}{r_2} - \frac{x^2}{2r_2} - \frac{y^2}{2r_2} + \ldots \left. \right\} - \frac{2\pi}{\lambda_L} \left\{ r_a - \frac{x_a x}{r_a} \right.$$

$$\left. - \frac{y_a y}{r_a} + \frac{y^2}{2r_a} + \frac{x^2}{2r_a} + \ldots \right\} = \frac{2\pi}{\lambda_L} \left\{ r_b - \frac{x_b x}{r_b} + \frac{x^2}{2r_b} + \ldots \right\} \tag{5}$$

The simulated receiver position (ξ, η) can be defined in terms of the actual and simulated velocities V_x, V_y, V_ξ, V_η, where

$$\xi = \frac{V_\xi}{V_x} x \quad \text{and} \quad \eta = \frac{V_\eta}{V_y} y \quad . \tag{6}$$

If we allow only parallel motion of the source and receiver, then the expressions for the source position are:

$$\alpha = a_0 + a_1 x \quad , \tag{7}$$

$$\beta = b_0 + b_1 y \quad . \tag{8}$$

The velocities of the source and receiver are related by

$$\frac{d\alpha}{dt} = a_1 \frac{dx}{dt} \quad , \quad \frac{d\beta}{dt} = b_1 \frac{dy}{dt} \tag{9}$$

where $\frac{d\alpha}{dt}$, $\frac{d\beta}{dt}$ and $\frac{dx}{dt}$, $\frac{dy}{dt}$ are the velocity components of the source and receiver, respectively. The final expressions for α and β (assuming $a_0 = b_0 = 0$) are

$$\alpha = \frac{V_{0\alpha}}{V_\xi} \xi \quad \text{and} \quad \eta = \frac{V_{0\beta}}{V_\eta} \eta \quad . \tag{10}$$

After substituting Eqs. (6) and (10) into Eq. (5) and equating coefficients, we obtain the following general expressions for image location:

$$\frac{x_b}{r_b} = \pm \frac{\lambda_L}{\lambda_S} \frac{V_x}{V_\xi} \left\{ \frac{x_1}{r_1} + (x_1 - x_0) \frac{V_{0\alpha}}{V_x} \frac{1}{r_0} - \frac{x_2}{r_2} \right\} - \frac{x_a}{r_a} \quad , \tag{11}$$

$$\frac{y_b}{r_b} = \pm \frac{\lambda_L}{\lambda_S} \frac{V_y}{V_\eta} \left\{ \frac{y_1}{r_1} + (y_1 - y_0) \frac{V_{0\beta}}{V_y} \frac{1}{r_0} - \frac{y_2}{r_2} \right\} - \frac{y_a}{r_a} \quad , \tag{12}$$

$$\frac{1}{r_{bx}} = \pm \frac{\lambda_L}{\lambda_S} \left(\frac{V_x}{V_\xi}\right)^2 \left\{ \frac{1}{r_1} + \left(\frac{V_{0\alpha}}{V_x}\right)^2 \frac{1}{r_0} - \frac{1}{r_2} \right\} - \frac{1}{r_a} \tag{13}$$

$$\frac{1}{r_{by}} = \pm \frac{\lambda_L}{\lambda_S} \left(\frac{V_y}{V_\eta}\right)^2 \left\{ \frac{1}{r_1} + \left(\frac{V_{0\beta}}{V_x}\right)^2 \frac{1}{r_0} - \frac{1}{r_2} \right\} - \frac{1}{r_a} \tag{14}$$

The image location equation for the composite (focused line source linear array) scanning configuration reduces to the following form:

$$\frac{x_b}{r_b} = \pm \frac{\lambda_L}{\lambda_S} \frac{V_x}{V_\xi} \left\{ \frac{x_1}{r_1} - \frac{x_2}{r_2} \right\} - \frac{x_a}{r_a} \tag{15}$$

$$\frac{y_b}{r_b} = \pm \frac{\lambda_L}{\lambda_S} \frac{V_y}{V_\eta} \left\{ \frac{2y_1}{r_1} - \frac{y_2}{r_2} \right\} - \frac{y_a}{r_a} \tag{16}$$

$$\frac{1}{r_{b_x}} = \pm \frac{\lambda_L}{\lambda_S} \left(\frac{V_x}{V_\xi}\right)^2 \left\{\frac{1}{r_1} - \frac{1}{r_2}\right\} - \frac{1}{r_a} \tag{17}$$

$$\frac{1}{r_{b_y}} = \pm \frac{\lambda_L}{\lambda_S} \left(\frac{V_y}{V_\eta}\right)^2 \left\{\frac{2}{r_1} + \frac{1}{r_2}\right\} - \frac{1}{r_a} \tag{18}$$

where the source velocity ($V_0\alpha$) in the "x" direction is zero and the source-receiver velocities are equal in the "y" direction (i.e., $V_0\beta = V_y$).

In order for r_b to be the same for each coordinate (i.e., stigmatic), then

$$2\left(\frac{V_y}{V_\eta}\right)^2 = \left(\frac{V_x}{V_\xi}\right)^2 \quad . \tag{19}$$

The scanned aperture to display aperture ratios can be expressed in terms of the velocities (Eq. (19))

$$\sqrt{2}\,\frac{Ly}{Lx} = \frac{L\eta}{L\xi} \quad . \tag{20}$$

If the scanned aperture ratio is unity (i.e., Ly = Lx), the display aspect ratio must be

$$L\eta = \sqrt{2}\, L\xi \quad . \tag{21}$$

If the display aspect ratio is unity, then the scanned aperture dimensions must satisfy the following relationship: Lx = $\sqrt{2}$ Ly.

Thus, one must increase L_η or Lx by the $\sqrt{2}$ in the hologram display or scanned aperture to construct a stigmatic image with the composite holographic scanning configuration. This correction is only valid for paraxial imaging. Figure 4a shows the point object hologram exhibiting astigmatism (i.e., elliptical rings). Figure 4b is the corrected stigmatic hologram. The display aspect ratio was corrected using Eq. (21).

LATERAL AND LONGITUDINAL RESOLUTION

The image resolution, as a function of object to hologram distance, effective aperture and frequency, is an extremely important parameter to verify experimentally. The maximum obtainable theoretical lateral and longitudinal (i.e., depth) using a focused source is given by the following equations:

$$\text{Lateral Resolution } (\Delta r) \simeq \frac{\lambda f}{a} \tag{22}$$

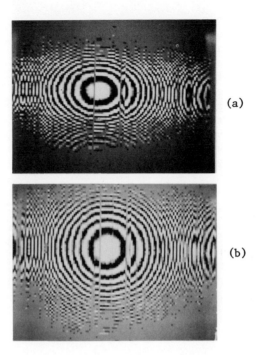

(a)

(b)

FIGURE 4.   ACOUSTIC HOLOGRAMS OF A POINT OBJECT ON AXIS
            (a) ASTIGMATIC HOLOGRAM
            (b) CORRECTION ASPECT RATIO ($\sqrt{2}$) HOLOGRAM

$$\text{Longitudinal Resolution } (\Delta z) \simeq \lambda \left(\frac{f}{a}\right)^2 \tag{23}$$

where

λ = acoustical wavelength in the medium
f = focal length in the medium
a = transducer aperture

The maximum lateral and longitudinal resolution in steel (L-wave)
for the 2.54 cm diameter, 5.7 cm focal length transducer operating
at 2.31 MHz is 1.45 mm and 3.3 mm, respectively. The focused line
source and point receivers will determine the holographic resolutions in one direction. If we assume the focus is in the "y"
direction, the holographic lateral resolution can be expressed in
the following form:[5]

$$\text{Lateral Resolution } (\Delta y) \simeq \frac{\lambda r_1}{2Ly} \tag{24}$$

where r_1 = flaw to hologram distance and Ly = effective aperture
length in the "y" direction.

These expressions assume the source and receive transducers
are ideal (i.e., perfect points). This, of course, is not true.
Thus, as in any imaging system, the weakest link dominates. The
focused line source or receiver resolution will dominate until the
holographic resolution degrades to this value or exceeds it. This
will occur at a predetermined depth for the focused line source
called the crossover point. The crossover point is easily determined by equating the focused source resolution with the holographic
resolution and solving for the depth, (r_1).[6] The result expresses
the crossover depth as a function of three parameters: 1) holographic aperture, 2) transducer focal length, and 3) transducer
aperture.

$$\text{Crossover Depth } (r_1) \simeq \frac{2fLy}{a} \tag{25}$$

For the composite configuration, $f \simeq 5.7$ cm and $a = 2.54$ cm, the
crossover depth occurs at 4.5 holographic aperture widths. This
depth is seldom ever reached in pressure vessel imaging. Thus,
the dominating resolution in the "y" direction is usually the
transducer resolution (1.45 mm) unless dimensions of the effective
receiver element exceeds this value. The configuration in the "x"
direction simulates a stationary source/point receiver system.

The holographic resolution in the "x" direction is

$$\text{Lateral Resolution } (\Delta x) = \frac{\lambda r_1}{L_x} \quad . \tag{26}$$

Equations (24) and (26) indicate the "y" holographic resolution
exceeds the "x" by a factor of two. In the above analysis, the
difference between source-to-hologram and receiver-to-hologram
distances is assumed negligible. The holographic resolution is
valid in the "x" direction unless the point receiver dimension
exceeds this value.

HOLOGRAPHIC IMAGING EXPERIMENTS IN WATER

Holographic Reconstruction of the Metal Letter "H" (5 cm x 5 cm)
at 24 cm Depth

Figure 5 shows the simultaneous source-receiver scan system
and the composite configuration imaging the letter "H" in water.
The object or target is located 23 cm from the hologram. Both
holograms contain 16,384 sampling points with 4,096 dynamic
signal range.

Figure 6a is the acoustic hologram (simultaneous source-
receiver scan). The spatial frequency content is graphically
illustrated with various fringe spacings across the hologram. The

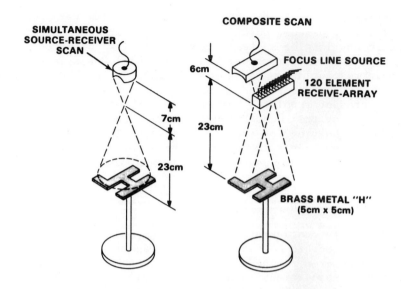

FIGURE 5.   HOLOGRAM CONSTRUCTION GEOMETRICS OF THE
METAL LETTER "H"

edges are defined by the high frequency fringes and the low contrast
interior with the low frequency fringes.

Figure 6b is the reconstructed image (simultaneous source-
receiver scan) of the letter "H".   The image exhibits extremely
sharp defined edges illustrating the high resolution capabilities
of the system.   The predicted theoretical resolution is approximately 1.8 mm in water at 2.31 MHz.

Figure 6c is the reconstruction image (composite scan) of the
letter "H" under identical conditions.   The edge sharpness is
decreased and the overall image quality has deteriorated as
compared with the simultaneous source-receiver scan image.

HOLOGRAPHIC IMAGE RESOLUTION TESTS IN THICK METAL BLOCKS

"y" Pattern (L-Wave) Holographic Resolution Tests (1 mm, 2 mm and
4 mm) at 12.7 cm Depth in Aluminum

Figure 7 is the L-wave hologram construction geometry of a "y"
pattern array of ten flat-topped holes positioned 12.7 cm in depth.
The edge-to-edge hole separations in each appendage are 1 mm, 2 mm
and 4 mm, respectively.

Figure 8 illustrates the results of the L-wave resolution
tests with the aluminum block "y" pattern.   Figure 8a is the

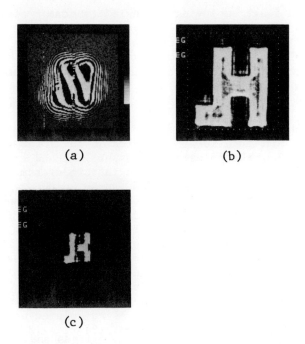

(a)   (b)

(c)

FIGURE 6.   HOLOGRAM, RECONSTRUCTIONS OF THE METAL LETTER "H";*
             (a)   HOLOGRAM
             (b)   SIMULTANEOUS SOURCE-RECEIVER SCAN IMAGE
             (c)   COMPOSITE SCANNING IMAGE
             *HOLOGRAPHIC DIGITAL RECONSTRUCTION (128 x 128)

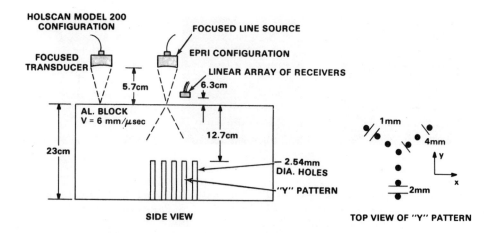

FIGURE 7.   SCHEMATIC OF ALUMINUM BLOCK CONSTRUCTED FOR LONGITUDINAL
             WAVE LATERAL RESOLUTION TESTS (1 mm, 2 mm and 4 mm)

reconstructed image employing a single transducer operating in the
simultaneous source-receiver configuration. All the holes in each
leg are resolved using this scanning technique. The theoretical
lateral resolution is approximately 1.8 mm. The upper left and
right legs contain the 1 mm and 4 mm hole separations. The frequency, focal length and transducer aperture are identical to the
focused line source.

Figure 8b is the "y" pattern image constructed with the composite scanning configuration. The holes in the lower vertical leg
are resolved, but the holes in the upper left and right legs have
merged together. The "y" resolution in this system is 1.45 mm and
aligns with the vertical leg. Thus, these holes are easily
resolved, but appear distorted in the image. The upper legs extend
outward at approximately 30° with respect to the horizontal and,
thus are dominated by the "x" resolution (i.e., 4 mm). The 1 mm
and 4 mm hole separations in the upper legs exceed the system
resolution with the effective 7.62 cm aperture. Thus, the holes
appear as a single integrated image and are not resolved. This
three directional image illustrates very nicely the variable
lateral resolution exhibited by the composite scanning configuration. The 4 mm lateral resolution was calculated using the
following parameters: (1) λ = 2.5 mm, (2) r_1 $\simeq$ 12.7 cm, and
(3) L = 7.62 cm.

(a)   (b)   (c)

FIGURE 8.   HOLOGRAPHIC LONGITUDINAL WAVE RESOLUTION TEST (1 mm,
             2 mm & 4 mm HOLE EDGE-TO-EDGE SEPARATIONS); (a) SIMUL-
             TANEOUS SOURCE-RECEIVER SCAN IMAGE (HOLSCAN MODEL 200),*
             (b) COMPOSITE SIMULTANEOUS AND STATIONARY SOURCE RECEIVER
             SCAN IMAGE* AND (c) PHOTOGRAPH OF "Y" HOLE (2.5 mm DIA.)
             PATTERN 12.7 cm DEPTH IN ALUMINUM

             *DIGITAL RECONSTRUCTION (128 x 128)

"y" Pattern (L-Wave) Holographic Resolution Tests (2 mm, 4 mm and 8 mm) at 10.2 cm Depth in Aluminum

Figure 9 is the hologram (L-wave) construction geometry with edge-to-edge hole separation of 2 mm, 4 mm and 8 mm, respectively. This pattern should be resolvable using the composite holographic scanning configuration. The predicted "x" and "y" lateral resolutions at 2.31 MHz are 3.33 mm and 1.45 mm, respectively. The "x" resolution (3.33 mm) will dominate the 30° and 120° "y" pattern appendages with 4 mm and 8 mm separations. The vertical row of holes is aligned in the "y" direction utilizing the higher 1.45 mm resolution.

Figure 9b is the "y" pattern image construction with the composite configuration. All holes in each of three legs are resolvable as predicted. The 1.45 mm and 3.3 mm resolutions are sufficient to resolve the 2 mm, 4 mm, and 8 mm hole separations in each leg. The "y" pattern in Figure 9b appears to exhibit an incorrect aspect ratio between the height and the width. This is probably related directly to the static aspect ratio correction.

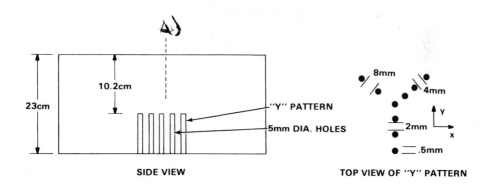

SIDE VIEW   TOP VIEW OF "Y" PATTERN

FIGURE 9.   SCHEMATIC OF ALUMINUM BLOCK FOR LONGITUDINAL
             WAVE LATERAL RESOLUTION TESTS (2 mm, 4 mm & 8 mm)

Figure 10a is the holographic reconstruction (simultaneous source-receiver scan) for comparison with the composite system. The "y" pattern image is of excellent quality and graphically illustrates the high resolution capabilities of the simultaneous source-receiver scanning configuration. The holes even appear circular as, for example, the hole at the extreme left side of the pattern. This image uniquely demonstrates the great potential inherent in holographic process.

"y" Pattern (L-Wave) Holographic Resolution Test (3 mm, 4 mm and 5 mm) at 36 cm Depth in Aluminum

Figure 11 is the hologram construction geometry for deep imaging resolution tests. The flat-top holes are positioned 36 cm below the upper surface. This depth exceeds most pressure vessel thicknesses.

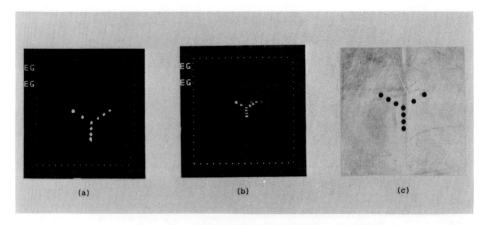

(a)   (b)   (c)

FIGURE 10.   HOLOGRAPHIC L-WAVE RESOLUTION TEST (2mm, 4mm and 8mm
              HOLE EDGE-TO-EDGE SEPARATIONS); (a) SIMULTANEOUS
              SOURCE-RECEIVER SCAN IMAGE*; (b) COMPOSITE SIMULTANEOUS
              AND STATIONARY SOURCE RECEIVER SCAN IMAGE*; AND (c)
              PHOTOGRAPH OF "Y" HOLE (5 mm DIA.) PATTERN 10 cm
              DEPTH IN ALUMINUM

              *DIGITAL RECONSTRUCTION (128 x 128)

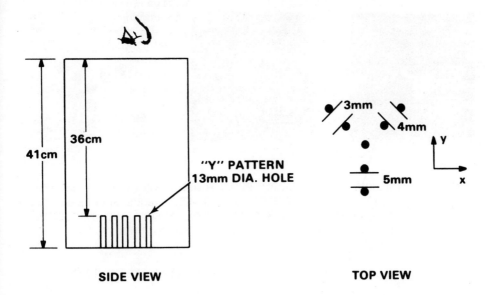

SIDE VIEW   **TOP VIEW**

FIGURE 11.   SCHEMATIC OF ALUMINUM BLOCK FOR DEEP IMAGING
RESOLUTION TESTS (3 mm, 4 mm & 5 mm)

Figure 12 is the reconstructed hologram obtained from the
simultaneous source receiver scanner system.   The predicted
resolution at 36 cm depth is 3 mm, with 15 cm aperture.   The
wavelength at 2.31 MHz is 2.6 mm.   The holes in all three legs are
resolved as predicted with simultaneous source-receiver scanning.

Figure 12b is the reconstructed image from the composite
configuration.   The "x" direction resolution is approximately 6 mm
and, thus not adequate to resolve the pattern.   All holes have
merged in the image as shown in Figure 11b.

Figure 12c is the photograph of the hole pattern viewed from
the bottom of the block.

"y" Pattern (45° S-Wave) Holographic Resolution Tests (1 mm, 2 mm
and 4 mm) at Variable Depths in 23 cm Thick Metal Block

Figures 13a and b illustrate the transverse or shear wave
hologram construction geometry.   The simulated internal defect
consists of a "y" pattern with various edge-to-edge hole separations in each appendage.   This pattern has been used extensively
in holographic resolution tests over the last few years.

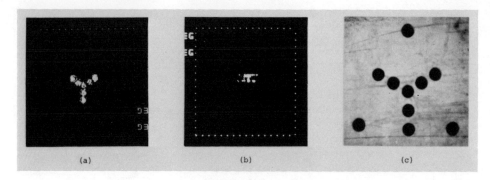

FIGURE 12.   HOLOGRAPHIC (L-WAVE) RESOLUTION TEST (3 mm, 4 mm AND
             5 mm) AT 36 cm DEPTH*; (a) SIMULTANEOUS SOURCE-RECEIVER
             SCAN IMAGE, (b) COMPOSITE SIMULTANEOUS AND STATIONARY
             SOURCE-RECEIVER SCAN, AND (c) PHOTOGRAPH OF "Y"
             PATTERN HOLES

             *HOLOGRAPHIC DIGITAL RECONSTRUCTION (128 x 128)

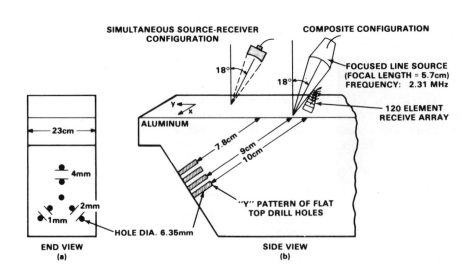

FIGURE 13.   SCHEMATIC OF ALUMINUM BLOCK FOR SHEAR WAVE LATERAL
             RESOLUTION TESTS VARIABLE DEPTH

The "y" hole tops vary in depth, 7.8 cm to 10 cm, from the
upper surface. The deeper holes lie at greater distances from the
hologram plane (i.e., metal surface) and, thus will exhibit a lower
resolution.

Figure 14a is the holographic reconstruction of the "y" pattern
obtained from the simultaneous source-receiver system. The 1 mm
(edge-to-edge) holes are aligned in vertical "y" direction. The
1 mm separations exceed the resolvable distance or resolution in
the "y" direction for both systems. The vertical row of holes in
Figure 14a (simultaneous source-receiver scan) are slightly merged
together as predicted. The vertical row of holes in the composite
system image, Figure 14b, also shows merging and blurring in the
image. The lower appendages of the "y" pattern are 30° from the
horizontal and, thus dominated by the "x" resolution. The effective
aperture of the composite configuration in this direction was
approximately 3.8 cm. Using Eq. (26) and the 3.8 cm aperture, the
stationary source-scanned receiver resolution is 3.24 mm. The two
horizontal legs in Figure 14b are very blurred and not resolvable.
The simultaneous source-receiver scan image clearly resolves the
two horizontal rows as predicted by theory.

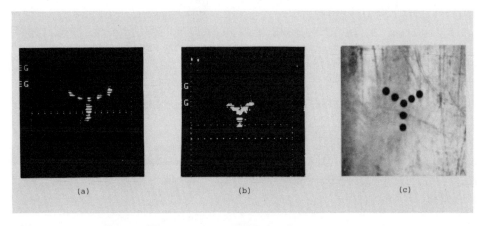

FIGURE 14.   HOLOGRAPHIC S-WAVE RESOLUTION TEST AT 15 cm (1 mm,
              2 mm, 4mm)*; (a) SIMULTANEOUS SOURCE-RECEIVER SCAN
              IMAGE, (b) COMPOSITE SIMULTANEOUS AND STATIONARY
              SOURCE-RECEIVER SCAN IMAGE, AND (c) PHOTOGRAPH OF
              THE "Y" PATTERN FROM THE BOTTOM SIDE OF THE BLOCK

              *DIGITAL RECONSTRUCTION (128 x 128)

CONCLUSIONS

Focused line source and linear receiver array holographic technique successfully imaged simulated defects in thick metal sections. This unique scanning configuration provides rapid inspection by electronically scanning along the receiver array and mechanically scanning perpendicular to it.

This hybrid scanning configuration produces directional image resolution and correctable astigmatism. The digital reconstruction system used with the rapid scanning technique is capable of processing 128 x 128 holograms in approximately three minutes. The system also has the control console options of multi-image integration, three dimensional viewing and spatial frequency filtering.

REFERENCES

1. Hall, Thomas and Wilson, Richard, Holosonics, Inc., Final Progress Report, EPRI, Project RP606-2-1, "Development of the Rapid Acoustical Holography Computer Reconstruction Device", 1979.

2. Hildebrand, B. P., "Development of an Ultrasonic Analyzing System", EPRI Report NP606-1, December 1979.

3. Keating, P. N., Acoustical Holography, Vol. 5, (edited by P. J. Green), Plenum Press, NY, 1973.

4. Collins, H. D., "Various Holographic Scanning Configurations for Under-Sodium Viewing", AEC Research and Development Report (BNWL-1558), 1971.

5. Collins, H. D., and Brenden, B. B., Acoustical Holography, Vol. 5, (edited by P. J. Green), Plenum Press, NY, 1973.

6. Collins, H. D. and Hildebrand, B. P., Evaluation of Acoustical Holography for the Inspection of Pressure Vessel Sections, book, (Periodic Inspection of Pressure Vessels), Institution of Mechanical Engineers, 1972.

ELECTRONIC ANALOGICAL DEVICES OF SECTOR SCAN IMAGING*

F. Haine, R. Torguet, C. Bruneel, E. Bridoux, G. Thomin
and B. Delannoy.
Université de Valenciennes
59326 Valenciennes Cedex FRANCE

INTRODUCTION

The sector scan imaging devices give good results in the medical applications. This fact may be explained by the possibility of obtaining images through narrow windows, since the space between two ribs for the observation of the heart.

Two kinds of sector scan apparatuses are actually developed:
- the mechanical devices (1,2)
- the computer assisted electronical scan devices (3,4,5)

The latter gives good results, due to the moving focus that is used. In these apparatuses two kinds of delay lines are used (fig.1):
- the first ones give the scanning whose delays are varied between 0 and 10 μS.
- the second ones give the focusing with delays lower than 1 μS.

So the scanning delay lines have to be very accurate in order to not disturb the focusing operation. This explains the high cost of these apparatuses.

In order to avoid this problem we have conceived two new devices where the scanning operation is obtained by using an analogical calculator (6,7,8). This analogical calculator is made of a cylindrical array of tranducers. The focus operation is then obtained :
- by an acoustical lens set in the center of curvature of the cylindrical array.
- or else by a delay line electronical lens.

* This work was supported by the DGRST FRANCE

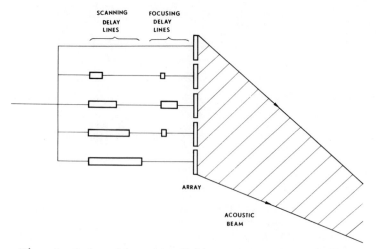

Fig. 1. Delay lines used in a sector scan device.

ANALOGICAL SCANNING DEVICE WITH ACOUSTICAL LENS

The set up is described on the figure 2. The scanning is obtained easily by sequentially switching the transducers of the cylindrical array. Each ultrasonic beam insonifies the lens set in the center of curvature of the cylinder. The focal length of the lens is calculated in order to focus the beam in the center of the observation zone.

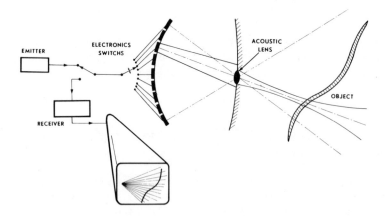

Fig. 2. Principle of the scanning device with an acoustic lens.

By using two cylindrical arrays, one at the emission stage and the other at the reception one, we may use two different lens with appropriate focal lengths in order to maximize the depth of focus.

For exemple, at a two megahertz operating frequency in water, a 5 mm
resolution is obtained between 5 cm and 15 cm.

Six 2 mm rods regularly spaced on a 4 cm circle have been imaged
by such a device, the obtained image (fig. 3) is very close to
the original object according to the theoretical predictions. However,
this apparatus doesn't give the expected results for medical applications, the reason is that parasitical reflections appear at the
same level that the useful echos of the biological structures. So
we develop a second version using an electronical lens, allowing
then a complete separation between the emission stage and the reception one and avoiding this parasitical reflections.

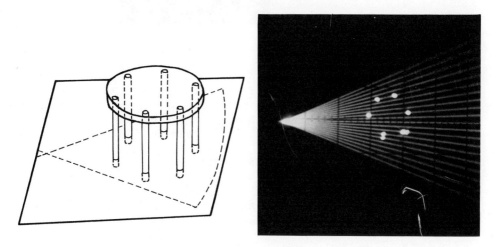

Fig. 3. Sector scan of model targets.

ANALOGICAL SCANNING DEVICE WITH ELECTRONICAL LENS

The principle of the device is illustrated on the figure 4
where we notice :
- the cylindrical array of transducers, each being sequentially connected via the switching system.
- a first linear array set in the center of curvature of the
circular array. This linear array picks up signals with the appropriate delays characterizing the direction Θ of the incident beam.

These two first parts of the device constitute the analogical
calculator which appears very simple and which is better and works
at least as good than the well known complex computer and delay
lines system.

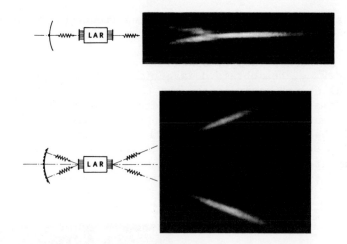

Fig. 4. Principle of the scanning device with electronical lens.

The last part is identical to the other sector scan system i.e delay lines with a quadratic law variation of the delays versus the order of the considered line with respect to the central one and a linear array which generate an acoustic beam in a direction making the same angle θ as the incident beam in the particular case of identical arrays and identical wavelengths.

The focusing effect and the scanning effect have been tested. The visual observation of the beams using a Schlieren's detection method has been performed as shown on the figure 5.

We may notice that such an analogical calculator allows the classical sequential operating mode but also a parallel operating mode i.e the different transducers of the cylindrical array are excited at the same time and at the reception stage a fast commutation is used in order to separate the informations corresponding to the different lines. Very high image rate may so be obtained (up to 1000 images per second) but, in counterpart, the resolution will decrease because the lack of focusing at the emission stage.

With this experimental set up we have observed images of the six rods target (figure 6). The matching of the system to medical applications is now under progress.

GENERALIZATION OF THE ANALOGICAL CALCULATOR

Other set up of this analogical calculator may be derived. In a first time the analogical calculator has been made using the same ultrasound charasteristics as these of the operating ones : same

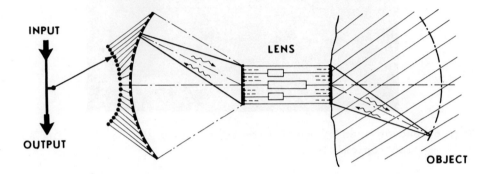

Fig. 5. Schlieren's pictures of the acoustic beams.

medium (water), same frequency, same longitudinal mode.However it is possible to use an other type of analogical calculator and one using Rayleigh wave mode has been made (9).

The phase law distribution on the receiving array is then :

$$\Phi_n = n\ 2\pi\ a\ \sin\ \theta/\lambda$$

where a sets for the step of the array, λ the incident wavelength, n the order of the considered transducer and θ the incident angle.

Using mixers and a reference signal this phase law may be trans-fered at the reemetting frequency and the generated beam will pro-pagate in a direction θ' such as :

Fig. 6. Sector scan of model targets given by the device using an
electronical lens.

$$\Phi_n = n2\pi a' \sin \Theta'/\lambda'$$

or $\sin \Theta' = \dfrac{\lambda'}{\lambda} \dfrac{a}{a'} \sin \Theta$

where λ' is the reemitting wavelength and a' the step of the reemitting array. If the steps of the arrays are chosen **proportional to** the corresponding wavelength, the receiving and reemitting angles are identical.

This analogical calculator may also be used for other applications and specially to generate the appropriated signals applied to the array of scanning radar devices.

CONCLUSION

We have described a new simple and low cost analogical calculator which is able to advantageously replace the digital calculator and delay lines commonly used in the electronic sector scan devices. The images of targets has been shown and confirm the feasabilities of such a system. The expected performances are even better than the one obtained by the other apparatuses. Other applications of the analogical calculator have also been derived.

References

1. Holm HH, Kristensen JK, Pedersen JF, Hanche S, and Northeved
   A (1975). "A new mechanical real time ultrasonic

   contact scanner". Ultrasound in Medecine and Biology 2,19.

2. Shaw A, Paton JS, Gregory NL, and Wheatley DJ (1976) "A real time 2-dimensional ultrasonic scanner for clinical use". Ultrasonic 14,35.

3. Somer JC. (1977) "Phased array systems". Echocardiography (N. Bom Ed, Rotterdam, 1977), p. 325-333.

4. Thurstone FL, and von Ramm OT (1973). "A new ultrasound imaging technique employing two-dimensional electronic beam steering". Acoustical Holography Vol. 5, Ed. Newell Booth, Publ. Plenum Press, New York and London, p. 249-259.

5. von Ramm OT, Thurstone FL, and Kisslo J. (1975) "Cardiovascular diagnosis with real time ultrasound imaging", Acoustical Holography, Vol. 6, Ed. Newell Booth, Publ. Plenum Press, New York and London, p. 91-102 (1975).

6. F. Haine, G. Thomin, R. Torguet, E. Bridoux and C. Bruneel (1977) "Nouveau dispositif d'imageur à balayage électronique sectoriel", 4ème colloque de la Société Française pour l'application des ultrasons à la médecine et à la biologie. Paris.

7. F. Haine, R. Torguet, C. Bruneel, E. Bridoux and G. Thomin (1979) "New sector scan echocardiography imaging devices" 3rd Symposium on echocardiography, Rotterdam.

8. F. Haine, C. Bruneel, R. Torguet and E. Bridoux (1979). "New sector scan imaging device" A.P.L. Vol. 34 n° 12.

9. P. Cauvard, P. Hartmann, C. Bruneel and F. Haine (1978) "Ultrasound beam scanning driven by surface acoustic waves" IEEE Ultrasonic.

NEW ARRANGEMENTS FOR FRESNEL FOCUSING

Bruno Richard[+], Mathias Fink[++], Pierre Alais[++]

+ Service de Biophysique, CHU Cochin, 75014 Paris, France
++ Laboratoire de Mécanique Physique,
   Université Pierre et Marie Curie, Paris, France

INTRODUCTION

Medical results of a 2 state Fresnel focusing, presented at the 7th International Symposium on Acoustical Imaging [1] , proved that a very good lateral resolution (1.5 mm) was obtained in vivo due to the use of a large aperture. The need for very low sidelobe levels in medical imaging had led us to associate to a Fresnel lens at reception a non focused transmission. Far off-axis sidelobes are then reduced (-40 dB) at the expense of the lateral resolution which is multiplied by 1.5. This apodizing technique gave good results for medical diagnosis [2] .

However, it has been theoretically proved that a better phase sampling of the Fresnel law with more than two states may reduce significantly the sidelobe level, with no loss in resolution [3] .

In CW mode, when focusing with a Fresnel aperture sampled by only two different states of phase $(0, \pi)$, parasitic converging and diverging waves are associated to the main converging beam and give rise to a high side lobe level. A better phase sampling of the Fresnel law reduces strongly the contribution of parasitic waves. It has been shown that for a p-phase sampling $(0, 2\pi/p , 4\pi/p , ...)$, the mean side lobe level is reduced by a factor of $\dfrac{1}{\sqrt{p} \ (p-1)}$.

For a wideband signal, such an effect is still observed when using different kinds of p-state sampling [4] . The arrangements that have been theoretically tested and experimentally used are described in fig.1 . For p = 8 we use either 8 different delays within the

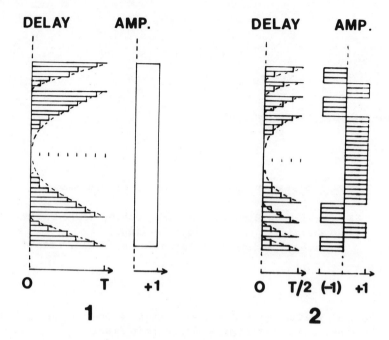

Fig.1 - This figure shows how a Fresnel law may be approximated
        by eight different states.
        - in 1, eight different delays are used within the
          period
        - in 2, four different delays are used within the half pe-
          riod and signal is transmitted either in phase (ampli-
          tude + 1) or in antiphase (amplitude - 1) realizing
          the eight different states.

fundamental period, or 4 different delays within the half period associated with an amplitude modulation by + 1 or - 1 (Fig.1). For practical reasons, the first one has been used for transmission and the second one for reception. We present here experimental measurements allowing comparisons to be done between these different focusing techniques.

EXPERIMENTAL METHOD

     For this purpose, we built a new experimental device using a
160 elements linear array, 12 cm long, working at 2.5 Mhz.For transmission as well as for reception, up to 64 transducers may be used
to synthetize the aperture. Fresnel lenses are sampled by eight dif-

ferent states and may be selected separately for transmission and for
reception. The echographic response is studied with a simple target
made of a copper wire (.3mm in diameter) in a water tank. Received
signals are logarithmically compressed so that a wide dynamic range
may be studied. The transverse echographic diagram is then obtained
by acting the Y deflection of a scope by this logarithmic signal while the X deflection corresponds to the position of the array relatively to the target.

EXPERIMENTAL RESULTS

- " Phase " sampling of the aperture

        We first checked the role played by the number of states used
for sampling the focusing aperture.Results are shown on figure 3. The
same wide aperture was used for transmission and for reception (f:2)

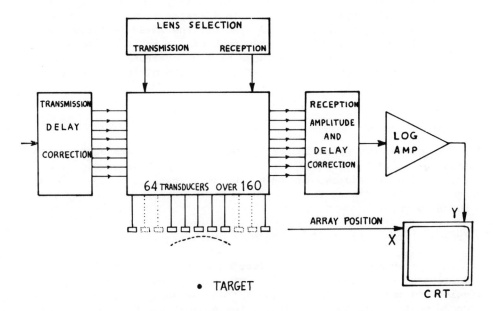

Fig. 2 - Block diagram of the experimental device

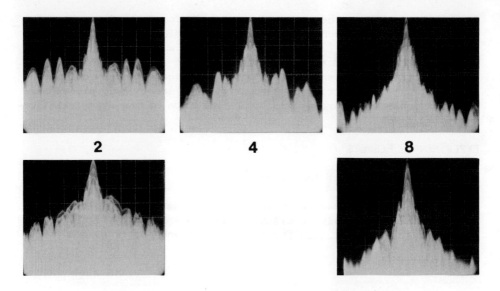

Fig. 3 - Sampling the aperture by 2, 4 and 8 different states.
        The top three pictures are experimental results obtained
        with a large aperture focusing (f:2) at the focal distance
        when using a 2, 4 or 8 state sampling of the aperture. The
        scale factors are in the Y direction 8 dB/div and in the
        X direction 5 mm/div. With two states, the mean sidelobe
        level is around - 30 dB, with 4 states around - 40 dB and
        with eight states around - 50 dB. The lower pictures are
        the patterns obtained when combining a small aperture in
        transmission with the large Fresnel aperture in reception
        with two states (lower left) and eight states (lower right).

The Fresnel law was approximated by two states, four states and eight
states. Experimental results are very similar to those predicted by
theory : with two states, the mean side lobe level is in the - 30 dB
range, with four states in the - 40 dB and with eight states in the
- 50 dB range.

- Fresnel large aperture focusing of a short signal

        The validity of Fresnel focusing when using a short signal has
been tested by varying the aperture of the lens used both for transmission and reception . Fig. 4 shows that the gain in lateral resolution is quite significant, at least for an aperture number of f:2.4.
Moreoever, the use of larger apertures don't increase the sidelobe

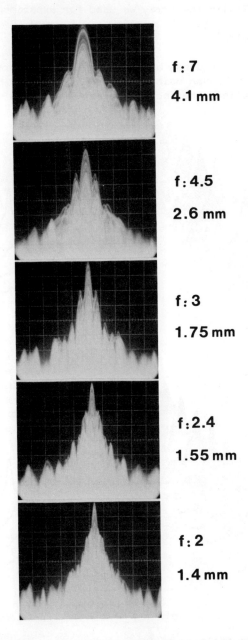

f : 7
4.1 mm

f : 4.5
2.6 mm

f : 3
1.75 mm

f : 2.4
1.55 mm

f : 2
1.4 mm

Fig. 4 - Transverse diagrams at the focal distance for different
apertures. For each case, the f number and the 6 dB lateral resolution at 75 mm of the array (focal distance) are
given aside the pictures. (Same scale factors as fig.3).

level. As for axial resolution, the 6 dB value is 1   mm for f:7 and
1.4 mm for f:2 with the array we used for experimentation.

- Depth of field

        Depth of field has been studied with different focusing confi-
gurations. First, when using the largest aperture both for transmission and reception, the focusing diagram exhibits of course the thin-

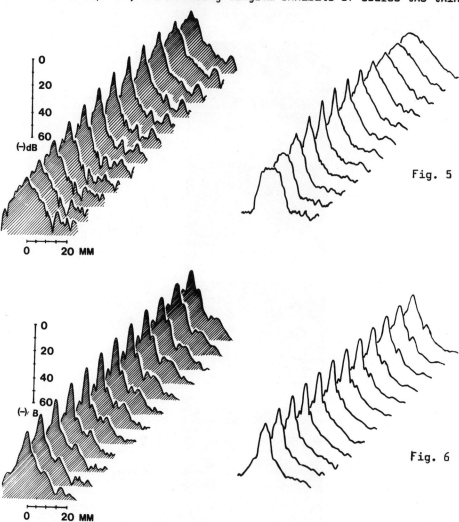

Fig. 5

Fig. 6

Fig.5,6 - Beam patterns of Fresnel focusing apertures from 50 to
105 mm. Transverse diagrams have been recorded every 5 mm. Fig. 5
correspond to a large aperture f:2,Fig.6 to a smaller aperture f:4.5
The hachured figure represents experimental results and on the
right a theoretical computed prediction in a similar scale.

nest peak at the focal distance but this focusing effect is restricted
to a limited depth of field. On the right of the figures 5, 6 has
been plotted a result of a numerical computation which shows good
agreement with experimental results for lateral resolution, depth of
field and side lobe level.

On the second figure, results obtained with a smaller aperture
(f:4.5) are presented. The loss in lateral resolution and the gain in
depth of field appear clearly always in accordance with the computed
beam pattern corresponding to this case.

A mixed technique, using a small aperture for transmission and
a large aperture for reception seems to be of interest (fig.7) as it
gives at the same time good lateral resolution (1.8 mm et 75 mm), appreciable depth of field and very good apodization.Moreover it should
be very well adapted to a dynamic focusing mode.

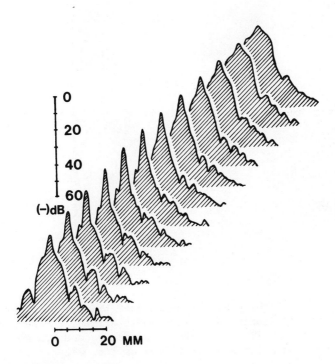

Fig. 7 - Experimental results obtained when using a small aperture
         for transmission combined to a large aperture in reception

- Spatial sampling of the aperture

        The same experimental mounting has also been used to study the
grating lobes in two different configurations (fig.8). In the first
one the spatial sampling of the aperture is . 75 mm (frequency 2.5
MHz), and the first order grating lobe is indicated by the white arrow. In the second case the pitch is 1.5 mm and the first order grating lobe appears, as expected, at half the distance of the previous
one with a much higher level. Both recordings were made with the same large aperture (f:2) using eight different states.

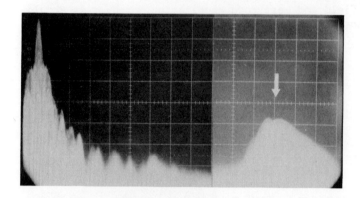

P= .75 mm

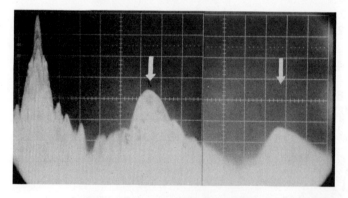

P= 1.5 mm

Fig. 8 - Effect of the spatial sampling of the aperture. Grating
        lobes are indicated by the white arrows, the sampled pitch
        being of . 75 mm for the upper figure and 1.5 mm for the
        lower one. (Same scale factor as fig. 3).

CONCLUSION

An experimental technique has been developed which may be useful to check the focusing properties of a B - scan device. It has been applied to a new Fresnel focusing method where any Fresnel aperture may be approximated by eight different states.

Results are analyzed in terms of lateral resolution, sidelobe level, and depth of field. Good agreement has been found between experimental measurements and theoretical predictions.

This approach may be an interesting alternative method compared to the classical delay line focusing in medical applications.

BIBLIOGRAPHY

1. P. ALAIS and M. FINK, " Fresnel zone focusing of linear arrays applied to B and C echography ", Acoustical Holography, vol 7, 1977, p. 509-522.

2. P. ALAIS and B. RICHARD, " Progress in Fresnel imaging, clinical evaluation ", Acoustical Imaging, Vol. 8, 1978

3. M. FINK, " Theoretical aspects of the Fresnel focusing technique", Acoustical Imaging, Vol. 8, 1978.

4. M. FINK and M.T. LARMANDE, " Wideband Fresnel focusing array response " . Proceedings of the IEEE ultrasonic symposium, New Orleans, Sept. 1979.

ULTRASONIC IMAGING USING TRAPPED ENERGY MODE FRESNEL LENS

TRANSDUCERS*

P. Das, S. Talley, R. Kraft, H.F. Tiersten
and J.F. McDonald

Electrical and Systems Engineering Department
Rensselaer Polytechnic Institute
Troy, New York 12181

ABSTRACT

A focusing transducer utilizing trapped energy modes can be
easily fabricated by plating a number of concentric rings of electrodes on a suitable piezoelectric plate of uniform thickness. The
concentric ring structure acts as a Fresnel lens and can be used to
obtain excellent lateral focusing of ultrasonic waves. Several
transducers operating in the 2-5 MHz range have been produced using
PZT-7A as the piezoelectric material. The near field radiation
pattern has been observed directly in front of the plate using the
first order diffraction peak from a laser probe. It is confirmed
that there is good acoustic isolation between the rings and that
they radiate in the trapped energy mode. The ultrasonic radiation
pattern of the Fresnel lens has been observed at various distances
from the plate and compared with computed results which show excellent agreement. Furthermore, the axial diffraction pattern of the
lens can be optimized by adjusting ring spacings. During the course
of this study a new, previously unreported mode of energy trapping
has been discovered. The explanation for this method of trapping is
briefly discussed. Finally an acoustic through-transmission imaging
system incorporating one focusing transducer is used for imaging of
flaws in composite materials.

INTRODUCTION

Successful application of the phenomena of energy trapping in

_____
*Partially supported by NASA under Grant No. NGL33-018-003.

the construction of ultrasonic transducer arrays has now been extensively reported in the literature [1,2,3,4,5,6] by the authors. In
this paper some recent results concerning the characteristics of
thickness-extensional trapped energy modes are discussed. These new
results relate to efforts of the authors to further explore the
bandwidth limitations imposed by the basic trapped energy mode
effect. In the course of this study a new mode of thickness extensional trapping has been discovered.

The basic trapped energy mode is achieved in a frequency range
between the cutoff frequencies of thickness vibration of the electroded and unelectroded regions of a piezoelectric plate of appropriate material [1]. As in previous papers we assume the coordinate system illustrated in Fig. 1. The piezoelectric plate which
is composed of hexagonal crystalline material in class C_{6v} is shown
with the X_3 axis along the hexagonal axis of symmetry.

As reported earlier, thickness-extensional energy trapping
will be obtained between the frequencies

$$\omega_A = \frac{\pi}{2h}(\frac{\overline{c}_{33}}{\rho})^{1/2} \tag{1}$$

and

$$\overline{\omega}_c = \eta_1(\frac{\overline{c}_{33}}{\rho})^{1/2} \tag{2}$$

where $\eta_1 h$ is the lowest root of

$$\tan \eta_1 h = \eta_1 h/(k_{33}^2 + R\eta_1^2 h^2) \tag{3}$$

and

$$\overline{c}_{33} = c_{33} + e_{33}^2/\epsilon_{33} \tag{4}$$

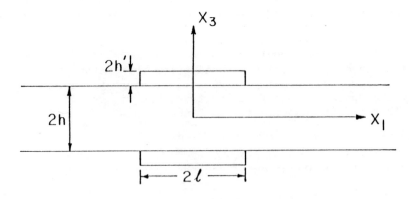

Fig. 1 Geometry of the trapped energy mode device.

$$k_{33}^2 = e_{33}^2 / (\bar{c}_{33} \varepsilon_{33}) \tag{5}$$

$$R = 2(\rho'h')/(\rho h) \tag{6}$$

Here c_{33}, e_{33}, and ε_{33} are the elastic constant, the piezoelectric constant and the dielectric constant for the x_3 axis, and $\bar{c}_{33}$ is the piezoelectrically stiffened elastic constant and R measures the relative mass loading of the metal electrode.

However, the basic range over which the trapped energy mode effect can be observed can be extended by lowering the resonance frequency of the electroded region of the plate further. This can be accomplished by introducing nondissipative external circuitry as shown in Figure 1 which can affect the frequency at which the admittance of the resulting circuit goes to infinity [6,7]. This is the condition for resonance in a lossless circuit. We deliberately do not include circuit resistance and ignore radiational losses in making such calculations. It has been shown [7] that the external reactance introduced must be inductive to lower the effective frequency ω_c which can be obtained approximately for low modes of high coupling material by solving a transcendental equation which replaced equation (3) as follows

$$\tan \bar{\eta}h = \frac{\bar{\eta}h}{k_{33}^2} - \frac{\bar{\eta}^{-3}h^3 \bar{c}_{33} LA\varepsilon_{33}^s}{2\rho h^3 k_{33}^2} \tag{7}$$

where A is the area of the electrode, L is the inductance, and again ω is related to $\bar{\eta}$ by

$$\bar{\omega}_c = \bar{\eta}(\frac{c_{33}}{\rho})^{1/2} \tag{8}$$

Technically the result is only valid down to a frequency of ω_B

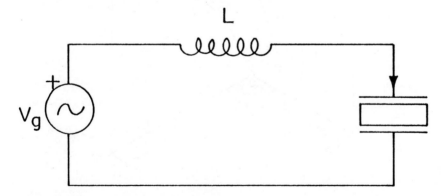

Fig. 2   Use of a tuning inductor to lower $\bar{\omega}_c$ and thereby increase bandwidth.

where

$$\omega_B = (\frac{\pi}{h})(\frac{c_{55}}{\rho})^{1/2} \ . \qquad (9)$$

This is the point at which the dispersion curve for the unelectroded region crosses back from the region of attenuation where the
wave number is imaginary along the length of the plate.

Nevertheless, if the value of the tuning inductance L in
figure 2 is increased beyond the point such that $\bar{\omega}_c \cong \omega_B$ then
further extensional trapping can be observed below the frequency
of ω_B. This is due to a complex branch of the dispersion curve
not shown in reference [1]. This is illustrated in figure 3. The
additional frequency range over which extensional trapping can be
observed can be appreciable when compared with ω_A-ω_B, the maximum
range of the basic extensional trapped energy mode effect. Unfortunately in the vicinity of ω_B there is a substantial resonance of
untrapped energy and any practical exploitation of trapping involving the complex branch must avoid exciting frequencies in a narrow
band near ω_B. This need not cause a severe distortion in image
quality.

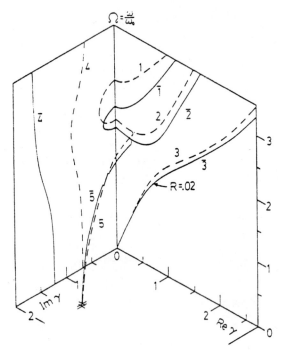

Fig. 3 Dispersion curves for the trapped energy mode device
        (with R≠0.02) showing the complex branch.

A complete analysis of the trapped energy mode device should take proper account of the finite length of the electroded region, explicitly include the effect of the complex branch and the inductor. To date an outline for the analysis covering the finite electrode length including coupling with the complex branch has been published [7] but this work is still in progress. Explicit inclusion of the tuning inductor in the dispersion curve calculation has not yet been attempted for the infinite plate.

APPLICATION OF ENERGY TRAPPING IN A FRESNEL LENS

One powerful advantage of the Trapped Energy Mode is the arbitrary shape that can be chosen for the electroded region. In particular we have chosen a concentric structure of ring electrodes. The result is in the form of a Fresnel zone plate lens if the rings are of proper width and location [8,9] as shown in Figure 4. The radii of the circular boundaries between the electroded and unelectroded regions of the piezoelectric Fresnel zone plate are given approximately by

$$r_n = \sqrt{n\lambda f} \qquad (10)$$

where f is the principal focal depth of the lens. Actually the full zone plate (with all of its rings present) acts simultaneously as a positive and negative lens with multiple "virtual" focal points at axial positions given by the distances [10]

$$z_m = \frac{f}{m} \qquad [m=0,\pm1,\pm2,\ldots] \qquad (11)$$

away from the plate. If the positive values of z_m correspond to

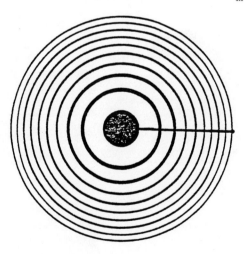

Fig. 4 Mask for Fresnel zone electrode configuration.

positions in front of the plate then the negative virtual foci
produce ring shaped side lobes at off-axial positions in any plane
perpendicular to the axis of the lens. These virtual spots and
rings are analogous to grating lobes in unfocused arrays since the
ring spacings of equation (11) undersample the aperture of the
plate. This is illustrated in Figure 5 where the diffraction
pattern is shown in the principal focal plane of the lens. As
observed by other authors [11] the energy in these ring shaped
side lobes can be substantial.

Similarly, if one examines the virtual foci along the axis as
shown in Figure 6 their effect can be equally unacceptable if some
of these foci happen to lie inside the volume to be examined. If
the zone plate focus is dynamically adjusted by varying λ or introducing additional "steering" phases for each ring then the virtual
foci will also move.

To compensate for these undesirable phenomena one can resort
to various techniques used in antenna array optimization [12]. By
varying the spacing of the rings a given region of the zone plate
diffraction pattern can be swept free of strong virtual lobes.
The technique was originally proposed by King [15] and modern
refinements of the array optimization problem are still actively

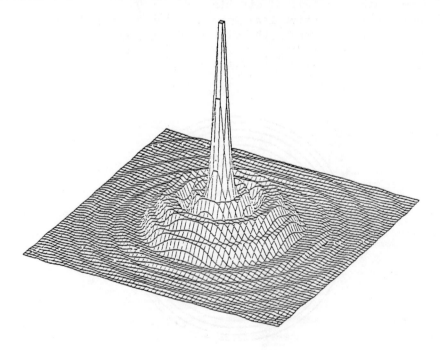

Fig. 5   Fresnel zone plate diffraction pattern in the principal
         focal plane.

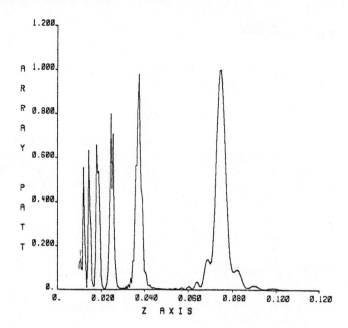

Fig. 6 Axial diffraction pattern showing false focal spots.

being investigated. The method of nonuniform ring spacing is in
contrast to the techniques discussed by Wild [16], Vilkomerson [17]
and Burckhardt, Grandchamp and Hoffman [18] which can also be employed in conjunction with spacing adjustment where a single discrete focal spot is desired.

The approach chosen by the authors has been to adjust the ring
spacings using nonlinear optimization. In particular, the procedures outlined by Madsen [19] have proved invaluable.

For the purposes of this discussion we limit our attention to
suppressing false focal spots along the zone plate's axis, although
joint sidelobe suppression in the principal focal plane is feasible.

Consider an electroded ring generating ultrasound $U(x_1, y_1)$ in
an annulus with inner radius r and outer radius R.

$$U(x_1, y_1) = \begin{cases} a & r \le \sqrt{x_1^2 + y_1^2} \le R \\ 0 & \text{otherwise} \end{cases} \tag{12}$$

The diffracted ultrasound at a point $x_0 y_0$ in an observation plane
a distance z from the plane of the annulus is approximately given
by the Fresnel integral [17]

$$U(x_0,y_0;z) = \frac{\exp(jkz)}{j\lambda z} \qquad \exp\{\frac{jk}{2z}(x_0^2 + y_0^2)\}$$

$$\int_{-\infty}^{\infty}\int U(x_1 y_1) \exp[\frac{jk}{2z}(x_1^2 + y_1^2)]$$

$$\exp\{-j\frac{2\pi}{\lambda z}(x_0 x_1 + y_0 y_1)\} \, dx_1 dy_1 \tag{13}$$

where $k = \omega/c$. We have ignored the Rayleigh-Sommerfield obliquity factor and made simplifying assumptions which amount to modeling the spherical propagation spreading loss with the single factor of $1/z$ shown in this equation. Since we are interested only in axial observation points: $x_0 = 0$ and $y_0 = 0$

$$U(0,0;z) = \frac{\exp(jkz)}{j\lambda z}$$

$$\int_{\infty}^{\infty}\int U(x_1 y_1) \qquad \exp\{\frac{jk}{2z}(x_1^2 + y_1^2)\} \, dx_1 dy_1 \tag{14}$$

Utilizing the symmetry of the annulus this integral may be transformed to polar coordinates

$$\eta = \sqrt{x_1^2 + y_1^2} \tag{15}$$

$$\phi = \text{Tan}^{-1}(x_1/y_1) \tag{16}$$

in the usual manner. Furthermore, by introducing a phase difference variable [10]

$$\psi = \pi \eta^2/\lambda z \tag{17}$$

we quickly obtain

$$U(0,0;z) = a \exp(jkz)[\exp(\frac{jkr^2}{2z}) - \exp(\frac{jkR^2}{2z})] \tag{18}$$

For a sequence of N rings driven with amplitude a_i and phase θ_i we then have

$$U(0,0;z) = \exp(jkz) \sum_{i=1}^{N} a_i \, e^{j\theta_i} [\exp(\frac{jkr_i^2}{2z}) - \exp(\frac{jkR_i^2}{2z})] \tag{19}$$

where r_i and R_i are the inner and outer radii of each annulus.

We are now free to adjust r_i and R_i subject to the constraints

$$r_i < R_i < r_{i+1} < R_{i+1} \tag{20}$$

for all i. We are also free to adjust the a_i and θ_i to try to achieve any desired axial behavior for $|U(0,0;z)|^2$ with the given number of rings. In particular, we can endeavor to adjust r_i, R_i, a_i and θ_i to suppress the false focal spots. Note however that only the dependence of $U(0,0;z)$ on a_i is linear. All other dependencies are very nonlinear. Hence, the optimization problem may be cast in the form of minimizing the maximum error, $E(\underline{V})$:

$$E(\underline{V}) = \max_{z} |U(0,0;z) - U_D(z)| \qquad (21)$$

over the z axis by adjusting the vector of parameters

$$\underline{V} = \{r_i, R_i, a_i, \theta_i; \quad i = 1,2,\ldots,N\} \qquad (22)$$

The Madsen procedure requires the specification of explicit expressions for the derivatives of the objective function $E(\underline{V})$ with respect to each parameter to be adjusted. These derivatives are given in Appendix A. Initially the ring spacings can be taken as Fresnel zone plate values. The optimization proceeds via a series of iterations similar to those of the Newton Raphson method except with step sizes adjusted to insure that the actual error reductions are close to those predicted by a locally linearized objective function.

As an example, consider the unoptimized diffraction pattern of a Fresnel array with N = 14 elements as was shown in Figure 6. After optimization of the gains and spacings (but not the phases), the axial diffraction pattern is shown in Figure 7.

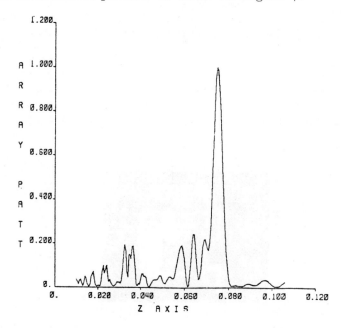

Fig. 7 Axial diffraction pattern after optimization

Two difficulties occur. First, due to the oscillatory nature
of the objective function (21) multiple extrema arise and it becomes necessary to examine them all to obtain the true global
optimum. Second, optimization of even a sparse array tends to
require that the spacing and thickness of at least a few of the
rings be quite small. Hence the interelectrode isolation provided
by the trapped energy mode is essential. Failure to achieve such
isolation will result in the ineffectiveness of the most tightly
spaced rings resulting in substantial changes in the diffraction
pattern.

EXPERIMENTAL TRAPPED ENERGY MODE BANDWIDTH STUDIES

In order to verify that the tuning inductor of Fig. 2 could
be utilized in a practical sense to extend the bandwidth of a
trapped energy array a series of measurements was performed on a
system of 3 electrodes on PZT as shown in Figure 8. Many induc-
tances in the range of 1.0µH to 30µH were examined. For this
particular plate material the optimum inductance with respect to
bandwidth broadening was 6.2µH, and the band pass characteristic
is shown in Figure 9. To verify whether the energy trapping was
present a laser optic scan was made using the system described in
earlier papers [1]. In the absence of any inductor trapping would
occur between the frequencies 3.5 MHz and 4.0 MHz, with the peak
at 3.6 MHz as shown in Figure 9. With an inductance of 6.2µH the
frequency response was observed as shown in Figure 10. Laser
spatial scans 1.8mm in front of the plate for points G,H,K,M and
P are shown in Figure 11. Trapping is evident in G,K and P but
not at H or M. Evidently trapping is present due to the complex
branch in Figure 3. There is also evidence of trapping on a higher
branch of the dispersion curve not drawn in Figure 3 which produces
trapping above 4.0 MHz. Further evidence of this is supported by
the multipeaked modal structure in the electroded region [6].
Extensional trapping on the complex branch has not been previously
reported, although much of the available bandwidth exists where it
was predicted to be in earlier papers.

Fig. 8 Experimental trapped energy mode array for bandwidth
        measurements.

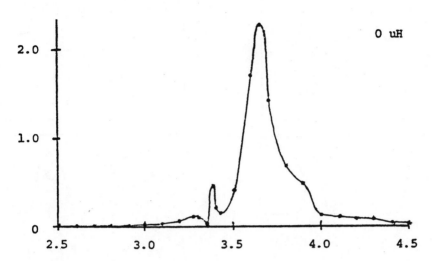

Fig. 9   Frequency response of the trapped energy mode device
with no tuning inductor.

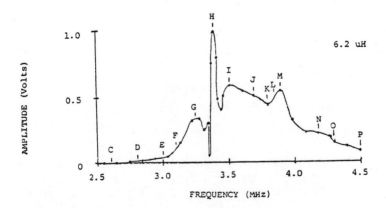

Fig. 10   Frequency response corresponding to Figure 9 but with a
6.2μH tuning inductor.

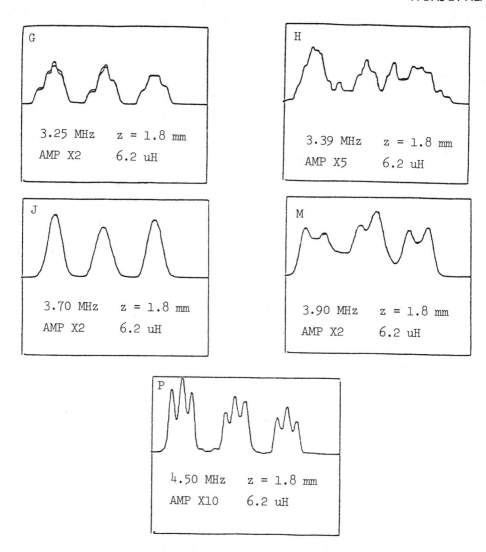

Fig. 11 Laser spatial scans to verify energy trapping at points
         G,H,J,M,P in Figure 10.

IMAGING EXPERIMENTS WITH THE FRESNEL LENS

     In an effort to gain some appreciation of the imaging quality
obtainable with energy trapping a Fresnel lens was constructed
using the mask shown in Figure 4. A laser scan of an unoptimized
lens in the principal focal plane is shown in Figure 12 at 16cm.
The lateral width of the primary peak is very close to the theoretical minimum of one wavelength as expected.

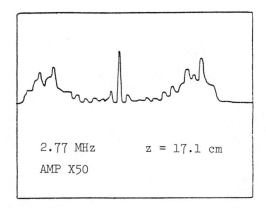

2.77 MHz   z = 17.1 cm

AMP X50

Fig. 12 A spatial laser scan in the principal focal plane.

In order to test the image quality without elaborate electronic scanning methods a mechanical scan was performed. The application in question involved detection of delamination or other flaws
in composite materials.

The graphite-epoxy composite materials studied have about 16
layers of fibers oriented in various directions to improve their
tensile resistance. The materials were subjected to tensile
stresses of about 70,000 psi before imaging them. The objective
of the experiment was to study what damage the tensile stress
caused on the materials. The materials also had holes drilled in
them with either a carbide tipped drill or a high speed steel (HSS)
drill and the damage density around the holes was also of special
interest. The images of a crossply composite with fiber orientation of 0° and 90° is shown in Figure 13. The dark spots on the
image correspond to high transmitted intensity as this when applied
to the Z-axis of the scope suppresses the intensity. The bright
spots result from low received intensity and can be a result of
acoustic damage in the material. The dark spots are either due to
a hole in the object or due to a clean, undamaged area on the
object. Figure 14 shows two images of a composite with fiber
orientations at 0° and ±45° to the axis of tensile stress. The
results show a damage pattern oriented at a 45° angle. This is
understandable as the composites have fibers at that orientation.

CONCLUSION

The thickness-extensional trapped energy mode array is particularly suitable for implementation of a two dimensional Fresnel
zone plate over a wide range of frequencies. Lateral beam resolutions on the order of one wavelength are achievable, indicating
excellent interelectrode isolation. The bandwidth of the device

Figure 13   Image of a crossply composite

Fig. 14 Ultrasonic image of a ± 45° orientation composite.

can be considerably enlarged with a tuning inductor, although some of this is achieved by using branches of the dispersion not discussed in the original papers on this subject by the authors. The trapping is sufficiently strong to permit optimization of electrode spacings to suppress spurious virtual foci and ring sidelobes.

ACKNOWLEDGMENT

It is a pleasure to acknowledge the technical assistance of Mr. S. Joshi and L. Esser for Figures 13 and 14 and fruitful discussions with Dr. B.K. Sinha.

REFERENCES

[1]   H.F. Tiersten, J.F. McDonald, M.F. Tse and P. Das, "Monolithic Mosaic Transducer Utilizing Trapped Energy Modes", in Acoustical Holography, (L. Kessler, Editor) Vol. 7, Plenum Press, New York 1977, pp. 405-421.

[2]   H.F. Tiersten, J.F. McDonald and P.K. Das, "Two-Dimensional Monolithic Transducer Array," in Proc. 1977 I.E.E.E. Ultrasonics Symposium, 1977, pp. 408-412.

[3]   H.F. Tiersten, B.K. Sinha, J.F. McDonald and P.K. Das, "On the Influence of a Tuning Inductor on the Bandwidth of Extensional Trapped Energy Mode Transducers", Proc. 1978 I.E.E.E. Ultrasonics Symposium, 1978, pp. 163-166.

[4]   P. Das, G.A. White, B.K. Sinha, C. Lanzl, H.F. Tiersten and J.F. McDonald, "Ultrasonic Imaging Using Monolithic Mosaic Transducer Utilizing Trapped Energy Modes", in Acoustical Imaging, Vol. 8, (A.F. Metherell, ed.) 1978, pp. 119-135.

[5]   H.F. Tiersten and B.K. Sinha, "An Analysis of Extensional Modes in High Coupling Trapped Energy Resonators", Proc. 1978 I.E.E.E. Ultrasonics Symposium, 1978, pp. 167-171.

[6]   P.K. Das, S. Talley, H.F. Tiersten and J.F. McDonald, "Increased Bandwidth and Mode Shapes of the Thickness-Extensional Trapped Energy Mode Transducer Array", Proc. 1979 I.E.E.E. Ultrasonics Symposium, 1979, pp. 148-152.

[7]   H.F. Tiersten and B.K. Sinha, "Mode Coupling in Thickness-Extensional Trapped Energy Resonators", Proc. 1979 I.E.E.E. Ultrasonics Symposium, pp. 142-147.

[8]   M.V. Klein, Optics, J. Wiley & Son, New York, 1970, Chapter 8.

[9]   J.W. Goodman, Introduction to Fourier Optics, McGraw Hill, New York, 1968.

[10] S.A. Farnow and B.A. Auld, "An Acoustic Phase Plate Imaging Device", in Acoustical Holography, Vol. 6, (N. Booth, Ed.), 1975, pp. 259-273.

[11] D. Vilkomerson and B. Hurley, "Progress in Annular-Array Imaging", in Acoustical Holography, Vol. 6, (N. Booth, Ed.), 1975, pp. 145-164.

[12] R.P. Kraft, F. Ahlgren and J.F. McDonald, "Minimax
      Optimization of Two-Dimensional Focused Nonuniformly Spaced
      Arrays", Proc. 1979 I.E.E.E. Int. Conf. on Acoust., Speech
      and Signal Processing, 1979, pp. 286-289.
[13] D.D. King, R.F. Packard and R.K. Thomas, "Unequally Spaced
      Broad Band Antenna Arrays", IRE Trans. on Antennas and
      Propagation, July 1960, pp. 880-884.
[14] J.P. Wild, "A New Method of Image Formation with Annular
      Apertures and an Application to Radio Astronomy", Proc. Roy.
      Soc. 286A, 499 (1965).
[15] D. Vilkomerson, "Acoustic Imaging with Thin Annular Apertures",
      in Acoustical Holography, Vol. 5, (P.S. Green, Ed.), 1973,
      Plenum Press, pp. 283-316.
[16] C.B. Burkhardt, P.A. Grandchamp and H. Hoffman, "Methods for
      Increasing the Lateral Resolution of B-Scan", in Acoustical
      Holography, (P.S. Green, Ed.), Vol. 5, 1973, Plenum Press,
      pp. 391-413.
[17] K. Madsen, "An Algorithm for Minimax Solution of Over-
      Determined Systems of Nonlinear Equations", J. Inst. Math.
      Applications, Vol. 16, pp. 321-328, Dec. 1975.

APPENDIX A

We can rewrite equation (19) as

$$F(\underline{V}) = \sqrt{(R-R_D)^2 + (I-I_D)^2} \tag{A-1}$$

where

$$U = R + jI$$
$$U_D = R_D + jI_D \tag{A-2}$$

Then the derivative of $F(\underline{V})$ with respect to an arbitrary parameter $\dot{v}$ is given by

$$\frac{\partial F(V)}{\partial v} = \frac{2(R-R_D)\frac{\partial R}{\partial v} + 2(I-I_D)\frac{\partial I}{\partial v}}{\sqrt{(R-R_D)^2 + (I-I_D)^2}} \tag{A-3}$$

The individual derivatives used in (A-3) are given by

$$\frac{\partial R}{\partial a_i} = [c_i - C_i] \tag{A-4}$$

$$\frac{\partial I}{\partial a_i} = [s_i - S_i] \tag{A-5}$$

$$\frac{\partial R}{\partial \theta_i} = a_i[c_i s_i - C_i S_i] = \frac{\partial I}{\partial \theta_i} \tag{A-6}$$

$$\frac{\partial R}{\partial r_i} = a_i c_i s_i r_i k/z = - \frac{\partial I}{\partial r_i}$$ (A-7)

$$\frac{\partial R}{\partial R_i} = a_i C_i S_i R_i k/z = - \frac{\partial I}{\partial R_i}$$ (A-8)

and

$$c_i = \text{Cos} \left(\frac{kr_i^2}{2z} + \theta_i \right) \qquad C_i = \text{Cos} \left(\frac{kR_i^2}{2z} + \theta_i \right)$$

$$s_i = \text{Sin} \left(\frac{kr_i^2}{2z} + \theta_i \right) \qquad S_i = \text{Sin} \left(\frac{kR_i^2}{2z} + \theta_i \right)$$ (A-9)

NONLINEAR EFFECTS IN ACOUSTIC IMAGING

Thomas G. Muir

Applied Research Laboratories
The University of Texas at Austin
Austin, Texas 78712

INTRODUCTION

The fundamental mechanisms of sound propagation are nonlinear. At low frequencies and/or intensities, linearity can usually be assumed but this is often not the case in ultrasonic imaging.

The classical nonlinear process is one of waveform distortion, shock formation and saturation of the radiated field. These effects result from the fact that the speed of an acoustical disturbance is not a constant, but depends on the intensity distribution within the waveform. The velocity v at a point x on a waveform should really be expressed as

$$v(x) = c_o + \left(1 + \frac{1}{2}\frac{B}{A}\right) u(x) \tag{1}$$

where c_o is the infinitesimal sound speed constant, B/A is the parameter of nonlinearity of the medium[1] and u(x) is the particle velocity at position x on the waveform.

If absorption is not dominant, a cumulative distortion in the propagating waveform will occur as shown in Fig. 1. The distortion is equivalent to the generation of new frequency components, which are harmonically related to the original frequency through a simple Fourier series[2]. In terms of the pressure p, this series may be expressed as

$$p = p_o \left(\frac{r_o}{r}\right)^a \sum_{n=1}^{\infty} B_n(\sigma) \sin n \left[\omega t - k (r - r_o)\right] \quad , \tag{2}$$

93

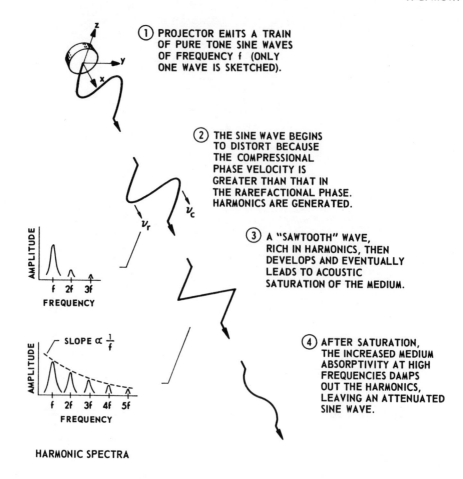

FIGURE 1

CASE HISTORY OF A HIGH INTENSITY SOUND TRANSMISSION ILLUSTRATING
FINITE AMPLITUDE DISTORTION AND HARMONIC GENERATION

where p_o is the pressure at the source, r_o is the radius of curvature of the source, r is range from the center of curvature, B_n (σ) is a Fourier coefficient, dependent on the shock parameter σ (to be defined below), n is the harmonic number, ω the original (angular) frequency, and k is the wave number. Here $a = 1$ for spherically converging as well as diverging waves and $a = o$ for plane waves. The Fourier coefficients may be calculated from formulas given by Blackstock.[2]

If the original wave is of sufficient intensity, shocks will form, as shown in Fig. 1. Very intense waves may experience acoustic saturation,[3] a form of amplitude limiting in the propagating radiation.

In saturation, an increase in source intensity produces no further increase in intensity arriving at remote points in the field.

Some of these effects may be deleterious in acoustic imaging, while others may be used to advantage. The purpose of the present paper is to comment on nonlinear effects in the light of imaging and to offer experimental demonstrations of their utilization in practical applications. Two recent papers[4,5] may be cited as a tutorial basis to the material to be presented here.

ESTIMATING NONLINEAR EFFECTS

How does one know if it is necessary to consider nonlinear effects in an imaging system? Several simple criteria exist and can be quite helpful in answering such questions.

One such criteria is the shock parameter σ, which may be expressed[6] for plane waves as

$$\sigma = \beta \epsilon k \int_{0}^{X} e^{-2\alpha r} dr \quad , \tag{3}$$

for spherically diverging waves as,

$$\sigma = \beta \epsilon k r_o \int_{r_o}^{X} \frac{\exp\left[-2\alpha\left(r-r_o\right)\right]}{r} \, dr \quad , \tag{4}$$

and for spherically converging waves as,

$$\sigma = -\beta \epsilon k r_o \int_{r_o}^{X} \frac{\exp\left[-2\alpha\left(r-r_o\right)\right]}{r} \, dr \quad , \tag{5}$$

where the nonlinearity parameter, $\beta = 1 + \frac{1}{2}\frac{B}{A})$, the acoustic Mach number $\epsilon = u_o/c_o$, where u is the particle velocity at the source, and α = the attenuation coefficient of the medium at the frequency f of the original radiation. For fresh water at 25° C, $\beta = 3.6$,[1] and $\alpha = 24 \times 10^{-17} f^2$, cm^{-1}sec$^{2.7}$ If the peak source intensity I_o is known, ϵ becomes $(2I_o \cdot 10^7/\rho_o c_o^3)^{1/2}$, where ρ_o is the static density and I_o is expressed in Watts/cm^2.

For many applications, the absorption is not significant over the range of interest. The shock parameters are then stated for plane waves as

$$\sigma = \beta \epsilon k x \quad , \tag{6}$$

for specifically diverging waves as

$$\sigma = \beta \epsilon k r_o \ln \frac{r}{r_o} \quad , \tag{7}$$

and for spherically converging waves as

$$\sigma = -\beta \epsilon k r_o \ln \frac{r}{r_o} \tag{8}$$

In practice, it is a good approximation to use the plane wave formulas in the immediate nearfield, spherically diverging formulas in the far field, and spherically converging results for focused transducers, including those self-focused at intermediate nearfield ranges.

When σ is $<<1$, nonlinear effects are small and may be neglected.

When $\sigma = 1$, the wave will have distorted to the point at which a vertical discontinuity will have formed at the zero axis crossing of the pressure waveform, thus creating a weak shock. The fundamental (original frequency) component will have lost 1 dB to harmonic conversion and the 2nd, 3rd, and 4th, harmonic amplitudes are 8, 12, and 15 dB below the fundamental, respectively.

When $\sigma = 3$, a strong "sawtooth" shock will have formed. The same respective harmonic amplitudes will be 6, 10 and 12 dB below the fundamental. The distortion mechanism may remain active beyond this point, but since multivalved waveforms are not permitted in nature, this tendancy must be matched by nonlinearly induced absorption at the shock fronts. The wave then greatly acclerates its attenuation in a self-destructive fashion.

NONLINEARITY OF EXISTING SYSTEMS

Some numerical examples of the degree of nonlinearity in several practical applications are appropriate. First, consider an offshore seismic source which produces a 100 msec transient signal with an energy of one megajoule (≈ 288 W/cm^2, equivalent to 1 lb of TNT). Use of Eq. 7 for the equivalent continuous wave parameters of the transient source leads to a value of $\sigma = 1.4 \times 10^{-3}$, for source to sediment separations around 1000 meters. Thus, the seismic source should not produce nonlinear acoustic effects over the path of propagation to the ocean bottom.

Second, consider the sonar systems introduced on destroyers and frigates in the era following the second World War. Frequencies near 10 kHz were emitted from sources about a meter in diameter, with radiated acoustic powers of several tens of kilowatts. Use of Eq. 7 at ranges of about 1 kM (where ordinary absorption begins to be significant) leads to σ parameters of a few tenths. Thus, the post WW II sonars were only approaching nonlinearity.

Next, consider the side scan sonars used commercially to image bottom objects, pipelines, offshore equipment, and shipwrecks. Many of these utilize rectangular transducers a meter or so long and

several centimeters high, with frequencies in the neighborhood of
100 kHz and source levels on the order of 225 dB re 1μPa at 1 m.
It can be shown (by use of the cylindrical wave equivalent[2] of Eq. 7)
that the σ parameter approaches unity at a range of 20 m and becomes
twice that near the Rayleigh distance around 60 m. These devices are
definitely nonlinear in that they will produce harmonics and will
experience some nonlinearly induced attenuation.

    Finally, consider the biomedical instruments used in 3D imaging
for diagnostic work. A recent survey by Carson, et al,[8] reports
comparative measurements on these devices. The parameters of a
typical system used in fetal scanning, echocardiology, etc., are
illustrated in Fig. 2. As can be seen, the beam is focused to a
finite spot some 3.7 mm in diameter, as can be estimated from the

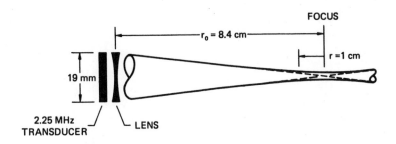

FIGURE 2
FOCUSED IMAGING SYSTEM USED IN
MEDICAL DIAGNOSTICS

intensity focusing gain G which is about 30. Since the theory em-
braced by Eq. 5 assumes an infinitesimal focus, it is appropriate
to compute the shock parameter σ at a range r of 1 cm, where geometrical focusing (dashed line) has an equivalent spot size.

    Some results for a water medium are shown in Fig. 3 which plots
σ as well as the peak intensity near the focus, I_f, as a function of
peak source intenisty, I_o. For values of I_o in the neighborhood of
5-15 Watts/cm^2, we should expect shock parameters of 1 to 2, which
is also well into the nonlinear range for a water medium. The departure from linearity in the input/output intensity curves is also
evident, and is manifested as a loss in the focusing gain of the
system. Special problems associated with this type of focused system are discussed in a subsequent section.

THE AMPLITUDE-FREQUENCY-ABSORPTION TRADEOFF

    The trend in the previous comparisons of system nonlinearity
showed the progressive increase in the shock parameter as we went

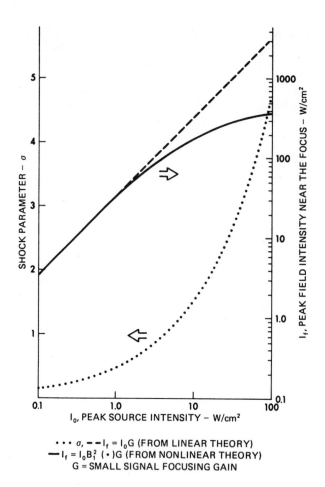

• • • σ, — — $I_f = I_0 G$ (FROM LINEAR THEORY)
— $I_f = I_0 B_i^2$ (•)G (FROM NONLINEAR THEORY)
G = SMALL SIGNAL FOCUSING GAIN

FIGURE 3
SHOCK PARAMETER versus INPUT/OUTPUT INTENSITY

from low to high frequency devices. This is due to the cumulative
nature of nonlinear effects. One simply accumulates a little nonlinearity with each wavelength of propagation path and these add up
fast at short wavelengths until absorption takes its toll and reduces
the radiation back to a linear entity.

Actually, the tendancy toward nonlinearity is summed up in the
βεk product, which appears in most solutions to problems of this
type. Increasing the magnitude of this product accentuates nonlinear
effects. This can be done with media of high nonlinearity constant
(β parameter), by using high amplitudes (ε parameter, which = $p/\rho_0 c_0^2$)
or by going to high frequencies (k parameter, = $2\pi f/c_0$).

The tendancy to abate nonlinearity involves diminishing the $\beta\epsilon k$ product as well as increasing the absorption. Nonlinear effects are effectively precluded when the radiation has propagated far enough to be absorbed. For plane waves, this occurs at $r = \alpha^{-1} - (\beta\epsilon k)^{-1}$.[2]

Although high frequency radiations are more quickly absorbed, there are many high frequency applications having substantial Mach numbers, that experience nonlinearity over the short range paths for which they were designed. On the other hand, low frequency radiations with small Mach numbers can produce nonlinear effects at extremely long ranges, since the medium absorption is usually quite low.

In the final analysis, these effects have to be considered quantitatively before the potential nonlinearity of an application can be ascertained.

EXPERIMENTAL TESTS OF NONLINEARITY

Although calculations are useful, the proper measurement of the nonlinear state of an application is always the ultimate proof. Some quite simple experimental tests can be made fairly easily, depending on the availability of measurement tools.

The best measurement tool is the wideband probe hydrophone. It should be small compared to both the dimensions of the beam and, if possible, to the acoustic wavelength. This includes the wavelength of the harmonic radiations, or at least several of the lower harmonics. Preferably, the hydrophone should have a fairly flat response, although variations in sensitivity with frequency can usually be accounted for by other means. The probe hydrophone, in conjunction with standard laboratory instruments such as amplifiers, oscilloscopes, filters, etc., can be used in the following tests illustrated in Fig. 4.

Amplitude Response Test

Locate the probe at some remote point in the acoustic field of the projector (near the focus, for example) and monitor the waveform as a function of input amplitude in terms of drive current and/or voltage. If the system is linear, there should be a 1:1 ratio in the change of input to output quantities. A departure from linearity will show up in the fundamental component (processed through a band-pass filter) as something less than 1:1 linearity between input and output, as the drive to the transducer is increased. Note that this portion of the test does not require a wideband hydrophone. If one is available, however, it can be used with bandpass filter(s) to mea-sure the amplitudes of harmonic components. If the system is non-linear, these should begin to increase with the input power before they saturate. With a wideband receiver system, the existence of harmonics will be noticed as a distortion dependent on the driver amplitude.

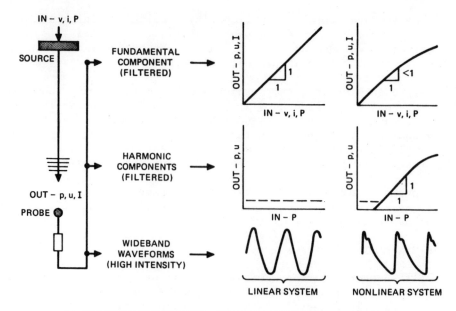

(a) AMPLITUDE RESPONSE TEST – DETERMINES EXIXTENCE OF NONLINEARITY

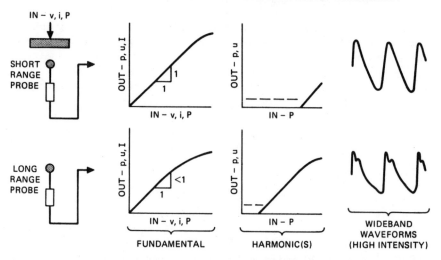

(b) AMPLITUDE RESPONSE versus RANGE TEST – DETERMINES SOURCE OF NONLINEARITY

FIGURE 4
EXPERIMENTAL TESTS OF NONLINEARITY

Amplitude Response vs Range Test

The preceding test may not be self-sufficient if the power amplifier is itself nonlinear. Since this if frequently the case, and since many transmitters directly radiate harmonic components, it is often necessary to measure the amplitude response at more than

one point in the field. Since nonlinear acoustical effects are
cumulative with range, a comparison of the amplitude response at a
short range (near the transducer) with that at other ranges should
identify the source of nonlinearity. Nonlinear acoustic distortion
and its harmonic components should increase with distance at short
ranges from the source. Measurements of this effect at frequencies
near 0.5 MHz have been reported.[9]

Some of these measurements can be conducted with sensors other
than hydrophones. Radiometer techniques, for example, have been used[5]
in amplitude response measurements of nonlinear phenomena. The radiometer approach normally precludes the preferential separation of
individual harmonics, so one must consequently deal with the total
field of all harmonic components. Departure from linearity can nonetheless be measured, especially in high frequency/intensity fields.[5]
Here, the departure from linearity represents the energy lost to nonlinearly induced absorption at the harmonic frequencies. In contrast,
hydrophone measurements on, say the fundamental component may represent the prior stage of the nonlinear process, namely the conversion
of energy to the harmonics which are eventually absorbed.

Measurements of nonlinear effects with thermocouples are perhaps
more complicated and are not well established. The thermocouple
measures temperature and infers acoustic intensity at an assumed
frequency, when the thermoviscous absorption and heat conduction
constants are known. The multiplicity of frequency components in
nonlinear waveforms with their accelerated rates of heat deposition
would appear to necessitate a special formulation of thermocouple
measurements in this case.

SPECIAL PROBLEMS INVOLVING FOCUSED SYSTEMS
WITH APPLICATIONS TO MEDICAL DIAGNOSTICS

The survey by Carson, et. al.,[8] reports some very high peak
intensities near the focus of imaging instruments used in medical
diagnostics. Many of these range from 10^2 to 10^3 Watts/cm^2. Problems associated with sharp "spikes" transmitted by impuse type
pulsers and accentuated by hydrophone resonances were mentioned as
possible sources of measurement error.[8] Some typical waveforms
resulting from this survey are shown in Fig. 5.

It could be that the measurements were quite accurate and that
these as well as other explanations for unusually high peak intensities are warranted. It is known, for example, that the waveform
at the focus undergoes a phase shift.[10-13] For converging spherical
waves, the phase shift is 90° at the focus and 180° beyond. Some
experiments on transient signals[13] indicate an increase in peak
intensities near the focus that are much higher than expected from
focusing gain alone. Presumably, these anomalous increases are associated with the phase shift. It is tempting to speculate that the

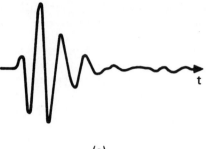

(a)

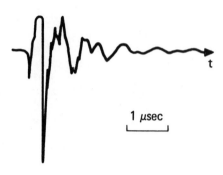

1 μsec

(b)

FIGURE 5
WAVEFORMS FROM IMPULSE SOURCE USED IN MEDICAL DIAGNOSTICS
(a) Generic Type Pulse; (b) Pulse Measured at Focus
(Taken from Carson, et al., Ultrasound in Med. and Biol., 1978)

waveform might be differentiated at the focus, thereby sharpening
its peaks, which may already have been distorted and steepened by
nonlinear effects. If nonlinear processes act on pulses like that
of Fig. 5, they will become evident in the peak excursion of the
transient-like signal.

In any case, measurements of peak intensities near the focus
should be dealt with carefully. Unfortunately, almost all nonlinear
acoustic calculations deal with peak intensities. However, it is the
peak intensity at the source that is input into most computations.

For reasons just stated, it may be inaccurate to extrapolate
the peak focal plane intensity back to that at the source. It is
probably better to use average intensity measurements at the source
as a starting point. We can obtain the average source intensity
during the pulse by dividing the time average source intensity by
the duty cycle, $\Delta t/\Delta T$, where Δt is the pulse length and ΔT is the
pulse repetition period. The peak source intensity during the pulse

must then be found. For an impulse type transmitter, like those popular in diagnostic instruments, there is a finite ratio between peak and average intensity within the pulse. For pulse shapes like that of Fig. 5, this ratio may be on the order of 3 to 4.

With this approach, one can argue that the most powerful instruments in the 2.5-3.5 diagnostic band previously surveyed[8] may have peak source intensities ranging from a few Watts/cm^2 to around 15 Watts/cm^2. With this sort of source intensity, values of σ up to 2 or so seem possible when the instruments are operated in a water medium. This would lead to extrapolated peak intensities at the source up to 200 Watts/cm^2, which is somewhat lower than that measured[8]. Although the waveform of Fig. 5(b) is rather complicated, it may contain nonlinear components that could be identified by application of the tests of Fig. 4.

Recalling the frequency tradeoff, it is worthwhile to comment on the calculated nonlinear properties of focused diagnostic instruments in the ophthalmic band, in the vicinity of 10 to 20 MHz. Two such instruments were included in the aforementioned survey, but incomplete data prevents a fair estimate of their peak intensity properties. An earlier instrument, reported by Giglio, et. al.[14], would appear to have σ parameters in excess of 10, which indicates very strong nonlinearities. Distorted waveforms were not reported[14], due perhaps to the finite bandwidth of the hydrophone.

Comments on the biological significance of nonlinear ultrasonic exposure are given in the discussion section.

IMAGING WITH THE HARMONIC RADIATIONS

Up to now, we have discussed nonlinear effects in the light of their existence in existing systems, some of which are unwittingly nonlinear. More interesting, perhaps, are those systems which were deliberately designed to utilize nonlinear effects. Although many of these have been used as sonars in ocean depth surveys as well as ocean bottom mining, we will concentrate here on experimental systems designed to explore imaging applications.

One of the best of these is the use of the harmonic radiations in pulse echo scanning. The harmonics are attractive for this type of work because of their great directivity. It has been shown[9] that the directivity function of the nth harmonic is equal to that of the fundamental raised to the nth power, or

$$D_n(\theta) = D_1^{\,n}(\theta) \tag{9}$$

Thus, as one progresses to the higher harmonics, the width of the beam decreases and the minor lobe suppression increases. These are desirable features in imaging applications.

Several experiments have been conducted along these lines, at Applied Research Laboratories' Lake Travis Test Station. The experimental apparatus is depicted in Fig. 6. A transducer measuring 1.0 x 0.1 m is energized at a fundamental frequency of 100 kHz in

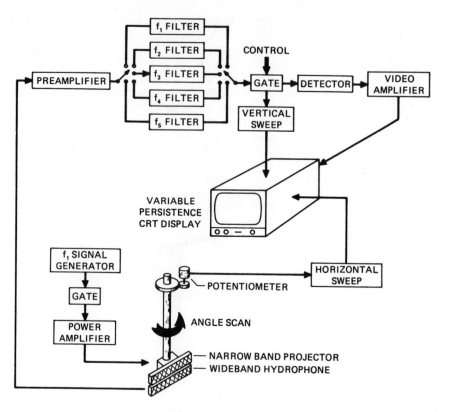

FIGURE 6
HARMONIC IMAGING SONAR

200 μsec cw pulses with 1.3 k Watts of pulse power. With a transducer efficiency of 50%, this yields a source intensity of 0.7. Watts/cm^2. Since the beam is only 0.8º x 8º in half power width, it has a high directivity and source level, which means that harmonic radiations are created as the high intensity pulse propagates through the water. Echoes from targets at ranges of 50 to 100 m are received by a wideband hydrophone, amplified, and passed through any of a series of narrow band pass filters tuned to the harmonic frequencies. The harmonic signals are processed and displayed in a range/bearing (B scan) format.

A typical illustration of some of the advantages of harmonic echo scanning is shown in Fig. 7. A point target at a range of 100 m

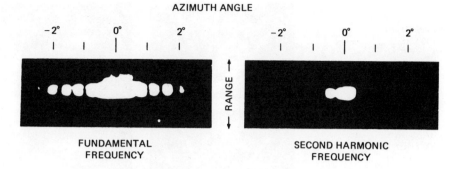

FIGURE 7
NONLINEAR HARMONIC SONAR DISPLAY RANGE VS
AZIMUTH ANGLE SCAN OF A POINT TARGET

is depicted with both the fundamental and its second harmonic component. The fundamental radiation has a half power resolution of
$0.8°$ or 1.4 m in bearing and 0.3 m in range, with a minor lobe suppression of 13 dB. A scan taken at the fundamental frequency shows
the target in a condition of display saturation which is a quite
common occurrence in pulse echo displays. In principle, it is possible to reduce the video gain to avoid the resulting blooming effect
when the target is in the major lobe as well as to diminish the tendancy for minor lobes to display false data as they are scanned
across the target. In practice, however, perfect electronic adjust-
ments are not always made, especially without some a priori knowledge
of the target size, strength, shape and orientation.

The second harmonic data demonstrate the alleviation of these
undesireable effects, even though the video gain was normalized to
the same peak input voltage. With the second harmonic, the half
power beamwidth was reduced to $0.5°$, yielding a resolution cell 0.9 m
in azimuth by 0.3 m in range. Also, the minor lobe suppression was
increased to around 20 dB (it should have been 26 dB). The target
display is much more compatible with the instrument's resolution
capability and the adjustment of the display is much less critical.

Further demonstration of the dynamics of harmonic echo scanning
is afforded by the data of Fig. 8 which depicts the viewing of larger
targets. Here the beampatterns are shown at left for the 1st through
5th harmonics of the experiment at hand. Range/bearing images of a
cylinder are shown at right, with each display corresponding to a
particular harmonic echo scan, each made at a range of 100 m. With
the fundamental radiation, the beam is almost as large as the target.
As a result, one really only sees some large blobs, probably due to
echos from each end of the target. Reception of these echoes with
the minor lobes is also evident. The sidelobe echoes are not a

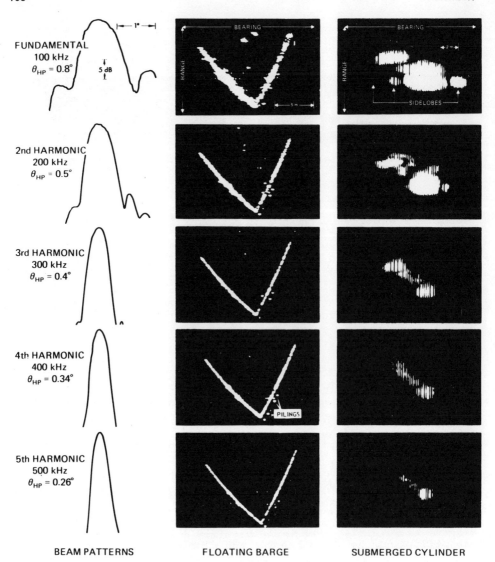

FUNDAMENTAL
100 kHz
$\theta_{HP} = 0.8°$

2nd HARMONIC
200 kHz
$\theta_{HP} = 0.5°$

3rd HARMONIC
300 kHz
$\theta_{HP} = 0.4°$

4th HARMONIC
400 kHz
$\theta_{HP} = 0.34°$

5th HARMONIC
500 kHz
$\theta_{HP} = 0.26°$

BEAM PATTERNS   FLOATING BARGE   SUBMERGED CYLINDER

FIGURE 8
IMAGING WITH THE HARMONIC RADIATIONS

problem in the second harmonic data. With the third harmonic, the
30° orientation of the target becomes clearly evident. Data for the
fourth harmonic shows a fair representation of the size of the target. This trend continues to the fifth harmonic, although the display has become somewhat sparse, perhaps for nonacoustic reasons.

The data displayed in the center column of Fig. 8 are for the viewing of an object much larger than the original beam. These are range/bearing scans of a barge having a diving platform with pilings attached. As can be seen, the definition in the field of view becomes progressively better with increase in harmonic number, angular resolution, and minor lobe suppression. At about the fourth harmonic, the four pilings (15 cm in diameter) become clearly evident.

Although it can be argued that the harmonic radiations are subject to more absorption than the fundamental, it is nonetheless true that many applications are not precluded by absorption limited effects. It appears reasonable, for example, to acquire crude images with the low frequency fundamental for initial study and then develop sharper images with the harmonics, as the range-dependent absorption factor permits.

DISCUSSION

We have treated the nature and identification of most nonlinear effects pertinent to imaging systems from both theoretical and experimental approaches. The amplitude-frequency-absorption tradeoff was advanced as a crucial factor in the significance of nonlinear effects for any given application.

It was argued that existing ultrasonic systems such as offshore, side-scan sonars and medical diagnostic devices are more prone to nonlinearity than are low frequency seismic sources and deep ocean sonars. Examples of the advantageous use of nonlinearly created harmonic radiations were offered. Some of the disadvantages of nonlinearity (saturation, loss of focusing gain) were also mentioned and are discussed in greater detail elsewhere[4,5]. The significance of nonlinear entities other than shocks and harmonics is emerging as a viable topic, also discussed elsewhere[15,16].

Although nonlinear acoustics offers new avenues for discovery and exploration, it should be seriously evaluated and approached with caution in certain areas. Medical applications are sensitive in this respect as they appear to require special consideration in terms of health hazard dosimetry. As shown in this paper, some existing medical systems may be unwittingly nonlinear, and may likely produce shock waves. Whether or not shock waves are produced in human tissue insonified by these devices and whether or not these shock waves are harmful, emerges as a worthwhile question. We will not be able to answer this question without further research. We are in a state of infancy in this area as is illustrated by the simple fact that we do not, for example, have any knowledge of the nonlinearity of animal tissue. If nonlinear effects pose a health hazard, it may be necessary to reduce the the intensities of medical ultrasonic devices, by placing more emphasis on the signal processing of echoes, as is done in both radar and sonar.

ACKNOWLEDGEMENTS

This work was supported in part by the U.S. Navy Office of Naval Research and by the Bureau of Radiological Health, U.S. Food and Drug Administration. L. L. Mellenbruch is thanked for his assistance in acquiring the harmonic imaging data.

References

1. R. T. Beyer, Parameter of nonlinearity in fluids, J. Acoust. Soc. Am. 32:719 (1960).

2. D. T. Blackstock, Connection between the Fay and Fubini solutions for plane sound waves of finite amplitude, J. Acoust. Soc. Am. 39:1019 (1966).

3. J. A. Shooter, T. G. Muir, and D. T. Blackstock, Acoustic saturation of spherical waves in water, J. Acous. Soc. Am. 55:54 (1974).

4. T. G. Muir and E. L. Carstensen, Prediction of nonlinear acoustic effects at biomedical frequencies and intensitites, submitted to J. Ultrasound in Med. & Biol. (1980).

5. E. L. Cartensen, W. K. Law, N. D. McKay and T. G. Muir, Demonstration of nonlinear acoustical effects at biomedical frequencies and intensities, submitted to J. Ultrasound in Med. & Biol. (1980).

6. T. G. Muir, L. L. Mellenbruch, and J. C. Lockwood, Reflection of finite-amplitude waves in parametric array, J. Acoust. Soc. Am. 62:271 (1977).

7. J. J. Markham, R. T. Beyer, and R. B. Lindsay, Absorption of sound in fluids, Rev. Mod. Phys. 23:353 (1951).

8. P. J. Carson, P. R. Fischella, and T. V. Oughton, Ultrasonic power and intensity produced by diagnostic ultrasound equipment, J. Ultrasound in Med. & Biol. 3:341 (1978).

9. J. C. Lockwood, T. G. Muir, and D. T. Blackstock, Directive harmonic generation in the radiation field of a circular piston, J. Acoust. Soc. Am. 53:1148 (1973).

10. M. Born and E. Wolf, "Principles of Optics," Pergamon Press, New York (1965).

11. E. P. Cornet, Focusing of an N wave by a spherical mirror, M. A. Thesis, Univ. Texas at Austin (1972).

12. R. D. Essert and D. T. Blackstock, Axisymmetric propagation of
     a spherical N wave in a cylindrical tube, J. Acoust. Soc. Am.
     66(S1):S68(A)(1979).

13. W. D. Beasley, J. D. Brooks, and R. J. Barger, A laboratory
     investigation of N-wave focusing, NASA Langley Report TN
     D5306 (1969).

14. E. J. Giglio, W. M. Ludlam, and S. Whittenburg, Improvement in
     the measurement of intraocular distances using ultrasound, J.
     Acoust. Soc. Am. 44:1359 (1968).

15. T. G. Muir, C. M. Talkington, B. S. Shaw, R. S. Adair and
     J. G. Willette, Parametric echoscanner for biomedical diag-
     nostics, J. Acoust. Soc. Am. 53:382(A)(1973).

16. J. Bjørnø and S. Grinderslev, Investigation of a parametric
     echoscanner for medical diagnosis, Proc. 8th Int. Symp.
     Nonlin. Acous. - Paris 1978, J. de Physique (in publication).

NOTE ADDED IN PROOF

     In the discussion of Fig. 5 of this paper, a speculation on
the differentiation of nonlinear waveforms at the focus was suggested as a possible reason for anomalous increases in measured peak
intensity. After preparation of this paper, the author encountered
a paper by Sutin[17] which presents theory as well as calculations
on this problem. The calculations support the aforementioned specu-
lation in the $\sigma < 1$ region, although there appears to be a singu-
larity in the solution offered.

17. A. M. Sutin, Influence of nonlinear effects on the properties
of acoustic focusing systems, Sov. Phys. Acous. 24; 334 (1979).

LOW FREQUENCY IMAGING TECHNIQUE
FOR ACOUSTIC AND ELASTIC WAVES

Bill D. Cook

Cullen College of Engineering, University of
Houston, Central Campus
Houston, Texas 77004

INTRODUCTION

Radar, active sonar, and ultrasonic pulse-echo non-destructive testing have the common goal of extracting as much information about the target as possible from backscattered signals.

Ranges, locations, velocities, etc. are easily established; however, size, shape, orientation and identification of the target are more difficult to obtain especially if the viewpoint is restricted to a single transducer acting as both sender and receiver.

This paper is concerned with demonstrating the possibility of acquiring shape and size information about isolated targets from selected time domain, acoustic and elastic backscattered signals. Ideally this time domain information displayed on an oscilloscope could be rapidly interpreted, say be a NDE technician, without elaborate analysis.

Specifically, the concept is to analyze the time-domain ramp response of backscattering for information of targets cross-sectional area as a function of distance through the target along the line-of-sight between transducer and target. Research is experimental

111

radar indicates that the ramp response is more easily
interpreted than the impulse response although it is
possible to obtain each response from the other.
Quoting from Young "the amplitude of the waveform,
(the ramp response) versus time is approximately proportional to the 'target profile function' is defined
as an artificial time-domain waveform equal to the
target cross-sectional area intersected by an imaginary
transverse plane moving along the line-of-sight at one
half the velocity of the interrogating transient radar
signal."[1]

The time-domain signature technique utilizing the
ramp function was first suggested for radar by Kennaugh
and Moffett.[2]   They developed a theory based on the
physical optics approximation which applies if the target is a perfect conductor, reasonably smooth and as
large

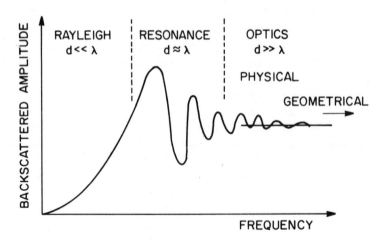

Fig. 1.   Typical frequency dependence of backscatter
          illustrating regions of nomenclature.

as a few wavelengths. However in reality, the interrogating frequencies composing the ramp must lie in the
upper Rayleigh region and lower resonance regions of
scattering; these regions do not correspond to the physical optics approximation. Figure 1 illustrates
typical frequency dependence of backscattering of electromagnetic, acoustic and elastic scattering with common
designations of regions of scattering.

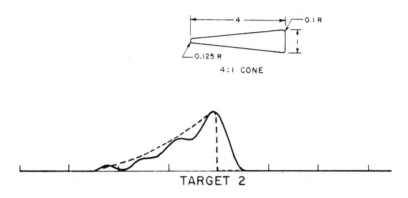

Fig. 2.   Result for electromagnetic waves. Top - Metal
          cone with dimension in inches. Middle- Dashed
          line, theoretical ramp response; solid line -
          experimental after some processing. Bottom -
          Image of cone constructed from measured profile
          functions (ramp responses) using incident angles
          of 0° and 30°.

The physical optics approximation establishes the
relationship between the target profile function and
cross-sectional area for the region directly illuminated
by the interrogating signal, or in other words, the region
ahead of the target shadow boundaries. However, experiments in radar indicate that some information can be
obtained from behind the shadow boundary.

If it is possible to radiate the target from more
than one direction, there are possibilities of inferring
more information. Viewing from one direction, it is
possible to infer these geometrical characteristics from
the ramp response:

    a)   Cross-sectional area versus distance along
         the line-of-sight

    b)   Target volume

    c)   Target length

at least up to the shadow region. For assimulating data
from more than one direction Das and Boener[3] have investigated tomographic algorithms that utilizes this type of
data. Young has also developed algorithms which utilizes
information from more than one direction.[4] The results
of one of these algorithms is illustrated in Figure 2.
These are experimental results. In this case, however,
the orientation of the body is known, and a true ramp
function was not generated but rather the results were
synthesized from scattering at harmonic frequencies.

The above mentioned concepts appear to be adaptable
to acoustic and elastic scattering provided that the
ramp responses can be similarly interpreted. To theoretically test the concepts, it is possible to calculate
backscatter signal using the exact theory based on the
appropriate wave equation and proper boundary conditions
for a few geometries. In acoustic scattering, the wave
equation is separable in several geometries and many
cases can be calculated. In elastic scattering, exact
solutions are restricted to cylindrical and spherical
geometries. All other geometries require approximate
techniques such as method of moments, T-matrix,

quasi-static solutions, etc.

NUMERICAL RESULTS

Two exact cases rather easy to calculate are the
rigid sphere in a fluid and a spherical cavity in a
solid.

Figure 3 shows a composite of ramp responses of
different sizes of rigid spheres. The ramp function in
the analysis was generated by Fourier synthesis. The
Ka is the product of the wave constant of the lowest
frequency and the radius of the sphere. The curves
shown have been normalized by the area of the scattering
sphere. The fact that the responses have approximately
the same height illustrates that maximum return is proportional to the cross-sectional area. The increasing
length with the increasing Ka indicates that length of
scatter is increasing in accordance with the physical
optics theory.

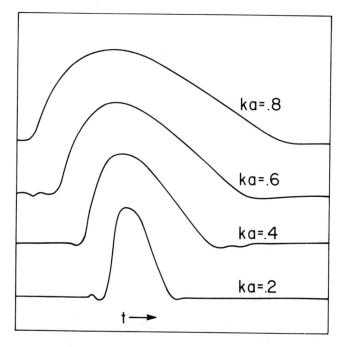

Fig. 3.   Theoretical calculations of ramp backscattered
          acoustic pulses from hard rigid spheres.

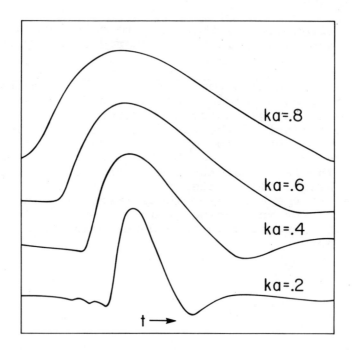

Fig. 4.   Theoretical calculation of ramp backscattered
          elastic pulses from spherical voids.

     Similar curves have been calculated for a void in
a solid. These are shown in Fig. 4. In both the acoustic and elastic cases, the asymmetry can be associated
with scattering from the illuminated and shadow areas of
the target.

EXPERIMENTAL RESULTS

     To test the concept, Shelford Wilson and I have
scattered low frequency acoustic waves from a large
aluminum body whose shape is approximately a prolate
spheroid (major axis 6 in., minor axis 3 in.). The
experimental arrangement is shown in Fig. 5. The acoustic signal generated by an overdamped speaker as detected
by the microphone is shown in Fig. 6. Fig. 7 shows the
amplified and expanded backscattered signals from the target. The angles are measured from nose-on, (along the
major axis.) When viewed from nose-on, the target pro-

file function is smaller in height and greater in length

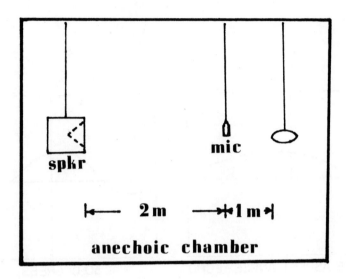

Fig. 5.   Experimental arrangement for scattering from
           spheroid.

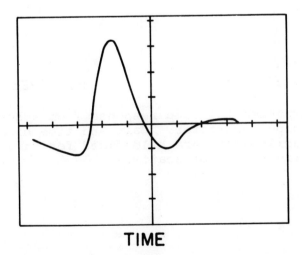

Fig. 6.   Acoustic signal directly from speaker as de-
           tected by microphone.

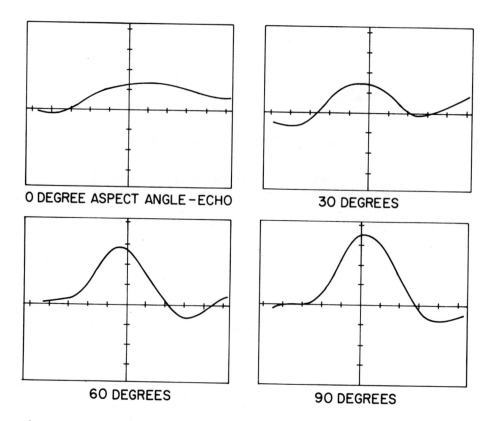

O DEGREE ASPECT ANGLE-ECHO   30 DEGREES

60 DEGREES   90 DEGREES

Fig. 8.   Experimental ramp response from spheroidal
          shaped scatterer. Gain and time base of
          oscilloscope are held fixed for varying
          orientation of scatterer.

than when viewed from the beam (along the minor axis).
The figures with zero and ninety degree aspect angles do
show that the backscattered signals from these two view-
points agree with the profile functions. It is obvious
that one can distinguish the viewing angles along the
axis, however, such simple interpretation is not possible
for the off-axis echos as demonstrated by the photography
for thirty and sixty degrees. It remains to be tested
whether algorithms such as those given by Young, Das
and Boerner can generate a rough construction of the
object.

REFERENCES

1.   J. D. Young, Radar Imaging from Ramp Response, IEEE
      Trans. on Antennas and Propagation, Vol. AP-24,
      pp 276-282 (1976).

2.   E. M. Kennaugh and D. L. Moffett, Transient and Im-
      pulse Response Approximations, Proc. IEEE, Vol. 53,
      p 893-901 (1957).

3.   Y. Das and W. M. Boener, On Radar Target Shape Es-
      timation Using Algorithms for Reconstruction from
      Projections, IEEE Trans. on Antennas and Propaga-
      tion, Vol. AP-26, pp 274-279 (1978).

4.   J. D. Young, et. al. "Basic Research in Three-dimen-
      sional Imaging From Transient Radar Scattering
      Signatures", Annual Status report to Ballastic
      Missle Defense Systems Command by the Electro-
      science Laboratory, The Ohio State University,
      Columbus, Ohio, (July 1978).

ACOUSTIC SWITCHING RATIOS OF PIEZOELECTRIC AND ELECTROSTRICTIVE OATS

Khalid Azim and Keith Wang

Department of Electrical Engineering
University of Houston
Houston, Texas   77004

ABSTRACT

The conventional piezoelectric Opto-Acoustic Transducer (PEOAT) has been studied theoretically and experimentally. The results indicate that due to the characteristics of the present photoconductive layer the Acoustic Switching Ratio (ACSWR) in the megahertz range of frequencies is in general inadequate for beam focusing or beam shaping applications. The physical properties of the present photoconductive layer must be successfully altered to bring about an improvement in the acoustic switching ratio.

We have recently started work on an alternative approach towards improving the ACSWR. This approach involves the use of an electrostrictive layer in the place of the piezoelectric layer in previous OATs. Electrostrictive layers have been prepared from some commercially available powders with very high dielectric constants. With these layers electrostrictive OATs (ESOATs) have been built. Preliminary experimental measurements indicate improved ACSWR and demonstrate the potential usefulness of this new class of OATs.

INTRODUCTION

The potential use of an optically controlled acoustic transducer for a scanning-focused beam acoustic imaging system has been under investigation for a few years.[1,2,3,4,5] Most of the work has been centered around the PEOAT (Piezoelectric Opto Acoustic Transducer) consisting basically of a piezoelectric layer and a photoconductive layer sandwiched between the two electrodes as shown in Figure 1. Due to material limitations,

however, these conventional OATs have not been able to produce
sufficient acoustic switching ratios (ratio of the acoustic output
from the device under illuminated conditions to that under dark
conditions) to be useful in the megahertz range of operating frequencies. Recent theoretical and experimental work on these
piezoelectric layer OATs has been reported.[6] These findings indicate that the characteristics of the present photoconductive
layer must be successfully altered to achieve good acoustic
switching ratios.

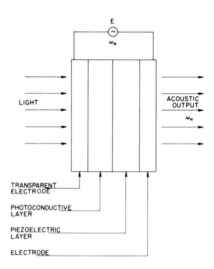

Fig. 1   The conventional PEOAT structure.

        To overcome these drawbacks, an OAT using an electrostrictive
layer in the place of the piezoelectric layer was proposed[7] as an
alternative approach. This new type of OAT, the Electrostrictive
OAT or ESOAT, utilizes the non-linear characteristics of the
electrostrictive layer to generate the acoustic output at the operating frequency. Several preliminary, small size ESOATs have
been built and tested. The acoustic switching ratios observed in
these ESOATs were in general found to be more than 20 dB above
those obtained with conventional PEOATs.

PIEZOELECTRIC OAT (PEOAT)

        The ACSWR (acoustic switching ratio) characteristics of the
PEOAT has been theoretically studied using the Active Model
Equivalent Circuit[6] shown in Figure 2. This is the complete
equivalent circuit which takes into account the piezoelectric
effect present in the photoconductive layer. This is necessary
because in all our recent studies Cadmium Sulphide has been used

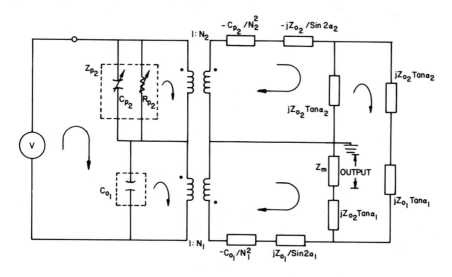

Fig. 2   The Active Model Equivalent
          Circuit for PEOAT.

as the photoconductive layer. Another equivalent circuit which
assumes no piezoelectricity in the photoconductive layer is the
Passive Model.[6,8] Under certain conditions,[6] these circuits can
be simplified. The simplified circuits, however, are valid only
over a limited range of frequencies that are close to the resonance frequency of the piezoelectric layer. The acoustic
switching ratio frequency response of the PEOAT has been theoretically calculated using the Active Model Equivalent Circuits.
The effect of various device parameters on the frequency response
have also been studied. The device parameters include the photoconductive layer dark resistivity (ρ_d), the photoconductive layer
thickness, (T_2), the photoconductive switching (S), etc. Some of
the important results are summarized below:

(1)   The ACSWR frequency response predicted by the active
       model and the passive model show significant differ-
       ences in the frequency range of 4.0 MHz - 10.0 MHz
       (resonance frequency, f_0 = 3.0 MHz) for the case of
       low dark resistivity of the photoconductive layer.
       This is demonstrated in Figure 3 where the ACSWR is
       plotted as a function of frequency for a particular
       set of parameters. The dark resistivity of the photo-
       conductive layer, ρ_d, is less than 10^7 Ω-cm which is
       considered to be low.

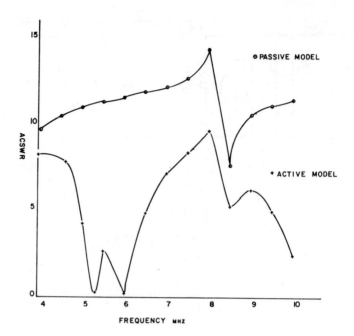

Fig. 3   PEOAT Active Model and Passive Model ACSWR
frequency response. Frequency range 4.0 MHz
to 10.0 MHz, f_0 = 3.0 MHz, ρ_d = $2 \times 10^5 \Omega$-cm,
s = 400, Sc = 1.7, T_2 = 75 μm.

(2)   The ACSWR frequency response peaks at around 0.83 f_0
      and dips at 0.9 f_0. This is clear from Figure 4 where
      ACSWR is plotted as a function of frequency for several
      values of ρ_d. The resonance frequency f_0 is taken as
      3.0 MHz. The ACSWR goes to a peak at 2.5 MHz and dips
      to a minimum at 2.7 MHz.

(3)   The above pattern of peak and dip is repeated just be-
      low 9.0 MHz. This may be viewed as a periodic behavior
      of the equivalent circuit since 9.0 MHz is equal to
      3 f_0. This can be seen in Figure 3 where a second peak
      can be observed at 8.0 MHz and a subsequent dip at
      8.5 MHz.

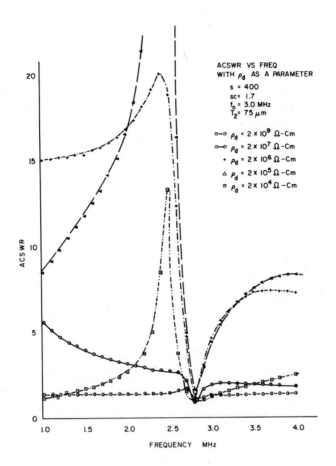

Fig. 4 Active Model predictions of ACSWR versus
frequency for several values of ρ_d.
Frequency range 1.0 MHz to 4.0 MHz,
f_0 = 3.0 MHz, s = 400, Sc = 1.7, T_2=75 μm.

(4) Under certain conditions the PEOAT shows N-type behavior (ACSWR less than unity) over a small range of
frequencies.

(5) The optimum dark resistivity (giving rise to the maximum ACSWR) is found to be a function of the operating
frequency.

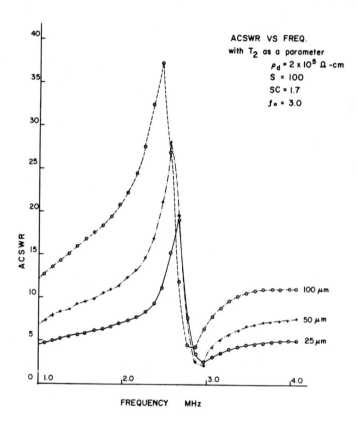

Fig. 5   Active Model predictions of ACSWR frequency response
for several values of photoconductive layer thickness,
t_2. Frequency range 1.0 MHz to 4.0 MHz, f_0=3.0 MHz,
s=100,Sc=1.7, ρ_d = $2\times10^5\Omega$-cm.

(6)   In general, the ACSWR increases with increasing thickness, T_2, of the photoconductive layer. This is illustrated in Figure 5 where the ACSWR frequency response is plotted for three different values of photoconductive layer.

It is important to point here that in the theoretical
studies, some of whose results are shown in Figures 3, 4, and 5,
the device parameters have been allowed to vary over a wide range
of values. However, in an actual OAT these device parameters,
like photoconductive switching S, photosusceptive switching Sc,
photoconductive layer dark resistivity ρ_d, etc. are a function
of the material characteristics and cannot be chosen at will.
Therefore a computer simulation of an actual OAT would require

the use of experimentally measured device parameters. In the theoretical studies when the photoconductive switching is chosen close to those experimentally observed in CdS, the ACSWR predicted is very low. This low ACSWR in PEOATs is also experimentally verified as shown in Figure 6. In this figure the experimentally measured values of ACSWR for a PZT5-CdS OAT is plotted (indicated by shaded triangles) at several discrete frequencies. The computer simulated values of ACSWR for a similar OAT is also plotted (indicated by shaded circles) on the same figure. The ACSWRs do not exceed 1.3 or 2 dB.

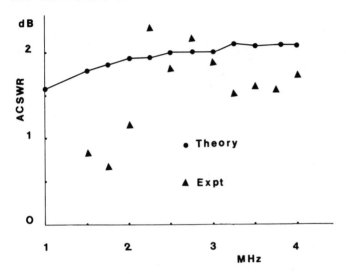

Fig. 6 ACSWR frequency response of a PZT5-CdS OAT. Experimental and theoretical values, f_0 = 3.0 MHz.

All the PEOATs we build recently use sintered CdS slurry as the photoconductive layer. The piezoelectric layer is usually $LiNbO_3$, PZT5 or $BaTiO_3$ ceramics.

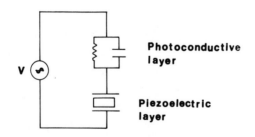

Fig. 7 Oversimplified circuit model for PEOAT.

Although for any detailed analysis the complete equivalent
circuit must be used, the low ACSWR can be explained by an oversimplified equivalent circuit shown in Figure 7. The voltage
applied to the device divides itself across the photoconductive
layer and the piezoelectric layer in accordance with their relative impedances. Under illuminated conditions, the impedance of
the photoconductive layer is lowered and the voltage drops across
the piezoelectric layer increases, resulting in a higher acoustic
output from the device. At high frequencies, however, most of
the signal feeds through the capacitor.

Since the capacitive switching from dark conditions to illuminated conditions is usually less than 2 around 1.0 MHz, the
overall switching is also restricted to about this value. Figures 8 and 9 show how the photoconductive layer (CdS) impedance
switching varies with frequency. The figure also indicates that
the use of a low frequency, e.g. 50 KHz, could give a high switching ratio, but such frequencies might be far below the desired
operating frequency.

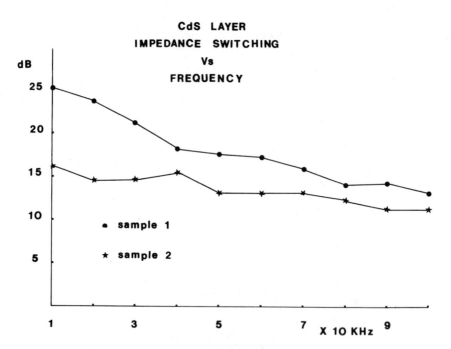

Fig. 8   Impedance switching of two CdS samples
         as a function of frequency. Frequency
         range 10 KHz to 100 KHz.

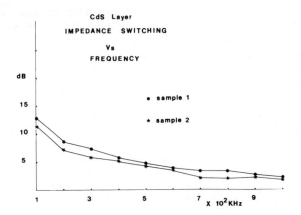

Fig. 9   Impedance switching of two CdS samples as a function
          of frequency. Frequency range 100 KHz to 1 MHz.

ELECTROSTRICTIVE OATS (ESOAT)

The ESOAT[7] accomplishes high switching at high frequencies
by utilizing the high photoconductive switching of CdS at low
frequencies.

The strain S produced by an applied electric field E due to
the electrostrictive effect is given by

$S = \gamma E^2$ where γ is the electrostrictive coefficient.

Suppose we have two electric fields applied to the ESOAT. Let
the voltage drops across the electrostrictive layer be represented by e_L and e_H where

$e_L = E_L \cos\omega_L t$ ω_L is a low frequency (around 50 KHz)

$e_H = E_H \cos\omega_H t$ ω_H is a high frequency (around 1 MHz)

Thus

$$S = \gamma(e_L + e_H)^2$$

$$= \frac{\gamma[E_L^2 + E_H^2]}{2} + \frac{E_L^2}{2}\cos2\omega_L t + \frac{E_H^2}{2}\cos2\omega_H t + E_H E_L \cos(\omega_H - \omega_L)t$$

$$+ E_H E_L \cos(\omega_H + \omega_L)t]$$

By suitably choosing the mechanical Q and the resonance frequency
of the transducer we can essentially avoid unwanted frequency
ranges and have radiation of acoustic energy only in the desired
frequency range.* Let us assume that our device allows only the
high frequency component (i.e. $\omega_L + \omega_H$) to radiate. Therefore, the
acoustic output is proportional to

$$E_{H_{dark}} E_{L_{dark}} \; Cos(\omega_L + \omega_H)t \quad \text{under dark conditions}$$

$$E_{H_{illum}} E_{L_{illum}} \; Cos(\omega_L + \omega_H)t \quad \text{under illuminated conditions}$$

The acoustic pressure amplitude switching ratio is thus
given by

$$ACSWR \; \alpha \; E_{H_{dark}} E_{L_{dark}} / E_{H_{illum}} E_{L_{illum}}$$

$$= \left(\frac{E_{H_{dark}}}{E_{H_{illum}}} \right) \left(\frac{E_{L_{dark}}}{E_{L_{illum}}} \right)$$

$$= \beta_H \beta_L$$

where β_H and β_L are the impedance switching in the photoconductive
layer at high frequencies and at low frequencies respectively.
For present CdS samples the impedance switching at high frequencies is small; β_H is usually just about 2 dB. However, β_L can be
quite large; for example, impedance switching at 50 KHz has been
measured to be over 20 dB in sintered CdS slurry layers. This
suggests that fairly high overall ACSWR in ESOATs could be
achieved using these CdS layers. The block diagram in Figure 10
illustrates the working principle of the ESOAT. A high frequency
source ω_h and a low frequency source ω_l are connected to a photoconductive layer and an electrostrictive layer next to it. The
low frequency source sees only a resistive path while the high
frequency source sees a resistor and a capacitor in parallel in
the photoconductive layer. Inside the electrostrictive layer
both the signals are summed and squared in the conversion into
acoustic energy.

---

*Note that such a choice may result in bandwidth limitation in
 this device.

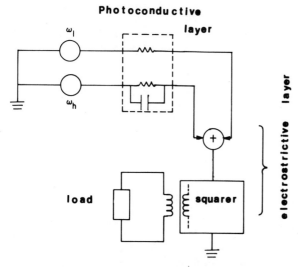

Fig. 10 A simpli-
fied model explain-
ing the working
principle of ESOAT.

PREPARATION OF ELECTROSTRICTIVE LAYERS

All materials, crystalline or amorphous, exhibit varying
degrees of electrostriction.[9,10] For good electro-acoustic
energy conversion the electrostrictive coefficient, γ, of the
material should be high. Significant electrostriction is generally observed in materials with high dielectric constant, ε_r.
The coefficient, γ can be related to ε_r for isotropic solids and
cubic crystals by the following relation.[11]

$$\gamma = - \frac{1}{2} (\varepsilon_r^2/\varepsilon_0)\pi \ m^2/v^2$$

where π is the photoelastic constant.

Materials such as Strontium Titanate ($SrTiO_3$) and Barium
Titanate ($BaTiO_3$) have been known to possess good electrostriction.[10,11] Electrostrictive coefficient in $SrTiO_3$ is on the
order of 10^{-19} m^2/v^2 while in $BaTiO_3$ it is around 10^{-17} m^2/v^2.
The dielectric constant is about 330 for $SrTiO_3$ and around 1000
for $BaTiO_3$. Certain commercially available powders have high
dielectric constants (6,000-12,000) and thus are good candidates
for making electrostrictive layers if the temperature stability
is satisfactory. They are listed below:

Product	Dielectric Constant
H402	6000
H602	7000
H702	8000
H123	12000

Considerable amount of effort was spent in making suitable electrostrictive ceramics with good mechanical strength out of these powders. The first problem was to find a suitable binder with which to mix the powder. Then a suitable substrate had to be found which could withstand the high firing temperatures and yet prevent the material from sticking to it or fusing into it. Finally, a proper heating schedule had to be experimentally established. After much experimentation the following procedure was found to yield the optimum results.

Binder used: Glycerine or Carboxy methyl cellulose.
Substrate: Mullite refractory shelf with a coating of
                 Koaline $(1/2\ Al_2O_32SiO_2)$ and Silica $(1/2SiO_2)$.
Firing temperature: Fired in a kiln to its max temperature
                 of about $2200°F$ over a period of approximately
                 12 hours and then allowed to cool.

The electrostrictive layers were then polished down to proper shape and size with fine grain sand papers. Electrodes were formed on the surface either by applying a thin coating of silver paint or by gold sputtering.

To test for their electrostrictive effect the output of a r.f. power source was directly applied across the samples inside a water tank. Necessary precautions were taken to prevent short circuit between the two electroded surfaces through the water. A Panametrics V301 transducer (center frequency 500 KHz) was used as a receiver. Since the received acoustic signal had several frequencies in it, a Kronhite variable band filter was used.

The results obtained with a H702 sample is shown in Figure 11. The applied electric field f_i as shown in the upper waveform is at 250 KHz and the received signal at 500 KHz $(2f_i)$ in the acoustic output is shown in the lower waveform.

Our preliminary studies have indicated that the transducers prepared with glycerine and electroded with gold have a stronger acoustic output. Also, at least from qualitative observations, H123 samples generate the strongest acoustic radiation, as expected from its high dielectric constant.

MEASUREMENTS OF ACSWRs IN ESOATs

In an earlier experiment[7] performed to demonstrate the feasibility of ESOATs, the device was physically modeled using discrete components. The photoconductive layer was modeled with a commercial photo resistor in parallel with a 0.001 µF capacitor. They were connected in series with an Erie red cap miniature capacitor, of 0.01 µF, representing the electrostrictive layer. Since the Erie red capacitor has a dielectric with a very high

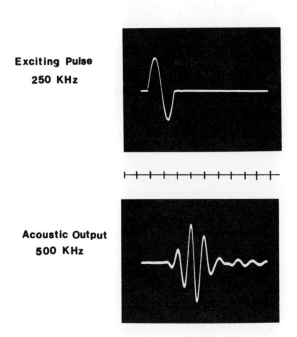

Fig. 11   Applied electrical pulse and acoustic
          output from an electrostrictive layer.

dielectric constant it served as a good electrostrictive transducer. In the first stage of our experimental investigations,
physical modelling using discrete components was experimented
again with the important difference of replacing the Erie Red
capacitor with an actual electrostrictive layer (approximately
1/2 in diameter and 0.5 in thickness we prepared. This provided
a more accurate modelling of an actual ESOAT and also served to
test the performance of our electrostrictive layer. The electrostrictive layer was electroded on both faces to provide the
necessary electrical connections.

The ACSWR obtained from these experiments were quite encouraging. As shown in Figures 12 and 13, ACSWR of over 20 dB

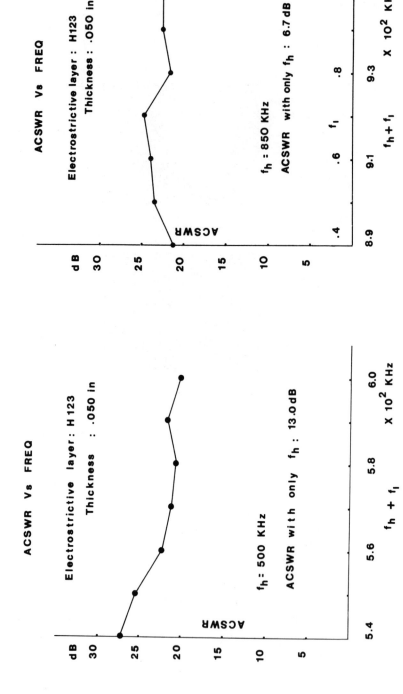

Fig. 13 ACSWR frequency response obtained with a discrete component ESOAT model.

Fig. 12 ACSWR frequency response obtained with a discrete component ESOAT model.

has been observed.   In these experiments the high frequency field
was kept at a constant frequency (500 KHz and 850 KHz) while the
low frequency field was varied from 30 KHz to 100 KHz.   The role
of the non-linear effect is demonstrated by turning off the low
frequency field and applying only the high frequency field; because of the lower photoconductive switching at high frequencies
the overall ACSWR also drops.

The next step was to build an actual ESOAT and measure its
ACSWR.   Figure 14 shows the basic structure of an ESOAT.   The ESOAT
consisted of an electrostrictive layer (H123) with only one side
electroded and a baked CdS slurry layer on a NESA glass substrate.
Both layers were mechanically clamped together as shown in Figure
15.   The CdS layers used were very similar to those used in conventional PEOATs.   The ACSWR measured for this ESOAT is shown in
Figure 16.   Again the high frequency component was kept constant
(at 800 KHz).   These results show that even with these preliminary
ESOATs the ACSWR has improved our similarly built PEOATs.   The impedance switching of CdS at high frequencies (around 1 MHz) is just
above 2 dB which means that the high ACSWR observed in ESOATs at
high frequencies must be due to the contribution of the low frequency switching as has been discussed theoretically in earlier
sections.   However, the relative impedances of the photoconductive
layer and the electrostrictive layer at different frequencies also
play an important role in the switching mechanism.   A more detailed study beyond these preliminary experiments is underway to enhance our understanding of the ESOAT switching mechanism and
characteristics.

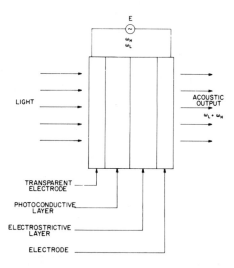

Fig. 14 An ESOAT structure.

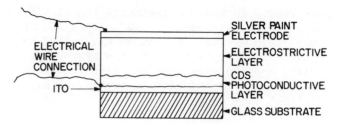

Fig. 15   Figure showing construction
          details of an ESOAT.

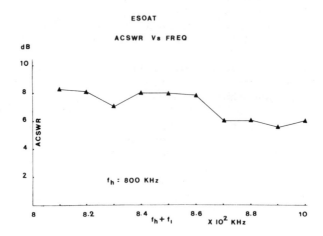

Fig. 16   ACSWR frequency response of
          an actual ESOAT.

The ACSWR, on these 'first generation' ESOATs, can be further
increased by improving the contact between the electrostrictive
layer and the CdS layer. This may be done by polishing the
electrostrictive layer surface making contact with CdS to a smooth
finish. The current method of smoothening the electrostrictive
layer with a fine sand paper still leaves unevenness on the surface. This results in small air gaps between the two layers which
in turn affect the voltage drops across the layers. The elimination of air gaps should help in improving the ACSWR.

CONCLUSION

Computer simulation studies as well as experimental measurements of the acoustic switching ratio of PEOATs show that the
photoconductive layer characteristics must be altered successfully
to improve the switching ratios at high frequencies. The ACSWR
presently obtainable with typical PEOATs is usually less than

2 dB in the MHz range.

An alternative approach for improving ACSWR is by using an electrostrictive layer in the place of the piezoelectric layer. Several electrostrictive ceramic layers were prepared using powders with high dielectric constants. The electrostrictive OATs built with these layers show promise of significant improvement in the ACSWR. However, further theoretical and experimental studies are necessary to enhance understanding of the switching mechanism and the switching characteristics of the ESOATs. These ESOATs should increase the probability of beam focusing and beam shaping applications of OATs.

ACKNOWLEDGEMENT

The work on PEOATs and ESOATs is supported by the National Science Foundation (NSF ECS 7913259   ).

REFERENCES

1. K. Wang and Glen Wade, "A Scanning Focused Beam System for Real-Time Diagnostic Imaging," Acoustical Holography, Vol. 6, N. Booth, Ed., p. 213 (1975).

2. C. Sabet and C. W. Turner, "Parameteric Transducer for High Speed Real-Time Acoustic Imaging," Electronics Letters, 22nd January 1976, Vol. 12, No. 2, p. 44.

3. Keith Wang, Vance Burns, Glen Wade and Scott Elliot, "Opto-Acoustic Transducers for Potentially Sensitive Ultrasonic Imaging," Optical Engineering, September/October 1977, Vol. 16, No. 5, p. 432.

4. T. Hu, Luis Basto and Keith Wang, "A Potential High Performance Ultrasonic Imaging System," Record of Conference Papers, 1978 IEEE Region Five Annual Conference, p. 206.

5. S. Elliott, G. Wade, T. Hu and K. Wang, "Characteristics of Opto Acoustic Transducers," 1978 Ultrasonics Symposium Proceedings, p. 250.

6. K. Azim, K. Wang and K. Bates, "Recent Investigations on OAT Structures," 1979 Ultrasonics Symposium Proceedings (to be published).

7. Ken Bates and Keith Wang, "PEOATs and ESOATs," 1979 Ultrasonics Symposium Proceedings (to be published).

8. S. Elliot, V. Domarkas and Glen Wade, "Frequency Character-
    istics of Opto-Acoustic Transducers," IEEE Transactions on
    Sonics and Ultrasonics, Vol. SU-25, No. 6, November 1978,
    p. 346.

9. Jaffe, Cook and Jaffe, Piezoelectric Ceramics, (Academic
    Press, 1971).

10. K. N. Bates, "A Highly Electrostrictive Ceramic," 1977
    Ultrasonics Symposium Proceedings, p. 393.

11. K. Iamsakun, W. Elder, C. D. W. Wilkinson and R. M. De La Rue,
    "Surface Acoustic Wave Devices Using Electrostrictive Trans-
    duction," J. Phys. D: Appl. Phys., Vol. 8, p. 266 (1975).

SPATIAL FREQUENCY CHARACTERISTICS OF OPTO-ACOUSTIC TRANSDUCERS

Behzad Noorbehesht and Glen Wade

Department of Electrical and Computer Engineering
University of California, Santa Barbara
Santa Barbara, California 93106

An Opto-Acoustic Transducer (OAT) is a multi-layer, planar device which produces a focused acoustic field distribution at its focal plane by transforming a pattern of optical intensity into a corresponding pattern of pressure amplitude. A typical OAT consists of four layers: a transparent electrode, a photo-conductive layer, a piezoelectric layer, and an opaque electrode. One measure of the effectiveness of such a device as an acoustic transducer is the degree of resemblance of the acoustical output to the optical input.

To study the resolving ability of the OAT, we have made use of linear systems theory by considering the OAT to be a linear spatial system. The spatial characteristics of the OAT are then uniquely determined by the Opto-Acoustic Spatial Transfer Function (OASTF), which is defined as the normalized Fourier transform of the acoustical focal spot distribution. Due to the linearity of the system, each layer is regarded as a subsystem and the OASTF is determined by the product of the individual transfer functions due to each subsystem (layer). We have derived the OASTF by including both the diffraction effects and the aberrations due to each layer. The effects of various device parameters and material properties on the resolution can be readily calculated from the OASTF.

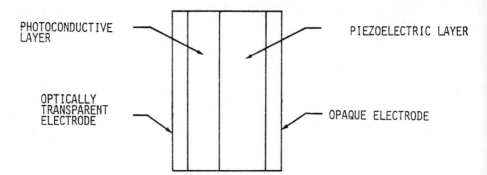

Fig. 1   Physical structure of the OAT.

INTRODUCTION

An Opto-Acoustic Transducer (OAT) is a multi-layer, planar
device which produces a focused acoustic field distribution at
its focal plane by transforming a focus-inducing optical pattern
into a corresponding acoustic pattern [1].   A typical OAT consists
of four layers: a transparent electrode, a photoconductive
layer, a piezoelectric layer, and an opaque electrode, as shown
in Fig. 1.   One measure of the effectiveness of such a device is
its inherent resolving ability.   This may be determined by the
sharpness of the focal-spot distribution of the device.   The
resolution is controlled by two factors:

1.   The degree of resemblance of the acoustic output immediately
      in front of the device, to the optical input [2].

2.   The size and nature of the focus-inducing optical pattern.

To study the resolving ability of the OAT, we have made use
of linear systems theory techniques for deriving a spatial
transfer function for the OAT.   This transfer function, which we
will call the Opto-Acoustic Spatial Transfer Function (OASTF),
turns out to be the normalized Fourier transform of the acoustical
focal spot distribution as we will show.   It consists of two
parts: one due to aberrations, which correspond to the first
factor given above, and the other representing the diffraction
effects, which correspond to the second factor.   The total
aberration effects may be determined by the product of the
individual transfer functions due to each of the four layers of
the OAT, each considered separately.

We will present computer plots of both the diffraction and
the aberration parts of the OASTF.   By comparing the diffraction
and aberration effects we will be able to determine the optimum

size of the OAT for the best resolution. Experimental measurements of the acoustic outputs due to simple optical input patterns
will be presented and compared with the corresponding theoretical
predictions.

DEFINITION OF THE OAT TRANSFER FUNCTION OASTF

A potential application of the OAT is in a real-time acoustic
imaging system in which the OAT is used as a scanning focused
transmitter [1, 3]. Several schemes have been proposed to achieve
the desired scanning and focusing of the OAT output [1]. However,
since our emphasis in this paper is on the focusing ability of the
OAT only, we will disregard scanning considerations and represent
the focusing mechanism by the simplified arrangement shown in Fig.
2. Although the actual scanning will be achieved by a different
and more sophisticated scheme, the one shown in Fig. 2 is complete
enough to demonstrate the basic principles of the focusing mechanism, yet it is adequately simplified to enable us to concentrate
on those aspects of focusing which are relevant in determining a
spatial transfer function for the OAT.

Fig. 2 shows an optical point source u_i placed in the front
focal plane of a perfect converging lens L (that is, one with no
aberrations or diffraction limitations). This lens produces a
plane wave which projects the focus-inducing optical pattern M
onto the transparent electrode of the OAT. The OAT thus generates
an acoustic field which converges to a point a distance f from the
device, to produce the focal-spot distribution u_o. For an ideal
device u_o would be an exact acoustical replica of u_i. In reality,
however, u_o has finite width and only approximates u_i.

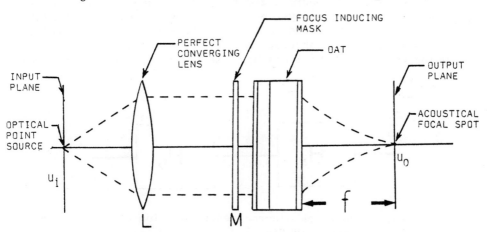

Fig. 2   A simplified scheme to achieve a focused sound beam,
         using the OAT.

For the OAT system shown in Fig. 2 we will define the Opto-Acoustic Spatial Transfer Function (OASTF), as the normalized ratio of the Fourier transform of u_o to that of u_i. This will relate u_o and u_i through their Fourier transforms. To simplify the analysis we will consider one dimensional variations only. Generalization to two dimensions may be achieved by simple mathematical manipulations. Let x and f_x represent the spatial variable and the corresponding spatial frequency respectively. The OASTF, denoted by $H(f_x)$ may then be written as [4]:

$$H(f_x) = \frac{\left[\int_{-\infty}^{+\infty} u_o(x)\ \exp(-j2\pi f_x x)dx\right] / \left[\int_{-\infty}^{+\infty} u_i(x)\ \exp(-j2\pi f_x x)dx\right]}{\left[\int_{-\infty}^{+\infty} u_o(x)dx\right] / \left[\int_{-\infty}^{+\infty} u_i(x)dx\right]} \qquad (1)$$

Since $u_i(x)$ is a point source, it may be represented by:

$$u_i(x) = \delta(x) \qquad (2)$$

where $\delta(x)$ is the Dirac delta function. Eqn. (1) may then be written as:

$$H(f_x) = \frac{U_o(f_x)}{U_o(0)} \qquad (3)$$

Where $U_o(f_x)$ is the Fourier transform of $u_o(x)$. Thus the OASTF may be regarded as the normalized Fourier transform of the focal plane distribution. The system shown in Fig. 2 can be represented by a block diagram as shown in Fig. 3. H_D represents the diffraction effects while the aberrations are described by H_A. In the following sections we will discuss how H_D and H_A may be calculated.

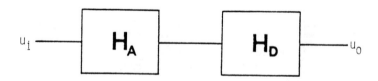

Fig. 3   A block diagram representation of the system shown in Fig. 2.

ABERRATION LIMITED TRANSFER FUNCTION H_A

    The degree of resemblance of the acoustical output immedi-
ately in front of the opaque electrode to the optical input at
the transparent electrode is determined by the aberrations introduced by the four layers of the OAT. If there were no aberrations, the acoustical output and the optical input would have the
same spatial distribution, exactly.

    To study aberrations we have represented the OAT structure
of Fig. 1 by the block diagram configuration shown in Fig. 4(a).
Each block represents a layer and its corresponding transfer
function. H_A is given by the product of the individual transfer
functions. Typical layer thicknesses are also indicated in the
figure. Since the piezoelectric layer is by far the thickest
layer of the device, its aberration effects can be expected to
influence H_A by far the most. In fact, the other three layers
have a combined thickness which is very much smaller than the
acoustic wavelength at the operating frequency of 3 Mhz and we
may neglect aberrations due to them. This is because each of the
four transfer functions is defined by an expression similar to
that in Eqn. (1), each starting at unity and eventually falling

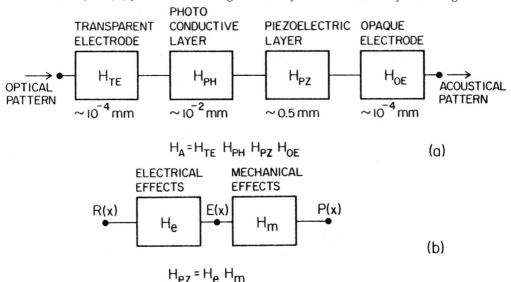

Fig. 4 a) A complete block diagram representation of the OAT.
        b) A simplified block diagram representation of the
            OAT, obtained by neglecting the aberrations due
            to all the layers except the piezoelectric one.

to zero at a high spatial frequency. Due to the small thickness
of the photoconductive and the two electrode layers, their transfer functions are much broader than H_{pz} and are practically equal
to unity over the non-zero range of H_{pz}. Consequently they may be
neglected. H_A will, therefore, be given by H_{pz}, the transfer
function of the piezoelectric layer.

    To derive the transfer function of the piezoelectric layer we
will consider it to be a linear spatial system with the electrode
configuration regarded as its input and the acoustic field immediately in front of it in the propagation medium, as its output.
The transfer function H_{pz}, may be treated as consisting of two
parts: H_e, representing the electrical nature of the aberrations,
and H_m, representing the mechanical aberrations, as shown in Fig.
4(b). H_e has the electrode configuration R(x) as its input and
the corresponding generated electric field E(x) as its output. H_m
has E(x) as its input, and the corresponding acoustic pressure
P(x) immediately in front of the device, as its output. H_e is in
general determined by the degree of resemblance of E(x) to R(x).
This resemblance is determined by the amount of spreading of
electric fields. H_m, on the other hand, represents the degree of
resemblance of P(x) to E(x) and is determined by the amount of
mechanical wave spreading.

    Piezoelectric plates used in OAT structures are generally
made of PZT ceramics having thicknesses of about 1/2 mm. Due to
the high dielectric constant of these ceramics and their small
thickness, electrical fringing is usually much smaller than the
corresponding mechanical fringing resulting in H_e being much
broader than H_m. Consequently, H_A may be represented to a first
approximation by H_m, the transfer function representing the
mechanical fringing effects in the piezoelectric layer.

    Mechanical fringing, i.e. mechanical wave spreading beyond
the electroded regions of the piezoelectric layer, may be determined by considering the details of mechanical wave generation and
propagation in the transducer, the resonance effects and the
nature of the wave transmission and reflection at the plate boundaries. We have determined H_m by solving the piezoelectric equations for the case of an infinite, air backed and water loaded
transducer excited with a single line electrode, as shown in Fig.
5. The Fourier transform of the output due to such an electrode
pattern is, by definition, the one dimensional transfer function
H_m. Here we will present the final results of our analysis. The
details of the analysis will be published elsewhere [5].

    In the following numerical calculations we will assume the
piezoelectric material to be, a PZT-5 ceramic polarized in the

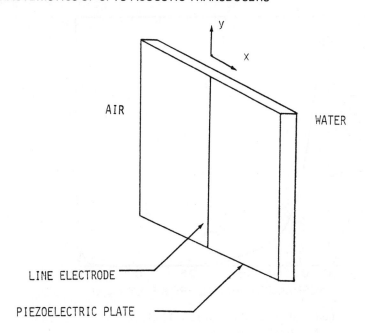

Fig. 5   Piezoelectric layer of the OAT, with a line electrode
         on the front side to simulate an input impules.

thickness direction having a thickness equal to half the wavelength (of the compressional waves) at an operating frequency of 3
Mhz.

    Fig. 6 is a plot of the magnitude of H_A as a function of the
normalized spatial frequency ν_x. The main lobe represents the
contributions due to the compressional waves in the plate, while
the second peak is due to the shear waves. The peaks represent
resonance. For compressional waves the resonance occurs for
normal incidence ($\nu_x=0$). For the shear waves, due to their shorter wavelength, resonance occurs at a large angle of incidence
corresponding to $\nu_x \approx 0.89$. The shear wave contribution is smaller,
partly because the plate thickness is such that the compressional
waves are resonant while the shear waves are off resonance and
partly because of the poling direction of the piezoelectric
ceramic. The spatial frequency at which the main lobe goes to
zero is given by

$$\nu_{xCA} = f_{xCA}\lambda = \frac{v}{V_{pz}} \tag{4}$$

where V_{pz} is given by [6]

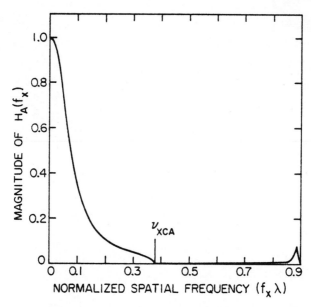

Fig. 6   Magnitude of the aberration limited transfer function
H_A, as a function of spatial frequency.

$$V_{pz} = (c_{11}^E/\rho_{pz})^{1/2} \qquad (5)$$

where λ and v are the sound wavelength and velocity in water
respectively, c_{11}^E is one of the elastic constants of the piezoelectric material, and ρ_{pz} is its density. The aberration limited
line spread function $h_A(x)$ is obtained from $H_A(f_x)$ by the inverse
Fourier transform relationship:

$$h_A(x) = \int_{-\infty}^{+\infty} H_A(f_x) \, \exp(j2\pi f_x x) df_x \qquad (6)$$

Fig. 7 shows a plot of $|h_A(x)|^2$ as a function of x/λ. The resolu-
tion of the OAT is closely related to the width of $|h_A(x)|^2$.
However, to obtain a complete measure of the resolution capabilities of the OAT we must also study the diffraction effects.

DIFFRACTION LIMITED TRANSFER FUNCTION H_D

The diffraction limited transfer function H_D, is defined as
the normalized ratio of the Fourier transform of the acoustical

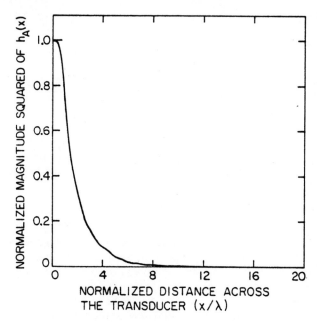

Fig. 7   Magnitude squared of the aberration limited linespread
         function h_A, as a function of the distance across the
         transducer.

focal plane distribution u_o to that of the optical input u_i in
Fig. 2, under the conditions of no aberrations, i.e. when the
acoustical output pattern at the opaque electrode is an exact
replica of the optical input pattern at the transparent electrode.
H_D is determined by the nature of wave propagation from the input
plane to the output plane in Fig. 2, as well as the nature of the
focus inducing mask M.

Several different masks may be used as the focus inducing
pattern M, among them are: Gabor Zone Plate (GZP), Gabor Phase
Plate (GPP), Fresnel Zone Plate (FZP), and Fresnel Phase Plate
(FPP) [7]. Each of these patterns has certain advantages and
disadvantages associated with it, but it can be said that in
general, in terms of their overall performance as focusing elements they are inferior to a diffraction limited converging lens
of the same dimensions.

Since our emphasis in this paper is on determining the inherent resolving ability of OATs rather than the effects of various focus inducing patterns, we will assume the mask M in Fig. 2
to be an ideal one i.e. one which produces a focal spot equivalent
to that of a diffraction limited acoustic lens. Under these
conditions the diffraction limited transfer function H_D will have

the characteristics shown in Fig. 8 [8].   The cut-off frequency ν_{xcD} is given by:

$$\nu_{xcD} = \sin\theta = \frac{D/2}{[f^2+(D/2)^2]^{1/2}} \qquad (7)$$

where θ is the angle subtended at the focal point by the aperture, D is the width of the mask, f is its focal length, and λ is the acoustic wavelength in water.   The corresponding normalized line spread function $h_D(x)$ is shown in Fig. 9.

OPTIMUM OAT SIZE

It can be seen from Fig. 8 that the cut-off frequency of H_D for a given f is a function of D the width of the mask M which in turn is determined by the size of the OAT.   A comparison of the plots in Figures 8 and 6 shows that no gain in resolution is obtained by making $\nu_{xcD} > \nu_{xcA}$, where ν_{xcA} is given by Eqn. (4). Therefore, by solving Eqn. (7) for D and equating ν_{xcD} to ν_{xcA} we obtain the following upper limit on D

$$D \leq \frac{2f\nu_{xcA}}{(1-\nu_{xcA}^2)^{1/2}} \qquad (8)$$

Any further increase of D beyond the value given above is useless as far as resolution is concerned.   The optimum thickness D_o is given by the equality sign in (8):

$$D_o = \frac{2f\nu_{xcA}}{(1-\nu_{xcA}^2)^{1/2}} \qquad (9)$$

As an example, consider an OAT having PZT-5 as its piezoelectric layer, radiating into water and with f=10 cm.   We get:

$$D_o \simeq 8.2 \text{ cm}$$

as the optimum width.   Fig. 10 shows several focal plane distribution curves corresponding to an OAT with the same parameters as those given above but for different values of D.   As expected the focal spot deteriorates as D is decreased below D_o.

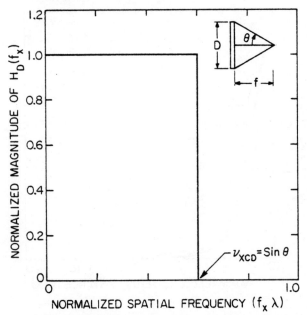

Fig. 8   Magnitude of the diffraction limited transfer function H_D, as a function of the spatial frequency.   The relationship between D, f and θ is shown in the inset.

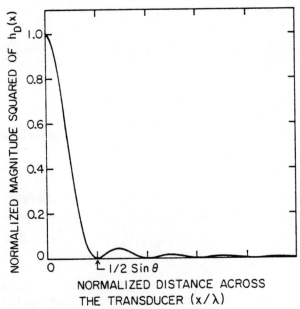

Fig. 9   Magnitude squared of the diffraction limited line-spread function $h_D(x)$.

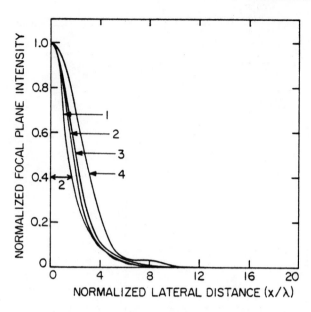

Fig. 10 Focal plane intensity distributions versus distance
        for several values of D.

    1. $D \geq 8.2$ cm   3. $D = 3$ cm
    2. $D = 4$ cm   4. $D = 2$ cm

    Using a Rayleigh-like criterion [9], i.e. taking a resolution
element to be equal to the width of 40% intensity points, we get
from curve 1 in Fig. 10:

$$\Delta x \simeq 4 \lambda$$

as the minimum resolvable distance for an optimum-sized OAT.

EXPERIMENTAL RESULTS

    Experiments were performed to measure the aberration limited
transfer function H_A of the OAT. Since direct measurement of H_A
is very difficult, we have verified the theoretical results in-
directly by comparing the acoustical output patterns due to simple
inputs with the corresponding theoretical predictions.

    Recalling that H_A is approximated by H_m the transfer function
of the piezoelectric layer, we may determine the validity of the
theoretical H_A by comparing the output $u_0(x)$ of a piezoelectric
transducer due to an input (electrode pattern) $u_i(x)$, obtained
through the inverse Fourier transform relationship:

$$u_o(x) = \int_{-\infty}^{+\infty} U_i(f_x) H_A(f_x) \exp(j2\pi f_x x) \, df_x \tag{10}$$

with the corresponding experimental results. In the above equation $U_i(f_x)$ is the Fourier transform of the electrode pattern $u_i(x)$.

Three different electrode patterns were used as the inputs: a 5 mm strip (Fig. 11(a)), a 2 mm strip (Fig. 11(b)), and two 2 mm strips (Fig. 11(c)). The piezoelectric transducer used in the experiments was a 5×5 cm plate of PZT-5 resonant at 3 Mhz.

Figures 12-14 show the resulting experimental plots together with the corresponding theoretical predictions. The corresponding input (electrode patterns) are also shown in the figures. The acoustic output was measured at a distance of 0.5 mm from the transducer by detecting the resulting Raman-Nath diffraction of a laser beam [10, 11]. The effect of the aberrations is to broaden the output relative to what it would be if there were no aberrations present. As it can be seen from the figures, experimental and theoretical results are in excellent agreement. This confirms the validity of H_A in Fig. 6, as the aberration-limited transfer function of the OAT.

CONCLUSIONS

Spatial frequency characteristics of the OAT have been studied by considering it to be a linear spatial system. The Opto-Acoustic Spatial Transfer Function (OASTF) of the OAT has been shown to consist of two parts: H_A the aberration-limited part, and H_D the diffraction-limited part. H_A determines the degree of resemblance of the acoustic output at the opaque electrode to the optical input at the transparent electrode. H_D, on the other hand, determines the sharpness of the focal plane distribution under the conditions of no aberrations. H_A has been shown to be essentially equal to H_m, the transfer function of the piezoelectric layer representing mechanical effects.

By comparing the aberration and diffraction components of the OASTF, the optimum size of an OAT having PZT-5 as its piezoelectric layer, was found to be 8.2 cm. Also the maximum resolution of an optimum sized OAT was found to be 4λ.

Experimental measurements of the output of a PZT-5 transducer due to simple electrode patterns were compared with the corresponding theoretical results, obtained by using the theoretical aberration-limited transfer function H_A. The agreement between the

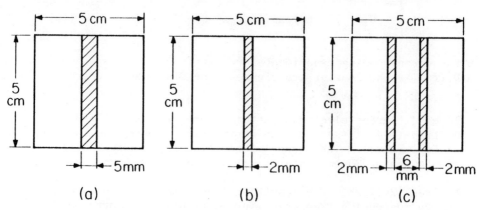

Fig. 11 Electrode geometries for the three transducers used
         in the experiments. Shaded areas represent electroded
         regions.

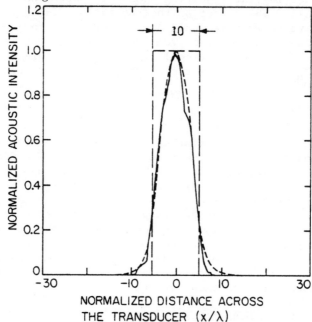

Fig. 12 Acoustic output immediately in front of the transducer,
         versus the distance across the transducer, for the
         transducer shown in Fig. 11(a)

         — — — : Input
         -------- : Output (theory)
         ———————— : Output (experiment)

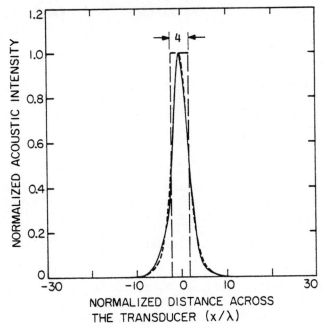

Fig. 13   Same as Fig. 12, but corresponding to the transducer
          in Fig. 11(b).

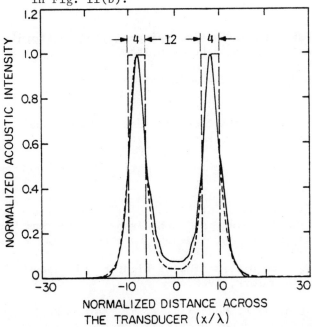

Fig. 14   Same as Fig. 12, but corresponding to the transducer
          in Fig. 11(c).

experiment and theory verifies the validity of the theoretical model.

ACKNOWLEDGEMENT

This work has been supported generously by the National Science Foundation under the auspices of the Automation, Bioengineering, and Sensing Systems Section of the Engineering Division.

REFERENCES

1. K. Wang and G. Wade, "A Scanning Focused-Beam System for Real-Time Diagnostic Imaging," Acoustical Holography, Vol. 6, N. Booth Ed., Plenum Press, New York, pp. 213-228, 1975.

2. S. Elliott, Ph.D. Thesis, University of California at Santa Barbara, 1979 (unpublished).

3. K. Wang, V. Burns, G. Wade and S. Elliott, "Opto-Acoustic Transducers for Potentially Sensitive Ultrasonic Imaging," Opt. Eng., Vol. 16, No. 5, pp. 432-439, Sept./Oct. 1977.

4. W. K. Pratt, "Digital Image Processing," John Wiley and Sons, New York, p. 352, 1978.

5. B. Noorbehesht, Ph.D. Thesis, University of California at Santa Barbara, currently being written.

6. H. F. Pollard, "Sound Waves in Solids," Pion Limited, London, p. 33, 1977.

7. W. E. Kock, "Engineering Applications of Lasers and Holography," Plenum Press, New York, 1975.

8. J. W. Goodman, "Introduction to Fourier Optics," McGraw-Hill, New York, 1968.

9. M. Ahmed, "The Response of Piezoelectric Face Plates Used in Ultrasonic Imaging Systems," IEEE Trans. Sonics and Ultrasonics SU-25, No. 6, pp. 330-339, November 1978.

10. M. V. Berry, "The Diffraction of Light by Ultrasound," Academic Press, New York, 1966.

11. H. F. Tiersten, J. F. McDonald, M. F. Tse and P. Das, "Monolithic Mosaic Transducer Utilizing Trapped Energy Modes," Acoustical Holography, Vol. 7, L. W. Kessler Ed., Plenum Press, New York, pp. 405-422, 1977.

TWO-DIMENSIONAL MEASUREMENT OF ULTRASOUND BEAM PATTERNS

AS A FUNCTION OF FREQUENCY

Paul P. K. Lee and Robert C. Waag

Departments of Radiology and Electrical Engineering
University of Rochester, Rochester, New York 14642.

INTRODUCTION

In ultrasonic imaging and tissue characterization, it is desirable to know the the pressure field of transducers. For a simple disc radiator operating in a piston mode, the field for continuous wave excitation may be calculated using diffraction theory. However, in practice, the field may be affected by transducer parameters such as backing and front surface matching. These effects are dependent on fabrication technique and sometimes difficult to control. Therefore, in order to ascertain imaging resolution or scattering volume dimensions which are determined by the beam profile, it is desirable to measure the actual pressure distribution.

It is common to obtain a beam profile by recording the signal response of a transducer as it is transversing a small target ball (1). The beam profile can also be measured by scanning a small piezoelectric hydrophone over the field (2). Transducer fields have been visualized in two dimensions using schlieren systems that employ an acousto-optical interaction (3). Two-dimensional measurement of transducer fields have been made by detecting the particle displacement at the field plane using laser scanning of a thin pellicle that moves in the sound field (4). This report describes the two-dimensional measurement of a beam pattern as a function of the excitation frequency.

In this study, an ultrasonic diffraction apparatus contained in a large water tank with precise spatial positioning has been used to measure the frequency-dependent continuous wave

two-dimensional beam patterns of ultrasound transducers. Both
the mechanical movement and the sampling of the electrical
signal representing pressure amplitude were under the control of
a minicomputer-based system that was also used in processing and
display of the data. Results are presented for a transducer
used in tissue characterization studies. Comparisons with
continuous wave calculations made using the Fresnel
approximation show good agreement but asymmetries at certain
frequencies are also demonstrated in the measured data.

METHODS

The two-dimensional spatial distribution of pressure
emitted by the ultrasonic transducer under test was measured by
scanning a commercially available microprobe in the plane of
interest while exciting the transducer. The microprobe, a small
piezoelectric hydrophone affixed to the tip of a hypodermic
needle, was mounted on a fixture allowing angular adjustment in
two planes. The fixture was attached to a gantry positioned to
rotate the fixture in an arc around the center point of the
transducer under test. The transducer under test was affixed to
a holder on a platform which moved perpendicular to the plane of
rotation (Figure 1). The motions are produced by stepping
motors.

Tone burst signals long compared to periods (32 to 128
cycles) were gated from an oscillator to a power amplifier and
applied to the transducer. The frequency of the oscillator was
stepped between 2 and 8 Mhz in increments of 23.5 KHz to obtain
measurement at 256 frequencies for each position of the
microprobe in the scan plane. The signal received by the
microprobe was amplified logarithmically with a dynamic range of
60 dB, gated, rectified and integrated over the duration of the
time gate (Figure 2). The output signal proportional to log of
the pressure field at the front surface of the microprobe was
then digitized into 8 bit samples and stored in a disk file that
also contained motor positional and other housekeeping
information.

A commercially available piezoelectric transducer used in
tissue characterization studies with ultrasound (5) was tested
with this system. The transducer was made with an active
circular disc having a radius of 0.238 cm and was specially
encapsulated with a thin layer of silicon to prevent water from
leaking in. The receiver microprobe was scanned in an arc with
an increment of 0.25 degrees over 128 steps with a fixed radius
of 13.5 cm around the center of the transducer. A total of 64
levels spaced 1.16 mm and centered on the transducer axis were

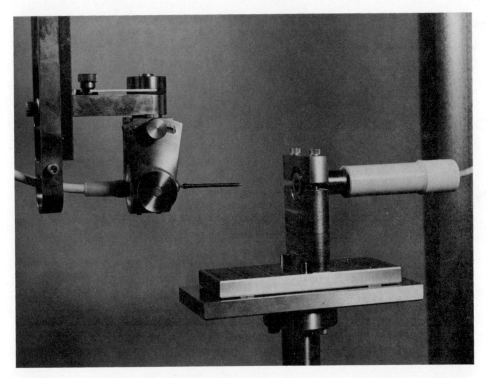

Figure 1.   Transducer Beam Pattern Scanning Geometry.   The
transducer under test is affixed to a holder which moves
vertically while the receiver microprobe is scanned in a
horizontal plane.

scanned.   The angular scan resulted in a spatial sampling
interval of 0.58 mm or one half of the spacing between scan
planes.

    The measured pressure   profiles   were   assembled   into
two-dimensional matrices with each representing a beam pattern
at a different frequency. Since the number of samples within an
angular scan is twice the number of scan levels, linear
interpolation was used to produce matrices having the same
number of points (128) in both directions.

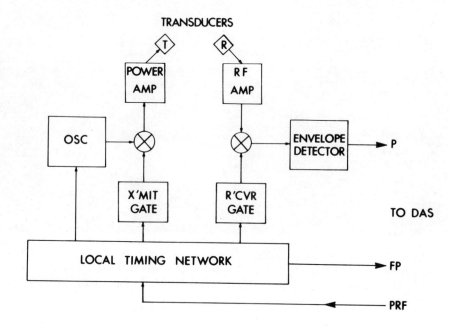

Figure 2. Block Diagram of Signal Electronics. The
transmission of a gated sine wave is initiated by a strobe (PRF)
sent from a remote clock in the data acquisition system (DAS).
The local timing network enables the oscillator (OSC), the
transmitting and receiving gates, as well as generates a framing
pulse (FP) that triggers digitization of the envelope-detected
ultrasound signal (P).

Two-dimensional beam patterns were calculated to compare with the measured values. The caculations were made for a circular piston radiator mounted in a rigid baffle and radiating into a homogeneous medium (Figure 3) . The exact expression for the pressure in this case is given by (6)

$$P_\omega \left(\underline{r} \right) = - j \, \frac{1}{2\pi} \, k\rho \, c \int_A u_\omega \left(\underline{r}' \right) \frac{e^{jk|\underline{r}-\underline{r}'|}}{|\underline{r} - \underline{r}'|} \, d^2 r'$$

where

k = wavenumber = ω/c
ω = angular frequency
c = sound speed
ρ = medium density
u_ω = harmonic velocity amplitude

Using the Fresnel approximation in which the term $|\underline{r} - \underline{r}'|$ in the exponent is expanded to include up to second order terms and the same term in the denominator is replaced by r yields the expression

$$P_\omega = -j \, \frac{1}{2\pi} \, k\rho \, cA_\omega \, \frac{e^{jk(r-\frac{r_1^2}{2r})}}{r} \int_0^a e^{-jk\frac{\eta^2}{2z}} J_0 \left(\frac{kr_1\eta}{z} \right) \eta \, d\eta$$

where

a = radius of radiator
A_ω = uniform velocity inside radiator
J_0 = the zero order Bessel Function

It has been shown that this equation may be stated in terms of Lommel functions (7). When the field point is sufficiently far away from the piston, and the quadratic terms in the exponent are negligible, the pressure amplitude reduces to

$$P_\omega = -j \, \frac{1}{2\pi} \, k\rho \, cA_\omega \frac{e^{jkr}}{r} \, \pi \, a^2 \left[\frac{2 \, J_1(ka \, \sin\theta)}{ka \, \sin\theta} \right]$$

where

J_1 = the Bessel Function of the first kind.

The experimental and theoretical results were displayed using a high resolution graphic display system . A 35mm cine camera under the control of the computer system was employed to photograph the sequence of displays corresponding to different frequencies of excitation. The resulting movie demonstates dramatically the narrowing of the beam at a fixed range as the frequency increases and any beam asymmetries.

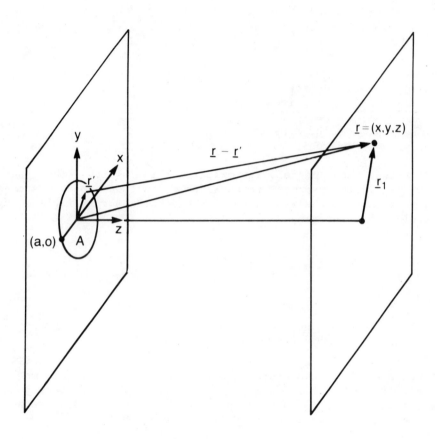

Figure 3. Geometry for Beam Pattern Calculations. Circular
aperture (A) centered at coordinate origin produces a field at
the point $\underline{r}$. The field results from summing contributions as
the vector $\underline{r}'$ is moved throughout the aperture.

RESULTS AND DISCUSSION

Beam narrowing at a fixed range when the frequency of excitation increases is shown in both experimental and computed results (Figure 4). Since in normalized units of distance, S(z), defined by

$$S(z) = z \ / \ (a^2/\lambda),$$

an increase in frequency is the same as moving towards the transducer, the narrowing of the beam pattern may also be viewed as a movement of the measurement plane closer to the transducer (Figure 5).

For the configuration employed in this study, the product of wavenumber and radius, ka, varies from 20.21 to 80.35. The normalized distance corrrespondingly varies from 17.64 to 4.44. Comparison of the measured beam profiles with the calculations shows good agreement in the shape, null positions, and side lobe locations (Figure 6). Since the generally recognized near-far field transition point is taken as a normalized distance of 1.0, the beam patterns correspond to far-field results. The position of the first null point or local minimum in the beam pressure profile predicted from the far-field theory compares very well with the measured and the computed result using Fresnel approximation (Figure 7). The error is less than the spatial sampling interval.

CONCLUSION

Two-dimensional continuous wave beam patterns at a fixed distance have been measured as a function of excitation frequency and compared to predictions of a piston radiator model. Overall agreement between the measurement and the calculations is excellent but asymmetries at specific frequency in the beam patterns are observed. Although the average radial distribution of pressure closely followed prediction based as a piston model, an asymmetry was observed in the frequency interval around 3 MHz. This asymmetry, which is not predicted by the model, may be due to another vibratory mode of the radiator.

ACKNOWLEDGEMENTS

We wish to acknowledge the assistance of Paul M. Collins and Jeffrey P. Astheimer who developed programs and Peter H. Helmers who provided hardware support. Dr. Raymond Gramiak supplied the clinical motivation for this work and gave encouragement as we proceed.

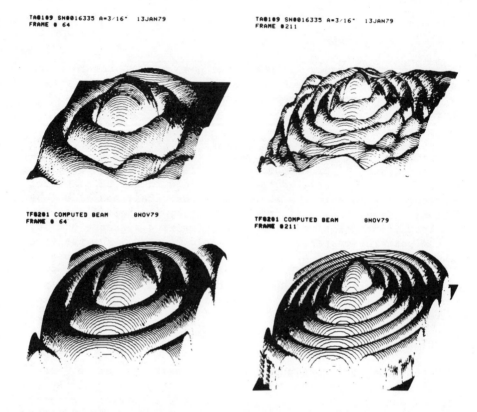

Figure 4. Representative Two Dimensional Beam Patterns at Low and High Frequency. The measured pressure amplitude produced by a transducer with radius of 0.238cm at a range of 13.5cm is displayed in a log scale axonometrically for an area that measures 7.4cm square. The main beam narrows as the frequency increases from 3.5 MHz (upper left) to 7.0 MHz (upper right) and the number of sidelobes within the measurement plane increases. This data agrees well with the pattern computed using the Fresnel approximation for both 3.5 MHz (lower left.) and 7.0 MHz (lower right).

This work was supported in part by the National Science Foundation through grant #DAR78-17849 and the National Institutes of Health through grant #HL15016.

TA0109 SN0016335 A=3/16" 13JAN79

TF0204 COMPUTED BEAM 090 D20NOV79

Figure 5. Measured and Computed Beam Profiles as a Function of Normalized Distance. Field amplitude along a 7.4 cm arc centered on the transducer axis and located 13.5 cm from a 0.238 cm radius radiator is plotted axonometrically on a log amplitude scale as frequency is varied from 8 to 2 MHz (upper). The beam broadening as frequency decreases is equivalent to a computation of the beam profile from a range of 8.5 cm to 33.8 cm at a fixed frequency of 5.0 MHz (lower).

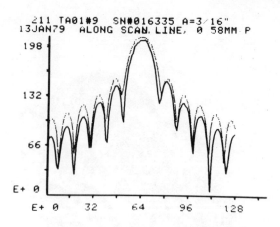

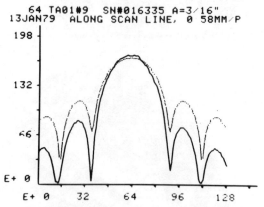

Figure 6. Comparison of Measured Beam Profiles with Results
Computed using the Fresnel Approximation. The measured beam
profiles (heavy lines) exhibit the same general shape as the
computed results (dotted lines) and have the same maxima and
minima locations at both 7.0 MH⁀ (upper) and 3.5 MHz (lower).

REFERENCES

(1) J.   T.   McElroy, "Identification and Measurement of
     Ultrasonic   Search   Unit   Characteristics",   TR   66-5,
     Automation Industries Inc., Boulder, Colorado, 1966.
(2) N. Bom, C. T. Lancee and G. van Zwieten, "Calibration
     of an Ultrasound Sensor", Ultrasonics 71 Conference Papers,
     73-75.
(3) W. G. Neubauer, "Observation of Acoustic Radiation from
     Plane   and   Curved   Surfaces," in Physical Acoustics,
     Principles and Methods, Vol.X, W.   P.   Mason and R. N.
     Thurston, eds. Academic Press, New York, 1973.

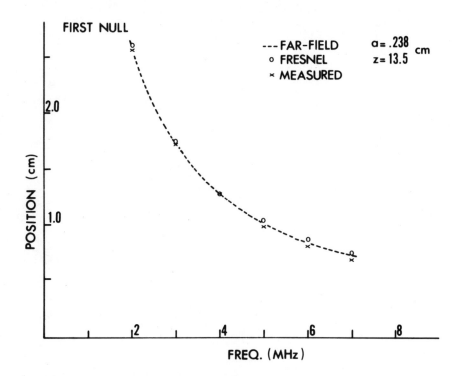

Figure 7. Measured and Computed Distance between the First Null and the Radiator Axis versus Frequency. The measured values are in close agreement with those computed using either the Fresnel or Far-field approximation.

(4) R. S. Mezrich, K. F. Etzold, D. H. R. Vilkomerson, "System for Visualizaing and Measuring Ultrasonic Wavefronts," Acoustical Holography, Vol. 6, N. Booth, ed., Plenum Press, New York, 1975.

(5) R. C. Waag, P. P. K. Lee, R. M. Lerner, L. P. Hunter, R. Gramiak, and E. A. Schenk, "Angle Scan and Frequency-swept Ultrasonic Scattering Characterization of Tissue," Ultrasonic Tissue Characterization II, M. Linzer, ed., National Bureau of Standards, Spec. Publ.525, 1979.

(6) P. M. Morse and K. U. Ingard, Theoretical Acoustics. McGraw-Hill Company, New York, 1968.

(7) H. Seiki, A. Granato, R. Truell, "Diffraction Effects in the Ultrasonic Field of a Piston Source and their Importance in the Accurate Measurement of Attenuation," J. Acoust. Soc. Am., 28(2), 230-238, 1956.

AN INTEGRATED CIRCUIT IMAGE

RECONSTRUCTION PROCESSOR

David E. Boyce

General Electric Co.
Electronics Park
Syracuse, NY 13221

INTRODUCTION

The reconstruction of an ultrasonic image is an ambitious undertaking because it is a complicated process if the reconstructed image is to be any good. To do it using integrated circuits would therefore be expected to require a large complicated integrated circuit. Nevertheless, there seems to be a principle in integrated circuits that, as time goes on, individual transistors get smaller, but circuits become more intricate. With this constant improvement in mind, the idea that the reconstruction of ultrasonic images can be done almost entirely in special integrated circuits seems practical. We have been attempting to develop a general method of image reconstruction that exploits the advantages that integrated processing can have. These advantages include small size, the ability to reproduce elaborate interconnection patterns easily, and improvement in frequency response and reduction in noise pickup, due to decreased stray capacitance. Integrated amplifiers and processors would make it possible to build a low cost imaging system that is contained almost entirely in the transducer probe.

We have concentrated on the development of a test integrated circuit to serve as an example of the general processing method. This integrated circuit will take signal currents from integrated transducer preamplifiers and combine these currents, depending on their received position and time, into a display signal. The preamplifiers can be derived from our previous work (1,2). Processing is done mostly with arrays of bipolar transistors, with the transistors acting as current switches. Bipolar is favored over MOS because a minimally sized transistor can have a low impedance

167

current sink node at its emitter. To get a low sink impedance from MOS, a much larger transistor would be needed. The low value of sink impedance allows the processor to maintain sufficient frequency response with fairly large internal parasitic capacitances from inactive transistor switches.

The processor first selects a group of signal inputs out of a larger number of signal inputs. This group may be shifted by applying a pulse to a clocking input. Shifting produces transverse scanning. The selected signal currents are next sorted into time interval bins. These bins are determined by the expected time of arrival of an acoustic wave from a given focal point and the frequency spectrum of the signal waveform. This sorting is what will be called "focusing." The time sorted signal currents are then appropriately delayed and combined in a bipolar charge transfer "bucket-brigade" (3,4), to yield a display signal. MOS charge transfer techniques are not used, mainly to be compatible with the rest of the circuit. In the test processor eight focal zones are provided. These may be switched as echoes from a single transmitted pulse are being received. If this works without excessive switching noise being introduced, the reconstructed image will have nearly continuous focusing.

The preamplifier-processor combination is a receiving system. The transmitter is separate, and for first experiments will probably be a single transducer located near the center of receiving transducer array, although oblique insonication also may be tried. So far, the test processor has been designed and drawn into a computer-aided drawing system that is used to produce the photo-lithographic masks that are then used to make the integrated circuit. Computed simulations of the reconstruction process have been done; typical results for the test processor will be given later.

SOME FORMALITIES

The structure and operation of a processor can be put into an organized form if some mathematical terms are used. Fig. 1 is a block diagram of a transversely scanned linear array reconstruction system, similar to the test processor. If the set of signals at a certain stage are represented as a signal vector, the operations performed on this vector may be represented as a matrix operator acting in the conventional way on the vector. For example, if the signals from all the transducer amplifiers are represented as the vector (a) of dimension L, the operation of selecting a subset of the vector's elements for scanning may be represented as:

$$[T](a) = (b) \tag{1}$$

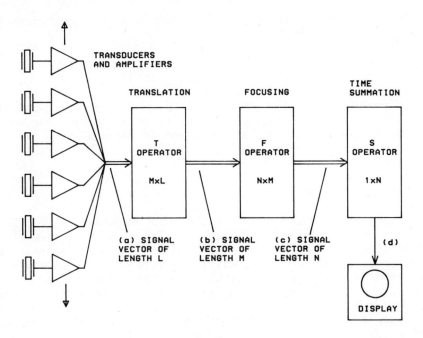

Fig. 1.   Block Diagram of a Transversely-Scanned Imaging System.
Amplified signals from the transducers are regarded as a
signal vector which is acted on by operators that produce
scanning, focusing, and time domain summation. The final
result is a reconstructed display signal.

where $[T]$ is a translation matrix operator of dimension MxL, and (b)
is a new signal vector of length M. Likewise, the "focusing"
operation can be represented as:

$$[F] (b) = (c) \tag{2}$$

where $[F]$ is the focusing operation matrix operator of dimension
NxM, and (c) is a new signal vector of length N. The final operation yields a single quantity, which means that the final operation,
the summation over time, is really a vector product. However, for
symmetry this operation may be represented as:

$$[S](c) = d \tag{3}$$

where $[S]$ is the time summation matrix operator of dimension 1xN,
and d, a scalar, is the display signal.

It is not difficult to determine the form of the matrices $[T]$ and $[F]$; they contain 1's and 0's in their elements. A "1" at a particular position would represent an active transistor in an actual integratedl transistor array. The $[S]$ operator has delay operators as its elements. In terms of the commonly used z-transform, elements of $[S]$ are of the form z^{-n}, where n is an integer from 1 to N. For the test processor to be described, the transistor arrays and circuits will be referred to by their matrix operator letter. That is, a "focusing" transistor array will be called an F array, and so on.

THE TEST PROCESSOR

A layout drawing of the test processor is shown in Fig. 2. A partial schematic is shown in Fig. 3. It would be impossible to show all the details, but the essentials are in these figures. The processor is a 500 mil by 450 mil integrated circuit, a very large one. Most of the area is inactive, so the effect of defects is minimal. It has 63 signal input connection pads around its periphery, and receives power and operating pulses at the rest of its pads. These pads may be made as "solder bumps" for bonding to a ceramic substrate. The display output is the second pad down in the upper right corner in Fig. 2. Voltages are applied at various points in the circuit. In Fig. 3 these voltages are shown as multiples of Vf, which is the forward voltage of a silicon junction at moderate bias current, or about 0.7 volt. This is a convenient voltage scale because these voltages can be obtained referenced to each other from a single external regulator circuit based on a string of diodes. The processor's estimated power dissipation is about 250 mW.

DC currents with added ac signal components flow into the signal inputs from the preamplifier circuits. The negative peak of the ac signal current must always be less than the dc current to maintain linearity. The preamplifiers may have variable gain for acoustic attenuation compensation. The dc current will be about $100\,\mu A$, which should produce a current sink impedance low enough to maintain frequency response to a few megahertz.

The square area at the bottom center of Fig. 2 is the T array. It is a 32x32 array made from bases and emitters of NPN bipolar transistors placed in a common collector region in each horizontal row. .The bases are connected vertically, and the emitters are connected along a diagonal. In operation, a single logical "1" is placed in the T shift register located above the T array, and this "1" is shifted one position for each T Clock cycle. The "1" raises the voltage of one vertical column of bases above all the other bases. This passes 32 of the preamplifier currents into the common collector rows. These collectors are connected together in pairs

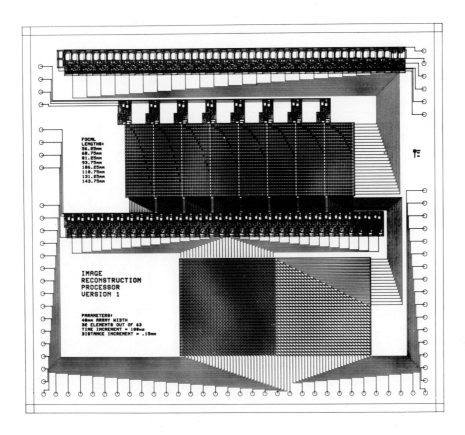

Fig. 2.   Layout Drawing of the Test Processor Integrated Circuit.
           Full detail is impossible to show at this scale, and
           certain layers have been omitted for clarity.   The con-
           nection pads are the circles around the periphery.   Sig-
           nal currents enter around the bottom and work their way to
           the top, eventually coming out near the top right corner
           recombined as a display signal current.

around the center because of transducer array symmetry.   There are
therefore 16 T array output lines.

     Near to the center of the processor shown in Fig. 2 are eight
F arrays that are selected one at a time by a set of shift register
circuits placed just above them.   The F arrays are 24 rows of
bipolar transistors that have common bases and collectors.   There
are individual emitters at each transistor site.   These emitters are
connected into metal lines that run vertically across the array.
These lines carry the 16 incoming currents.   The array is programmed

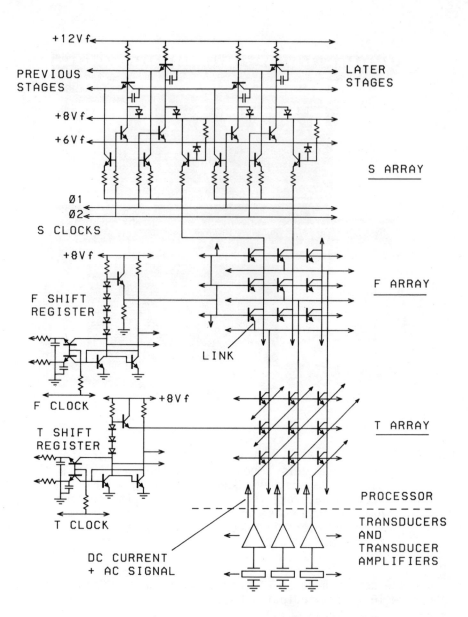

Fig. 3.   Partial Schematic of the Test Processor. Small portions
           of the T and F arrays are shown along with one of their
           respective shift register circuits. Two sections of the
           S array are shown. The arrows at the end of a line
           indicate that that line continues to the next similar
           circuit.

by linking one emitter into one current line with a short length of metal. The placement of these links is determined by the desired focal distance. One F array is selected over the others because the shift register that contains the "1" raises the base voltage of all the transistors in that particular array above the base voltages of all the other arrays. Clocking the shift registers with F Clock pulses causes the "1" to move to the next F array. The F registers may be clocked as echoes are being received. This permits focusing to follow with the distance of the reflecting points. Some things should be noted in regard to this array switching: when F arrays are changed rapidly very little signal information is lost, and there should be no need to throw away image information when switching focal zones. Also, the total current passing through the F arrays remains constant, which tends to reduce switching transients.

Summation in the time domain is next done in a 24 stage charge storage and transfer circuit located along the top of the processor in Fig. 2. This is called the S array. This circuit samples the incoming currents for a certain time interval, and then transfers the accumulated charges to the next stage. The currents from the F array are first reduced by a current divider because the current at that point is larger than is required by the S array to function properly, and larger capacitors would be needed in the S array stages. There are two parallel transfer strings in the array. These are clocked in opposite phase by the two-phase S Clock, so that one is accumulating charge while the other is transferring charge. The currents are alternately steered into either string at the current divider. The two strings add at the output to produce a nearly constant flow of charge that is low in clocking frequency components. The time interval for the S array is 100 ns. Each section of the S array accumulates or transfers charge for slightly less than this time to prevent overlap. The two strings in the array operating in opposite phase give an effective sampling frequency of 10 MHz, which allows frequency components up to 5 MHz to be processed. The capacitors that are charged in the S array are about 25 pF and are made by an ion implantation into the base diffusion in the integrated circuit fabrication. The small capacitors shown in the T and F shift registers of Fig. 3 are made this way also.

To evaluate how well the processor can reconstruct an image, a program was written for a microcomputer that does waveform reconstruction using numerical simulation of the operations described. The effective reflecting point object could be moved in relation to the point of focus. Some of the results for transverse reflecting point movement are shown in Fig. 4. The original waveform was taken to be a single cycle of a 2.5 MHz sine wave. For the test processor as it is configured, that is, for a 40 mm 32 element receiving array, the results show the reflection amplitude dropping to about

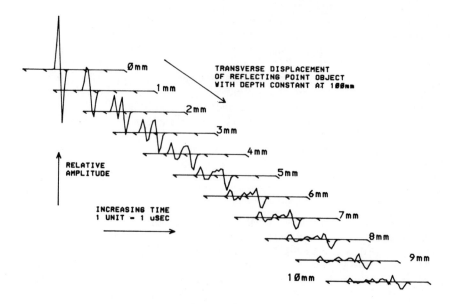

Fig. 4.   Computed Reconstructed Waveforms for the Test Processor.
Data points were computed at each 100 ns, and the points
were linearly connected.   Focus was held at x=0,
z=100 mm.

half at about 1 mm off the point of focus, and then the waveform
gradually dissipating for greater movement. Not shown in the figure
is that in the depth direction focusing changes slowly, as might be
expected, so that the eight focal zones in the test processor should
give adequate focusing over depths of 5 to 15 cm.

CONCLUSIONS AND COMMENTS

The processor has been called a test processor because it is
not intended to be an optimum design.   With further work, it
probably could have been made smaller. The circuits used in the
test processor also have not been made before in integrated form,
and may require some changes. When built, it probably will be put
into some standard integrated circuit package, and perhaps not all
the signal inputs will be used if there is a shortage of package
pins. Input currents may be obtained from preamplifiers made from
standard integrated circuits, or simulated inputs may be obtained
from pulse generators made from standard digital integrated

circuits. When an integrated preamplifier is developed, a complete imaging system will probably be assembled on a ceramic interconnection substrate. The processor and preamplifiers will be connected to the substrate either by solder bumps or by tiny leads; eventually, the solder bumps should allow easiest construction, unless the leads are put in place by an automatic wire bonding machine. Also to be developed are the transducers. It is intended that these will be located near to the input of the preamplifiers, and therefore on the proposed interconnection substrate. This might be a good application for an integrated transducer-preamplifier combination.

The test processor was made for translational scanning. It is hoped that it will stimulate ideas for other types of imaging systems. For example, with slightly different architecture, sector scanning could be done. The T array would be eliminated, but the set of F arrays would have to be expanded so that there is an array for each scan angle as well as each focal distance. Quite a few F arrays might then be needed if the F array is permanently programmed. One possibility is to make an electrically programmed F array, which would have a memory element associated with each transistor switch. Between each acoustic transmission pulse the F array could be reprogrammed to focus at a different angle. A set of programmable F arrays would be needed, with one array for each focal distance, because reprogramming could not be done fast enough for zone switching. For sector scan the F arrays would have to be at least twice as large as those used in translational scanning, and would probably be made in several integrated circuits.

Another possible application of integrated processing, and a truly ambitious undertaking, is the building of a three-dimensional imaging system. This system uses a two-dimensional transducer array. Translation of the active receiving elements is done in two stages; first by a set of Tx arrays that selects a transducer group in the x-direction, and then by a set of Ty arrays that does translation in the y-direction. The set of F arrays is expanded in an additional dimension because of the two-dimensional transducer array. The S array is almost the same as in the test processor, but a little longer. A three-dimensional processor would require the development of a three-dimensional interconnection technique, and would be too complicated to be a single integrated circuit. If an acoustic transmission is done for each scanning position of a 32x32 position receiving array, then 5-10 complete volume images could be done in a second. This same system could produce reduced volume images at a higher rate, and could also produce C-scan images.

We have been promoting a concept in the preceding examples and throughout this paper. The concept is this: good quality real-time image reconstruction can be done using specially designed integrated circuits. In our test processor a single integrated circuit does the reconstruction operations. In general, an imaging system would have a highly compact central part consisting of a set of dedicated integrated circuits, each performing an operation in the sequence of operations required for the reconstruction of an acoustic image.

REFERENCES

1.   H.A.F. Rocha, D.E. Boyce, C.E. Thomas, "Real-Time Acoustical Imaging with Two-Dimensional Arrays," paper presented at the 8th International Acoustical Imaging Symposium, June, 1978.
2.   H.A.F. Rocha, D.E. Boyce, C.E. Thomas, "A Two Dimensional Array for Real-Time Acoustic Imaging," paper presented at the 3rd International Symposium on Ultrasonic Imaging and Tissue Characterization, National Bureau of Standards, Gaithersburg, MD, 1978.
3.   F.L.J. Sangster, K. Teer, "Bucket-Brigade Electronics — New Possibilities for Delay, Time-Axis Conversion, and Scanning," IEEE Journal of Solid-State Circuits, Vol. SC-4, No. 3, June, 1969.
4.   C.N. Berglund, "The Bipolar Transistor Bucket-Brigade Shift Register," IEEE Journal of Solid-State Circuits, Vol. SC-7, No. 2, April, 1972.
5.   P.R. Grey, R.G. Meyer, Analysis and Design of Analog Integrated Circuits. New York: John Wiley and Sons, 1977.
6.   H.F. Harmuth, Acoustic Imaging with Electronic Circuits. New York: Academic Press, 1979.

ADAPTIVE ARRAY PROCESSING FOR ACOUSTIC IMAGING

Gregory L. Duckworth

Department of Electrical Engineering and Computer Science
Massachusetts Institute of Technology
Cambridge, MA   02139

INTRODUCTION

The need for high resolution acoustic image formation under the constraints imposed by small ($\tilde{}$ 100 wavelengths) apertures, long wavelength radiation, and sparsely sampled discrete apertures is encountered in many applications. Most techniques currently in use require large arrays for resolution and uniform sampling of the array aperture for sidelobe control to yield adequate performance when imaging specular reflectors. This paper examines the applicability of the data-adaptive array processing technique known as the Maximum Likelihood Method to image formation in the undersea environment, where acoustics and acoustic imaging techniques have long been of interest as a result of their utility in probing where electromagnetic radiation will not penetrate. The approach taken is to suppress the usual deterministic outlook wherein the propagation phenomenon is "undone", and to look at the problem in a statistical sense. The result is an imaging technique that is essentially spatial and temporal spectral density estimation for a space/time random process which is sampled at a small number of discrete spatial locations.

Using the appropriate data acquisition techniques, a spatial covariance characterization of the received acoustic reflections from the object to be imaged is obtained. This statistical characterization of the received data is then used to generate the image that is "optimal" in the sense that the influences of uncorrelated sensor noise and reflections from other directions are minimized when constructing the image value for each point. Simulation results will show that this technique yields substantially higher resolution images than conventional techniques for the same aperture size, wavelength, and signal to noise ratio when some realistic assumptions

regarding the imaging scenario are made. It also will allow the
use of arrays with arbitrarity spaced sensors.

IMAGING GEOMETRY

     Several different source/object/receiver geometries have been
proposed for use in acoustical imaging systems.[1,2,3] Consideration
of the specifics of the undersea imaging problem narrows the practical choices available, and of the three imaging geometries that
emerge, the one considered in this paper is illustrated in figure 1.
The other viable alternatives are systems that have resolution in
the transmitter and an omnidirectional receiver, or both directional
receivers and transmitters. Specifically excluded from consideration
for this undersea application are scanned or synthetic aperture systems that rely on the relationship between the three components being
fixed for substantial periods of time. In practice the aperture is
not a continuous detector, but consists of a number of point receivers sampling the reflected pressure field. Since it is impractical to sample densely enough to render all propagating waves in
the positive z half-space unique, it is assumed that the source element is directional enough to insonify only that region of space for
which the receiving aperture is capable of uniquely discriminating
differing propagation directions.

     The insonification will consist of continuous wave pulses so
that the region imaged can be confined to an approximately planar
area in space of thickness equal to half the length of the pulse in
the propagating medium, and of lateral extent equal to the field of
view. The generation of one image at a given range will typically
utilize the data from many of these pulses, or "pings", with equal
round trip times.

RELATING CLASSICAL IMAGE RECONSTRUCTION TO THE STOCHASTIC FORMULATION

     In general, the imaging problem is viewed deterministically -
the Kirchoff diffraction integral at one or more frequencies is used
to relate the received signal field to the object reflectance function and source/object/receiver geometry. The imaging problem is
then solved each time the object is insonified by attempting an inverse operation to the generating integral, usually using the Fresnel
approximation. In this frame of mind, randomness is dealt with only
insofar as some thought is given to the uncorrelated measurement
noise that corrupted the observations of the output of the diffraction integral. The compensating operation is to average several
noisy images to eliminate spatially uncorrelated noise effects,
such as locally generated sensor noise.

IMAGING GEOMETRY

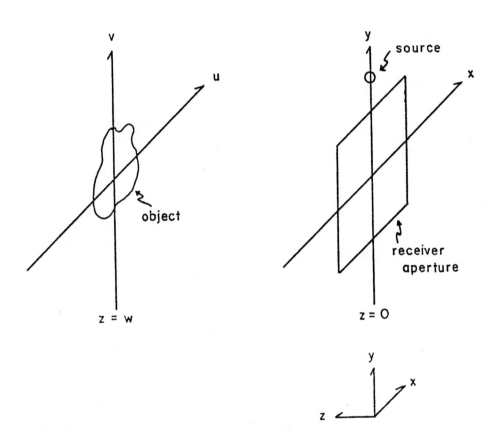

Figure 1.

The approach of this paper will be to consider the reflection process on each insonification ping to be a random experiment. With every ping the experiment is repeated and a new sample function in space and time is received at the array. As a quick argument for the validity of this assumption, it is obvious that in all but the most carefully controlled experimental setup, the source/object/ receiver geometry and the reflectance function are time dependent. Because of this, it should not be expected that each time the transmitter is pulsed that the identical results of the diffraction integral appear at the receiving aperture corrupted by some uncorrelated noise. Instead, the signal in space and time should be viewed as a random process which is meaningful only in a statistical sense as the output of a deterministic system, expressed by the diffraction integral, driven by another random process, the reflectance function of the object. It has been shown[4] that the changes in the imaging geometry may be lumped into random phase variations of the reflection coefficient as a function of location on the object, with the source/ object/receiver geometry held fixed. This output may then be corrupted by additive noise before observation. Thus the imaging procedure becomes one of the determination of the statistics of the received data, followed by the use of these estimated statistics to obtain a characterization of the reflecting object that bears some resemblance to what one would call an image.

To first motivate this statistical viewpoint, it is easy to show the equivalence between the classical method of acoustic imaging and a non-adaptive stochastic approach. Assume that the signal from one ping is received at an N sensor array and frequency analyzed to yield the magnitude and phase (relative to an arbitrary reference) of the signal at each frequency. This may be done by quadrature demodulation if only one frequency is of interest, or by digitizing and applying the discrete Fourier transform if the Fourier coefficients are desired for a broad band of frequencies. For clarity it will be assumed that all imaging is done in the far field, thus focusing the array will not require quadratic phasing. This is not restrictive since by looking only for the power reflected from an object point, a phase correction term which is only a function of sensor location and object to receiver distance is required, and may therefore be applied to the data before any spatial processing is done. We consider the set of Fourier coefficients at one frequency over all channels on ping i to be concatenated in the N-vector $\underline{X}_i$. For each point in the image plane (u,v) for which it is desired to know the reflected signal power, a beam is directed by steering the array with a steering vector $\underline{E}(u,v)$ and squaring the sum of the steered Fourier coefficients. Thus the image on the ith ping is:

$$I_i(u,v) = ||\underline{E}^H(u,v)\underline{X}_i||^2$$

where

$$\underline{E}(u,v) = \begin{bmatrix} e^{j\frac{2\pi}{\lambda w}(ux_1 + vy_1)} \\ e^{j\frac{2\pi}{\lambda w}(ux_2 + vy_2)} \\ \vdots \\ e^{j\frac{2\pi}{\lambda w}(ux_N + vy_N)} \end{bmatrix} \tag{1}$$

(x_n, y_n) = coordinates of sensor n in the z = 0 plane
w   = z range to object plane
H denotes Hermitian, or conjugate transpose

For the case of uniformly sampled linear or rectangular arrays, this operation may be efficiently implemented by 1 or 2-dimensional Fast Fourier Transforms. The final image is constructed by averaging several one ping images:

$$I(u,v) = \frac{1}{M} \sum_{i=1}^{M} I_i(u,v) = \frac{1}{M} \sum_{i=1}^{M} \underline{E}^H(u,v) \underline{X}_i \underline{X}_i^H \underline{E}(u,v) \tag{2}$$

$$= \underline{E}^H(u,v) \left[\frac{1}{M} \sum_{i=1}^{M} \underline{X}_i \underline{X}_i^H \right] \underline{E}(u,v)$$

The stochastic approach involves various statistical characterizations[4,5] of the space/time data. Space does not permit a complete discussion here so the needed quantitites will merely be defined. The "spectral covariance function", $K(f:\underline{z}_1, \underline{z}_2)$ of the signals received at two points in space, $\underline{z}_1$ and $\underline{z}_2$ is equivalent to the expected value of the product of their Fourier coefficients at a given frequency, f. If E{ } is the expectation operator then:

$$K(f:\underline{z}_1, \underline{z}_2) = E\{\underline{X}(f:\underline{z}_1) \underline{X}^*(f:\underline{z}_2)\} \tag{3}$$

This is simply the spatial analog of the time domain autocorrelation function. Additionally, if the received process has the attribute of homogeneity, the spatial equivalent of stationarity, then one may define on the spectral covariance function a "frequency-wavenumber function", equivalent to the spectral density in time series analysis, and implemented by a spatial Fourier transform over all location differences. In two dimensions this yields:

$$P(f:\underline{v}) = \int\int_{-\infty}^{\infty} K(f:\underline{z}) e^{j\frac{2\pi f}{c}(\underline{v}\cdot\underline{z})} d\underline{z} \tag{4}$$

$$\underline{z} = \underline{z}_1 - \underline{z}_2 \qquad\qquad \underline{v} = (v_x, v_y)$$

This corresponds to a decomposition of the received wave-field into uncorrelated plane wave components, characterized by their expected power as a function of direction, or wavevector, $\underline{v}$, and frequency f. This is analogous to the spectral analysis characterization

of a time series by its expected power as a function of frequency.
Note that the spatial quantities just defined already assume temporal
stationarity, since they have been defined on the time domain transforms of the received data.

To characterize a sampled wave-field, a spectral covariance
matrix is defined which contains samples of the spectral covariance
function for all pairs of sensors. Henceforth, for notational simplicity a single frequency, f, will be implicit in all quantities
as it was when describing the conventional imaging technique above.

$$\underline{K} = \left[E\{X_{ni}\, X^{*}_{mi}\} \right] = E\{\underline{X}_i\, \underline{X}^{H}_i\} \tag{5}$$

where $\underline{X}_{ni}$ is the nth component of $\underline{X}_i$ and $E\{\ \}$ is the expectation
operator. Since the received wave-field is assumed to be homogeneous
the values of the spectral covariance function are dependent only on
the separation of the two sensors in space, and not their absolute
location. Thus, corresponding to applying a discrete Fourier transform to a sampled autocorrelation to obtain a spectral density estimate, the steering vectors are applied to the spectral covariance
matrix in quadratic form – the sensor location differences appearing
in the net complex exponential applied to each element. The transforming variables are the wavenumbers:

$$\underline{k} = \frac{2\pi f}{c}\, (\nu_x, \nu_y) = \frac{2\pi}{\lambda}\, \underline{\nu}$$

where ν_x and ν_y are the directional cosines of the plane wave,
forming the "wavevector", $\underline{\nu}$. Thus the sampled frequency-wavenumber
function is:

$$\tilde{P}(\underline{\nu}) = \underline{e}^{H}(\underline{\nu})\, \underline{K}\, \underline{e}(\underline{\nu})$$

where

$$\underline{e}^{H}(\underline{\nu}) = \left[e^{j\frac{2\pi}{\lambda}\, (\underline{\nu}\cdot\underline{z}_1)} \quad e^{j\frac{2\pi}{\lambda}\, (\underline{\nu}\cdot\underline{z}_2)} \quad \ldots \quad e^{j\frac{2\pi}{\lambda}\, (\underline{\nu}\cdot\underline{z}_n)} \right] \tag{6}$$

$$\underline{\nu}\cdot\underline{z}_n = \nu_x x_n + \nu_y y_n$$

Recalling the definitions of $\underline{E}(u,v)$ in equation 1, and $\underline{e}(\underline{\nu})$ in
equation 6:

$$\nu_x = u/w \qquad\qquad \nu_y = v/w \; . \tag{7}$$

And we may redefine:

$$\underline{E}(u,v) = \underline{e}(\underline{\nu}) \triangleq \underline{E}(\underline{\nu}) \tag{8}$$

Inserting what is known to an observer of the process into the
defining equation for $\tilde{P}$, we are led to an estimate for the frequency-wavenumber function:

$$\hat{P}(\underline{\nu}) = \underline{E}^H(\underline{\nu}) \; \hat{\underline{K}} \; \underline{E}(\underline{\nu}) \tag{9}$$

where $\hat{\underline{K}}$ is an estimate of $\underline{K}$. The estimation is carried out just as in periodogram spectral analysis in time series characterization generalized to auto and cross-spectra:

$$\hat{\underline{K}} = \frac{1}{M} \sum_{i=1}^{M} \; \underline{X}_i \; \underline{X}_i^H \tag{10}$$

i.e., by averaging over the pings, each corresponding to one time domain Fourier transformed block of data. Inserting $\hat{K}$ into $\hat{P}$ yields:

$$\hat{P}(\underline{\nu}) = \underline{E}^H(\underline{\nu}) \; \left[\frac{1}{M} \sum_{i=1}^{M} \underline{X}_i \; \underline{X}_i^H \right] \; \underline{E}(\underline{\nu}) \tag{11}$$

which is simply $I(u,v)$ indexed differently:

$$\hat{P} \; ((u/w, v/w)) = I(u,v) \quad .$$

This method of imaging suffers from the same problems as the periodogram method of spectral analysis: the finite aperture (time window) causes the expected frequency-wavenumber estimate (spectral density estimate) to be the convolution of the beampattern (transform of the time domain windowing function) with the true value of the quantity. The classical solution to this has been to apply a tapering function to the array elements to obtain the a priori "best" beampattern, usually suppressing sidelobes and increasing the main lobe width, favoring a decrease in resolution to spurious images. The problem is further compounded, however, in imaging situations by the vast increase in the number of samples needed to uniformly fill a two-dimensional sample field. This can lead to imaging systems with large but sparsely sampled apertures or to small uniformly sampled apertures. Both of these aperture types exhibit poor imaging properties. In general, sparsely sampled apertures lead to beampatterns with very high sidelobe structures which generate poor images, especially when confronted with specular targets. Small, uniformly sampled arrays have more controllable sidelobes and lend themselves to the application of fixed array tapering functions, but have generally poor resolution due to their small size when constructed with practical numbers of sensor elements. The next section will deal with the application of the data adaptive frequency-wavenumber estimation algorithm known as the Maximum Likelihood Method to the imaging problem. It has been successfully used in both time series analysis and geophysical frequency-wavenumber estimation to overcome the problems described above.

THE ADAPTIVE ESTIMATOR

The data adaptive frequency-wavenumber estimator commonly

known as the "Maximum Likelihood Method"[6,7,8] has seen a great
deal of use in seismic signal processing for frequency-wavenumber
estimation of wave-fields consisting primarily of isolated plane
waves.   The primary difficulty in applying the method to imaging,
or distributed source frequency-wavenumber estimation, stems from
the fact that the estimator is formulated for estimating the power
in uncorrelated plane waves, i.e., a homogeneous wave field.   The
instantaneous received field for one ping in an imaging application
is anything but homogeneous, as indicated by the diffraction pattern
present at the receiving aperture.   However, if certain constraints
are met by the imaging geometry[4] the averaged spectral covariance
matrix defined by equation (9) will appear as that arising from a
homogeneous spatial process.   The adaptive estimator is then defined
on this quantity, and we may achieve a superior frequency-wavenumber
estimate to that of equation (11).   These constraints will be discussed briefly later.

The adaptive estimator is formulated as the minimum variance
unbiased estimate of the power in a plane wave received at the array.
Assuming that a unit amplitude monochromatic plane wave described by
the frequency coefficient vector $\underline{E}$ is incident on the array buried
in the noisy received vector $\underline{X}$.   We hypothesize some optimum steering
vector $\underline{W}$ to apply to $\underline{X}$ such that:

$$\underline{W}^H \underline{E} = \underline{E}^H \underline{W} = 1 \qquad\qquad \text{(unbiased constraint)}$$

and

$$P_A = \min_{\underline{W}} E\{||\underline{W}^H\underline{X}||^2\} = \min_{\underline{W}} \underline{W}^H E\{\underline{X}\,\underline{X}^H\}\,\underline{W} = \min_{\underline{W}} \underline{W}^H \underline{K}\,\underline{W} .$$

Carrying out the constrained minimization using Lagrange multipler 2λ

$$0 = \nabla_{\underline{W}}\left[\underline{W}^H\underline{K}\,\underline{W} - 2\lambda(\underline{W}^H\underline{E} - 1)\right]$$

$$\underline{K}\,\underline{W} = \lambda\underline{E}$$

$$\underline{W} = \lambda\underline{K}^{-1}\,\underline{E}$$

but, multiplying the last equation by $\underline{E}^H$ yields

$$\lambda = \frac{1}{\underline{E}^H\underline{K}^{-1}\,\underline{E}}$$

and thus:

$$\underline{W} = \frac{\underline{K}^{-1}\,\underline{E}}{\underline{E}^H\underline{K}^{-1}\,\underline{E}}$$

Substituting the estimate of $\underline{K}$, $\hat{\underline{K}}$, into $\underline{W}$ and applying it to all the data yields:

$$\hat{P}_A(\underline{\nu}) = \left[\underline{E}^H \; \hat{\underline{K}}^{-1} \; \underline{E}\right]^{-1} \tag{12}$$

where the $\underline{E}$ vectors are the steering vectors directed for $\underline{\nu}$.

Intuitively what has been done is to generate a tapering function and steering vector, $\underline{W}$, that varies with the point of estimation, $\underline{\nu}$, and is a function of the statistics of the received wavefield through $\hat{\underline{K}}$. This combination "adapts" the beampattern to be the optimum for each direction, and varies the positions, amplitudes, and widths of the main and sidelobes to achieve a unity response in the "look-direction", $\underline{\nu}$, and to suppress non look-direction sources and uncorrelated sensor noise. Figures 2 and 4 show how the beampattern adapts to the received wave field.

Figure 2 illustrates the beampatterns for a thirty two element .186 meter linear array receiving two plane-waves with directions $\nu = 0$ and $\nu = .1$ and wavelengths of 2.5 mm. (a) and (b) are the conventional and adaptive beampatterns when the array is directed at $\nu = 0$. In 2c the look-direction has been moved slightly off one of the sources. When the array is directed at a given wavenumber, the conventional processor merely shifts the beampattern shown in (a) to that location. However, the adaptive processor optimally weights the array data to try to place a null in the direction of the offending source, as shown in (b) and (c). Note from (c) that the look-direction response may not be the largest, and that large sidelobes may exist in areas where there is no energy to fall into them. Figure 3 indicates the resulting conventional and adaptive processor output for the point sources. Note the greater resolution of the adaptive processor.

The response is similar for distributed objects when a small amount of source/object/receiver motion is assumed and care is taken to average adequately to achieve a homogeneous spectral covariance estimate. The adaptive beampatterns are shown in figure 4 for a distributed object subtending $||\underline{\nu}|| = \pm .1$ in normalized wavevector space. The suppression in the high energy regions is obvious. Figure 5 compares the adaptive and conventional outputs for the distributed reflector. Note that the improvement in sidelobe suppression yields sharper edge definition and less blurring.

Figure 6 indicates the degradation that can occur in the adaptive processor when the spectral covariance is incorrectly formed, i.e. the waves incident on the array from disparate directions appear to be correlated and the spatial field is not homogeneous. This condition would occur when the source/object/receiver geometry is fixed and small perturbations cannot occur, or when insufficient averaging is used in the estimate of the spectral covariance matrix[4]. That

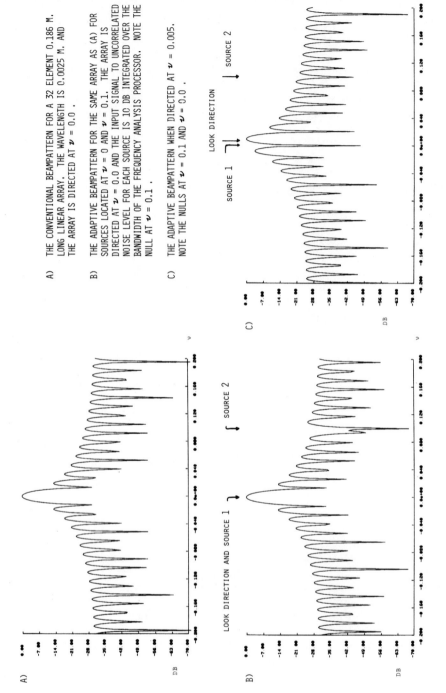

A) THE CONVENTIONAL BEAMPATTERN FOR A 32 ELEMENT 0.186 M. LONG LINEAR ARRAY. THE WAVELENGTH IS 0.0025 M. AND THE ARRAY IS DIRECTED AT $\nu = 0.0$.

B) THE ADAPTIVE BEAMPATTERN FOR THE SAME ARRAY AS (A) FOR SOURCES LOCATED AT $\nu = 0$ AND $\nu = 0.1$. THE ARRAY IS DIRECTED AT $\nu = 0.0$ AND THE INPUT SIGNAL TO UNCORRELATED NOISE LEVEL FOR EACH SOURCE IS 10 DB INTEGRATED OVER THE BANDWIDTH OF THE FREQUENCY ANALYSIS PROCESSOR. NOTE THE NULL AT $\nu = 0.1$.

C) THE ADAPTIVE BEAMPATTERN WHEN DIRECTED AT $\nu = 0.005$. NOTE THE NULLS AT $\nu = 0.1$ AND $\nu = 0.0$.

Figure 2.

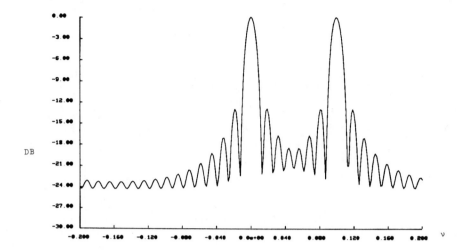

A)   THE OUTPUT OF THE CONVENTIONAL PROCESSOR FOR POINT SOURCES
      LOCATED AT ν = 0.0 AND ν = 0.1.   THE INPUT SIGNAL TO NOISE
      RATIO IS 10 DB FOR EACH SOURCE WHICH LEADS TO A TOTAL SIGNAL
      TO NOISE RATIO AT THE SENSORS OF 13 DB.

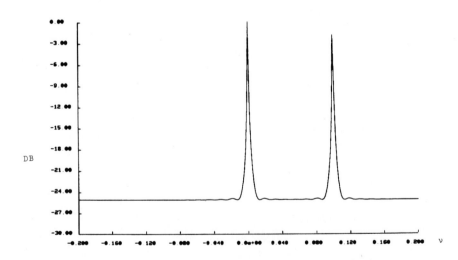

B)   THE OUTPUT OF THE ADAPTIVE PROCESSOR APPLIED TO THE SAME
      DATA USED IN (A).   THE BIAS INDICATED IN THE PEAK HEIGHT
      IS DUE ONLY TO INSUFFICIENT SAMPLING OF THE HIGH RESOLUTION
      OUTPUT.

Figure 3.

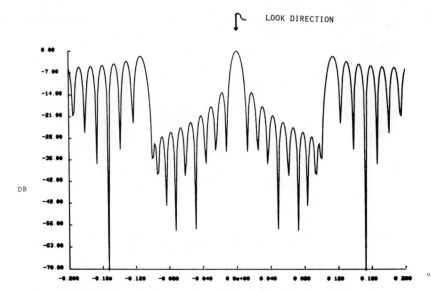

A) THE BEAMPATTERN FOR A 32 ELEMENT ARRAY ADAPTED ON DATA FROM AN
    INCOHERENT REFLECTOR SUBTENDING ± 0.1 IN NORMALIZED WAVENUMBER
    (ν) SPACE.
    THE SIGNAL POWER IN ONE STANDARD RESOLUTION WIDTH (λ/L) TO
    UNCORRELATED NOISE POWER RATIO IS 10 DB. FOR THIS EXAMPLE
    THE REFLECTOR IS 15 RESOLUTION WIDTHS, WHICH LEADS TO A TOTAL
    SIGNAL TO NOISE RATIO AT THE SENSORS OF 22 DB.

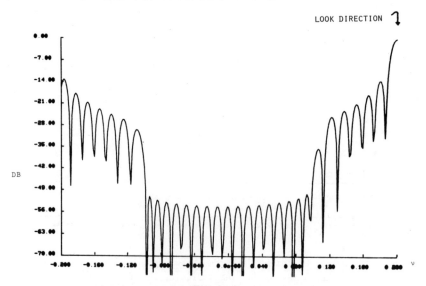

B) THE BEAMPATTERN FOR THE SAME DATA AS IN (A) WHEN THE ARRAY IS
    DIRECTED AT ν = 0.2.

Figure 4.

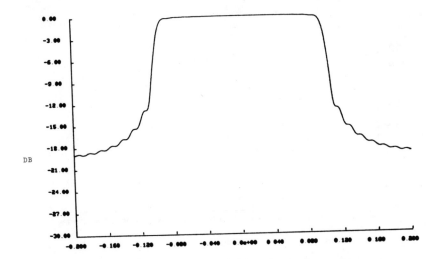

A)   THE CONVENTIONAL PROCESSOR OUTPUT FOR THE DATA DESCRIBED IN
     FIGURE 4(A).

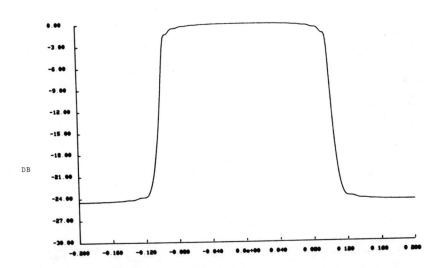

B)   THE ADAPTIVE PROCESSOR OUTPUT FOR THE DATA DESCRIBED IN
     FIGURE 4(A).   THE EQUIVALENT POINT SOURCE ARRAY GAIN IS
     24.5 - 10 = 14.5 DB.   THUS, THE ADAPTIVE TAPERING FUNCTION
     APPLIED TO THE ARRAY HAS ONLY DECREASED THE ARRAY GAIN FROM
     15 = LOG(32) TO 14.5 = LOG(28).   THE IMPROVEMENT IN SIDE-
     LOBE SUPPRESSION OVER THE CONVENTIONAL PROCESSOR IS CLEARLY
     EVIDENT.

Figure 5.

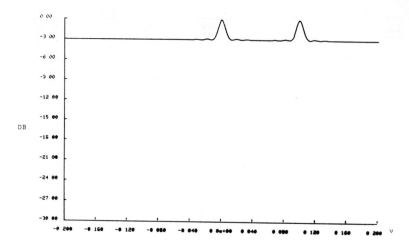

A)   ADAPTIVE PROCESSOR OUTPUT FOR DATA FROM TWO PERFECTLY
     CORRELATED POINT SOURCES INCIDENT ON THE 32 ELEMENT LINEAR
     ARRAY.   NOTE THAT THE PEAK OUTPUT SIGNAL TO NOISE RATIO IS
     ONLY ABOUT 3 DB.   AN ANALYSIS OF THIS PHENOMENON APPEARS IN
     REFERENCE 4.

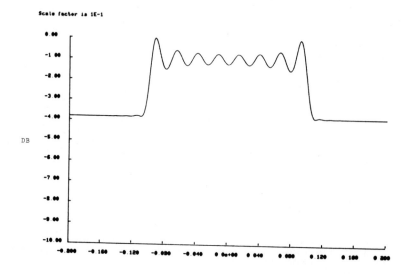

B)   ADAPTIVE PROCESSING OUTPUT FOR REFLECTIONS FROM A COMPLETELY
     COHERENT DISTRIBUTED REFLECTOR WITH NO COMPENSATING
     SOURCE/OBJECT/RECEIVER MOTION.   THE RESULTING WAVE-FIELD
     AT THE RECEIVERS IS INHOMOGENEOUS AND THE ADAPTIVE ALGORITHM
     FAILS, YIELDING AN OUTPUT SIGNAL TO NOISE RATIO OF ONLY
     0.4 DB.

Figure 6.

small object motions with respect to the source/receiver will allow
successful spectral covariance estimation can be seen from the following argument: at a single frequency for a given ping each
incremental reflector will generate a wave with a certain initial
phase. If the object moves even slightly between pings (on the order
of a wavelength) the initial phases for all the incremental reflectors
will be different, and in general for an arbitrary motion the initial
phases at each reflector will be uncorrelated. Thus, sufficient time
averages of the outer products of the frequency coefficient vectors
will cause the cross terms, or interference patterns, between the
incremental reflectors to disappear, and the spectral covariance
estimated will appear as that of a homogeneous random process. Note
that since each incremental reflecting area has been modelled as a
point reflector, the required random phase shifts may be produced by
randomly translating the insonifying source between shots. In essence
all that is required for correct covariance estimation is that the
bandwidth of the reflection process be negligible on a pulse (ping)
duration scale so that the frequency decomposition can be done
correctly, but yet significant on an interpulse period scale. Simple
calculations[4] show that this constraint is commonly met in undersea
imaging situations.

Before presenting the results of the adaptive processing applied
to two dimensional distributed source data it is important to consider the computational intensity of the algorithm. Disregarding the
frequency analysis, which is common to all frequency domain imaging
techniques, the algorithm requires approximately $8MN^2$ real operations
to compute an M averaged inverse spectral covariance, $\hat{K}^{-1}$, matrix via
the matrix inversion Lemma[11]:

$$\left[\frac{1}{M} \sum_{i=1}^{M} X_i \, X_i^H + \frac{\varepsilon}{M} I \right]^{-1} \quad \triangleq \quad \hat{\underline{K}}^{-1}$$

$$\hat{\underline{K}}_1^{-1} = \frac{1}{\varepsilon} I$$

$$\hat{\underline{K}}_{i+1}^{-1} = \hat{\underline{K}}_i^{-1} - \frac{\left[\hat{\underline{K}}_i^{-1} \, \underline{X}_{i+1} \right]\left[\hat{\underline{K}}_i^{-1} \, \underline{X}_{i+1} \right]^H}{1 + \underline{X}_{i+1}^H \, \hat{\underline{K}}_i^{-1} \, \underline{X}_{i+1}} \qquad i = 1, 2, \ldots, M$$

$$\hat{\underline{K}}^{-1} = M\hat{\underline{K}}_M^{-1}$$

Note that this method avoids the direct inversion, requiring an
order N^3 computation. Next, the quadratic form of the steering
vectors with the inverse spectral covariance in general requires
$4QN^2$ real operations after taking advantage of the conjugate symmetry
of the matrix, where Q is the number of pixels in the output image.
However, there exists a much faster algorithm if the array has uniform spacings, i.e. a filled rectangular array is used. In this
case the elements of the steering vectors $\underline{E}$ have uniform phase

increments in each direction. Consider first a one dimensional example – a linear array. The ith element e_i of $\underline{E}$ is simply $e^{j\phi i}$, where ϕ is the phase increment for the desired steering direction. Thus, the evaluation of a quadratic form for a general matrix $\underline{A} = (a_{i\ell})$ can be done with an order N operation via:

$$\underline{E}^H \underline{A}\, \underline{E} = \sum_{i=0}^{N-1} \sum_{\ell=0}^{N-1} e_i^* \, a_{i\ell} \, e_\ell = \sum_{i=0}^{N-1} \sum_{\ell=0}^{N-1} e^{-j\phi(i-\ell)} a_{i\ell}$$

$$= \sum_{n=0}^{N-1} e^{j\phi n} \sum_{i=n}^{N-1} a_{(i-n),i} + \sum_{n=1}^{N-1} e^{-j\phi n} \sum_{i=n}^{N-1} a_{i,(i-n)}$$

$$= \sum_{n=0}^{N-1} e^{j\phi n}\, d_n + \sum_{n=1}^{N-1} e^{-j\phi n}\, d_{-n}$$

where the d_n are the sums over the main and superdiagnals, and the d_{-n} are the sums over the subdiagnals. Finally, we see that when the $\underline{A}$ matrix is conjugate symmetric this may be evaluated for all directions simultaneously via a Fast Fourier Transform on the sequence:

$$\{d_0, d_1, d_2, \ldots d_{N-1}, 0, 0, \ldots, 0, d_{-(N-1)}, d_{-(N-2)}, \ldots d_{-1}\}$$

where the zeros are used to pad the sequence for the desired length transform. This may be evaluated by "pruned" FFT's for further savings when the number of zeros is large[9],[10]. The generalization to two dimensional rectangular arrays is straightforward: Consider an array of samples in the x and y directions which is mxn points, mn = N. The samples are ordered by concatenating them row wise, i.e., n is the most quickly varying index. An NxN (inverse) covariance matrix, $\underline{A}'$, that can be partitioned into m^2, nxn sub-matrices corresponding to groupings over the rows results.

$$\underline{A}' = \begin{bmatrix} A_{00} & A_{01} & \cdots & A_{0,(m-1)} \\ A_{10} & & & \\ & & & \\ & & & A_{(m-1),(m-1)} \end{bmatrix}$$

The steering vectors have the form:

$$\underline{E}^{H} = \left[\underline{e}_{x0} \; \underline{e}_{y}^{H} \; \vdots \; \underline{e}_{x1} \underline{e}_{y}^{H} \vdots \cdots \cdots \vdots \; \underline{e}_{x(m-1)} \underline{e}_{y}^{H} \right]$$

$$\underline{e}_{y}^{H} = \left[e_{y0} \; e_{y1} \cdots \cdots e_{y(n-1)} \right] \qquad e_{yi} = e^{j\phi_{y}i}$$

$$\underline{e}_{x}^{H} = \left[e_{x0} \; e_{x1} \cdots \cdots e_{x(m-1)} \right] \qquad e_{xi} = e^{j\phi_{x}i}$$

Thus

$$\underline{E}^{H} \; \underline{A}' \; \underline{E} = \underline{e}_{x}^{H} \left[\begin{array}{cccc} \underline{e}_{y}^{H} A_{00} \underline{e}_{y} & \cdots \vdots \cdots & \vdots \underline{e}_{y}^{H} A_{0,(m-1)} \underline{e}_{y} \\ \vdots & \vdots & \vdots \\ \vdots & \vdots & \vdots \\ \underline{e}_{y}^{H} A_{(m-1),0} \underline{e}_{y} \vdots & \cdots & \cdots \end{array} \right] \underline{e}_{x}$$

and each of the block operations may be carried out by "condensing"
the submatrices by summing over the diagonals as described above.
The reduction in dimensionality at each stage greatly reduces the
amount of computation required, and in general only half the operations above need to be carried out due to symmetries. Even when not
applying FFT's to evaluate the submatrices at all angles simultaneously , since in general we do not desire a complete hemisphere
of wavevectors, the number of operations is greatly reduced - a factor
of 10^{3} for the N = 144 sensor examples used in the simulations to be
seen later. The real operation count is[9]:

$$6q\left(\frac{m^{2}+m}{2}\right) (2n-1) + 6\ell qm + (m^{2}+m) (n^{2}-2n+1)$$

$$+ 2q\left[\left(\frac{m^{2}-m}{2}\right) + (m^{2}+m)(n-1)\right] + 2\ell q(m-1)$$

where ℓq = Q are, respectively, the numbers of unique wavevectors in
the x and y directions, and the total number of output samples.

SIMULATION RESULTS

     In keeping with the assumptions outlined above, the wave-fields
generated for the simulations were done by discrete evaluations of
Kirchoff's integral, with the appropriate approximations used when

applicable. The source/object/receiver motions were simulated by
1) Gaussian distributed random tilting motions of the object with
translational standard deviations on the order of one to five wavelengths of the edge of the object field, and 2) assuming a new
distribution of uniform random phase reflection coefficients on each
incremental radiator of the hypothesized object for each ping. These
operations were found to yield the same results. Care was taken to
sample the reflectance function adequately in order to simulate the
continuous integral. The two-dimensional arrays used in the examples
to follow are plotted in figure 7. In all examples the wavelength
of sound used is 0.0025 meters, corresponding to 600 kHz insonification in 1500 m/s water.

To show the increased performance on point targets, two reflectors were simulated at 10 meters, separated by 0.29 meters, exactly
the Rayleigh resolution of the 12 x 12 sensor array with a 0.086
meter aperture and imaged with 2.5 mm wavelength sound. The signal
to uncorrelated noise ratio at the array input was 0dB for each
reflector, and was simulated by adding a diagonal matrix to the
theoretical spectral covariance matrix. Note that the points are
just barely resolved by the conventional processor in figure 8(a),
but are clearly separated by the adaptive processor in figure 8(b).

Figure 9 shows the distributed objects used to compare the processors. The "man" object is a uniformly reflecting silhouette and
subtends the same space occupied by a 6 foot tall human at 10 meters.
The second object consists of the letters "RAT", some resolution
test bars, and an area in the lower right corner for which the
reflection coefficient gradually increases from 0 to 64 times that
of the letters or test bars. This was used to test the algorithm's
performance on specular reflectors.

Figure 10 indicates the processing results for the RAT object
subtending a total of 1.5 x 1.5 meters at 10 meters from the receiving
aperture. One hundred data pings were used assuming random phase
reflection coefficients and 1% total uncorrelated sensor noise. (See
reference 4 for typical expected signal to noise ratios, note that
this actually corresponds to a very low equivalent point source
signal to noise ratio, since the signal power in any resolution bin
on the diffuse areas of the object is quite small.) Figures 10a
and 10b are for the array of figure 7b, and figure 10c corresponds
to the sparse array of figure 7a. Observe that the largest resolution bars are resolved only for the adaptive cases, and that the
strong source in the lower right is much more confined in the adaptive
images. Note that the larger aperture of the sparse array shows
slightly better resolution, however its peak to background signal
to noise ratio ("definition") suffers due to the fact that parts of
the aperture are spatially undersampled. One property of the adaptive algorithm is that for irregular arrays very little spatial

occurs as long as the spatial distribution of sample points contains some "lags" or separations that lead to phase increments (mod 2π) between the Fourier coefficients that can resolve all wavenumbers unambiguously. In fact, the more random the distribution of lags the better the aliasing rejection, and the less the decrease in array gain to uncorrelated noise. For this example the non-adaptive image formed with the sparse array was unrecognizable.

Adaptive and conventional results for the 16 x 16 square array of figure 7c appear in figure 11. For this larger array, the size of the object was reduced to 1 x 1 meters, again 10 meters from the array. The increased resolution exhibited by the adaptive pro-cessor is apparent.

Finally, the 6 foot "man" oject was imaged at 10 meters by the small 12 x 12 sensor array. Figure 12 shows the increase in resolution on a practical scale. Note that the excellent sidelobe behaviour of the adaptive processor leads to a very low and flat background level.

CONCLUSIONS

The comparisons shown between the data adaptive and conven-tional processing techniques on simulated acoustic returns show that the adaptive technique leads to better images in all examples. The adaptive estimator also has the property that when extremely poor signal to noise ratios are encountered, the adaptive estimate converges to the conventional and therefore does no worse. For expected signal to noise ratios in the undersea environment for the frequency studied[4], the resolution performance of the adaptive estimator is superior to that of the classical estimator.

The disadvantages of the technique are basically two: first, there is no guarantee that the data will allow a homogeneous spectral covariance estimate easily, and a great deal of averaging may be required in the spectral covariance estimator. Second, since the arrays used in imaging applications generally have a large number of sensors, the time and space required for the spectral covariance estimate necessitates the use of a powerful processor. It should be possible, however, to construct an inverse spectral covariance estimation algorithm that uses space and computer time in proportion to the instantaneous rank of the matrix. If this is done, the technique should be very feasible computationally.

ACKNOWLEDGEMENTS

The author wishes to thank his advisor, Professor A. B.

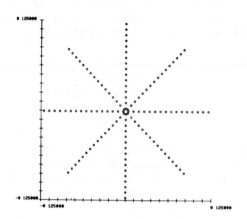

(A) 128 ELEMENT SPARSE ARRAY. INTER-ELEMENT
SPACINGS ARE 7.8 MM.

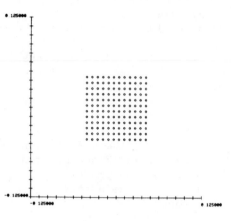

(B) 12 BY 12 FILLED ARRAY. INTER-ELEMENT
SPACINGS ARE 7.8 MM, YIELDING A
FIELD OF VIEW OF ± 10 DEGREES.

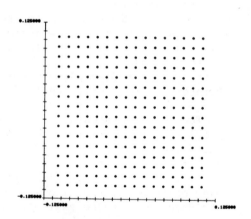

(C) 16 BY 16 FILLED ARRAY. INTER-ELEMENT
SPACINGS ARE 14 MM, YIELDING A FIELD
OF VIEW OF ± 5 DEGREES.

Figure 7. The Two-Dimensional Arrays Used in The Simulations

A) CONVENTIONAL PROCESSING OUTPUT.
THE POINT SEPARATION IS 0.29
METERS AND THE FIELD OF VIEW IS
± 1 METER IN EACH DIRECTION.
THE SIGNAL TO NOISE RATIO IS
0 DB FOR EACH SOURCE.
NOTE THAT THE POINTS ARE JUST
RESOLVED.

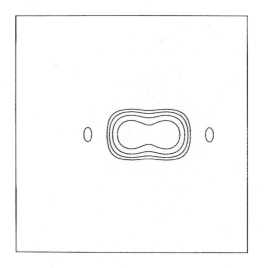

CONTOUR INTERVAL = 3 DB

SLICE THROUGH Y = 0

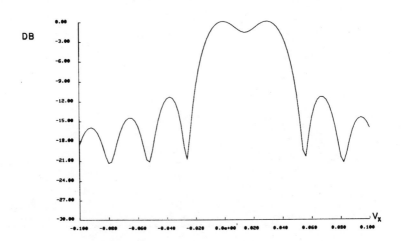

Figure 8A.

B) ADAPTIVE PROCESSING OUTPUT.
    THE POINT SEPARATION IS 0.29
    METERS AND THE FIELD OF VIEW IS
    ± 1 METER IN EACH DIRECTION.
    THE SIGNAL TO NOISE RATIO IS
    0 DB FOR EACH SOURCE.
    THE INDIVIDUAL POINTS ARE CLEARLY
    RESOLVED.

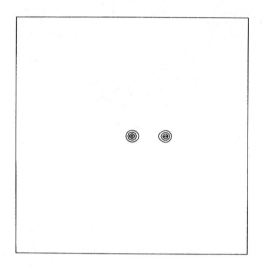

CONTOUR INTERVAL = 3 dB

SLICE THROUGH Y = 0

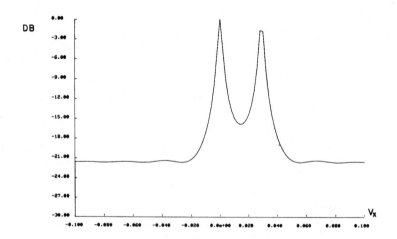

Figure 8B.

"MAN" TEST OBJECT.
6 FEET TALL, SURFACE
IS DIFFUSELY REFLECTING.

"RAT" TEST OBJECT.
THE LETTERS AND BARS
ARE DIFFUSELY REFLECTING,
THE SPECULAR AREA AT
LOWER RIGHT REFLECTS
64 TIMES AS STRONGLY
PER UNIT AREA.

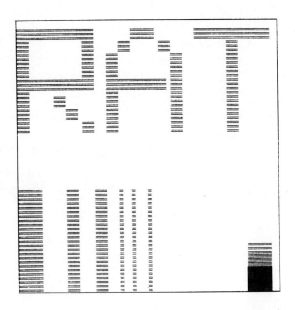

Figure 9.

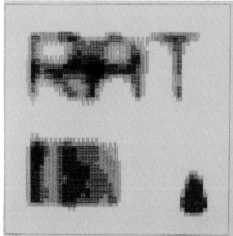 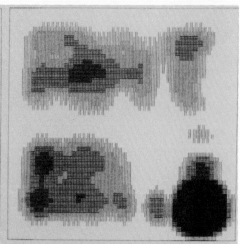

(A) ADAPTIVE PROCESSING   (B) CONVENTIONAL PROCESSING

FIGURE 10.

(A) & (B) RAT OBJECT USING 12 BY 12 ELEMENT
FILLED ARRAY. INSONIFICATION WAVELENGTH
= 2.5 MM AND INPUT SNR = 20 DB. PLOTS
USE 14 GREY-SCALE INCREMENTS OF 1.5 DB.
OVERALL IMAGE SIZE IS 2 BY 2 METERS AT A
10 METER RANGE.

(C) SAME PARAMETERS AS A&B, BUT WITH 128
ELEMENT SPARSE ARRAY. NOTE THE SLIGHTLY
BETTER RESOLUTION BUT POORER PEAK TO
BACKGROUND LEVEL. THE SMEARING ON THE
RIGHT IS DUE TO AN UNTIMELY COMPUTER
FAILURE.

(C) ADAPTIVE PROCESSING WITH SPARSE ARRAY DATA.

FIGURE 11.

IMAGE OF "RAT" OBJECT USING A
16 BY 16 ELEMENT FILLED ARRAY
WITH TOTAL APERTURE OF 0.21
METERS. INSONIFICATION WAVE-
LENGTH = 2.5 MM, AND INPUT SNR
IS 20 DB. THE TOTAL IMAGE SIZE
IS 1.2 BY 1.2 METERS AT A
RANGE OF 10 METERS.

IMAGES PLOTTED WITH 14 GREY-
SCALE INCREMENTS OF 1.5 DB.

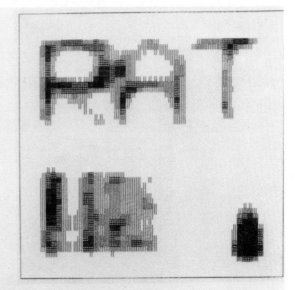

(A) ADAPTIVE PROCESSING

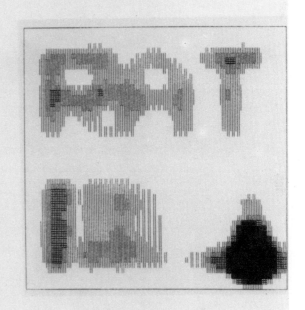

(B) CONVENTIONAL PROCESSING

(A) ADAPTIVE PROCESSING WITH
     BLOCK INVERSE COVARIANCE

(B) ADAPTIVE PROCESSING WITH FADING
     MEMORY INVERSE COVARIANCE ESTIMATE.

     SEE REFERENCE 4 FOR FURTHER DISCUSSION
     OF THIS TECHNIQUE.

FIGURE 12.

IMAGES OF 6 FOOT MAN AT 10
METERS WITH 12 BY 12 FILLED
ARRAY. (APERTURE .086 BY .086 M)
SNR = 20 DB ON 80 SHOTS OF DATA.
GREY LEVELS ARE 3 DB TO -21 DB.

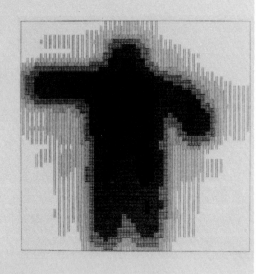

(C) CONVENTIONAL PROCESSING WITH
     BLOCK COVARIANCE ESTIMATE.

Baggeroer, and Dr. R. T. Lacoss of Lincoln Laboratories for their help and encouragement through the duration of this work. I am also indebted to the Seismology Group at Lincoln Laboratories for providing the unlimited computer time required for the simulations, and to the support of the Office of Naval Research under Contract No. N00014-77-C-0256.

REFERENCES

1.   Wade, G.; Wollman, M.; and Wang, K.; "A Holographic System for use in the Ocean", Acoustical Holography, Vol. 3; Plenum, Press, 1971.

2.   Wollman, M.; Wade, G.; "Experimental Results from an Underwater Acoustical Holographic System", Acoustical Holography, Vol. 5; Plenum Press, 1973.

3.   Flesher, G.T.; Wollman, M.; and Wade, G.; "Multichannel Under-Water Acoustic Holographic System", IEEE Ocean '75, IEEE Press, 1975.

4.   Duckworth, G.L.; Adaptive Array Processing for High Resolution Acoustic Imaging; Masters Thesis in the Department of Electrical Engineering and Computer Science, Massachusetts Institute of Technology, September 1979.

5.   Baggeroer, A.B.; Space/Time Random Processes and Optimum Array Processing; Naval Undersea Center document NVD TD 506, 1976.

6.   Capon, J.; Greenfield, R.J.; Kolker, R.J.; "Multidimensional Maximum Likelihood Processing of a Large Aperture Seismic Array", Proceedings of the IEEE, Vol. 55 #2, February, 1967.

7.   Capon, J.; "High Resolution Frequency Wavenumber Analysis", Proceedings of the IEEE, Vol. 57, #8, August, 1969.

8.   Capon, J.; Goodman, N.R.; "Probability Distributions for Estimators of the Frequency Wavenumber Spectrum", Proceedings of the IEEE, Vol. 58, pp. 1785-86, October, 1970.

9.   Duckworth, G.L.; "Array Processing for a Distributed Sensor Network Node", (A Lincoln Laboratories Technical note to be published).

10. Markel, J.D.: "FFT Pruning", <u>IEEE Transactions on Audio and Electroacoustics</u>, Vol. AV-19, #4, December, 1971.

11. Strang, G.; <u>Linear Algebra and its Applications</u>, Academic Press, N.Y., 1976.

COMPUTER HOLOGRAPHIC RECONSTRUCTION SYSTEM IMPLEMENTING

INTEGRATED THREE DIMENSIONAL MULTI-IMAGE TECHNIQUE

Thomas E. Hall,[1] Richard L. Wilson, and H. Dale Collins

Battelle, Pacific Northwest Laboratories

P.O. Box 999, Richland, Washington 99352

ABSTRACT

This paper describes a computer (Fresnel/Fourier transform) holographic reconstruction system used in the inspection of nuclear pressure vessels. The computer system can generate quasi-real-time (128x128) holographic reconstructions using an array processor and FFT program.

The display software has the unique capability of integrating up to ten holographic images taken of one defect at different viewing angles or positions, and displaying a single composite true three dimensional image. The image can then be rotated or tilted to assist the operator in interpretation and analysis.

Additionally, the computer reconstruction system provides an image spatial frequency filtering program for image enhancement and object characterization. This program has the ability to catalog and store in disk memory spatial frequency signatures which can be then used for classification and identification analysis.

Experimental results are presented that uniquely demonstrate the capabilities of the holographic reconstruction system.

INTRODUCTION

The mathematical nature of the holographic reconstruction process lends inself intuitively to existing computer array algorithms. However, until recent years, physical constraints have made impractical attempts to capitalize on this relationship for real-time field applications. Memory size, data storage,

205

computation speed and computer expense, among others, have been overwhelming hurdles. Due to the development of low cost and high density MOS memory, microcomputer systems, and high density mass storage devices, dedicated minicomputers can now be practically implemented for rapid large data array manipulations. Array processor technology (computer peripheral devices dedicated to data array computations) has contributed proportionally to this ability to handle large data arrays. In addition, CRT refresh display devices are improving dramatically, due to the availability of very fast semiconductor memories and microcomputers.

It is the aim of this paper to present a computer holographic reconstruction system that has taken advantage of the state-of-the-art by providing a minicomputer based system compatible with real-time field applications. The system was developed to perform the reconstruction portion of a complete acoustical holography imaging system for the inspection of nuclear pressure vessels. As will be apparent from the following text, applications for the device are not limited to pressure vessel scanning characteristics, but, rather, this device is adaptable to any acoustical holographic apparatus that can supply data in a digital form. The reconstruction system is intended to supply a mode by which the integrity of holographic imaging in the field is enhanced with respect to optical systems, and provide a rapid and reproducible field instrument for sizing defects.

Figure 1 shows a general block diagram of the computer reconstruction system illustrating the buss structure and peripheral devices. The hardware implemented includes all off-the-shelf items. The host computer is a Data General NOVA 3/12 with 128 KW of memory and hardware multiply and divide. The array processor selected for the number crunching task is a CDA MSP-2 single board array processor. Data is stored on 10 MB disk packs and displayed on a RGB color video monitor via a Lexidata graphics display unit.

Features of the completed system include:

1. Near Real-Time Reconstruction Processing - Image reconstruction of a 128x128 complex plane-wave hologram takes approximately 180 seconds.

2. Multi-Image Three-Dimensional Viewing - A sophisticated graphics display module uses an algorithm that has the capability of "over-laying" up to ten individual images of the same object. Thus, the interpretation of the object is greatly enhanced since multiple-angle views of the object may be incorporated into one composite image. Any viewing angle may be selected for a three-dimensional display. Also, alpha-numeric image information is presented on the video monitor alongside the image displayed.

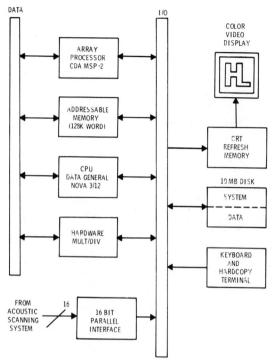

Fig. 1.   Block Diagram of Rapid Acoustical Holography Computer
          Reconstruction System

3.   Spatial Filter Module - A software module is incorporated that
     allows the operator to selectively filter spatial frequencies
     present in the plane wave hologram. Since spatial frequencies
     have a direct relationship to the edges or sharpness of the
     image, this module provides a mechanism for enhancement of
     related image characteristics.

4.   Improved Image Quality and Interpretation Capabilities - Because
     of the precise digital processing and display algorithms, the
     final color video display offers a calibrated image presenta-
     tion. The system supplies qualitative and quantitative data
     for flaw sizing and interpretation.

                    SYSTEM OVERVIEW AND RESULTS

     Figure 2 shows a general flow diagram of the reconstruction/
display system as viewed from the perspective of the operator. As
is evident from this diagram, the software was designed in a
modular fashion. This incorporates a very distinct set of algorithms that facilitate individual algorithm and software development with minimum impact on the complete system. Three modes

(1) DATA ACQUISITION MODULE

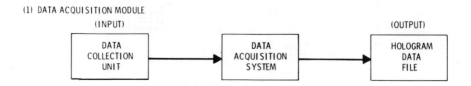

(2) RECONSTRUCTION MODULE

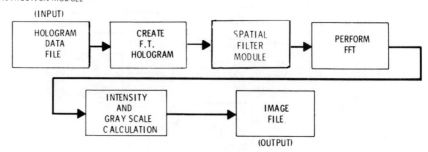

(3) DISPLAY MODULE

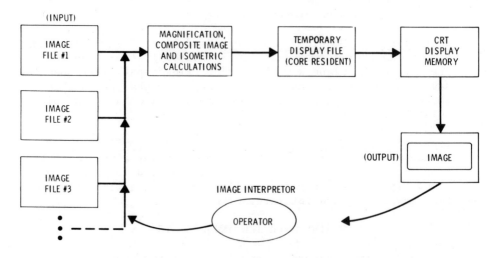

Fig. 2.   Reconstruction/Display System Flow Diagram

are available to the operator: data acquisition, reconstruction
and display.

The system is provided with complex Fresnel acoustical hologram
data in the form of a two-dimensional array (128x128 complex values).
Via the data acquisition module, a complex plane-wave hologram is
transferred to the reconstruction system mini-computer memory.   The
objective of the module is simply to create a data file in the
appropriate format such that the reconstruction module can operate
on it.   Data acquisition may occur concurrently with the scanning
operation.   Because of the modular construction of the software,
subsequent to data transfer the operator may choose to acquire
another hologram or reconstruct data currently on disk.

The reconstruction module has a minimum amount of operator
interaction, yet much of the complexity of the operation as a
whole is here.   To collect the necessary physical parameters (e.g.,
wavelength, material velocity, etc.) for reconstruction a header
on the data file is scanned.   The reconstruction calculations
(described later in this paper) are performed primarily by the
array processor.   Elapsed time for reconstruction of one hologram
with the present hardware is approximately 180 seconds.

Prior to reconstruction the operator has the option of spatially
filtering the hologram data for image enhancement.   This is
performed by masking portions of the Fourier transform hologram
itself.

Perhaps the most critical aspect of the holographic process
is the ability to interpret the reconstructed image accurately
with a minimum amount of confusion.   Therefore, the display module
must allow maximum flexibility to the user in a clear and concise
manner in order that he may concentrate his efforts on interpretation.   It was with this philosophy in mind that the display module
was designed.

The purpose of the display module is to provide a composite
isometric presentation of reconstructed holographic images.   That
is, the user must have the ability to "rotate and tilt" the displayed
image and, at the same time, be able to "overlay" more than one
image on any given display.   The algorithm chosen to perform this
task requires that the operator initially identify the related
image files to be "overlayed".   This establishes the input to the
isometric calculations.   At this point, use of the array processor
expedites the operation.   Temporary use of the CPU memory for a
buffer facility is utilized for the output of the arithmetic computations.   After all the defined image files have been mapped into
the new coordinate image space, the data is then dumped into the
display device for image display.

The user then enters into the system as the feedback device.
He may choose to add or subtract image files from the overlay list,
alter the isometric display angle, alter the overall display gray
scale, or possibly choose to go back to another module and optimize
physical parameters.   Thus, the operator is the "image interpreter".

The display module will require a significant amount of operator
interaction.   Image enhancement/interpretation requires that the
user be included in the feedback loop (note Figure 2).   In the
computer sense, the operator is being included as a peripheral
device that can be labeled as an "image interpreter".   This may
seem dehumanizing, but in an analytical sense it is precisely what
is taking place and, in fact, this is taking advantage of a very
sophisticated computer device.   So the operator or "image interpreter" must have control of variables significant to the image
display.

An example of a hologram and its respective reconstruction is
shown in Figure 3.   The object was a flat metal "H", 2 inches square,
placed at a depth of 9.72 inches in water.   The scanning frequency
was 3.2 MHz.   Note that a scale factor is displayed below the image
to provide a physical transformation from the image space to the
object space.

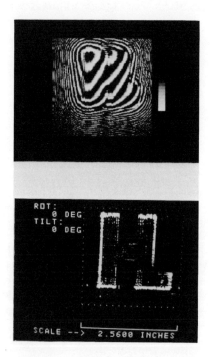

Fig. 3.   Example Hologram and Reconstructed Image

Figure 4 incorporates a simple multi-sided object to dramatize the three-dimensional multi-image feature of this system. The object was located 16.2 inches in depth in water and scanned at a frequency of 3.2 MHz. The dotted line on the display outlines the image space cube (i.e., the depth is presented at the same magnification as the lateral position). In this example the cube is approximately 4.6 inches on each side.

THE RECONSTRUCTION ALGORITHM

The following discussion is intended to illustrate the holographic technique (Fresnel/Fourier transform holography) incorporated in this computer reconstruction system.

The reconstruction system is supplied with a complex Fresnel hologram required with an on-axis reference wave. This can be represented as

$$H_c = \left[A_o^2 + R_o^2 + 2A_o R_o \cos\left\{\phi_o(x_o, y_o) - \phi_r(x_o, y_o)\right\}\right] +$$

$$i\left[A_o^2 + R_o^2 + 2A_o R_o \sin\left\{\phi_o(x_o, y_o) - \phi_r(x_o, y_o)\right\}\right] \quad (1)$$

where

A_o = amplitude of the object wave

R_o = amplitude of the reference wave

ϕ_o = phase of the object wave

ϕ_r = phase of the reference wave

(x_o, y_o) = defines the hologram plane.

If A_o and R_o are assumed to be slow varying functions over (x_o, y_o), then they can be ignored and since the plane wave reference is on axis, ϕ_r is constant. The hologram stored in memory is then

$$H_c = A \cos\left\{\phi_o(x_o, y_o)\right\} + i A \sin\left\{\phi_o(x_o, y_o)\right\} \quad (2)$$

Incidentally, it can be shown [2] that the "conjugate" image, a common annoyance with optical reconstruction systems, is eliminated from the reconstructed image since a complex hologram rather than a purely real hologram is implemented.

To take advantage of the Fast Fourier Transform algorithm available on today's computers (computed rapidly in an array processor), it is necessary to convert this expression into a form mathematically an inverse Fourier transform of the image. This may be accomplished by rotating H_c in the complex plane to a new H_c

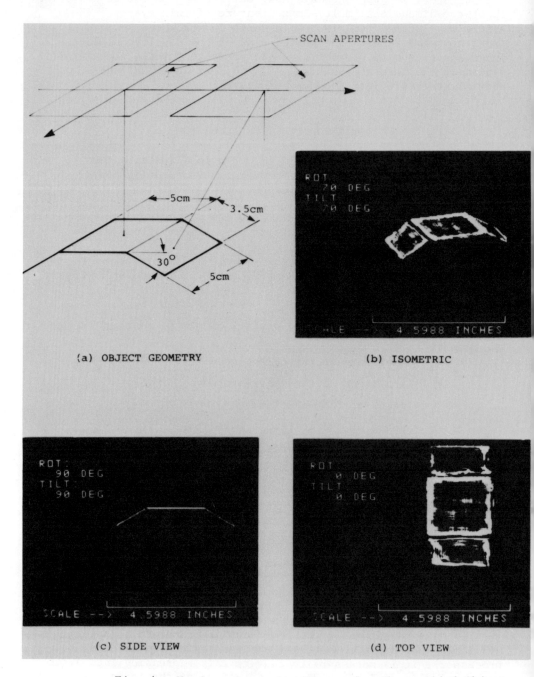

(a) OBJECT GEOMETRY   (b) ISOMETRIC

(c) SIDE VIEW   (d) TOP VIEW

Fig. 4.   Various Composite Views of a Three Sided Object

vector which is in the form of a Fourier Transform Hologram. Standard FFT software may then be used to supply a reconstructed image.

Holograms made with a point reference source rather than a plane wave can be called Fourier Transform Holograms where the object plane is located in the plane of this reference source. This is shown simply by observing the Fresnel diffraction equation [3] and noting that a point reference will eliminate the quadratic phase factor within the integrals. The expression then degenerates into the form of a Fourier transform of the aperture distribution (the hologram itself). In general, then, if one wishes to examine objects at depth Z, it is appropriate to place (mathematically) a point reference source at this depth. This will produce a complex hologram that is an inverse Fourier transform of the object.

The problem now is to convert the Fresnel hologram that was recorded into the form of a Fourier transform hologram. This is accomplished by recognizing that this hologram can be represented by equation (1) if the reference phase factor is replaced by a point source phase factor, i.e.,

$$\phi_r(x_o, y_o) = \frac{2\pi d}{\lambda} \tag{3}$$

where

$\quad\quad$ d = distance from the point source to the point (x_o, y_o)

$\quad\quad$ λ = acoustic wavelength.

Equation (1) then becomes

$$H'_c = A_o \cos\left(\phi_o - \frac{2\pi d}{\lambda}\right) + iA_o \sin\left(\phi_o - \frac{2\pi d}{\lambda}\right) \tag{4}$$

To calculate the H_c function then we need merely to perform a rotation operation at each location. That is

$$H'_c = P\,H_c \tag{5}$$

where

$$P = \begin{bmatrix} \cos\,(kd) & \sin\,(kd) \\ -\sin\,(kd) & \cos\,(kd) \end{bmatrix} . \tag{6}$$

This transforms the original Complex Plane Wave Hologram data into a Complex Fourier Transform Hologram Data.

Since the Fourier Transform Hologram is a Fourier Transform of the resultant image, it is possible to take advantage of this for image enhancement purposes. At this point a spatial filter may be introduced to enhance edges or for more extensive image analysis. Nonetheless, a spatial frequency "map" of the object is available for data manipulation and analysis purposes.

An FFT routine will transform the new, H'_c, hologram to a reconstructed image given by

$$A = FFT\ (H'_c) = R + iI\ .\tag{7}$$

The usual image is formed in analogy with the physical process
of optical reconstruction by forming the "intensity" of A or its
modulus square. This is also the equivalent of the Fourier
"spectrum" of the Fourier transform hologram. This Fourier spectrum
is identical with the object image intensity:

$$A*A = R^2 + I^2\ .\tag{8}$$

Unlike the case of optical reconstructions where only the
intensity A*A can be recorded on photographic film or a TV monitor,
we have recorded A which includes all relevant phase information
regarding the object. This means that we are at liberty to calculate any desired function of A and display this as an image. For
example, by taking $\sqrt{A*A}$ instead of just A*A and plotting this, one
can often improve certain aspects of the image contrast. Or calcu-
lating $|R| + |I|$ can produce an accurate image, while simultaneously
reducing computation time significantly.

To summarize the holographic reconstruction algorithm, the
stored complex hologram data characteristic of a plane wave, i.e.,
H_c, is "rotated" by subtraction of the phase, θ, where

$$\theta = \frac{2\pi d}{\lambda}$$

is the phase due to a point reference. The rotated complex hologram, H'_c, is then a Fourier Transform Hologram of the objects in
the plane (same depth as the point of reference). The hologram
H'_c may then be reconstructed using standard FFT routines capable of
operating on complex data of four quadrants.

HOLOGRAPHIC MAGNIFICATION CONSIDERATIONS

Holographic magnification has been well documented in the
past. [4] In the algorithm presently used, the reconstruction
illumination wavelength can be thought of as being equal to the
recording wavelength. In addition, the hologram size has not been
altered. Therefore, a unity magnification will always be realized
when considering exclusively holographic parameters.

In the case of Computer Holographic reconstruction, however,
consideration of the effects of sampling must be taken into account
to determine the true observed magnification. An evaluation of the
Nyquist Criterion for the present configuration and, subsequently,
the number of points across an image will be discussed. From this,
an expression for image magnification can be determined.

Given an object of circular shape (diameter D) at depth Z from the hologram plane, the fringe spacing produced by points at opposite ends of any diameter is

$$\delta = \frac{\lambda}{2 \sin \theta} \tag{9}$$

where θ is the opening angle of the cone whose base is the circular object of diameter D and the height of the cone is Z. From the laws of trigonometry

$$\sin \theta = 2 \sin \phi \cos \phi, \quad \text{if } \theta = 2\phi.$$

Therefore,

$$\sin \theta = 2 \; \frac{D/2}{Z^2 + D^2/4} \; \frac{Z}{Z^2 + D^2/4}$$

$$\sin \theta = \frac{ZD}{Z^2 + D^2/4} \tag{10}$$

The maximum spatial frequency of the fringes produced by the object is therefore

$$f_{OB} \leq \frac{1}{\delta} = \frac{2}{\lambda} \left\{ \frac{ZD}{Z^2 + D^2/4} \right\} \tag{11}$$

where it has been assumed that the point reference source was placed somewhere on the circumference of the circular object. A smaller maximum spatial frequency would have been obtained if the point reference had been placed at the center of the circular object. Thus, f_{OB} represents the maximum spatial frequency of fringes on the hologram when the point reference source, used to make the hologram, lies within the object.

If the hologram is a sampled N X N hologram array, then the reconstructed image can contain spatial frequency information up to and including the Nyquist spatial frequency maximum

$$f_{max} = \left(\frac{N - 1}{2a} \right) \tag{12}$$

The digital reconstructed image space is divided into four quadrants, the center of the N X N image array having zero spatial frequency and the "edge", f_{max}, is the Nyquist frequency. Thus, any image corresponding to spatial frequency information on the original hologram must fall in this range.

The foregoing circular object produced a maximum spatial frequency f_{OB} which corresponds to the image of an object of diameter D. Thus, the number of points across this image is

$$Ni = \left\{ \frac{f_{OB}}{f_{max}} \right\} \cdot N$$

or

$$Ni = \left\{ \frac{4aZD}{\lambda \ (Z^2 + D^2/4)} \right\} \left(\frac{N}{N-1} \right) \qquad (13)$$

Note that this result is essentially independent of the number of points in the array for large N, but does depend on the size of the aperture (a), the size of the object (D), the object depth (Z) and the wavelength (λ).

If the image magnification due to sampling is defined as follows:

$$M = \frac{\Delta d_i}{\Delta d_o} \qquad (14)$$

where

Δd_i = the distance between PIXELS in the image space,

Δd_o = the distance between samples in the hologram (object) space,

then it is a simple matter to derive a resultant expression.

An object is selected that has a diameter equal to the aperture size (D = a). Therefore, the expression for M becomes

$$M = \frac{Ni}{N} \bigg|_{D \ = \ a} \qquad (15)$$

The original hologram was acquired in a simultaneous source/receiver mode. From well documented holographic theory [5] it is known that the effective depth that is observed by the hologram in this mode is twice the physical depth of the object. Taking this into account, the final expression becomes

$$M = \frac{32 \ a^2 z}{N \ \lambda \ (16 \ z^2 + a^2)} \qquad (16)$$

where

a = aperture size
z = object depth
N = number of samples in one dimension
λ = acoustic wavelength

IMAGE RESOLUTION

Contributing factors to the final image resolution are the holographic resolution of the acoustic system and data sampling. Holographic resolution has been treated thoroughly in the past [6] and can be represented by:

$$\Delta x = \frac{\lambda d}{2A} \qquad (17)$$

where

λ = acoustic wavelength,
d = distance of object from hologram aperture,
A = aperture width.

In the sampled image, however, the expected resolution is Δx modulo, the image PIXEL size, Δd_i. Or if we allow T to be a trunkation operator, then

$$\Delta x' = T\left(\frac{\Delta x}{\Delta d_i}\right) \cdot \Delta d_i \tag{18}$$

CONCLUSIONS

This computer holographic reconstruction system was developed to encourage acoustical holography and digital reconstruction in field applications for accurately sizing material defects. An operator oriented device evolved with features to facilitate image interpretation, such as a near real-time reconstruction process, integrated multiple images and isometric color viewing.

A modular design lends itself directly to adaptability. The reconstruction system as a whole is independent of the acoustic processor, thus in general only the data acquisition module need be changed for separate applications. Other reconstruction algorithms (e.g., backward wave propagation [7]) may be readily adapted by altering the reconstruction module. The system's design inherently promotes future analysis and enhancements.

REFERENCES

1. The work described in this paper was performed in its entirety at Holosonics, Inc., in Richland, Washington, for the Electric Power Research Institute. Mr. Hall and Dr. Collins are currently employed at Battelle, while Mr. Wilson is self-employed at 20 Nuclear Lane, Richland, Washington 99352.
2. Keating, P. N., et al., in Acoustical Holography, vol. 5, (edited by P. S. Green), Plenum Press, NY, 1973, pp 515-528.
3. Goodman, J. W., Introduction to Fourier Optics, McGraw-Hill, 1968, pp 60.
4. Hildebrand, B. P. and B. B. Brenden, An Introduction to Acoustical Holography, Plenum Press, NY, 1972, pp 37.
5. Ibid., pp 85-87.
6. Ibid., pp 43.
7. Boyer, A. L. et al., Acoustical Holography, vol. 3, (edited by A. F. Metherell), Plenum Press, NY, 1971, pp 333-348.

ULTRAFAST ANALOGICAL IMAGE RECONSTRUCTION DEVICE WITH SLOW MOTION CAPABILITY*

B. Delannoy, C. Bruneel, R. TORGUET and E. Bridoux

Laboratoire O.A.E.

Université de Valenciennes 59326 Valenciennes Cedex FRANCE

GENERALITY

In the cardiological area, the increase of the image rate of echographic devices is a very suitable feature, owing to the very fast motion of cardiac valves. It would then be possible to follow precisely the valves during theirs motions, by using a slow motion picture system.

A first reconstruction imaging device was built in our laboratory by R. Torguet, C. Bruneel and al (1,2). The outpout images appear in optical form at rates up to 1000 per second and may be recordable on film using a cine camera. Such a technique seems however somethat involved because at time needed for film development and a magnetic record on tape would be more satisfactory. The quasi-real-time examination of the observed tomograph resulting in rapid diagnosis would then be possible. The matching of a magnetic recording system to the optical imaging apparatus would require the fast conversion of optical images into video-frequency electrical signals using a camera, whose characteristics would need to be well above the current technological limits. Another solution has therefore been tested, which consists of a purely electronic reconstruction of the acoustic images, thus giving the necessary video-frequency electrical signals directly. The problems encountered with the fast magnetic recording on a standard video tape recorder are examined. A possible solution is given for our particular case, whereby the information may be expanded in time, during two consecutive images. A solution for slow motion pictures images may be obtained using only the reduction of the magnetic tape speed in the reading stage.

* This work was supported by the DGRST FRANCE

ELECTRONIC RECONSTRUCTION OF THE ACOUSTIC IMAGES

The magnetic tape recording of an optical image, using a camera, may be performed using, either a photo-diode array with very fast multiplexing, or else a "Fanworth"-type image dissector. The electronic image reconstruction scheme enables a real image to be obtained using a group of N piezoelectric transducers (Fig. 1) as shown by B. Delannoy (3,4,5) without the need for any intermediate optical system. In each group, the electrical signals issuing from the different transducers of the acoustical field sampling array are so phased as to sum constructively for the signals arising from a point source lying on the axis of symmetry of the group, and destructively for off-axis point sources.

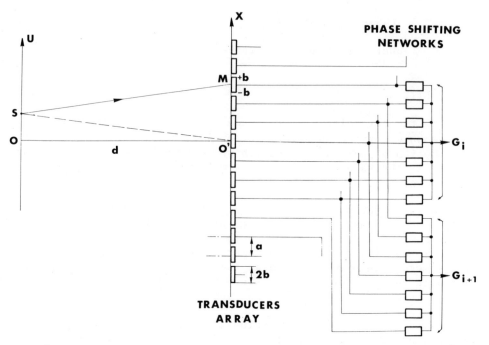

Fig. 1. Receiving array and transducer groups connections.

The reconstruction of the whole image requires a number of groups equal to the number of lines in the image. Each group is shifted from its neighbours by the spatial period of the array and all groups work in parallel, thus enabling very high image rates to be obtained. The amplitude level of the first order parasitic images introduced by the spatial sampling process is related to the aperture of each group of N transducers as shown by B. Delannoy and al (4,5), by the analytical relation :

$$\frac{I_1}{I_0} = \frac{Si\left(\frac{\pi b}{\lambda a}(Wa + 2\lambda)\right) + Si\left(\frac{\pi b}{\lambda a}(Wa - 2\lambda)\right)}{2Si\left(\frac{\pi b}{\lambda a}Wa\right)}$$

where Si is the Sine Integral function, b the half width of each transducer, λ the radiated acoustic wavelength, a the spatial sampling period and $W = Na/d$ is the angular aperture of each group, d being the observation distance.

In Figure 2, the first order parasitic image relative amplitude level versus the product W.a is shown for the values $\lambda = 0.5$ mm and $2b/a = 0.9$.

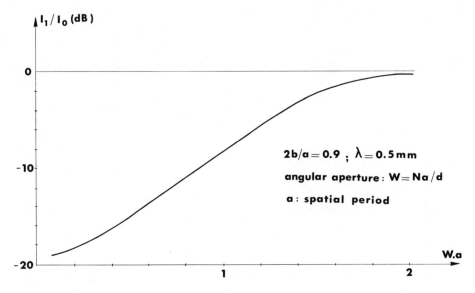

Fig. 2. First-order, secondary image level for the experimental
        system.

The system was designed to retain a constant angular aperture when the observation distance is varied. In this way, the parasitic image level is limited to a known fixed value. The moving focus feature is performed by varying the phase delays when the source distance changes, in order to get a zoom effect. The fast multiplexing of the group output signals gives the video-frequency signal, which is visualized, using a suitable monitor, at rates up to 1000 images per second, or else recorded, using a video tape recorder, after processing. The performance rated for this imaging device is comparable with that of the acoustico-optical processing system. The spatial resolution is nearly equal to 4 mm, if a 4 dB criterion is assumed and the apparatus has 20 processing channels and a spatial sampling period equal to 2 mm.

In order to increase the resolution without raising the parasitic image level, the reduction of the spatial sampling period, a, is necessary. This in turn must be accomplished while keeping the product "angular aperture W and spatial period a" constant. The number of transducers per group is then multiplied by a factor inversely proportionnal to a^2, and W by a factor inversely proportionnal to a.

REDUCTION OF BANDWIDTH

The cardiac ultrasound image corresponds to a maximum exploration depth of 15 cm and may be produced in 200 μsec with a frame rate of 1000 per second. The mean data rate for this type of echocardiographic imaging is given by 1000 frames per second, with 40 lines of 100 image points per line, giving 4×10^6 image points per second. The data rate is therefore lower than that for european standard television, where frames of 625 lines with 830 image points per line and a frame rate of 25 per second gives 12×10^6 image points per second. The acoustic image points are not, however, uniformly time-distributed but are concentrated in the 200 μsec image format time (fig. 3). Thus the instantaneous data rate is as high as 20×10^6 image points per second, requiring a minimum bandwidth of 10 MHz, and thus exceeding current technical capabilities. This problem may be solved by temporarily storing the data in buffer memories and then processing it before re-recording, using a constant 1 msec frame time. This then requires a minimum bandwidth of 2 MHz, which is available as a standard option.

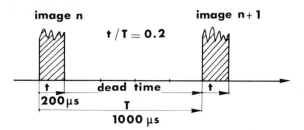

Fig. 3. Time stretching of the images without processing (1000 frames per second).

IMAGE RECORDING

The reduction of the required bandwidth by buffering enables a standard video tape recorder to be used as the recording device, even for a 40-channels imaging device, which seems to be the current technological limit imposed by the electronics needed for the reconstruction system and by the construction of the sampling transducer array. This technique has some definite advantages related to the helical scan used on the magnetic tape : a great quantity of data may be stored on reasonable lengths of tape and slow-motion observation is possible, while fixed bead rotation speed enables frequency-modulated recording and provides good fidelity at the reading stage.

The recording of ultrasonic images is performed at the standard head and tape motion speeds for a 50 half images per second magnetoscope. The image rate must then be one multiple of 50, say 50 P (with P integer). For a 1000 images per second rate, this gives P = 20, each with a 1 ms duration, after time stretching in order to reduce bandwidth. These images are written diagonally on the magnetic tape.

The very first results obtained using this method and a standard magnetoscope are shown in fig. 4, where the two extreme positions of the mitral valve are seen.

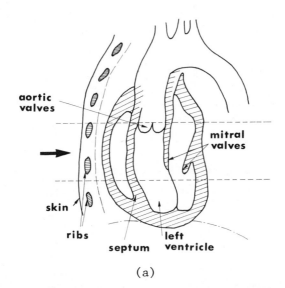

(a)

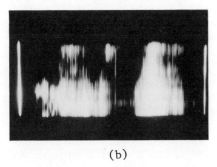   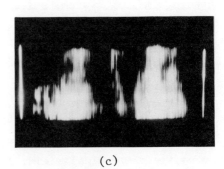

(b)   (c)

Fig. 4. Cut in the observation plane of the heart.
a) diagram, b) opening, c) closing.

IMAGE READING

The image reading may be performed at the same rate used for recording. A slow motion picture reading may be also implemented by winding the tape at a slower speed in the read mode than in the write one. The tape speed must then be slaved to the head motion in order to retrieve the recorded images. The principle of writing and slow motion reading is depicted in fig. 5 in the particular case P = 5.

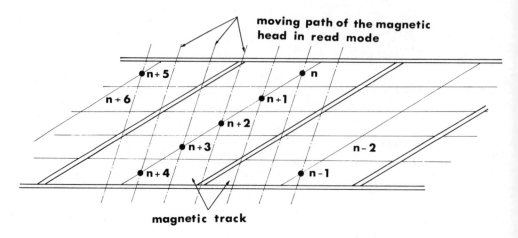

Fig. 5. Principle of slow motion reading of recorded ultrasonic images (P = 5).

CONCLUSION

In order to retain the advantages of the high image rate capability of some imaging devices, a recording and slow-motion picture system must be added to them. The recording may be performed using a cine camera and magnetic tape recording for the acousto-optic processing imaging system, but only if the optical images are rapidly converted into video-frequency signals. With the purely electronic reconstruction system currently developed in our laboratory, the optical-electronic conversion is bypassed and direct recording of the output video-frequency signals onto magnetic tape is possible, using time expansion of the information-carrying signals after buffering. The firts results have been obtained using CCD's. This technique doesn't however lead to very reproducible results., so that a digital solution is now devised. Moreover, a standard video tape recorder has been used for this purpose. A very simple speed control loop has allowed for the obtention of ultrasonic slow motion pictures. In the medical field, this device could not only provide as useful information as that to be obtained from a TM scanner but would provide this data for the whole cardiac valve instead for a single isolated point on the valve.

References

1. R. Torguet, C. Bruneel, E. Bridoux, JM Rouvaen, B. Nongaillard, 7th International Symposium on acoustical Holography and Imaging, 1976, pp. 79-85.

2. C. Bruneel, R. Torguet, JM. Rouvaen, E. Bridoux and B. Nongaillard, "Ultrafast echotomographic system using optical processing of ultrasonic signals", Appl. Phys. Lett., 30, 371, 1977.

3. B. Delannoy   , R. Torguet, C. Bruneel, E. Bridoux, Traitement analogique électronique d'images ultrasonores à grande cadence, Biosigma, Paris 1978, pp. 128-133.

4. B. Delannoy, R. Torguet, C. Bruneel, E. Bridoux and JM. Rouvaen, "Acoustical image reconstruction in parallel- processing analog electronic system", J. Appl. Phys. Vol. 50, n° 5, May 1979.

5. B. Delannoy, R. Torguet, C. Bruneel, E. Bridoux, Ultrafast electronical image reconstruction device, 3rd Symposium Echocardiology 1979, Rotterdam.

DISPLAY OF 3-D ULTRASONIC IMAGES

Lowell D. Harris, Titus C. Evans,* Jr., and
James F. Greenleaf

Biodynamics Research Unit, Mayo Clinic,
Rochester, Minnesota 55901
*Division of Cardiovascular Diseases
and Internal Medicine, Mayo Clinic,
Rochester, Minnesota 55901

ABSTRACT

Three-dimensional (3-D) ultrasonic images made up
of a "stack" of adjacent digitized B-scans are displayed as a volume image using the method of numerical
projection. This display method involves the generation by a computer, of two-dimensional digital projection images of the display volume viewed from any
desired direction. They are formed by summing the
"intensity" values of the individual volume picture
elements (voxels) of the 3-D B-scan along the set of
paths which project the 3-D image array onto a plane.
Stereo-pair projection images generated at two angles
of "view" differing by 2° to 8° are utilized to visualize the volume image in three dimensions. The 3-D
display of a "stack" of B-scans may result in the
obscuring of some portions of the image due to superposed "bright" structures unless this undesirable
effect is overcome by selective enhancement of·fea-
tures of interest in the 3-D B-scan. These enhancement methods are selective "tissue dissolution" and
"numerical dissection" whereby portions of the 3-D
B-scan are either partially "dissolved" or totally
eliminated from the volume before projection to enhance
the visibility of desired structure. Because they
operate on the 3-D image array before projection and
not on the resulting projection image, they are both

228 L. D. HARRIS ET AL.

effective and highly selective as visual enhancement
methods.

INTRODUCTION

The display of a "stack" of multiple, parallel,
ultrasonic B-scans as a three-dimensional (3-D) image,
utilizing any one of several methods (1-7), presents a
dilemma. The dilemma is that the successful display of
such a volume image (in order to appreciate 3-D anatomic morphology) may defeat a primary advantage of the
B-scan as an imaging modality which overcomes the
obscuring effect of superposition of bodily organs.
B-scans are two-dimensional (2-D) images of the reflective surfaces contained within 2-D sections* through an
organ. Imaged structures are, therefore, unobscured in
B-scans by superposed anatomy unlike in conventional
radiographic images which are 2-D projection images of
3-D structures. By simply displaying a stack of
B-scans as a 3-D image, anatomic structures at various
depths may be obscured by superposed structures. This
is especially true of 3-D ultrasonic B-scans which are
often high contrast (i.e., black and white) images and
therefore may contain very "bright" portions.

Described here are: 1) methods for displaying the
entire 3-D ultrasonic B-scan with each cross-sectional
image in its correct spatial position relative to the
other images, and 2) image enhancement methods which
successfully overcome the obscuring effects of superposition to visualize selected 3-D anatomy.

The display method, accomplished utilizing a digital computer, requires the analog-to-digital conversion (digitization) of the multiple parallel B-scans
which make up the 3-D image. The magnitude of each
picture element (pixel) in the digital array represents
the "intensity" of the B-scan at that position. The
2-D digital B-scans are assembled into a 3-D image
array, which is then mathematically projected by the
computer onto a plane to form a 2-D digital "projection"
image suitable for display on a television monitor.

---

* The assumption made here is that the path taken by
  the ultrasonic energy from the transducer to the
  reflecting surface and back is a straight line.
  Therefore, all reflecting surfaces seen in the B-scan
  are assumed to lie in the imaged plane.

Paired projection images of ultrasonic image volumes, generated at angles of view corresponding to binocular disparity (2° to 8° apart), are used as stereo image-pairs to generate a display of the volume which may be perceived in three dimensions.

The obscuring effect of superposition is effectively overcome by two enhancement methods. These computer-based techniques, referred to as "non-invasive numerical dissection" and "selective tissue dissolution", operate on the volume image data by selective alteration of image volume picture element (voxel) brightness values prior to their projection and display. Removing or selectively "dimming" obscuring anatomy enhances the visibility of the remaining portions of the image.

METHODS AND RESULTS

Digital B-Scan Image Generation

The system used to generate 3-D ultrasonic digital B-scans of isolated tissue specimens is shown in Figure 1. To the left of the dashed line is the scanning subsystem consisting of an ultrasound transmit/receive transducer (10 megahertz, focused), pulser, time variable gain amplifier, log compression circuit, rectifier, associated control logic, and an X-Y cathode ray tube (CRT) display device. The computer subsystem with its associated analog-to-digital conversion equipment used to digitize B-scans is on the right of the dashed line.

Digital B-scans are obtained by analog-to-digital conversion of the video output of the scanning subsystem as the transducer is swept across the specimen. Each row of the digital image is generated by pulsing the transducer and sampling the output signal of the scanner at a rate of 5×10^6 samples per second (8 bits/sample). The digitized row is transferred to the memory of the computer, the transducer is again pulsed, and the output of the scanner digitized for the next row. When the desired number of rows has been collected, the entire 2-D image array is transferred from the memory of the computer to digital tape for later processing and display. The B-scan of the adjacent section is obtained by positioning the transducer to the level of the next section (see inset) where the scanning and digitizing processes are repeated until

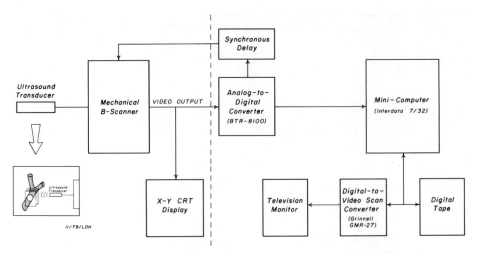

Figure 1   Diagram of system used to generate
digital ultrasonic B-scans of isolated tissue
specimens.   Mechanical B-scanner is on left of
dashed line.   Digital computer with associated
analog-to-digital conversion equipment is on
right.

the set of parallel digital B-scans covering the anatomic extent of the specimen are obtained.

Four digital B-scans of transaxial sections
through an excised human carotid bifurcation immersed
in saline are shown in Figure 2.   The images are of
parallel sections separated by 1.5 mm beginning 2.5 mm
cephalad to the bifurcation (Panel A) to 2.0 mm caudad
to the bifurcation (Panel D).   The external carotid is
located above the internal carotid artery in Panels A
and B.   The transducer scanning and analog-to-digital
sampling rates have been adjusted so that each pixel
represents a 0.15 x 0.15 mm region.   The magnitude of
the pixel represents the "brightness" of the B-scan
at that location.

Projection Image Generation

The process by which projection images of 3-D
B-scans are generated is illustrated diagrammatically
in Figure 3.   The voxels of the 3-D image at a single
level, k, are depicted as a two-dimensional array of
cubes contained within the perspective outline of the
image volume.   The outline of the projection image
plane is on the left of the volume, with the pixels

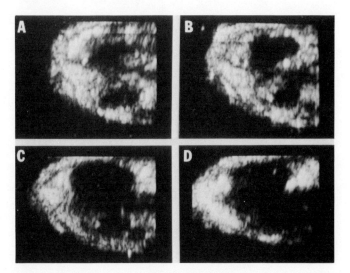

Figure 2 Digital B-scans of isolated human
carotid bifurcation immersed in saline. B-scans
are separated by 1.5 mm beginning 2.5 mm cephalad
to the bifurcation (Panel A) to 2.0 mm caudad
(Panel D).

corresponding to the level of this cross section shown
as a rectangular array of juxtaposed squares. The
values of the voxels located along the projection
paths, centered on each pixel of the projection image,
and passing through the volume image are summed to
compute the values of the array of pixels which form
projection image. Three representative summation paths
through the image volume are illustrated in this
figure. When all volume elements at a level k have
been projected onto the appropriate pixels in the projection image, the voxels at level k + 1 are projected,
and so on until all ultrasonic image voxels have been
projected onto the appropriate pixels. The resulting
two-dimensional array of pixel values can be scaled in
intensity, transferred to the memory of a digital scan
converter, and subsequently viewed and photographed as
an image on a television monitor. Before projection,
the image volume can be mathematically rotated to view
the 3-D B-scan from any desired angle. The location
of the center of rotation is arbitrary, although generally chosen near the center of the volume so that,
after rotation, the image volume does not pass out of

the effective viewing "window" defined by the projection image.

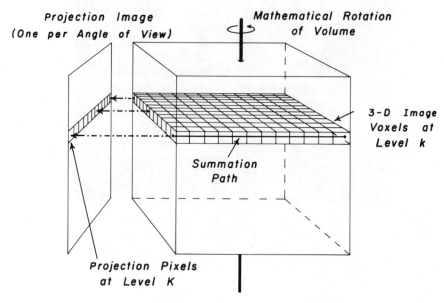

Figure 3 Diagram of three-dimensional (3-D)
image array projection process. (Reproduced
with permission from Harris et al., (5)).

Perspective is introduced into the projection
image by using a converging or conical set of summation
paths. The use of perspective adds a powerful monocular depth cue to images by making parallel surfaces or
lines appear to converge as they recede from the
observer.

Selective Enhancement of Imaged Anatomy

"Non-invasive numerical dissection" and "selective tissue dissolution" are terms used to describe two
methods used to overcome the obscuring effects of
superposition which often occurs when viewing 3-D
ultrasonic images, i.e., some structures are hidden by
overlying anatomy. Numerical dissection (5) is a process whereby selected portions of the image volume are
removed or "cut away" before projection in order to
more clearly visualize underlying portions of the
image. Numerical tissue dissolution (6) is a process

whereby obscuring portions of an image are selectively
dimmed but not totally removed. The effect of tissue
dissolution is to produce a "see through" or "ghosting"
effect so that the observer looks "through" obscuring
anatomy to visualize desired portions of the image.

Selective enhancement of the anatomic features
contained within a 0.36 mm thick section through a 3-D
B-scan of a human carotid bifurcation is shown from
two angles of view in Figure 4. The displayed image
volume is made up of 128 parallel transaxial digital
B-scans, each made up of 128 x 128 picture elements,

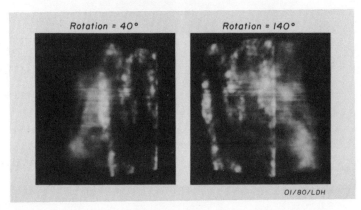

Figure 4 Projection images of 3-D ultrasonic
B-scan from two angles of view.

i.e., a 128 x 128 x 128 voxel image. The value of
each voxel represents the brightness of a small cubic
region with side dimensions of approximately 0.15 x
0.15 x 0.18 mm. Before projection, the volume was
rotated by 40° (left panel) and 140° (right panel).
The intersection of the vertically oriented lumen of
the vessel with the plane of dissection can be visualized as the dark "Y" shaped structure seen in the
highlighted plane.

Figure 5 illustrates the increased visibility of
a highlighted section through the 3-D B-scan achieved
by "dimming" the portions of the image "behind" the
plane and removing portions of the image "in front" of
the plane. In Figure 5, Panel A is a projection image
of the numerically dissected artery with no dissolution. Panels B-D are reprojections of the identical
portion of the image volume after the voxels behind

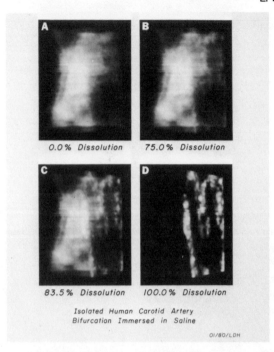

Figure 5   Illustration of enhancement effect of
0.36 mm thick section through 3-D B-scan by
selective numerical "tissue dissolution" and
"dissection".

the selected section have been dimmed or dissolved
by 75%, 83.5%, and 100%, respectively. Dimming is
accomplished by replacing the value of each voxel with
the product of the original voxel intensity value and
a constant which is less than unity (e.g., a constant
of 0.1 would result in a reduction of the magnitude
of each voxel by 90%). The result is that the structures within the slice are preferentially enhanced
because the voxels representing "bright" superposed
anatomy are either dimmed or totally removed. In
Panel A, it is not apparent exactly where the volume
has been sectioned. After dissolution by 75% (Panel
B), the intersection of the enhanced slice and the
arterial wall is just visible. After dissolution by
83.5%, the bifurcation of the artery is visible as the
dark "Y" shaped structure in the center of the enhanced
section. Finally, after the dissolution by 100%, the
highlighted section is clearly visible in perspective
but none of the remaining image behind the section
remains. In Panel C, with 83.5% dissolution, the

spatial relationships of the anatomic structures contained within the highlighted section to the anatomy
behind the plane can be visualized because the underlying arterial wall has been selectively dimmed but
not totally removed. These spatial relationships are
better understood when the images are visualized in
three dimensions.

Three-Dimensional Display of Reconstructed Volumes

The visualization of ultrasonic image volumes in
three dimensions can be accomplished by appropriately
viewing two projection images generated at angles of
view differing by a separation angle greater than 2°
but less than 8°. Two such images taken together are
referred to as a "stereo-pair". Stereo-pair projection images of an isolated human carotid arterial
bifurcation are shown in Figure 6. This image can be
perceived in three dimensions by cross fusing the
images, i.e., viewing the image on the right with the
left eye and the image on the left with the right eye.
In three dimensions, the orientation of the highlighted
sectioning plane and the spatial relationships of the
anatomy behind the plane to the structure in the plane
are clearly visualized.

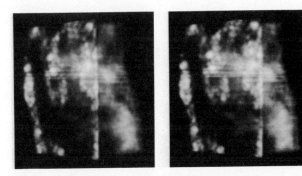

Figure 6 Stereo-pair projection images of 3-D
B-scan of isolated human carotid bifurcation
immersed in saline.

The image volume can also be visualized in three
dimensions by viewing sequences of projection images
which show the 3-D B-scan as it appears to rotate
before the observer. Rotation sequences are formed by
generating a series of projection images of the 3-D
B-scan at 2° angular increments around 360°. The

sequence, in the form of a series of video frames, is recorded on a video disc. When "played" back at normal or reduced rates, the powerful monocular depth cue of motion parallax occurs as portions of the image move relative to one another and the scene can be perceived as three-dimensional.

DISCUSSION

The obscuring effect of superposed anatomy which is overcome by ultrasonic B-mode tomographic scanning may reappear when the multiple parallel cross sections of a 3-D ultrasonic image are displayed simultaneously as a volume image. Straightforward image enhancement methods, such as numerical dissection and dissolution, can be utilized to increase the visibility of selected anatomy while maintaining the capability of observing important 3-D morphology. Visual enhancement accomplished using these methods is highly selective because these methods operate on the 3-D data set before the volume is displayed, i.e., before the 3-D to 2-D summation-projection transformation.

Additional feature enhancement methods might involve the selective dissolution or dissection of anatomy in a more structure-dependent manner. Ultimately, it would be desirable to control these processes operator interactively, i.e., to allow the observer to non-destructively "cut", "peel", and "dissect" the 3-D image in a manner which is similar to the methods utilized now only by the pathologist or surgeon.

ACKNOWLEDGEMENTS

The authors express appreciation to Dr. Erik L. Ritman for many stimulating conversations relating to the methods described in this paper, to Mr. Tak S. Yuen for his significant contribution to the design and implementation of the required computer algorithms and software, to Messrs. William F. Samayoa and Christopher R. Hansen for their assistance in operating the mechanical B-scanner, and to Ms. Marge C. Fynbo, Ms. Marge A. Engesser, and their colleagues for their assistance in the preparation of this manuscript and figures.

These investigations were supported in part by Research Grants HL-04664 and RR-00007 and Contract N01-HV-7-2928-2, National Institutes of Health, United States Public Health Service.

REFERENCES

1. Baum, G. and G. W. Stroke: Optical holographic
    three-dimensional ultrasonography. Science 189:
    994, 1975.
2. Greguss, P.: Holographic displays for computer
    assisted tomography. Journal of Computer Assisted
    Tomography 1:184-186, 1977.
3. Mark, H. and F. Hall: Three-dimensional viewing of
    tomographic data - the tomax system. Proceedings
    of the Society of Photo-Optical Instrumentation
    Engineers 120:192-194, 1977.
4. deMontebello, R. L.: The synthalyzer for three-
    dimensional synthesis and display of optical
    dissection. Proceedings of the Society of Photo-
    Optical Instrumentation Engineers 120:184-191,
    1977.
5. Harris, L. D., R. A. Robb, T. S. Yuen, and E. L.
    Ritman: Non-invasive numerical dissection and
    display of anatomic structure using computerized
    x-ray tomography. Proceedings of the Society of
    Photo-Optical Instrumentation Engineers 152:10-
    18, 1978.
6. Harris, L. D., R. A. Robb, T. S. Yuen, and E. L.
    Ritman: The display and visualization of 3-D
    reconstructed anatomic morphology: Experience
    with the thorax, heart, and coronary vasculature
    of dogs. Journal of Computer Assisted Tomography
    3(4):439-466 (August) 1979.
7. Ueda, M. and H. Nakayana: Three-dimensional dis-
    play of ultrasonic images using a Fly's-eye lens.
    Ultrasonic 17:128-131, 1979.

TOLERANCE ANALYSIS FOR PHASED ARRAYS

Kenneth N. Bates

Stanford University

Stanford, California 94305[*]

ABSTRACT

The effects of phase and amplitude quantization, random and periodic errors, channel to channel crosstalk, and missing radiators upon the line response of a phased array imaging system are discussed. A tolerance analysis is developed that allows the design engineer to evaluate the effects of those errors and to set limits as to their maximum allowable magnitudes.

INTRODUCTION

Acoustic phased arrays are being developed to meet the high speed imaging needs of medicine and nondestructive evaluation. In many respects the ideas and technology of phased array radar can be, and are being utilized in acoustic imaging systems. There is one major difference between these two imaging modalities. Radar imaging is almost always accomplished in the far field of the aperture formed by the array of antennas. In acoustic imaging, the object field routinely falls within the near field of the acoustic array. In this region, focusing of the acoustic radiation may be employed to improve the resolution of the system. Focusing is accomplished by the application of a quadratic phase shift across the array.

This phase shift may be introduced by the addition of a thin acoustic lens placed in front of the array. If one desires to steer the acoustic beam or to vary the focal distance, the phase

---
[*]The author is now at HP Corporate Labs., 1501 Page Mill Road, Palo Alto, California 94304.

shift will have to be introduced electronically. Due to the tremen-
dous signal handling capacity of digital electronics, there is a
well warrented tendency of those in the field to use digital devices
to control the electronic phase shift introduced into the signal
paths to and from the acoustic array. The price to be paid for the
use of digital processing is that the phase relationship between
elements of the array must be quantized. This will lead to quantization errors. Those errors will have some effect upon the line
response of the imaging system and hence upon the images formed by
that system.

Before designing such systems, it is necessary for the engineer
to understand what those effects are and how they are mitigated by
system parameters. Toward this end the following analysis is developed. The fundamental assumptions are those of linear propagation of acoustic fields in homogeneous isotropic media. The analysis is developed for narrow bandwidth systems. Within the framework of linear propagation, the results may easily be extended to
broad band (such as pulsed) systems.

PHASE QUANTIZATION

The analysis of this section is concerned with the effects of
phase quantization on the line response in the primary focal plane
of a one dimensional acoustic phased array imaging system. The
line response is, in the primary focal plane, the Fourier transform
of the pupil function that describes the complex acoustic field
generated at the array. The side lobe structure can be considered
a result of modulating the focusing or lens function with the pupil
function of the array. Therefore, any extraneous side lobe structure that appears from the quantization of the phase of the focusing
aperture must be accompanied by some modulation of the original continuous aperture function.

The quantization of the phase function can indeed be interpreted as a modulation of the original lens function. The phase
relationship of a thin lens is $\exp i\{kx^2/2f\}$.[1] The phase angle
$\phi = kx^2/2f$ is plotted in Figure 1. Figure 2 shows that quantizing
the phase relationship into steps $a_p = 1/2$ increment high produces
a "stair step" like variation of ϕ with respect to x. Subtracting this function from the original produces a third function shown
in Figure 3. This function is identified as error function $\xi(x)$.

The quantized lens function $L'(x)$ is therefore seen to de-
compose into the error free term $L(x)$ times a phase error term

$$L'(x) = L(x) \cdot \exp(-i2\pi\xi(x)$$

$$L(x) = \exp[i(k/2f)x^2]$$

(1)

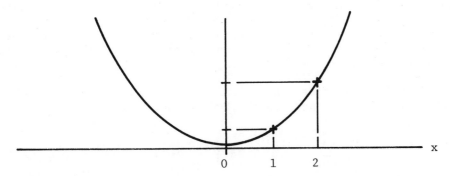

Figure 1.   Phase angle of the thin lens function.

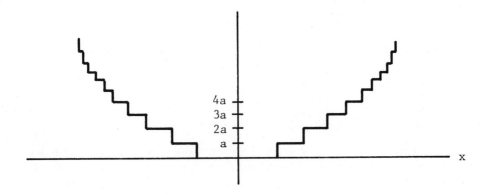

Figure 2.   Quantized thin lens phase relationship
a_p = quantization step size.

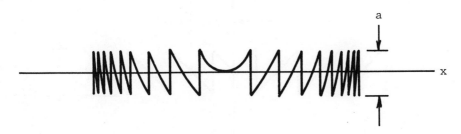

Figure 3.   Quantized phase error function.

Any side lobe structure, resulting from such a modulation, must arise from some periodicity in the modulating function itself. The function $\xi(x)$, however, is periodic in x^2 not in x. A function periodic in x^2, may exhibit periodicities in x when sampled at equal intervals. For this reason, it might be expected that the function $\xi(x)$ may also exhibit such periodicities when sampled in the same manner.

Changing variables, $X = x^2/2\lambda f$, $\xi(x)$ becomes $\xi(X)$ which is plotted in Figure 4 and has the Fourier series described in Eq. (2):

$$\xi(X) = \frac{a_p}{\pi} \sum_{m=1}^{\infty} \frac{(-1)^{m+1}}{m} \sin\left(\frac{2\pi m}{a_p} |X|\right) \tag{2}$$

Changing back, $x^2 = 2\lambda f X/k$, gives Eq. (3):

$$\xi(X) = \frac{a_p}{\pi} \sum_{m=1}^{\infty} \frac{(-1)^{m+1}}{m} \sin\left(\frac{\pi m}{\lambda f a_p} x^2\right) \tag{3}$$

Keeping the largest term, $m = 1$ allows $\xi(x)$ to be approximated by Eq. (4):

$$\xi(X) \simeq \frac{a_p}{\pi} \sin\left(\frac{\pi}{\lambda f a_p} x^2\right) . \tag{4}$$

Sampling the error function at 'b' intervals will produce a function that will replicate itself at interval Q, when Q meets certain conditions:

$$Q^2 = 2a_p \lambda f \ell \qquad \ell = 0, \pm 1, \pm 2, \ldots \tag{5}$$

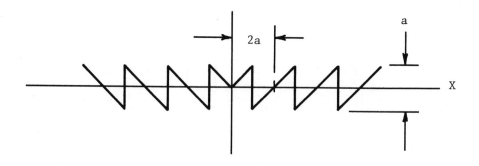

Figure 4. Quantized phase error function vs $x^2/2\lambda f$.

and

$$Q = \frac{a_p \lambda f}{b} P \qquad P = 0, \pm 1, \pm 2, \ldots \qquad (6)$$

As an example of this replication property, refer to Figure 5. Here, $\xi(x = nb)$ exhibits replication at $Q = 4.0$ cm when $2a_p \lambda f = 0.4$ cm and $b = 0.05$ cm . Notice that the sampled error function looks like a frequency modulated sinusoid. The modulation frequency is seen to have a period Q (i.e., the central pattern is replicated at Q intervals). An interesting question to ask is whether the other Fourier series components of $\xi(x)$ behave in a similar way. In point of fact, they do. However the conditions on Q may or may not be dependent on which term in the series is chosen. This means that under special circumstances, the sampled error function will behave more like a sinusoid of period Q than a Fourier series of sinusoids whose spacial frequencies are proportional to x . It is assumed, then, that the worst possible situation would occur when the sampled error function gives rise to a constant frequency sinusoid of period Q and amplitude a_p/π .

Under certain conditions the combination of the effects of phase quantization and spacial sampling will impress a sinusoidal phase modulation upon the original lens function. This phase modulation acts as a sinusoidal phase grating that will, in turn, diffract energy out of the main focus. The energy is diffracted into multiple diffraction orders whose position depend upon the modulation frequency $1/Q$, and amplitude upon the peak to peak phase variation, $2a_p$. When the magnitude of phase modulation a_p/π is shallow, the first diffraction order will be largest;

$$\frac{\text{Increase in Side Lobe Intensity}}{\text{Actual Main Lobe Intensity}} \simeq J_1^2(2a_p) \qquad (7)$$

When the phase step 'a_p' is determined by a digital word, 'a_p' becomes $1/2^N$, where N is the word length, Eq. (7) becomes Eq. (8);

$$\frac{\text{Increase in Side Lobe Intensity}}{\text{Actual Main Lobe Intensity}} \simeq J_1^2(2/2^N) \qquad (8)$$

$$\approx (1/2^N)^2 \qquad \text{for } N > 2 .$$

Therefore, it would be expected that large extraneous side lobes (first order diffraction lobe due to the phase modulation) will appear at interval $\lambda f/Q$ (when the restrictions upon Q are met) and have a maximum intensity of $J_1^2(2a_p)$.

There is yet another error function that can be ascribed to effect of phase quantization. In this case, the phase is not changed until the phase factor $x^2/2\lambda f$ equals some integer number

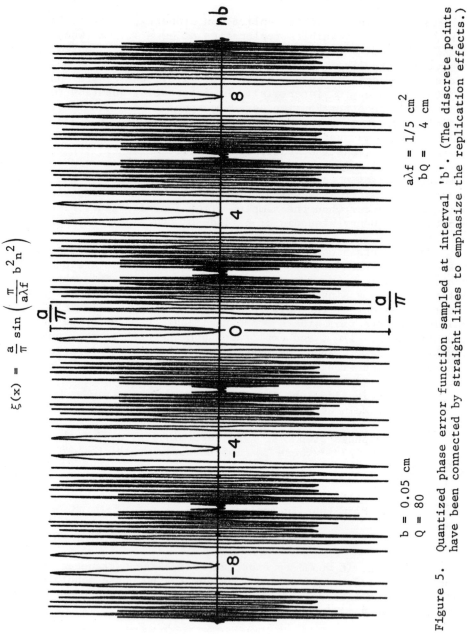

$$\xi(x) = \frac{a}{\pi} \sin\left(\frac{\pi}{a\lambda f} b^2 n^2\right)$$

aλf = 1/5 cm^2
bQ = 4 cm

b = 0.05 cm
Q = 80

Figure 5.   Quantized phase error function sampled at interval 'b'. (The discrete points have been connected by straight lines to emphasize the replication effects.)

times the quantization level a_p . This produces an error function $\xi'(x)$ shown in Figure 6. Changing variables, $X = x^2/2\lambda f$, $\xi'(x)$ becomes $\xi'(X)$ which is plotted in Figure 7. This sampled error function has the Fourier series described in Eq. (9):

$$\xi'(X) = \frac{a_p}{2} - \frac{a_p}{\pi} \sum_{m=1}^{\infty} 1/m \, \sin\left(\frac{2\pi m}{a_p}\,|X|\right) \tag{9}$$

Changing back to x , Eq. (9) becomes Eq. (10):

$$\xi'(x) = \frac{a_p}{2} - \frac{a_p}{\pi} \sum_{m=1}^{\infty} 1/m \, \sin\left(\frac{\pi m}{\lambda f a_p}\,x^2\right) \tag{10}$$

The constant $a_p/2$ will have no effect upon the response of the system. Since this merely represents a constant phase shift, it may be ignored. Now keeping the $m = 1$ term, $\xi'(x)$ is approximated by Eq. (11):

$$\xi'(x) \simeq - \frac{a_p}{\pi} \, \sin\left(\frac{\pi}{\lambda f a_p}\,x^2\right) . \tag{11}$$

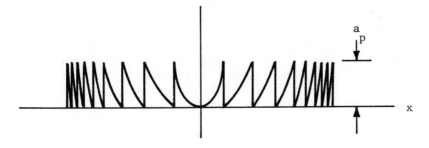

Figure 6. Quantized phase error function of the second kind.

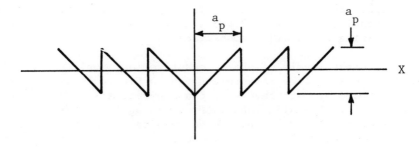

Figure 7. Quantized phase error function vs $x^2/2\lambda f$.

This is the same as $\xi(x)$, except for a minus sign. The significance of the sign of the error function is that it describes
the phase relationship of the extraneous side lobes with respect to
the error free response of the system. For $\xi(x)$, the phase of
each order alternates. All orders of $\xi'(x)$, however, are out of
phase with the error free response.

Since this result is based on several assumptions, numerous
focal plane responses were computed from a modeling program based
on first principles. Figure 8 is the computed response of a phased
array system whose phase was varied quadratically across the array
in a continuous manner. The aperture of this array was amplitude
weighted as $\cos^2(\pi x/A)$, where A is the width of the array aperture.

Figure 9 is the computed response of the same system with the
phase quantized in π radian increments. The second phase quantization method was used for this example. Notice the high level
background produced by the first diverging focus. A very intense
side lobe structure is also obvious. The spacing of the side lobes
is approximately 0.2 cm, instead of the predicted 0.1 cm interval.
The discrepancy is a result of the very large quantization step and
the phase quantization method used. The quantization step is large
enough that the higher orders of at least the first two sinusoidal
phase grating components of the error function must be included.
The combined effect is the coherent addition of the phase grating
side lobes expected at 0.1 cm with that at 0.2 cm, that at 0.3 cm
with that at 0.4 cm, and so on. The nature of the first extraneous lobe p = 1 , m = 1 is further altered by the coherent subtraction with the error free response.

For smaller phase quantization steps, these complicated effects
disappear, thus allowing the use of the simplified equations. In a
similar vein, the side lobe structure, resulting from both quantization methods, become virtually identical for phase steps less
than $\pi/2$ radian.

Figure 10 is a plot of the line response for the system with
phase quantized in $\pi/2$ radian steps. The periodic structure of a
phase grating is clearly demonstrated.

For eight levels of phased quantization, i.e., every $\pi/4$
radian, the side lobe structure shown in Figure 11 is similar to
that produced by four levels of phase quantization, albeit lower
in intensity. The side lobe interval is again 0.2 cm. This interval was correctly predicted as the requirement on the square of the
replication interval, Q , of the error function demands that
p = 2 .

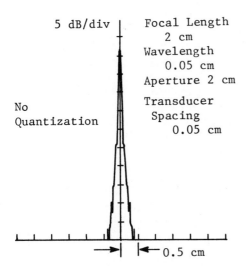

Figure 8.   Computed line response of focused phased
array (Hanning weighted).

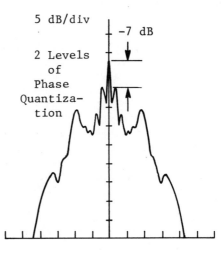

Figure 9.   Computed line response for $a_p = 1/2$.

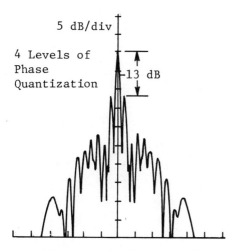

Figure 10.   Computed line
response for $a_p = 1/4$.

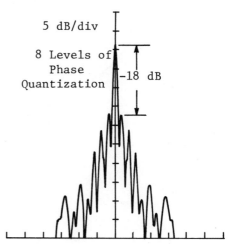

Figure 11.   Computed line response for $a_p = 1/8$.

Figure 12 is a plot comparing the predicted maximum extraneous side lobe level with the maximum side lobe level observed from the computer plots. The focal lengths and the wavelengths of each of the three series were different.

Though not exact, this analysis does predict the occurrence of large extraneous side lobes, their positions, and their intensity. The random variable analysis,[2] by comparison, predicted only their existence and their possible peak intensity. Of particular importance is the fact that the phase quantization of the focusing aperture did not generate large extraneous side lobes that were sensitive to the number of radiators. This is a direct contradiction to the prediction of the random variable analysis.[3] The cause is that the prime assumption, the indeterminacy of the phase error, is incorrect. The phase error is NOT random. Therefore, increasing the number of radiators cannot "average out" the phase error. As demonstrated, the phase error, introduced by phase quantization, can be determined. With some simplifying assumptions, the worst of these quantization effects in the focal plane can be readily calculated.

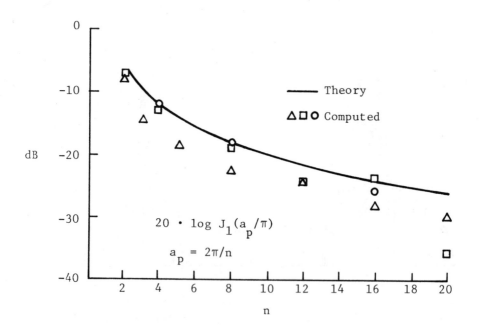

Figure 12.   Maximum extraneous side lobe level due to phase quantization.

AMPLITUDE QUANTIZATION

The effects of quantizing the amplitude of the aperture distribution are easier to determine. Since the aperture distribution
and its line response are Fourier transform pairs, the error introduced into the line response is the Fourier transform of the amplitude error introduced into the aperture distribution. This amplitude error will, of course, depend upon the actual amplitude relationship across the aperture as well as the magnitude of the
quantizing step. For example, only two quantum levels 0 and 1
would be needed to exactly reproduce the rectangular aperture
function shown in Figure 13. Having more levels would be of no
use. However, in acoustical imaging, it is desirable to amplitude
weight or apodize the aperture in a more complex manner in order to
reduce the original error free side lobe level.

As an example of the effects of amplitude quantization, consider the triangle apodization function shown in Figure 14.
Quantizating this function into steps a_A high produces the
"stair step" function in Figure 15. Subtracting this from the
original, produces the third function seen in Figure 16. This
function is identified as the amplitude error function $\eta(x)$.
Equation (12) is the Fourier series of this error function;

$$\eta(x) \;=\; \frac{4a_A}{\pi^2} \sum_{m=1,3,5,}^{\infty} \frac{(-1)^{(m-1)/2}}{m^2} \sin\left(\frac{2m\pi}{\sqrt{2}\;a_A}\,x\right) \qquad (12)$$

Dropping the higher order components gives Eq. (13);

$$\eta(x) \;\simeq\; \frac{4a_A}{\pi^2} \sin\left(\frac{2\pi}{\sqrt{2}\;a_A}\,x\right) \qquad (13)$$

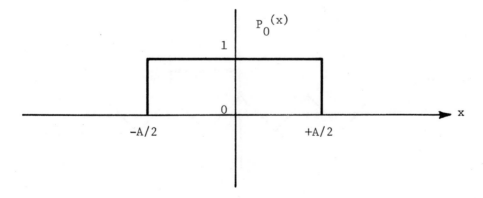

Figure 13.   Rectangular aperture weighting function.

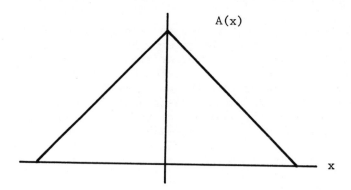

Figure 14. Triangular aperture weighting function.

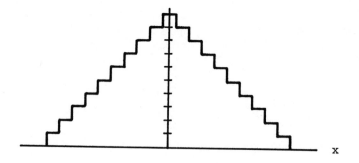

Figure 15. Quantized triangular aperture weighting function.

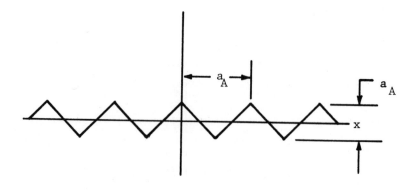

Figure 16. Quantized amplitude error function.

The amplitude error function is seen to be approximately the same as that of a sinusoidal amplitude diffraction grating. Therefore it will produce at least two diffraction orders $\pm$ f/a_A away from the origin of the focal plane with a relative amplitude of $2a_A/\pi^2$. When the step size is controlled by a digital word, a_A becomes $1/2^N$, where N is the word length;

$$\frac{\text{Peak Increase in Side Lobe Intensity}}{\text{Actual Main Lobe Intensity}} \simeq \left(\frac{2a_A}{\pi^2}\right)^2$$

$$\simeq \left(\frac{2}{\pi^2 2^N}\right)^2 \tag{14}$$

$$\text{Position of Extraneous Side Lobe} = \pm \frac{\lambda f}{\sqrt{2}\ a_A}$$

$$= \pm 2^N \lambda f / \sqrt{2} \tag{15}$$

Figure 17 is a computer generated line response similar to those in the previous section. Here, the phase was not quantized. The amplitude was quantized into four equal steps between 0 and 1. The triangle function was used as the amplitude apodization function. Notice that four levels of amplitude quantization produces a side lobe structure significantly lower than four levels of phase quantization.

Figure 18 is a plot of the extraneous side lobe levels observed in the computer generated line responses versus the amplitude quantization step size. Again, there is good agreement between the computed values and the ones predicted by analysis. The relative level of the extraneous side lobes generated by the amplitude quantization of other apodization functions can be determined in a like manner. It was observed, however, that the quantized triangle aperture generated the most intense side lobes of the apodization functions studied (Hanning and Hamming, i.e., $\cos^2$ and $\cos^2$ on a pedestal). Therefore, it seems likely that choosing the amplitude step size such that the extraneous side lobes generated by the triangle aperture are below the maximum acceptable level will produce satisfactory results.

As might be expected, the random variable analysis[4] predicts a side lobe level due to amplitude quantization that depends on the number of radiators, M ;

$$\frac{\text{Mean Increase in Side Lobe Intensity}}{\text{Actual Main Lobe Intensity}} \simeq \frac{1}{12}\frac{10}{M}\left(\frac{1}{2^N}\right)^2 \tag{16}$$

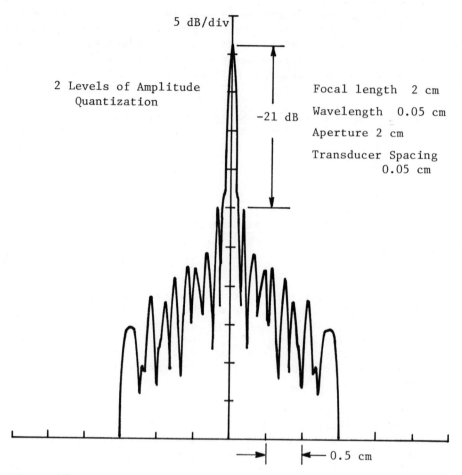

Figure 17.   Computed line response of focused phased array
             triangular aperture weighting function;
             a_A = 1/2 .

Again, the discrepancy lies in the random error assumption of the random variable analysis.

Comparing the results of the amplitude and phase quantization analysis it is seen that, for a given extraneous side lobe level, the number of quantization steps in phase must be. at least twice the number of quantization steps in amplitude. In other words, the extraneous side lobe structure is more sensitive to phase quantization than amplitude quantization. Therefore, when the step size is determined by a digital word, the word length of the amplitude control word can be one bit less than the phase control word.

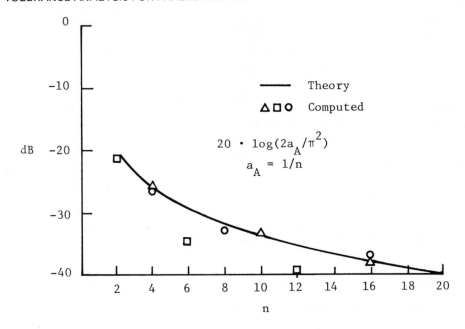

Figure 18.   Maximum extraneous side lobe level due to amplitude
             quantization.

ERRORS – PERIODIC AND RANDOM

   The operational parameters of any physical device will depart
from the optimum ones imparted by the designer. Therefore, in the
design of phased arrays, the effects of such "real world" problems
as random and periodic errors, channel to channel crosstalk, and
missing radiators should be considered.

   The idea behind such tolerance analysis is to pinpoint the
possible trouble spots and cure them at the design stage. Blithely
proceeding from theory to construction can cause great trouble if
the predicted performance is sensitive to variation in the design
parameters. Knowledge of the effects of possible variations will
enable the designer to make intelligent decisions as to the engineering trade offs incurred during any design procedure.

   In the following analysis, only those variations that are unpredictable will be termed errors. In turn, errors will be classified as random or periodic. Periodic errors in a phased array
are those errors in phase and/or amplitude that exhibit some
periodicity across the array. Conversely, random errors are those
that exhibit no discernable periodicity.

As demonstrated in the analysis on quantization effects, periodic variations in phase and amplitude give rise to phase and amplitude modulation of the error free aperture function. The result of such modulation is extraneous side lobes in the line response. Periodic errors produce similar results for the same reasons. However, the exact effects of such errors are undeterminable by definition. Fortunately the designer can ask what the maximum side lobe level could be for a given magnitude of error.

For a magnitude 'a', a periodic error describable as a square wave of period 'L' would most likely produce the largest extraneous side lobe. The square wave $\mathcal{M}(x)$ graphed in Figure 19 has the Fourier series:

$$\mathcal{M}(x) \;=\; \frac{2a}{\pi} \sum_{n=1,3,5}^{\infty} \frac{1}{n} \sin\left(\frac{2\pi n}{L} x\right) \tag{17}$$

Dropping the higher order terms,

$$\mathcal{M}(x) \;\simeq\; \frac{2a}{\pi} \sin\left(\frac{2\pi}{L} x\right) \tag{18}$$

If this error 'a_A' were described as some fraction of the amplitude distribution across the array, the largest extraneous side lobes produced would be

$$\frac{\text{Increase in Side Lobe Intensity}}{\text{Actual Main Lobe Intensity}} \;\simeq\; \left(\frac{2a_A}{\pi}\right)^2 \tag{19}$$

This demonstrates that a square wave of a given magnitude contains a sinusoidal component that is $4/\pi$ times greater than a sine wave having the same period and magnitude. This is the impetus behind the assumption that a square wave error would be responsible for the largest extraneous side lobe.

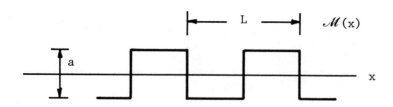

Figure 19.   Square wave error function.

If this error described that found in the phase distribution of the array, the largest extraneous side lobe intensity would be π^2 times greater if 'a_p' is described as some fraction of 2π ,

$$\frac{\text{Increase in Side Lobe Intensity}}{\text{Actual Main Lobe Intensity}} = J_1^2(4a_p) \qquad (20)$$

$$\simeq (2a_p)^2 \text{ for } a_p < 1/4$$

The period of these errors determines the position of the extraneous side lobes. Errors with periodicity approaching that of the width of the array will produce side lobes very close to the main lobe; so close that the major effect will be to broaden the main lobe. In acoustical imaging, main lobe broadening is preferred to extraneous structure in the line response. As a result, the allowable limits of the magnitude of the error can be relaxed for those errors of long period.

Notice that for a given magnitude of error, the worst case periodic errors produce extraneous side lobes four times more intense than those caused by quantization. Consequently for digital controlled systems, the maximum acceptable magnitude of the periodic errors should be less than one half that of a single quantum step. Or conversely, for a phased array having a periodic error of a given magnitude, it is of little use to quantize the phase or amplitude to less than twice that magnitude.

Random errors also produce extraneous structure in the line response of a phased array. These errors are best described by the random variable analysis.[5] Here, the prime assumption of indetermincy is met by definition. Assuming that the amplitude and phase errors are mutually independent:

$$\frac{\text{Expected Value of the Increase in Side Lobe Intensity}}{\text{Actual Main Lobe Intensity}}$$

$$\simeq \frac{\partial_\phi^2 + (\partial_a^2/a)}{M} \qquad (21)$$

where

∂_ϕ^2 = variance of phase error PDF

∂_a^2 = variance of amplitude error PDF

a = amplitude of radiation from single radiator

M = total number of radiators

PDF = Probability Density Function.

ERRORS - CROSSTALK

Crosstalk, as used in phased arrays, is the linear coupling of the signal from one radiator to the next. Due to crosstalk, the actual signal radiated from the nth radiator, S_n, will be a weighted sum of all the error free signals, U_m, of all the other radiators;

$$S_n = \sum_{m=-(M-1)/2}^{(M-1)/2} a_{nm} U_m \qquad (22)$$

where

a_{nm} = coupling coefficient

M = total number of radiators

This cross coupling is the predominant source of errors in both phase and amplitude. Here, the phase and amplitude errors are not mutually independent and hence cannot be treated by the random variable analysis. Nevertheless, the effects of this form of error should be studied to determine acceptable design limits.

As the focusing aperture is of prime interest to acoustic phased arrays, the following analysis will be limited to the effect of cross talk on the line response of such an aperture.

Assuming the coupling coefficients for the nth radiator can be described by a sampling of a continuous distribution function $C_n(mb - nb)$, the coupling coefficient a_{nm} becomes

$$a_{nm} = C_n(mb - nb) \qquad (23)$$

where 'b' is the spacing between samples. Since the samples represent contributions from each radiator, 'b' must be the same as the spacing between the radiators.

Figure 20 is a plot of the coupling functions for an array of seven radiators. Each radiator is shown to have the same coupling function. Fortunately for a given acoustic array, the environment of each radiator is approximately the same. It is likely that the distribution function is indeed the same for every radiator. Assuming this to be true, the subscripts may be dropped:

$$a_{nm} = C(mb - nb) \qquad (24)$$

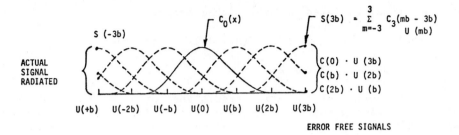

Figure 20.   Coupling functions.

Therefore

$$S(nb) \quad = \quad \sum_{m=-(M-1)/2}^{(M-1)/2} C(mb - nb) \; U(mb) \tag{25}$$

Let $q = m - n$ become $m = q + n$

$$S(nb) \quad = \quad \sum_{q=-n-(M-1)/2}^{n+(M-1)/2} C(qb) \; U(qb + nb) \tag{26}$$

The actual radiated signals, $S(nb)$, can therefore be written as a linear combination of weighted sets of error free signals. Each set, centered about qb , is weighted by $C(qb)$.

For example, if

$$C(qb) \quad = \quad \begin{cases} 1 & \text{for} \quad qb = 0 \\ \\ 0 & \text{otherwise} \end{cases}$$

then there is no crosstalk and the actual radiated signals must be the error free signals supplied:

$$S(nb) \quad = \quad U(nb)$$

If there is crosstalk between nearest neighbors,

$$C(qb) \quad = \quad \begin{cases} 1 & \text{for} \quad qb = 0 \\ \frac{1}{2} & \text{for} \quad qb = \pm b \\ 0 & \text{otherwise} \end{cases}$$

Now the actual radiated spacial signal becomes:

$$S(nb) = 1/2\ U(nb-b) + U(nb) + 1/2\ U(nb+b)$$

except at the ends of the array.

   If U describes the error free signal of a focusing aperture, then the resulting aperture can be described as a collection of M lenses weighted by C . Each 'lens' will produce its own characteristic response in the focal plane. As shown in Figure 21, these responses will be located a radiator spacing, b , apart and have a weighting C . Therefore, ignoring the end effects, the total response in the focal plane will be:

$$U'_f(x_f; z = f) \simeq \sum_q C(qb)\ U_f(x_f - qb; z = f) \tag{27}$$

   If there is no crosstalk, the coupling function C will be nonzero only at the origin and the resulting line response will become the error free response. As the cross coupling increases, the response will change. If the coupling function is smooth, real, and positive, the result would be a broadening of the line response. For such a coupling function, the main effect of crosstalk on the focusing aperture will be to broaden the error free line response by an amount dependent upon the distribution of the coupling coefficients. In fact, the general behavior is one of convolution. The overall response is approximately the error free response convolved with the coupling function. To see this, let $M \to \infty$, $b \to 0$, and * represent convolution:

$$U'_f(x_f; z = f) \simeq C(x) * U_f(x_f; z = f) \tag{28}$$

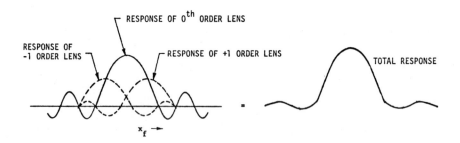

Figure 21.   Effect of nearest neighbor crosstalk in the focal
             plane.   (z = f)

For a phased array imaging system using the focusing aperture, crosstalk will degrade the line response of the system by broadening the main lobe. If the shape of $C(x)$ could be controlled, then the convolution effect might be used to smooth out the side lobe structure found in the line response. Unfortunately, it is difficult to identify all sources of crosstalk much less control them. The only practical solution is to keep the crosstalk to a minimum (i.e., $C(x)$ should be kept as narrow as possible).

ERRORS - MISSING RADIATORS

Missing radiators, those elements in the array that do not function, pose a very serious problem to phased array imaging. The net effect of the loss of two or more radiators is the introduction of a periodic "side lobe like" structure to the line response of the system.

To demonstrate this, the aperture, $U_0'(x)$, that results from the loss of one or more radiators can be decomposed into the error free aperture, $U_0(x)$, times a modulation function, $E(x)$, that is unity except at the position of the missing radiators where it is zero;

$$U_0'(x) = U_0(x) \cdot E(x) \tag{29}$$

For one missing radiator located at $x = r$, $E(x)$ becomes

$$E(x) = 1 - \delta(x-r) \tag{30}$$

For two missing radiators located at r and s

$$E(x) = 1 - \delta(x-r) - \delta(x-s) \tag{31}$$

and so on for more missing radiators. Now if $U_0(x)$ represents the focusing aperture, the actual response in the focal plane will be

$$U_f'(x_f; z=f) = U_s(x_f; z=f) * \int_{-\infty}^{\infty} E(x) \exp\left(i \frac{2\pi}{\lambda f} x_f x\right) dx$$

$$= U_f(x_f; z=f) * \left\{ \delta\left(\frac{x_f}{\lambda f}\right) - \exp\left(-i \frac{2\pi r x}{\lambda f}\right) \right.$$

$$\left. - \exp\left(-i \frac{2\pi s x}{\lambda f}\right) \right\} \tag{32}$$

Evaluating the convolutions,

$$U_f(x_f;z=f) * \delta\left(\frac{x_f}{\lambda f}\right) = U_f(x_f;z=f)^\dagger$$

$$U_f(x_f;z=f) * \exp\left(-i\frac{2\pi rx}{\lambda f}\right) = \exp\left(-i\frac{2\pi rx}{\lambda f}\right)\int_{-\infty}^{\infty} U_f[(\eta/\lambda f):z=f]$$

$$\times \exp\left(i\frac{2\pi r}{\lambda f}\eta\right) d\eta \qquad (33)$$

But

$$\int_{-\infty}^{\infty} U_f\left(\frac{\eta}{\lambda f};z=f\right) \cdot \exp\left(i\frac{2\pi r}{\lambda f}\eta\right) d\eta = |U_0(x_0=r;z=0)|$$

since $U_0(x)$ represents the focusing aperture. $|U_0(r)|$ is the magnitude of the aperture function describing the amplitude weighting or apodization at $x = r$. Therefore;

$$U_f(x_f;z=f) * \exp\left(-i\frac{2\pi rx}{\lambda f}\right) = \exp\left(-i\frac{2\pi rx}{\lambda f}\right)|U_0(x_0=r;z=0)|$$

$$(34)$$

Likewise

$$U_f(x_f;z=f) * \exp\left(-i\frac{2\pi sx}{\lambda f}\right) = \exp\left(-i\frac{2\pi sx}{\lambda f}\right)|U_0(x_0=s;z=0)|$$

$$(35)$$

The response in the focal plane becomes

$$U_f'(x_f) = U_f(x_f) - |U_0(r)| \exp\left(-i\frac{2\pi rx}{\lambda f}\right) - |U_0(s)|$$

$$\times \exp\left(-i\frac{2\pi sx}{\lambda f}\right) \qquad (36)$$

Thus each missing radiator introduces a phasor having spacial frequency proportional to its position and weighted by the magnitude of the error free aperture function at that position.

One missing radiator has the effect of adding a constant background $|U_0(r)|$ high. Two missing radiators introduce a cosinusoidal "side lobe like" structure that may be as large as $|U_0(r)| + |U_0(s)|$.

In general, the peak amplitude of this extraneous structure may be as large as the sum of all the weights of the phasors;

$\dagger$The scale factor of $1/\lambda f$ is understood.

Peak Amplitude of Structure $= \displaystyle\sum_{\substack{\text{All missing} \\ \text{elements}}} |U_0(x_0=\text{missing})|$ (37)

The main lobe or central response will be reduced by this amount;

$$\text{Actual Main Lobe} = \left\{ U_f(x_f=0) - \sum_{\text{missing}} |U_0(x_{\text{missing}})| \right\}^2$$

$$= \left\{ \sum_{\text{All elements}} |U_0(x_0=\text{all})| \right.$$

$$\left. - \sum_{\text{missing}} |U_0(x_{\text{missing}})| \right\}^2 \qquad (38)$$

The relative side lobe level becomes

$$\text{Relative Side Lobe Intensity} = \left\{ \frac{\displaystyle\sum_{\text{missing}} |U_0(x_{\text{missing}})|}{\displaystyle\sum_{\text{All}} |U_0(x=\text{all})| - \sum_{\text{missing}} |U_0(x_{\text{missing}})|} \right\}^2$$

$$(39)$$

which is the same relationship as obtained by Kino, et al.[2]

For an unweighted (not apodized) aperture, this simplifies to

$$\text{Relative Side Lobe Intensity} = \left(\frac{N}{M-N} \right)^2 \qquad (40)$$

where

N = the number of missing elements

M = the total number of elements.

In acoustical imaging normally 30 elements at a time are activated. If there is no apodization, 4 missing elements could be tolerated. However, for imaging purposes, the array is almost always apodized as to produce side lobes that are at least 23 dB down from the peak response. If triangle apodization is used, a single missing element (if it were in the center of the array) would produce a barely acceptable background level 23 dB below the peak. Two missing radiators would produce totally unacceptable "side lobe like" structure. Therefore, in the construction of a

phased array system, great care must be taken to insure that there
will be no more than one missing element in the array. For the
sake of uniformity of the image, there should be no missing radiators.

ACKNOWLEDGEMENTS

The author would like to thank H. J. Shaw and G. S. Kino for
their helpful discussions. This work was supported by Electric
Power Research Institute Contract No. RP609-1.

REFERENCES

1.   J. W. Goodman, "Introduction to Fourier Optics," MdGraw-Hill,
      New York, (1968).

2.   B. D. Steinberg, "Principles of Aperture and Array System
      Design," John Wiley and Sons, New York (1976), Chapter 7.

3.   Ibid., Chapter 13.

4.   Ibid., Chapter 13.

5.   Ibid., Chapter 13.

6.   J. Fraser, J. Havlice, G. Kino, W. Leung, H. Shaw, K. Toda,
      T. Waugh, D. Winslow, and L. Zitelli, "An Electronically
      Focused Two-Dimensional Acoustic Imaging System," Acoustical
      Imaging, Plenum Press, New York (1975), Vol. 6, pp. 292-299.

QUADRATURE SAMPLING FOR PHASED ARRAY APPLICATION

J.E. Powers, D.J. Phillips, M. Brandestini
R. Ferraro, D.W. Baker
Center for Bioengineering, University of Washington
Seattle, Washington, 98195

Receiver beam-forming in ultrasonic phased array systems typically demands very wide-band delay systems, having wide dynamic range and often complicated electronics. Sampled approaches to these delay lines generally employ sampling rates at least twice the highest frequency component in the returned RF signal. However, since the RF is bandlimited about some center frequency, by sampling in quadrature with respect to this center frequency the actual sampling rate may be substantially lowered. This produces two channels of information -- the in-phase or "real" component, and the quadrature or "imaginary" component. This "complex video" from each array element may be delayed and added to that from the other elements to provide receiver focusing. In addition to lowering the sampling rate, this technique retains all phase information for Doppler processing as well.

A 4 element, 5 MHz annular array has been constructed using both a lumped constant analog delay line system as well as the above sampling scheme for comparison. Results include beam plots, wire target scans, and tissue scans.

INTRODUCTION

Ultrasound phased array systems have been employed in diagnostic trials for several years with considerable success. Advantages of phased array systems include: ability to generate arbitrary scan formats; high rates of data acquisition; ability to electronically focus the beam in the receive mode; numerous possibilities for signal processing; and increased reliability since no moving parts are required to scan the sound beam through the ob-

263

ject volume of interest.[1,2,3] Probably the most significant
engineering trade-off in system design involves the requirements
for a large transducer aperture for good lateral resolution while
minimizing the number of active transducer elements comprising the
aperture. Each active channel requires its own transmitter and
receiving/ delay electronics in order to achieve the coherent
processing upon which phased array principles are based. Conventional receiver delay electronics generally use analog lumped constant delay lines,[2,3] or delay lines employing standard uniform
sampling techniques.[4,6,7] The latter approach requires rates that
meet or exceed the Nyquist limit allowing reconstruction of the
entire RF waveform by lowpass filtering. These conventional approaches result in a high cost and a large number of electronic
components required for each active channel.

Starting with the ultrasound transducer bandwidth, center frequency, and aperture geometry, one can specify the ultimate range
and lateral resolution achievable in a constructed two-dimensional
image. These parameters, therefore, set the video bandwidth required for image display. In particular, transducer bandwidth considerations can be used to simplify the delay line electronics required for each active channel.[8] This paper describes the theory
and application of a sampling technique which, when designed upon
bandwidth considerations, results in considerable simplification
over more conventional approaches used to date in diagnostic systems. In addition, the sampling approach to be described can be
readily adapted to different RF signal bandwidths and center frequencies through straightforward changes in system timing. The
technique also retains all phase (hence Doppler) information in a
form that lends itself to further processing.

Quadrature or "Complex Video" Representation

The design of a delay line system which minimizes the amount
of signal processing necessary can best be approached by analysis
of the required system output. Since many ultrasound 2D imaging
systems in use today employ some form of discrete time scan conversion prior to video display, the sampled video signal provides
a convenient starting point. Fig. 1 illustrates the detection
process typically used in present systems whether designed around
a single element transducer or a phased array. The returned RF
signal undergoes envelope detection and sampling prior to video
display; the sampling rate, generally no more than 1-3 MHz, determined by the required system axial resolution. Thus, the incoming
signal need be characterized only at these points in order to produce an output suitable for video display. In order to characterize the signal sufficiently to allow for coherent interference, as
is required for phased array application, the sampling method must
retain phase information in addition to amplitude. Since Doppler

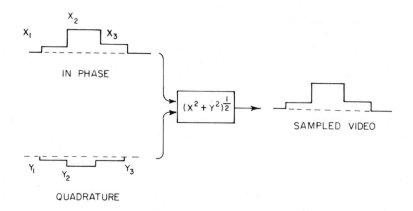

Fig. 1. Conventional Detection Method. The RF signal is detected
and sampled for discrete time video display.

Fig. 2. Coherent Detection Method. The RF has been demodulated
into its in-phase and quadrature components. Samples of
these signals are combined via the "hypotenuse function,"
or a linear approximation, to produce the same output as
conventional detection.

information is contained in the relative phase of consecutive
echoes,[9,10] this makes a straightforward method of determining
phase even more important.

Since the RF signal is typically band-limited about some
center frequency, f_o (usually, but not necessarily, the transducer resonant frequency), it may be represented in quadrature as

$$f(t) = X(t) \cos(2\pi f_o t) + Y(t) \sin(2\pi f_o t) . \quad (1)$$

Then the low frequency signals $X(t)$ and $Y(t)$ determine the amplitude and phase of $f(t)$ relative to f_o as

$$f(t) = A(t) \cos[2\pi f_o t + \Phi(t)] , \quad (2a)$$

where $A(t) = [X^2(t) + Y^2(t)]^{\frac{1}{2}}$ (2b)

and $\Phi(t) = -\tan^{-1} \dfrac{Y(t)}{X(t)} .$ (2c)

Thus, the signals X(t) and Y(t) contain the phase information required to process the signals of a phased array. Also, samples of X(t) and Y(t) taken at the video sampling rate are all that is required to produce the video output required for display, as illustrated in Fig. 2.

The similiarity between the quadrature representation and complex phasor notation has led our group to coin the phrase "complex video" to describe the signals X(t) and Y(t). By regarding these two signals as "real" and "imaginary" components of a phasor, the signals can be sampled and processed at the same rate as that required for video display without loss of information. The criterion for sampling rate then depends on video bandwidth and envelope rise time, rather than on transducer center frequency and highest frequency component.

The significance of this representation for single element systems is that by maintaining the phase information in the video signal, both tissue interface information and Doppler signals may be obtained from the same source. The application of this concept has been described by E. A. Howard.[11] For phased array systems, however, the implications are more far reaching. By maintaining phase information in the complex video from each element, coherent interference can still take place even though the samples are taken only at the video sampling rate.

Quadrature Sampling

One relatively common method of generating the signals X(t) and Y(t) is quadrature demodulation, or mixing the signal with $\cos(2\pi f_0 t)$ and $\sin(2\pi f_0 t)$ followed by low pass filtering. This approach is typically used in Doppler systems. For processing in sampled data systems such as an echo-Doppler processor,[11] the in-phase and quadrature signals are sampled simultaneously to produce the sampled complex video.

The method we chose, however, obtains samples of the quadrature components directly from the RF signal. Repeating eqn.(1),

$$f(t) = X(t) \cos(2\pi f_0 t) + Y(t) \sin(2\pi f_0 t), \qquad (1)$$

it can be seen that samples of X(t) can be obtained directly from f(t) by sampling at times $t = \frac{n}{f_0} = nT$ (n, a positive integer; T, the period of the carrier), thus

$$f(nT) = X(nT) = X_n. \qquad (3)$$

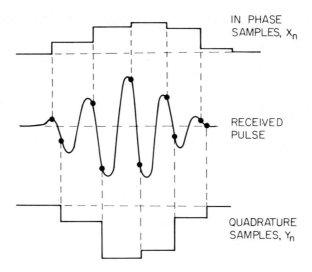

Fig. 3.   Quadrature Sampling Points. The timing for sampling
          in quadrature at the rate of f_o.

Similarly, samples taken at times $t = nT + T/4$ become

$$f(nT + T/4) = Y(nT + T/4) = Y_n. \qquad (4)$$

This is illustrated in Figure 3.

Calculation of amplitude and phase of the signal from eqn.
(2) requires that the signal be characterized for both $X(t)$ and
$Y(t)$ at the same instant in time; however, the samples generated
using eqns. (4) and (5) are $T/4$ seconds apart. Assuming a 50% RF
fractional bandwidth, a figure seldom significantly exceeded with
current diagnostic ultrasound equipment, the shortest rise time
possible will be on the order of $t_r = 1.4T$. Thus, the maximum
error in an estimate of the quadrature components at a point midway between the X sampling points and the Y sampling points will
be on the order of ±9%. In view of the wide dynamic range commonly encountered in diagnostic ultrasound applications, this error
can usually be neglected and the samples regarded as characterizing the signal at the same instant in time. If more accuracy is
required, an interpolation function has been derived whereby the
signal may be reproduced exactly (12).

The sampling rate assumed in obtaining eqns. (4) and (5) is
unnecessarily restrictive, however. It can be easily shown that
any two samples taken $T/4$ seconds apart characterize the signal
for amplitude and phase at a point midway between them subject to
the maximum error explained above. The amplitude is still obtain-

ed through eqn. (2b); however, determination of the phase requires
the addition to eqn. (2c) of a phase term to account for the arbitrary sampling time. Then a signal represented by eqn. (2a),

$$f(t) = A(t) \cos[2\pi f_o t + \Phi(t)], \qquad (2a)$$

may be described by the sampled complex video equations:

$$A(t) \simeq [\hat{X}^2(t) + \hat{Y}^2(t)]^{\frac{1}{2}} \qquad (5a)$$

$$\text{and} \quad \Phi(t) \simeq -\tan^{-1}\frac{\hat{Y}(t)}{\hat{X}(t)} + 2\pi m - 2\pi f_o t + \frac{\pi}{4}, \qquad (5b)$$

where m is the largest integer such that $(2\pi m - 2\pi f_o t) \leq 0$,

$$\text{and} \quad \hat{X}(t) \equiv f(t - T/8) \qquad (6a)$$

$$\hat{Y}(t) \equiv f(t + T/8) \qquad (6b)$$

are samples of the original RF signal taken relative to an arbitrary time t. In this representation, the signals $\hat{X}(t)$ and $\hat{Y}(t)$
still have the same properties as the actual in-phase and quadrature components, in that they characterize the signal in terms of
a phasor; the only difference is the additive phase factor due to
the arbitrary, but known, sampling time. This additive phase term
does not affect Doppler acquisition, however, as a pulse-Doppler
processor measures the change in phase of the signal returned from
a given object volume over several transmit/receive cycles, without
regard to the absolute phase.[9,10]

Complex Signal Processing

This representation is particularly handy for signal processing. Since complex addition merely requires separate summation
of the real parts and the imaginary parts, the same can be done
with the complex video from several array elements. Forgetting,
for a moment, the need for delay between elements, if two signals
are added vectorially -- real with real, imaginary with imaginary
-- then the result is identical to adding the signals in continuous time and sampling the result. If the signals were in phase
they will add constructively; if they were out of phase, the samples will cancel each other. The critical requirement is that the
samples be taken simultaneously.

Delay can easily be provided to the element signals as long
as the proper timing is observed, as illustrated in Fig. 4. In a
conventional delay line system, the received signals are delayed
the proper amount, τ, so that echoes from the target line up in
phase. Then they are added to produce a large RF signal which is
detected and sampled as mentioned previously.

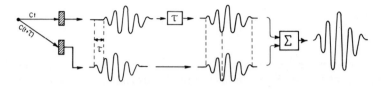

CONVENTIONAL DELAY LINE SYSTEM

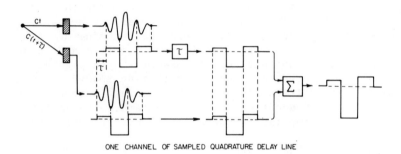

ONE CHANNEL OF SAMPLED QUADRATURE DELAY LINE

Fig. 4.   Conventional vs. Quadrature Delay Lines. In both systems
illustrated, an echo reflected from the target on the left
is received first by the nearest transducer element, then
later by the second, lower element. A conventional system
delays the entire RF waveform so that both echoes line up
in phase, then they are added. The quadrature delay line
samples the RF in quadrature (note only the in-phase channel is shown) at the rate of $2f_0/3$. The samples are
taken separated in time by the delay time τ, so that after
delay they line up exactly. The samples are then summed
vectorially to produce focused complex video. (Note that
the sampling rate of $2f_0/3$ produces a phase inversion
with each sample.)

    The same can be done using the sampled complex video. In
order for the sampling points to coincide after delay, however,
the original signals must be sampled at times that differ by the
amount of delay to be imposed. Then, only the samples need be
delayed, not the entire RF signal. Fig. 4 illustrates this process. The signal received by the element closest to the target
is sampled first. Then, τ seconds later the second signal is

sampled (also in quadrature), and the two sets of samples added vectorially. In the figure, a sampling rate of $2f_o/3$ is assumed,
producing the alternating sign of the samples.*

RESULTS

The quadrature sampling scheme that has been described has
been implemented to provide dynamic focusing for a 4 element, 5 MHz
annular array with a 15 mm aperture. The system also may be used
with a set of lumped constant analog delay lines for comparison.

Since the system will be used for Doppler as well as tissue
interface data acquisition, a 3-cycle gated burst is transmitted
from each element. The timing of each burst can be controlled in
increments of 50 nsec, or T/4, to provide focusing of the transmitted burst. The received signals are amplified and input to
either the lumped constant delay lines or the sampled delay lines.
The lumped constant delay lines have 10 taps spaced at 50 nsec.;
the sampled delay lines also have phase delay increments of 50
nsec., giving $\lambda/8$ accuracy for both systems. The latter system
operates with a sampling frequency of $2f_o/3$, or 3-1/3 MHz.

Fig. 5 shows beam plots made with the two systems. The received signal is used as the vertical deflection on an oscilloscope, the horizontal deflection is provided by a potentiometer
sensing the lateral translation of the target, a 4 mm plastic
ball. The beam plot is recorded via time exposure on Polaroid
film as the target is translated through the sound field. Peak
signal amplitude is indicated by the outermost envelope, interior
envelopes being generated by portions of the signal other than the
peak. Fig. 5a is a plot done with the lumped constant delay lines.
Fig. 5b was produced using the sampled delay lines, both with the
target at 60 mm in range. As can be seen, the two systems focus
equally well.

Some two-dimensional "B" scans were also done using the sampled delay lines. The wire target assembly used to produce Fig. 6a
was made from .25 mm monofilament line, and the individual targets
spaced as indicated. It was inclined at a 45° angle so that the
depth of field due to dynamic focusing can be seen. The array maintains good resolution throughout the area shown -- from 45-80 mm.

---

* It has been shown [12] that this sampling rate may be used to
reproduce a signal with a 66% RF fractional bandwidth with no loss
of information. In addition, it simplifies the expression for
relative phase.

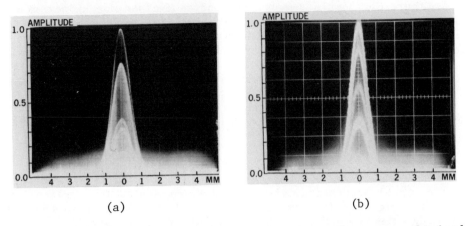

(a)   (b)

Fig. 5.   Annular Array Beam Plots. These beam plots were obtained
          by scanning a 4 mm ball reflector through the sound field
          at 60 mm in range. The received signal was focused by (a)
          the lumped constant, and (b) the sampled delay lines.

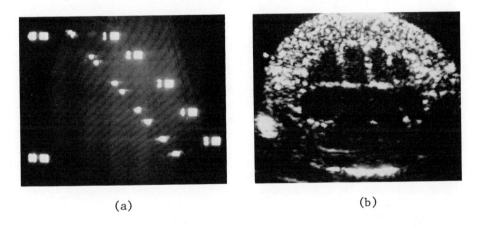

                (a)   (b)

Fig. 6.   2D B-mode Scans. (a) is a wire target scan with a spacing
          of 1 mm, 2 mm, 3 mm, 4 mm, and 5 mm (upper left to lower
          right) between the targets of each pair. The depth ranges
          from 45 mm to 80 mm. Part (b) is a scan of an excised dog
          kidney showing renal pelvis (bright line), papilla (echo-
          free circular regions), and renal cortex (peripheral gran-
          ular region). Length about 7 cm.

NOTE: Figures 5 and 6 have previously been submitted for publicat-
       ion in IEEE Trans. Sonics and Ultrasonics as part of ref. 12.

Figure 6b shows a B-mode image of an excised dog kidney made with the sampled delay lines. The overall length of the specimen was roughly 7 cm. Several basic structures are quite easily seen; the renal cortex is the granular region around the periphery, the papilla the echo free regions, and the renal pelvis the bright line down the center.

These scans demonstrate the utility of this sampling approach in obtaining ultrasound scans with a dynamically focused array.

SUMMARY

A signal processing scheme has been presented for application to ultrasonic phased array delay lines. It has been shown that the incoming signals need be sampled only at the video sampling rate if they are sampled in quadrature with respect to the center frequency. Thus, they are characterized as phasors and may be added vectorially to produce the required system output. Since the sampling process retains phase information, Doppler signals are easily acquired by standard pulse Doppler techniques without the need for mixers.

The advantages of this system over more conventional techniques include:

1. lower sample rates, hence reduced delay line bandwidth;
2. phase delay increments independent of sampling rate;
3. system flexibility -- different array frequencies and configurations require changing only sample timing;
4. phase information contained in complex video signal for possible further processing;
5. Doppler information obtained without mixing.

It is felt that the concepts outlined in this paper will prove to significantly reduce the complexity and improve the performance of phased arrays in the future.

ACKNOWLEDGEMENTS

The authors wish to express their appreciation to Vern Simmons, Bob Olson, John Ofstad, and Florencio Daquep for their technical assitance. We would also like to thank Carolyn Phillips for her help in preparing the manuscript and wish her luck on her African Safari.

Supported by NIH Grant HL-07293

REFERENCES

1. J. C. Somer: Electronic sector scanning for ultrasonic
    diagnosis. Ultrasonics 6:153-159, 1968.
2. F. L. Thurstone and O. T. Von Ramm: A new ultrasound imaging
    technique employing two dimensional electronic beam steering.
    Acoustical Holography. Vol. 5, 1973.
3. H. E. Melton, Jr.: Electronic Focal Scanning for Improved
    Resolution in Ultrasound Imaging. Ph.D. thesis, Department of
    Biomedical Engineering, Duke University, 1971.
4. P. Doornbos and J. C. Somer: An electrically variable analogue
    delay achieved by fast consecutively commutated capacitors.
    Progress Report 3, Inst. Med. Phys. TNO, Da Costakade 45,
    Utrecht, The Netherlands, 1972.
5. R. E. McKeighen and M. P. Buchin: Evaluation of novel dynam-
    ically variable electronic delay lines for ultrasonic imaging.
    Ultrasonics in Med., Vol. 4, 1977.
6. R. E. McKeighen and M. P. Buchin: New techniques for dynamical-
    ly variable electronic delays for real time ultrasonic imaging.
    Proceedings IEEE Ultrasonics Symposium, 1977.
7. J. T. Walker and J. D. Meindl: A digitally controlled CCD dyna-
    mically focused phased array. Proceedings IEEE Ultrasonics
    Symposium, 1975.
8. C. B. Burckhardt, P. A. Grandchamp, H. Hoffman, and R. Fehr: A
    Simplified Ultrasound Phased Array Sector Scanner. Echocardio-
    logy, C. T. Lancee, ed., pp. 385-393, 1979.
9. D. W. Baker: Pulsed Doppler Blood Flow Sensing. IEEE Trans.
    Sonics and Ultrasonics, SU-17, pp. 170-185, 1970.
10. M. A. Brandestini: Topoflow -- A Digital Full Range Doppler
    Velocity Meter. IEEE Trans. Sonics and Ultrasonics, SU-25, pp.
    287-293, 1978.
11. E. A. Howard, M. A. Brandestini, J. E. Powers, M. K. Eyer, D. J.
    Phillips, E. B. Weiler: Combined Two-Dimensional Tissue/Flow
    Imaging. Acoustical Imaging, this volume.
12. J. E. Powers, D. J. Phillips, M. A. Brandestini, R. A. Sigelmann:
    Ultrasound Phased Array Delay Lines Based on Quadrature Sampling
    Techniques. Submitted to IEEE Trans. Sonics Ultrasonics, Nov.
    1979.

ACOUSTIC MICROSCOPY-A TUTORIAL REVIEW

Lawrence W. Kessler and
Donald E .Yuhas
Sonoscan, Inc.
530 E. Green Street
Bensenville, Illinois 60106

ABSTRACT

Acoustic microscopy is emerging as an important analytical technique serving the needs of both biomedical and materials technology.
By means of very high frequency elastic waves, acoustic microscopes
reveal structural-mechanical properties of specimens at high magnification. A review of the techniques and applications is presented in
this reprinted article entitled, Acoustic Microscopy-1979 (37).

INTRODUCTION

The acoustic microscope is an analytic tool whose course of
development and importance to the scientific community may parallel
that of the scanning electron microscope. Though in a relatively
early stage of development compared with electron microscopy, the
acoustic microscopy field is demonstrating its importance in a broad
spectrum of applications, such as biomedical research, materials
technology, and quality control. In an earlier article (1), Kessler
reviewed the field to about 1973. The purpose of this article is to
update the review and provide the reader with an addition to the 1973
comprehensive bibliography. Examples are presented to illustrate the
new level of information provided by acoustic micrographs.

By way of contrast to other articles in this volume in which
low frequencies (1 to 10 MHz) are employed, acoustic microscopes use
frequencies ranging from 100-3000 MHz. In clinical medicine and nondestructive testing, relatively long wavelength (e.g., 0.5 mm) acoustic energy is used to penetrate the depth of the human body or many
inches of metal. On the other hand, microscopy, which discriminates
structural features down in the micrometer size range, requires

appropriately short wavelength ultrasonic energy. Unfortunately,
the shorter acoustic wavelengths are often associated with increased
absorption losses which decrease penetration depth. The increased
absorption is often so severe that in the "microscopy" regime the
conventional applications of low-frequency (1-10 MHz) ultrasound are
precluded. However, the increased resolution capabilities have lead
to variety of new applications.

Modern analytical instrumentation has played an important role
in the advancement of technology. Whenever images are formed of an
object with a different form of radiation, new information can be
gained because of the possible unique interactions which occur between the radiation and the object. An acoustic microscope reveals
localized changes in elastic properties of materials by measuring
point by point reactions of the specimen to a periodic stress wave.
Thus the physical properties of matter which govern sound propagation,
namely, the compressibility, the density, and the viscoelasticity are
translated into acoustic micrographs. From the pragmatic point of
view, the sonically transparent nature of most materials permits internal examination of samples which are opaque to light. Subsurface
defects, flaws, inclusions, and disbonded areas are particular areas
of materials science and quality control which are amenable to acoustic microscopy. In biomedical research, the capacity to differentiate
structure in live tissue without chemical staining makes a new level
of experimentation possible. The new structural information that is
revealed and the efficiency of visualization without chemical processing of the tissue make the technique of ultimate great value in clinical diagnosis of disease. In materials technology and nondestructive
testing, the elastic properties of the sample control its behavior
under stress. Acoustic microscopy is used to characterize microelas-
tic structure as well as to find hidden defects and flaws.

II. BACKGROUND

The notion of acoustic microscopy dates back to 1936 when
Sokolov (2) proposed a device for producing magnified views of structure with 3 GHz sound waves. However, due to technological limitations at the time, no such instrument could be constructed, and it
was not until 1959 that Dunn and Fry (3) performed the first acoustic
microscopy experiments, though not at very high frequencies.

The scientific literature shows very little progress toward an
acoustic microscope following the Dunn and Fry experiments up until
about 1970 when two groups of activity emerged, one headed by C. F.
Quate, and the other by A. Korpel and L. W. Kessler. The first
efforts to develop an operational acoustic microscope concentrated
upon high-frequency adaptations of low-frequency ultrasonic visualization methods. For example, one early system employed Bragg diffraction imaging (4)-(6), which is based upon direct interaction between
an acoustic-wave field and a laser light beam. Another example was

based on variations of the Pohlman cell (7). The original device is
based upon a suspension of asymmetric particles in a thin fluid layer
which, when acted upon by acoustic energy, produce visual reflectivity
changes. Cunningham and Quate (8), modified this by suspending tiny
latex spheres in a fluid. Acoustic pressure caused population shifts
which were visually detectable. Kessler and Sawyer (9) developed a
liquid crystal cell that enabled sound to be detected by hydrodynamic
orientation of the fluid. In 1973, the Quate group began the development of its present instrument concept (10) which utilizes a confocal
pair of acoustic lenses for focusing and detecting the ultrasonic
energy. Advancements of this instrument, a Scanning Acoustic Microscope (SAM), have to do with achieving very high resolution, novel
modes of imaging, and applications. In 1970, the Korpel and Kessler
group began to pursue a scanning laser detection system for acoustic
microscopy (11). In 1974, the activity was shifted to another organi-
zation under Kessler, where practical aspects of the instrument were
developed. This instrument, the Scanning Laser Acoustic Microscope
(SLAM), was made commercially available in 1975.

A great variety of methods can be thought of for ultrasonic
visualization. Some of these are adaptable for microscopy and there
may be many new ones which will be thought of in the future when it
becomes economical to justify the task. At the present time, in the
opinion of these authors, the optimum use of resources seems to be
in the area of applications of the technology. Out of these efforts
will come the needs and justification for improved techniques. Such
improvements may achieve higher resolution, different modes of operation, customized instruments for particular types of samples, etc.

The balance of this article presents theory and applications of
the presently employed techniques. Innovations to these techniques
have been constantly in process in order to make acoustic microscopy
viable as an analytical tool to serve the needs of industrial and
scientific communities in biomedicine, materials technology, and
quality assurance.

Out of the research efforts of the past five years, essentially
two techniques have emerged as viable practical acoustic microscopes:
The SLAM and the SAM. The end result of both is the production of
high resolution acoustic images; however, the imaging technology,
capabilities, and areas of applicability are quite different. In
the next section we describe and contrast the operational principles,
design philosophy, and capabilities of the two systems.

III. THE SCANNING ACOUSTIC MICROSCOPE

In the SAM, an incident acoustic wave is launched into a water
medium by a piezoelectric transducer bonded onto one end of a cylindrical sapphire rod. The other end of the rod has a concave spherical
surface ground into it, causing the acoustic beam to be focused a

short distance away. A receiver rod, geometrically the same as the
transmitter rod, is colinearly and confocally aligned with the transmitter to achieve maximum signal and resolution. For visualization,
a thin sample is placed at the focal zone and it is systematically
indexed mechanically. Variations in the sonic transmission are used
to brightness modulate a CRT display. The position output signal of
the sample drive mechanism is synchronized to the X and Y axes of
the display. Horizontal scanning is accomplished by a rod attached
to a vibrating loudspeaker cone. The vertical mechanism is a small
precision hydraulic piston. A block diagram of the acoustic system
is shown in Fig. 1 after Lemons and Quate (10).

Regarding the focusing of waves, in an optical system lens designs are usually quite sophisticated in order to compensate for
spherical abberations. Multielement lenses are usually necessary.
In the acoustic system, however, the situation is much less involved
for two reasons. In the first place, the lens in the SAM is always
used "on axis" rather than in an imaging mode; here, the sample is
moved through the focused beam. In the second place, the spherical
abberations which preclude the use of simple lenses in an optical system are minimal in the acoustic system because of the larger refractive index ratio between the lens material and the fluid space in
which the rays come to focus (12), (13). As a comparison, in typical
optical systems, a light beam is focused from glass of index 1.5 to
air of index 1.0. In the acoustic system the velocity of sound ratio
(the reciprocal of the index of refraction ratio) is 7.5 between the
sapphire rod and water. Thus the rays from the paraxial region and
the far region join focus in a more precise manner than they would
if the index of refraction ratio was smaller.

The frequency of operation of the SAM is limited by the ultrasonic attenuation in the fluid (usually water) between the transmitting and receiving rods. The attenuation of signal depends upon
the separation distance between the lens surfaces; this in turn depends upon the lens radius of curvature. The distance can be decreased if the lens radius is made smaller. The first reported experiments in 1973 were at 160 MHz (12) and the lens radius was 1.59
mm. Because of the high velocity of sound ratio, the focal zone is
located about 13 percent farther out than the center of curvature of
the lens. Therefore, the lens-to-lens separation is about 3.6 mm,
and the consequent attenuation loss in the water path, excluding that
due to the specimen, is 20 dB at room temperature. In the most recent experiments the reported operating frequency is 3 GHz (14)
(Sokolov's proposed frequency) where the wavelength of sound in water
(0.5 μm) is comparable to visible light. However, because of the extremely high ultrasonic absorption in water, viz. 8000 dB/cm at 60°,
the lens-to-lens distance must be considerably reduced. Thus Jipson
and Quate have achieved microscope operation at 3 GHz with a lens
whose radius is only 40 μm (14).

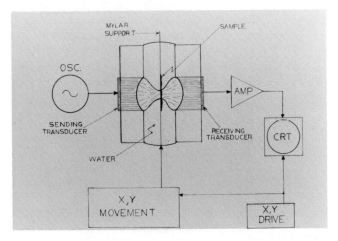

Fig. 1.   Schematic diagram of the Scanning Acoustic Microscope (SAM)
(10).

The basic advantage of piezoelectric acoustic detection is that
it is capable of very high sensitivity (11). Unfortunately, however,
the fluid space between the sample and the lenses must be long enough
to achieve acoustic focus, and significant signal is lost in travers-
ing the path. The possibilities of further increases in frequency
and, therefore, resolution are based upon finding suitable low-loss
fluids as well as fabricating precise lens surfaces with still smaller radii of curvature. A very interesting approach to achieving
high resolution in the SAM, suggested by Quate is to operate the instrument with the samples in liquid helium where the wavelength of
sound is much shorter than in water, and the absorption losses are
orders of magnitude lower (15).

IV.   THE SCANNING LASER ACOUSTIC MICROSCOPE

In the SLAM, a specimen is viewed by placing it on a stage where
it is insonified with plane acoustic waves (instead of focused waves
as in SAM) and illuminated with laser light (16), (17). The block
diagram of this system (18a), (18b) is shown in Fig. 2. Within the
sample, the sound is scattered and absorbed according to the internal
elastic microstructure. The principle upon which the laser beam is
employed as a detector is based upon the minute displacements which
occur as the sound wave propagates. As shown in Fig. 3, an optically
reflective surface placed in the sound field will become distorted
in proportion to the localized sound pressure. The distortions are
dynamic in that the pressure wave is periodic and the mirror displacements accurately follow the wave amplitude and phase. At every instant of time the mirror surface is an optical phase replica of the
sound field. The laser is used to measure the degree of regional
distortion. By electronically magnifying the area of laser scan to

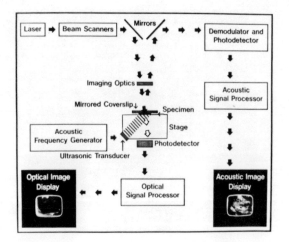

Fig. 2.   Schematic diagram of the Scanning Laser Acoustic Microscope (SLAM) (18).

Fig. 3.   Laser detection of acoustic energy at an interface (16).

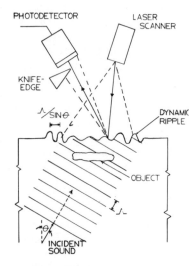

the size of the CRT monitor and by brightness modulating the display, the acoustic micrograph is made visible. If the sample is made of a solid substance which can be optically polished (as for metallurgical examination), the sample is viewed directly with the laser. However, if the sample is not polished (for example, biological tissue or un-prepared solid material), a plastic mirror (coverslip) is placed in contact with the sample to relay the sonic information into the laser beam.

The laser detection process is further explained with reference to Fig. 3. A light beam incident upon the mirror will be reflected

at an angle equal to the incident angle. When the surface is tilted
by an amount proportional to the sound pressure, the reflected light
is angularly modulated (spatially). If all the reflected light is
captured by a photodiode, its electrical output signal will be a dc
level only because the light power reaching the detector will not
change as a function of angle. However, if part of the light beam is
blocked by an obstacle (or knife-edge) then the amount of light reaching the photodiode will depend upon the instantaneous angular position of the beam. Thus the electrical signal output will now consist
of a dc component plus a small ac term coherent with the acoustic
amplitude.

As a byproduct of the laser scanning technique, a corresponding
optical image of the sample is obtained simultaneously as shown in
Fig. 2. In the case of optically translucent samples, such as biological material and some solid materials, a "partially silvered" coverslip is employed. Now, a fraction of the probing laser light can
penetrate the mirror and sample. This transmitted light is detected
and the resulting signal fed to an adjacent CRT monitor, thus constituting an optical image. In the case of a reflective polished specimen such as a metal prepared for metallographic examination, an optical reflection image is produced. The importance of the simultaneous optical image is very great to the user, for it permits newly obtained information, acoustically, to be immediately compared to a
familiar optical frame of reference. This is accomplished without
repositioning the sample or disturbing its environmental circumstances.

The laser beam scanning technique is based upon the use of an
acoustooptical diffraction cell originally developed for a "laser
TV" projection scheme (19) in the late 1960's. This was developed
because of the difficulties associated with linear mechanical scanning of mirrors for the horizontal deflection at "TV" frequencies,
viz. 15.75 kHz. In an acoustooptical cell a traveling ultrasonic
wave serves as a diffraction grating for the laser beam. If the
interaction distances are sufficiently long, efficient Bragg diffraction occurs and virtually all of the light is diffracted into a single side order whose angular position is a function of the grating
spacing. By changing the acoustic frequency to the cell, the output
laser beam is made to scan over a range of angles limited by the
bandwidth of the transducers. The vertical scanning of the laser
beam is accomplished by linearly moving a mirror which is attached
to a servo controlled galvanometer. The frame rate of the system
is identical to that of standard TV, i.e., 30 frames per second.

The frequency of operation of the SLAM is usually between 100
and 500 MHz, the frequency of choice being dictated by intrinsic
attenuation of signal in the sample. Comparing to the SAM, wherein
a lossy fluid path is needed to establish focus prior to entering
the specimen, the SLAM needs only a thin-fluid film to couple the

sound from the stage to the sample. Also thick samples are not as much of a problem with the SLAM except for losses within the samples themselves. Another basic advantage of the SLAM is that the sample remains stationary; the light beam does the scanning and produces acoustic and optical images simultaneously. Like the SAM, the SLAM can be pushed upwards in frequency and down in temperature to liquid helium in order to achieve higher resolution. Unlike the SAM whose resolution limit depends upon the quality of a lens, the limit of resolution in SLAM is governed by the wavelength of the laser light.

V. RESOLUTION

The ultimate resolution of an imaging system of any kind is governed by two factors, the wavelength (λ) of the insonification or illumination and the numerical aperture (NA) of the detection system. The NA is the sine of the half-angle of the response by the detector. For example, an optical system having a lens which is too small to accept (or focus) rays beyond an angle $\pm\theta_m$. The classical Rayleigh criterion for resolution states that two distant objects can be resolved when their separation distance d is greater than or equal to 0.66λ/NA. Therefore, with an ideal omni-directional detector, the maximum value of NA=1, and therefore, d = 0.66λ.

In the SAM, an acoustic wave is brought to a focus by a lens. The waist dimension of the spot is equal to d (above) in the focus zone. However, the receiver also embodies an acoustic lens, and the combination of lenses in this case actually improves the overall resolution by about a factor of 2. This occurs because the directivity patterns of transmitting and receiving transducers multiply, thereby decreasing the effective spot size. This discussion of resolution in the SAM is valid for focused spots produced in the fluid medium which occupies the sapce between the two lenses. Soft biological tissues have velocities of sound very close to that of water and, therefore, the resolution criteria are still valid. For solid materials, however, the wavelength of sound is considerably higher than that of a fluid; and, therefore, the resolution within the solid is degraded by the longer wavelength of sound as well as by the loss of focus due to aberrations which occur in media with high velocities of sound. If the solid specimens have a thickness limited to the depth of focus of the spot, then the resolution is close to that which would be obtained with biological material. For thick specimens, the resolution is degraded due to loss of focus.

Methods of resolution determination involve the use of test targets and the measurement of spatial frequency response of the system, the so-called modulation transfer function. Weglein and Wilson (20) have measured the modulation transfer function of a SAM operating in the reflection mode shown in Fig. 4. Here, one single lens rod is employed both for transmission and reception of the acoustic energy. This mode is quite useful for inspection of near surface phenomena

of a sample since, if penetration depth is limited, the ultimate resolution is governed by the focal spot size in the fluid couplant.
Based upon theoretical calculations for spatial frequency response
of the SAM, a surprisingly better performance was experimentally observed by Weglein and Wilson. These results, shown in Fig. 5, represent the normalized transfer function for the SAM operating at 375 MHz
in the reflection mode. The ultimate limiting resolution appears to
be consistent with the calculations; however, there is an enhancement of the low spatial frequency response.

In the SLAM, the effective numerical aperture of the system does
not depend upon acoustic lens performance since none are used.
Rather, the numerical aperture is limited by the response of the
knife-edge detector as a function of insonification angle. The effective numerical apertures for the SLAM is dependent upon the type of
samples as will be discussed below. Consider a solid sample that is
optically polished on the side being examined by the laser beam. Acoustic energy at any angle but $0°$ (normal incidence) or the critical angle
(if it exists), will produce surface tilt and thereby elicit a response by the knife-edge photodetector. Here the numerical aperture
can be close to 1. Consider a solid sample that is not polished. In
order to transfer the information onto the light beam, a mirror is
placed in contact with a thin film of fluid to promote effective energy transfer. If the velocity of sound in the mirror itself is higher
than that in the sample, a critical angle will occur at which the
transmitted wave is refracted to $90°$. Mirrors can be made from hard
materials such as glass as well as from plastic. It is advantageous
to use plastics since they characteristically have lower velocities
of sound than glass. As long as the velocity of sound in the mirror
is lower than the sample, no critical angle will occur. For soft
biological tissue, which has a typical velocity of sound of 1500 m/s,
and for a low velocity of sound plastic, typically 2200 m/s, the critical angle will be $42°$ thus restricting the effective numerical
aperture to about 0.6.

Aside from pure numerical aperture considerations, there is an
opportunity to achieve a factor of about 2 improvement in resolution
in the SLAM without a frequency increase. The method is to use
transverse acoustic wave illumination instead of compressional mode.
It turns out that there is a factor of about 2 smaller wavelengths
for the transverse mode for most materials. With plane-wave illumination of a specimen, the choice can be made between the 2 modes. However, in the SAM, since a focused beam is employed, the 2 modes simultaneously exist causing a confusing set of 2 foci to exist within
a thick sample.

In the SLAM, optimum resolution is obtained in the near-field
zone of the object area imaged. The near-field zone extends inward
from the laser viewed surface of the specimen by an amount which depends upon the zone diameter and the sonic wavelength. Outside of

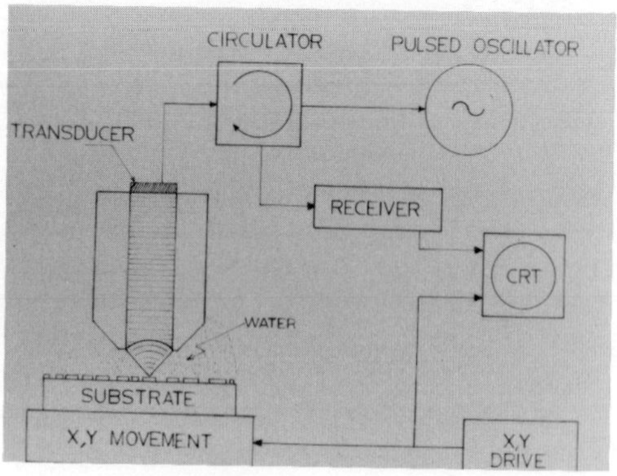

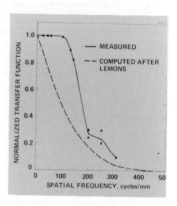

Fig. 4. SAM in the reflection mode examining a substrate material (20).

Fig. 5. Spatial frequency response of SAM (20).

the near-field zone, the resolution is degraded, in an analogous manner as with the SAM, when the plane of interest lies outside the focused zone of the acoustic lens.

VI. IMAGING TECHNIQUES

The usefulness of acoustic microscopy as an analytical tool depends on its adaptability to the problem in need of solution. The basic techniques described above form a good basis; however, numerous modifications and innovations have been necessary to achieve versatility. In this section a variety of imaging modes and acoustic microscopy techniques are described.

Acoustic amplitude transmission images were the first to be demonstrated during the course of the development of both SLAM and SAM. Transmission imaging is still one of the most useful acoustic imaging modes. In this image type, structure within a specimen is revealed by local variations in the transmission of acoustic energy.

Attenuation can arise from either absorption of sound by the sample or by reflection and scattering of the sound field by elastic inhomogeneities. Inhomogeneities larger in size than an acoustic wavelength will be individually resolved and will appear as distinct structures on the acoustic micrographs. The size, location, and morphology of these features can be measured and used to characterize the material. When inhomogeneities are smaller than an acoustic wavelength, the attenuation of the beam is the result of scattering. Scattering by smaller structures produces acoustic images with a unique "texture" which can be used to characterize these materials.

In addition to variations in amplitude transmission through a specimen, variations in transit time or phase of the acoustic wave are important. The phase of the transmitted acoustic wave is related to the sonic velocity which is in turn related to the compressibility bulk density. There are a variety of ways that phase information can be incorporated into acoustic images. Two methods—phase contrast imaging and the acoustic interferogram—are described below. Prior to the advent of acoustic microscopy, and, in particular, phase sensitive imaging techniques, measurements of density and compressibility variations at the microscopic level were not possible.

Acoustic phase contrast imaging is directly analogous with optical phase contrast imaging. In phase contrast mode, the brightness level of the monitor is modulated by the relative phase of the transmitted acoustic wave. Thus images are the result of variations in sonic velocity (compared with amplitude transmission images where brightness is controlled by attenuation variations). Aside from the physical properties being revealed, the use of a phase imaging mode may be useful from the pragmatic standpoint of making it easier to "see" low contrast structures. Fig. 6 (Marmor et al. 1977) shows an example of a 1-μm section of human retina (21) which is first visualized on an optical microscope (Fig. 6(a)), then mounted on a Mylar support and visualized in acoustic amplitude and acoustic phase. The lack of detail in the optical image is due to the fact that the section is very thin and it has not been histologically stained. (This staining process, though necessary for optical viewing, alters the state of the original tissue and introduces significant artifacts.) Acoustically the structures are differentiated very well without stains (as seen in Fig. 6(b)) and the phase image (Fig. 6(c)) reveals interesting variations of properties within the cell nuclear layers which were not evident in the optical or acoustic amplitude images.

In acoustic interferograms, sonic velocity data is incorporated into the images by superimposing a number of fringes on an acoustic amplitude micrograph. Fig. 7 shows a mouse embryo heart (22) which was maintained alive in the SLAM. This figure is a still frame of a real-time videotape recording of the cardiac contraction. The acoustic frequency is 100 MHz. Optically, the sample is totally opaque due to its thickness. The field of view is 3 mm horizontally and the

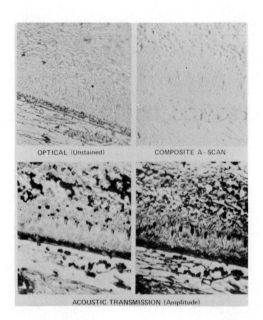

Fig. 6. Optical and SAM acoustic micrographs of a thin (1 um) un-
stained section of normal human retina at about 400 X. Both acoustic
amplitude (lower left) and phase (lower right) images are shown (21).

crossmarks represent 1 mm spacings. In this case the advantage of
acoustic microscopy is circumvention of optical opacity in a specimen
and differentiation of the tissue structure without introducing
stains or fixatives which would interfere with the viability of the
live material. The acoustic interferogram (Fig. 7(c)) consists of a
series of vertical fringes which arise from the spatial beat note between a 10° angular insonification field and an electronically simulated normal incident reference signal introduced into the detector.
Whereas a conventional phase image translates phase information to
grey levels, the interferogram yields phase information by the lateral shifts of the fringes as the sound propagates through regions with
different indices of refraction. In the interferogram shown here,
the fringe shifts result from two conditions of the sample: first,
the velocity of sound in cardiac tissue is higher than that of the
surrounding nutrient fluid bath, thereby causing fringe shifts to the
right by an amount dependent on the tissue thickness, and second, it
has been found (23) that as muscle contracts, the intrinsic velocity
of sound increased a small amount.

    Newer additions to the arsenal of acoustic microscopy imaging
techniques include reflection and acoustic darkfield imaging. In
contrast to transmission imaging schemes where the structures are
revealed when sound is scattered from main beam, these imaging techniques detect the scattered or reflected energy directly.

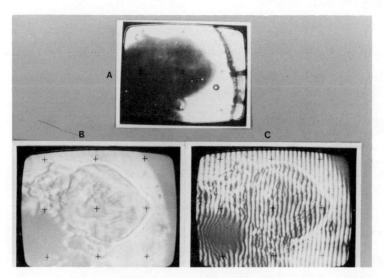

Fig. 7. Live mouse embryo heart. (a) Optical image. (b) SLAM
acoustic micrograph at 100 MHz. Cross-marks are placed 1 mm apart.
(c) Acoustic interferogram from which variations in elastic properties of the muscle are determined (22).

Acoustic reflection images have been made with the SAM (20),
(24) by eliminating one of the transducers and employing the other
in a pulse-echo mode. Except for the large difference in frequency,
the setup is similar to a conventional pulse-echo system used in nondestructive testing and clinical medicine. (Refer back to Fig. 4.)
The technique is very useful for examining near surface phenomena
within the reflecting object. An obvious example where this is important is in integrated-circuit technology. A typical sample is
shown in Fig. 8 (from (24)) in which the SAM brings out features
within the overlapping electrode layers that are not detectable optically.

Acoustic dark-field images in the SLAM can be produced in one
of two ways, by insonifying the specimen at an angle corresponding
to low knife-edge response, such as normal incidence, or by engaging
a sharp band reject filter in the RF output of the photodiode. The
latter method is made possible by the fact that when the laser rapidly scans the acoustic field, the perceived electrical frequency is
Doppler shifted from that driving the ultrasonic transducer by an
amount dependent on the sine of the angle of the incidence (27a),
(28). Thus acoustic energy scattered by an object is incident upon
the detector plane with a spectrum of angles of incidence, each angle
of which is perceived as a unique electrical frequency. An electrical filter can then be used to spatially filter the content of the

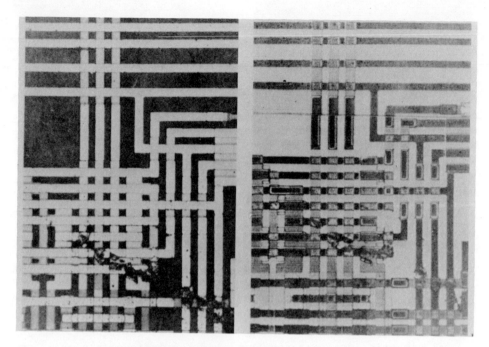

Fig. 8. Nomarski optical (a) and SAM acoustic (b) images of an inge-
    grated circuit with a silicon-on-sapphire wafer (24).

acoustic image and produce effects such as dark-field and edge enhancement. Going one step further, the filter can also be so positioned that the image consists only of forward scattered sound or
backscattered sound (28). In this case, forward-scattered and backscattered directions refer to rays with components which are parallel
and antiparallel to the direction of the laser scan. This capability
is of great value in analyzing wave patterns existing on surface-
acoustic-wave filters, delay lines, and other such devices (29).

   Two additional acoustic imaging schemes-nonlinear acoustic
imaging and polarization sensitive acoustic imaging-should be mentioned. These methods use the same basic format as normal transmission imaging but differ by the sonic mode used to produce the
image. The importance and usefulness of these imaging techniques
has only recently been explored.

   Nonlinear elastic behavior of materials is a very interesting
aspect of the field of mechanics and, with the acoustic microscope,
these nonlinear effects can be probed at a very fine level. The
basic technique for visualizing nonlinear effects is to excite the
specimen at one acoustic frequency while detecting acoustic energy
produced at harmonics of the applied frequency. Kompfner and Lemons
(30b) produced images with the SAM which displayed the second harmonic image of a biological tissue which was insonified at 400 MHz.

Although they found that the resolution was equivalent to that of an
800 MHz image, there were striking features of the image contrast
which were not revealed at 400 MHz or 800 MHz regular (nonharmonic)
images. Presumably, the detail enhancement was caused by differential nonlinearities of the tissue components; however, the investigators feel that there are still puzzling features of this type image
which cannot be explained fully.

For polarization sensitive imaging, samples were insonified with
linearly polarized transverse acoustic waves. Transverse acoustic
waves, also known as shear waves, cannot propagate in liquids; therefore, the SAM cannot take full advantage of this shorter wavelength
type of insonification. In the SLAM, a transverse wave transducer
is bonded onto the quartz stage block and the sample is affixed to
the block with a thin film of a viscous gel. Transverse waves transmitted through the sample may be acted upon differently than compressional waves by defects within a sample. For example, particular orientations of flaws may make them difficult to detect with compressional waves. One further aspect of the transverse mode is that similar to electromagnetic waves, there are two polarizations. In the
SLAM, only one of these will produce surface ripples. Recall that
the SLAM relies upon surface deformation for detection. Therefore,
SLAM is polarization sensitive, and if the specimen is insonified by
the polarization that produces no surface ripples, the acoustic image
will reveal anisotropic elastic properties of the material (26).

VII. ANALYTICAL TECHNIQUES

Acoustic microscopy, like all microscopic techniques, is primarily a qualitative technique. The information contained in acoustic
amplitude or phase micrographs constitutes characteristic material
signatures. Thus much of the useful information obtained with acoustic microscopy is purely morphological in nature. These data can be
used to classify and sort materials, detect and localize flaws and
defects in optically opaque samples, and map compressibility and density variations on a microscopic scale.

In addition to the qualitative aspects, techniques have been
developed to quantitatively measure elastic properties on the microscopic scale. These techniques include graphic analysis of the acoustic interferograms to obtain velocity data as well as measurement of
acoustic transmission levels through samples to obtain quantitative
attenuation data. Additional techniques include stereoscopy, acoustic line scan, and acoustic reflectivity profiling. The last two are
not true imaging modes or image analysis methods, but instead are new
characterization capabilities which have arisen from the microscope
technology.

Stereo viewing is an accepted method of depth determination; it
also enhances the information gathering process within the eye-brain

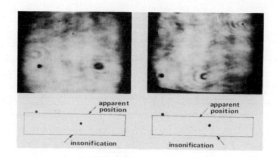

Fig. 9. Defect depth by stereoscopy. SLAM stereo pair of images in
which the apparent position of a deep lying defect is shifted by
changing the angle of insonification.

complex. The technique is commonly used for observing the terrain
from aerial photographs as well as for understanding the features of
a sample visualized in a scanning electron microscope. The acoustic
microscope employs relatively straightforward applications of the
standard technique; namely, a set of images is made with the source
of insonification at different angles. Thus Bond et al. (27b) get
the stereo effect on the SAM by rotating the axis of the transmitting
transducer-lens with respect to that of the receiving element by an
angle θ. The sample is rotated by half the angle θ for reasons of
symmetry. For small angles of rotation, say 10°, the stereo effect
appears to be significant in their images of onion skin. For greater
angles, however, the images produced in the SAM have accentuated object detail over that which is evident in the perfectly aligned
through transmission system. The effect that is occurring has to do
with the angular response curve of the system in which energy received at large angles produces a smaller signal than energy at normal incidence. Therefore, when the specimen and transmitter are
"off-axis," the low-amplitude scattered energy is detected near the
peak in the detector response, thereby producing dark-field enhancement effects.

     Depth determination can also be made by simple axial translation
of the sample. This method was used by Wang et al. (27c) to determine defect locations in joints of various materials.

     In the SLAM, the sample is usually insonified on an angle to
start with. Therefore, in order to produce a stereo image plane, the
sample can simply be rotated on the stage and the set of 2 images
recorded directly. An example is shown in Fig. 9 in which a deep
lying defect in a homogeneous ceramic material is imaged. In order
to record the apparent position shift, a fiducial mark was made on
the top surface of the sample. The image shift of the defect with
respect to the fiducial can be measured and the depth calculated.

     In addition to producing two-dimensional images in the X-Y plane as in the previous examples in this article, the SAM has been used to produce acoustic reflectivity profile. With this technique, graphical spectrum of several materials are produced by axial translation of the object along the axis of the acoustic lens (36).   In this mode the SAM is used in the reflection mode. It is suggested by Weglein and Wilson that in this mode the SAM may have the ability to reveal crystallographic information on a specimen which has otherwise been obtainable by X-ray diffraction alone.

The acoustic line scan mode is an option on the SLAM which permits measurements of sonic velocity variations on a millisecond time scale. This method involves measuring the lateral fringe shifts along a single horizontal line of an acoustic interferogram. This method arose from a need to investigate rapid variations in the elastic properties of viable muscle undergoing contraction (23).

VIII. APPLICATIONS OF ACOUSTIC MICROSCOPY-EXAMPLES

The acoustic microscope is an analytical instrument with problem solving capabilities as well as applications to basic and applied research. Illustrations presented below are taken from the biomedical sciences, materials technology, nondestructive testing, and from the electronics components technology.

A. Biomedical

The study of embryos is complicated by the fact that the whole specimen must be optically clarified and selectively stained in order to see internal tissues. These techniques modiy the natural state of the tissues and may not be acceptable. A 10.5 day mouse embryo (31) is shown optically in Fig. 10(a). The simultaneously produced acoustic micrograph is shown in Fig. 10(b) and it quite clearly

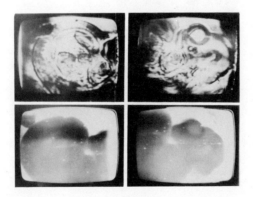

Fig. 10.   Mouse embryo at age 10.5 days. Top image set is 100 MHz acoustic transmission and lower images are optical (31).

demonstrates a high degree of differentiation for hard and soft tissues. The frequency of ultrasound is 100 MHz and the resolution in tissue is 20 μm. Although this particular specimen was fixed in ethanol, similar studies have been done in the live state, thereby revealing dynamic events such as cardiac activities which were recorded on video tape.

Collagen is an abundant protein in mammals and is responsible for the structural strength of tissues and organs. Acoustic microscopy of fresh collagen (32) carried out at 100 MHz is illustrated in Fig. 11(a) and (b) which are, respectively, a normal acoustic micrograph and an interferogram. Measurements of the fringe shifts in the collagen relative to the surrounding aqueous fluid indicate a velocity of sound which is at least 25 percent higher than soft tissue.

Pathological disease in human liver is shown in Fig. 12 which was obtained at 100 MHz on samples supplied by R. Waag of the University of Rochester Medical School, Rochester, N.Y. This particular sample shows a metastic adenocarcinoma. Some of these tumor cell nests have necrotic centers which appear darker than the surrounding tumor cells. A transmission acoustic micrograph of a thick section shows collagenous septae as dark bands between glandular nests of

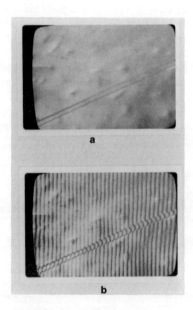

a

b

Fig. 11. Acoustic micrograph (a) and interferogram (b) of
        mammalian tendon at 100 MHz (32).

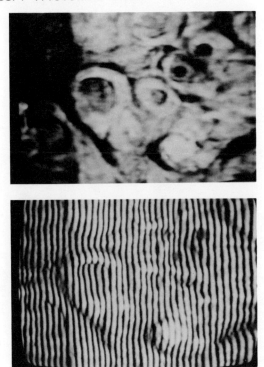

Fig. 12. Human liver with metastatic adenocarcinoma. Acoustic micrograph (a) and interferogram (b) at 100 MHz with SLAM. (Courtesy
of R. Waag, University of Rochester Medical School.)

tumor cells. The necrotic centers of the tumors appear darker than
the peripheral tumor cells. The reduced transmission through the
septas and necrotic centers indicates that they attenuate sound more
than the surrounding tissue. An acoustic interferogram with fringes
bending to the left as they traverse the collagenous septae and necrotic tissue demonstrates that the speed of sound is greater in
them than in the adjacent tissue. The distribution of collagen and
the architecture of the adenocarcinoma give a markedly different
appearance from normal liver.

B. Materials Technology and Nondestructive Testing

     Silicon nitride turbine blades were examined at 100 MHz to de-
fine the typical elastic microstructure and to detect the presence
of flaws (33a). Fig. 13(a) shows an acoustic micrograph which shows
small pores distributed throughout the sample as well as a much larger laminar defect which could not be detected by radiography even
after the flaw was spotted acoustically. Fig. 13(b) shows the flaw
optically after the sample was sliced apart to confirm the findings.

Fig. 13. Ceramic turbine blade showing large defect among a network
of porosity (a) acoustic micrograph and (b) sectioned sample revealing laminar flaw (33).

The seam welds on steel cans with a thickness of 0.010 in and
approximately 0.020 in. thick-weld sections were inspected by SLAM
at 100 MHz (33b). The acoustic micrograph Fig. 14(a) revealed major
defects (disbonded zones) along the seam, smaller imperfections within the junction regions and details of the adjacent heat affected
metal. Conventional metallographic data obtained by sectioning revealed only the large defects due to the wide separations between the
sections as seen in Fig. 14(b). Nine polished and etched sections
provided 29 visual data points and required many hours of preparation,
whereas the acoustic micrographs containing 40,000 image points each
were produced in 0.033 s without destroying the sample.

C. Electronic Components

Acoustic microscopy is a very sensitive method for detecting delaminations of bonded materials. An investigation into solder bonds
on a commercially available transistor was carried out at 150 MHz by
Tsai et al. (34). Fig. 15 shows a silicon wafer 2.5 mm square bonded
to a copper plate by means of Au-Sn solder. The regions of disbond
permit no sonic transmission and the corresponding dark areas of the
micrograph.

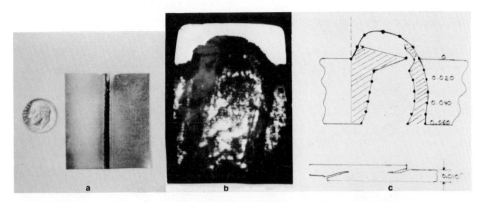

Fig. 14. Seam welded steel can with poor bonding at the lead edge.
(a) Optical photograph of sample, (b) 100 MHz SLAM acoustic transmission image with dark zones corresponding to disbonds, and (c)
confirming metallographic sectioning of sample revealed approximate
disbond areas.

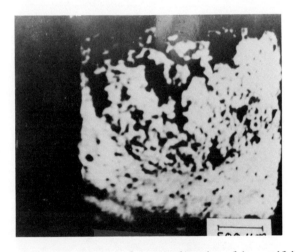

Fig. 15. Acoustic micrograph of a production line silicon-solder-
copper bond. Wafer thickness is 100 μm and copper plate is 300
μm. SAM transmission image (34).

Another example is shown in Fig. 16 which is an interferogram
(100 MHz) of a ceramic chip capacitor along with a schematic diagram
of its construction. The field of view is 4 mm horizontally, and
the dark zone in the center correlates with delamination of a layer.
In this study, Love and Ewell (35) pointed out that delaminations
such as these shorten the lifetime of the component. Each layer is

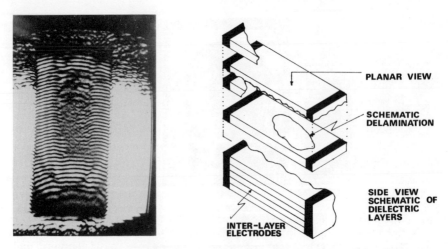

Fig. 16. Ceramic chip capacitor with interlayer delamination. Acoustic interferogram at 100 MHz and SLAM (35).

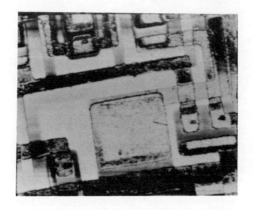

Fig. 17. SAM acoustic image at 3 GHz of an integrated circuit. The line widths are 7 μm (14).

constructed of a high dielectric constant material which has been electroded on both surfaces. Prior to failure, i.e., complete delamination and break of electrical contact, the capacitor behaves normal electrically.

At a frequency of 3 GHz, Jipson and Quate (14) have examined an

integrated circuit in the SAM. Fig. 17 shows a reflection mode image
of an IC circuit fabricated on silicon. The acoustic reflectivity is
a function of the layering structure beneath the aluminum lines connecting to the transistors; thus the dark spots appearing are attributed to subsurface structure such as regions where the oxide has
been removed and emitter contacts are formed.

IX.   CONCLUSIONS

Approximately 35 years after Sokolov first proposed the notion
of acoustic microscopy, the technology has now developed to the point
where it is practical. Current emphasis is on applications of the
technology as well as improvement on the techniques employed. It is
anticipated that the SLAM and SAM will soon take their place along
the electron microscope as standard analytical instruments.

REFERENCES

1. L. W. Kessler, "A review of progress and applications in acoustic
    microscopy," J. Acoust. Soc. Amer., Vol. 55, pp. 909-918, 1974.
2. S. Sokolov, USSR Patent No. 49 (Aug. 31, 1936), British Patent
    No. 477 139, 1937, and U.S. Patent No. 21 64 125, 1939.
3. F. Dunn, and W. J. Fry, "Ultrasonic absorption microscope,"
    J. Acoust. Soc. Amer., Vol. 31, No. 5, pp. 632-633, May 1959.
4. J. Havlice, C. F. Quate, and B. Richardson, "Visualization of
    sound beams in quartz and sapphire near 1 GHz," paper 1-4 pre-
    sented to the 1967 IEEE Ultrason, Symp., IEEE Trans. Sonics
    Ultrason., Vol. SU-15, 68, 1968. See also (15) and (17).
5. H. V. Hance, J. K. Parks, and C. S. Tsai, "Optical imaging of a
    complex ultrasonic field by diffraction of a laser beam," J. Appl.
    Phys. Vol. 38, pp. 1981-1983, 1967.
6. A. Korpel, "Visualization of the cross section of a sound beam by
    Bragg diffraction of light," Appl. Phys. Lett. Vol. 9, p. 425,
    1966.
7. R. Pohlman, "Material illumination by means of acoustic optical
    imagery," Z. Phys., 1133 697, 1939. See also Z. Angew. Phys.,
    Vol. 1, p. 181, 1948.
8. J. A. Cunningham, and C. F. Quate, "Acoustic interference in
    solids and holographic imaging," in Acoustical Holography, Vol.
    4, G. Wade, Ed. New York: Plenum, 1972, pp. 667-685.
9. L. W. Kessler, and S. P. Sawyer, "Ultrasonic stimulation of opti-
    cal scattering in nematic liquid crystals," Appl. Phys. Lett.
    Vol. 17, pp. 440-441, 1970.
10. R. A. Lemons, and C. F. Quate, "Acoustic microscope-scanning
    version," Appl. Phys. Letters, Vol. 24, No. 4, pp. 163-165, Feb.
    15, 1974.
11. A. Korpel, and L. W. Kessler, "Comparison of methods of acoustic
    microscopy," in Acoustical Holography, Vol. 3 by A. F. Metherell,
    Ed. New York: Plenum, 1971, pp. 23-43.
12. R. A. Lemons, and C. F. Quate, "A scanning acoustic microscope,"

13. ___, "Advances in mechanically scanned acoustic microscopy,"
    Proc. 1974 Ultrason. Symp., IEEE Catalog #74CHO 896-ISU, p. 41,
    J. de Klerk, Ed., 1975.

14. V. Jipson, and C. F. Quate, "Acoustic microscopy at optical
    wavelengths," Ginzton Laboratory rep. No. 2790 on Contract
    AFOSR-77-3455 and NSF APR75-07317; submitted to Appl. Phys.
    Letters, June 15, 1978 (in press).

15. "The promise of acoustic microscopy," in Mosaic, Vol. 9. Wash-
    ington, DC: National Science Foundation, Mar./Apr., 1978, p.35.

16. A. Korpel, L. W. Kessler, and P.R. Palermo, "Acoustic microscope
    operating at 100 MHz," Nature, Vol. 232, No. 5306, pp. 110-111,
    July 9, 1971.

17. L. W. Kessler, A. Korpel, and P. R. Palermo, "Simultaneous
    acoustic and optical microscopy of biological specimens," Nature,
    Vol. 239, No. 5367, pp. 111-112, Sept. 8, 1972.

18. a) L. W. Kessler, and D. E. Yuhas, "Structural perspective,"
    Indust. Rev., Jan. 1978, Vol. 20, No. 1, 53-56.
    b) SONOMICROSCOPE ^{TM} 100, manufactured by Sonoscan, Inc., Ben-
    senville, IL. 60106.

19. A. Korpel, R. Adler, P. Desmares, and W. Watson: "A television
    display using acoustic deflection and modulation of coherent
    light," Appl. Opt., Vol. 5, 1667-1675 1966.

20. R. D. Weglein and R. G. Wilson, "Image resolution of the scan-
    ning acoustic microscope," Appl. Phys. Lett., Vol. 31, No. 12,
    pp. 793-796, 1977.

21. M. F. Marmor, and H. K. Wickramasinghe, "Acoustic microscopy
    of the human retina and pigment epithelium," Invest. Ophth.
    Vis. Sci., Vol. 16, No. 7, pp. 660-666, July 1977.

22. R. C. Eggleton, and F. S. Vinson, "Heart model supported in
    organ culture and analyzed by acoustic microscopy," in Acousti-
    cal Holography, Vol. 7, L. W. Kessler, Ed. New York: Plenum,
    1977, pp. 21-35.

23. R. C. Eggleton, "Application of acoustic microscopy of the study
    of muscle mechanics," SPIE Proc. Vol. 104, Multidisciplinary
    Microscopy, R.L. Whitmen, Ed. 117-124 Published by Society of
    Photo-Optical Instrumentation Engineers, P.O. Box 10, Belling-
    ham, WA. 98225, 1977.

24. R. Kompfner, and C. F. Quate, "Acoustic radiation and its use
    in microscopy," Phys. Technol., pp. 231-237, Nov. 1977.

25. L. W. Kessler, and D. E. Yuhas, "Listen to structural differen-
    ces," Indust. Dev. Vol. 20, No. 4, Apr. 1978, pp. 102-106.

26. L. W. Kessler, U.S. Patent 4 012 951, Acoustic Examination Me-
    thod and Apparatus.

27. a) L. W. Kessler, P. R. Palermo, and A. Korpel, Practical High
    Resolution. New York: Plenum Publishing, 1972, pp. 51-71.
    b) W. L. Bond, C. C. Cutler, R. A. Lemons, and C. F. Quate,
    "Dark-field and stereo viewing with the acoustic microscope,"
    Appl. Phys. Lett., Vol. 27, No. 5, pp. 270-272, Sept. 1, 1975.

27c. S. K. Wang, C.C. Lee, and C. S. Tsai, "Nondestructive visualiza-
    tion and characterization of material joints using a scanning

acoustic microscope," in Proc. IEEE 1977 Ultrason. Symp., IEEE
Cat. 77-CH1-264.

28. D. E. Yuhas, "Characterization of surface flaws by means of acous-
tic microscopy," in Program and Abstracts: First International
Symposium on Ultrasonic Materials Characterization. National
Bureau of Standards, Gaithersburg, MD, p. 79.

29. L. W. Kessler, "Real time SAW visualization techniques to aid
device design," Paper F-9. 1975 Ultrason. Symp. (IEEE), Sept.
22-24, Los Angeles, CA. Also available as application note
SAW-1 from Sonoscan, Inc., 530 E. Green St., Bensenville, IL.
60106.

30a. H. K. Wickramasinghe, R. Bray, V. Jipson, C. F. Quate, and J.
Salcedo, Photoacoustics on a Microscopic Scale. Submitted to
Appl. Phys. Lett. 1978.

30b. R. Kompfner, and R. A. Lemons, "Nonlinear acoustic microscopy,"
Appl. Phys. Lett., Vol. 28, No. 6, pp. 295-297, Mar. 15, 1976.

31. W. D. O'Brien, and L. W. Kessler, "Examination of mouse embryol-
ogical development with an acoustic microscope," Amer. Zool.,
Vol. 15, p. 807, 1975.

32. W. D.O'Brien, and S. Goss, In preparation.

33. a) D. S. Kupperman et al., "Preliminary evaluation of several
non-destructive evaluation techniques for silicon nitride gas
turbine rotors," Argonne Nat. Laboratory Tech. rep., Materials
Science Division, Argonne, IL. 60439 (1978). Rep. #ANL-77-89.
b) D. E. Yuhas, and L. W. Kessler, "Acoustic Microscopy," Principles and Applications, Paper Summaries, Spring Conf. ASNT,
165-171 (Mar. 1977).

34. C. S. Tsai, S.K. Wang, and C. D. Lee, "Visualization of solid
material joints using a transmission-type scanning acoustic
microscope," Appl. Phys. Lett., Vol. 31, p. 317, Sept. 1977.

35. G. R. Love, and G. J. Ewell, "Acoustic microscopy of ceramic
capacitors," Proc. of 28th Electronic Components Conference,
IEEE/EIA, Apr. 24-26, 1978, Anaheim, CA.

36. R. D. Weglein and R. G. Wilson, "Characteristic signatures by
acoustic microscopy," Electron, Lett., Vol. 14, p. 352, 1978.

37. L. W. Kessler and D. E. Yuhas, "Acoustic Microscopy-1979",
Proceedings of the IEEE, Vol. 67, No. 4, April 1979.

DEFECT CHARACTERIZATION BY MEANS OF THE

SCANNING LASER ACOUSTIC MICROSCOPE (SLAM)

D. E. Yuhas and
L. W. Kessler
Sonoscan, Inc.
Bensenville, Illinois   60106

ABSTRACT

High frequency acoustic imaging represents a powerful technique for the nondestructive evaluation of optically opaque materials. In this report the Scanning Laser Acoustic Microscope (SLAM) is used to detect and characterize flaws in ceramics. SLAM micrographs showing typical examples of cracks, laminar flaws, porosity and solid inclusions are presented. The various flaw types are easily differentiated on the basis of their characterisitc acoustic signatures. The importance of an imaging approach to the nondestructive evaluation of ceramics is demonstrated.

INTRODUCTION

This report describes high frequency (100 MHz) acoustic imaging of ceramics. The current investigation deals primarily with the recognition and interpretation of a variety of different flaw types. In this and other studies the Scanning Laser Acoustic Microscope (SLAM) has been used to investigate cracks, delaminations, porosity, machine damaged surfaces and solid inclusions[1-5]. Examples of these are presented to provide a general overview of acoustic micrographs obtained with the SLAM and to gain an appreciation for the unique features of the various flaws. In another paper, attention is focused on implanted solid inclusions in silicon nitride discs[6]. The differentiation of solid inclusions from cracks, delaminations and porosity is based on comparison of the characteristics of flaws found in these discs with those observed in other samples where the exact nature of the flaw type has been confirmed by destructive analysis.

EXPERIMENTAL APPARATUS AND TECHNIQUE

Figure 1 is a photograph of the SONOMICROSCOPE 100[7]. The prin-
cipals of operation have been described in the literature[8]. Briefly,
the instrument produces full grey scale acoustic images at frequencies from 30-500 MHz. The results appearing in this paper are all at
100 MHz where the field of view is 2.3 mm x 3 mm, resolution is about
wavelength, or 30 microns in silicon nitride (shearwaves), and images
are produced in real time. The transmitting transducer is a piezoelectric element. A focused laser scans the insonified zone and acts
as the receiving transducer. By synchronizing the laser scan with a
television monitor, real-time acoustic images are obtained. As a byproduct of the technique, an optical image of the scanned surface is
produced on a separate TV monitor. This image is quite useful in
documenting the location of flaws.

Fig. 1.   Scanning Laser Acoustic Microscope (SLAM) used in this
          investigation [7].

SLAM DATA

Two types of acoustic micrographs, amplitude images and interferograms are commonly used. They are easy to distinguish and both
are useful for defect characterization. Figure 2 shows an amplitude
image (top) and interferogram (bottom) obtained in the same silicon
nitride test sample. The amplitude image shows the relative transmission level over the field of view. In the amplitude mode, bright

areas correspond to zones in the sample with good acoustic transmission, whereas the darker areas are attenuating. The interferogram
has a characteristic set of vertical fringes which are produced by
electrical mixing of the detected acoustic signal with a phase reference. In principle, the interferogram is an acoustic hologram containing all three dimensions of information throughout the sample
volume. In our employment of this mode, the hologram is not "reconstructed", rather graphical analysis of the fringes yield quantitative elastic property information. In particular, lateral shifts of
the fringes indicate sonic velocity variations within the sample or
thickness changes. In the interferogram (Fig. 2b) a lateral shift
of one fringe corresponds to a 1.0 percent velocity variation[9]. The
field of view in this sample is 3 mm across; the spacing between the
fringes in the interferogram is 85 microns. It can be determined
from this figure that the silicon nitride sample shown is fairly
clean and uniform in its elastic properties. The horizontal streaks
in the acoustic images are due to surface texture caused by a grinding operation. Though visible, the grinding marks do not interfere
with the visualization of buried flaws. By way of contrast, other
ultrasonic inspection techniques[10] require a high degree of surface
flatness and polish. Therefore, poorly prepared samples such as this
are ordinarily difficult to inspect.

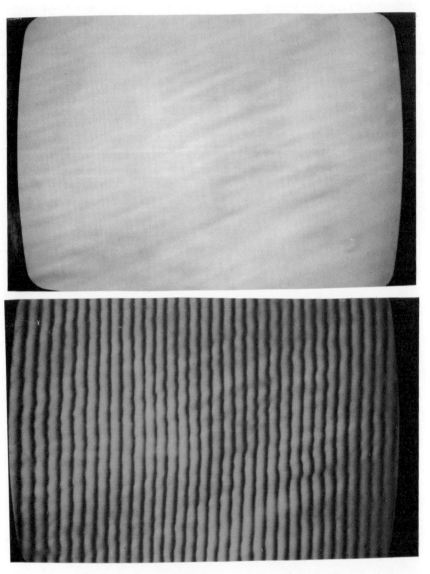

Fig. 2 – Acoustic amplitude (a) micrograph and interferogram (b) obtained on silicon nitride sample. This sample exhibits uniform elastic properties. The field of view is 2.3 by 3.0 mm.

ACOUSTIC CHARACTERISTICS OF FLAWS

The differentiation of various flaw types is made by comparing their characteristic signatures. In order to acquaint the reader with the various flaw characteristics, a series of micrographs illustrating a variety of flaw types is presented.

Influence of Porosity

Figure 3 shows an acoustic amplitude micrograph obtained on a reaction sintered silicon nitride turbine airfoil. This is a much more porous material than shown in Figure 2 as evidenced by the high attenuation regions. Destructive analysis of these samples indicates the presence of pores ranging in size from less than 1-50 um. The larger pores are directly resolvable acoustically while the smaller pores give rise to the texture seen in this micrograph. A qualitative correlation was found between the porosity distribution determined acoustically and that obtained destructively by sectioning[1].

Recent investigations of a variety of silicon carbide samples reveal substantial changes in the characteristics of acoustic micrographs as a function of grain size and percentage of unreacted material[3].

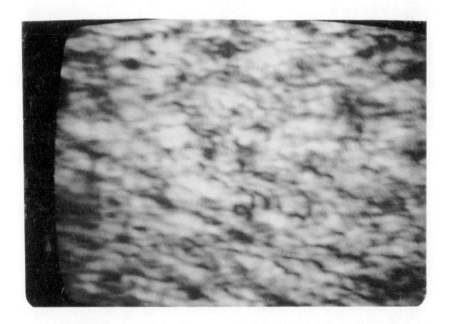

Fig. 3. Acoustic amplitude micrograph revealing porosity in reaction sintered silicon nitride turbine airfoil.

Laminar Flaws

    Figure 4 is an acoustic micrograph showing a laminar flaw found in a reaction sintered silicon nitride turbine blade[1]. The central dark zone (D) indicates an extended area of increased acoustic attenuation. This same region shows perturbed, discontinuous interference fringes indicating increased scattering (not shown). The region away from the flaw (the background structure) is similar to that presented in Figure 3 and is attributed to porosity in the blade.

Fig. 4 - Acoustic amplitude micrograph of laminar flaw found in a cera mic turbine airfoil.

    Figure 5 is an optical reflection micrograph confirming the acous tic results presented in Figure 4. The most apparent feature is a 100 micron diameter pore located near the center of the section (see arrow). Contiguous with the pore is a laminar crack-like flaw running horizontally in the micrograph. Both the crack and pore lead to high sound attenuation thus producing a dark region in the acoustic micrograph (see Fig. 4). The length of pore and crack is approximately 600 microns which is comparable dimensionally with the dark zone imaged acoustically. Remember that the acoustic micrograph provides a two-dimensional projection image of the defect. Several serial sections would be required to map the entire flaw optically.

Fig. 5 - Optical reflection micrograph on turbine airfoil section
confirming the flaw found in Figure 4.

Crack Detection (Acoustic Shadow)

      The fracture characteristics revealed by acoustic microscopy
depend on the crack orientation relative to the direction of sonic
propagation, the size of the crack opening, the roughness of the
fracture interface, and the extension of the fracture beneath the surface. The acoustic transmission level across the fracture interface
may vary from almost total attenuation (evidenced by a shadow, as
shown schematically in Fig. 6) to only slight variations in acoustic
contrast. Similarly, the perturbation of the interferogram fringe
spacing may range from complete disruption to an almost imperceptible
fringe shift.

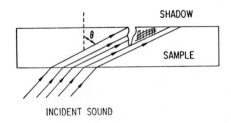

Fig. 6 – Schematic illustrating the effect of crack shadowing observed in the SLAM.

Figure 7 shows acoustic micrographs of a fracture opening to the surface of an alumina sample. The micrographs are oriented such that the sound field propagates out of the plane of the paper at an angle of 45° from left to right across the field of view. Sound propagating through the sample is attenuated at the fracture interface resulting in a shadow to the right of the fracture. The primary feature that distinguishes a surface opening crack is the abrupt and sharply defined onset of the shadow region. This is seen as a sharp boundary separating the light and dark portions of the micrographs in Figure 7. The interferogram fringes are almost obliterated in this area, indicating almost total sound attenuation. By measuring the length of the shadow and using the known propagation angle, measurement of crack extension beneath the surface is made. In Figure 7 the propagation angle was 45°, the shadow is approximately 1 mm in length, thus the crack extends 1 millimeter below the surface.

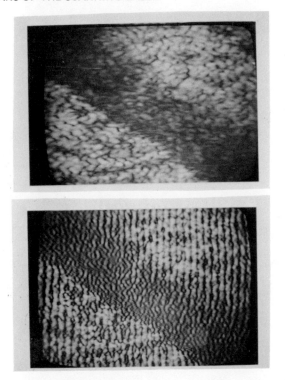

Fig. 7 - a) amplitude micrograph; b) acoustic interferogram showing a crack in an alumina substrate.

Microcracks (Mode Conversion)

     For large cracks which extend several wavelengths below the surface (greater than 200 microns) the primary micrograph feature is the shadow zone. Smaller cracks are rendered visible due to easily detected mode conversion at the flaw site.

     Figure 8 shows a micrograph obtained in the vicinity of a Vickers indent in a hot-pressed silicon nitride test bar[11]. The presence of the mode-converted surface-skimming bulk wave leads to the characteristic cone-shaped ripple pattern observed in the vicinity of the flaw. Explanation of the detection phenomenon and the image characteristics have been reported elsewhere[2].

     A nice feature of the intrinsic mode conversion at the site of small cracks is that it is easy to detect flaws at low magnification. Waves generated at the flaw site propagate several millimeters beyond the flaw site, thus leading to enhanced detection sensitivity and remote sensing capability.

Fig. 8 - SLAM detection of mode conversion at the site of a flaw
produced by a Vicker's indenter. (5Kg load)

Characteristics of Solid Inclusions

Figure 9 shows a typical silicon nitride sample with an implanted solid inclusion. The primary features which distinguish this flaw
type from other defects are the - 1) bright center with acoustic transmission greater than or equal to the background structure. 2) ring
pattern due to diffraction of sound by the inclusion, 3) and the well-
defined boundary of the flaw. These image characteristics seem to be
prevalent for solid inclusions observed so far and they are quite a
different type from porosity variations or laminar flaws. The micrograph was obtained on a sample containing an implanted 400 micron
diameter silicon inclusion. The image size, the diameter of the
first ring, is 160 microns and the flaw is 900 microns below the
surface.

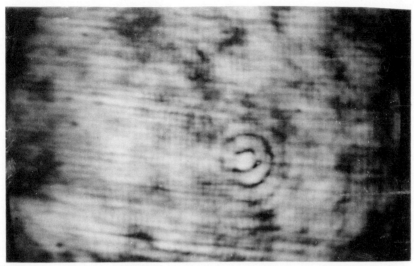

Fig. 9 - Acoustic amplitude micrograph showing a flaw with image
characteristic indicative of a solid inclusion.

Flaw Size and Depth Determinations

To obtain the relationship between SLAM image size and the actual
flaw size, it may be necessary to account for the effects of diffraction beam spreading, and refraction. In general, we expect the
sound field to diverge from the flaw site which leads to image sizes
typically larger than the actual flaw size. However, for some flaws,
focusing may lead to an image size smaller than the actual flaw size.

Flaw depth may be determined by stereoscopy which is illustrated
in Figure 10. This involves imaging the same flaw at two different
angles of insonification. By measuring the shift in the projected
image position of the flaw (relative to a fixed point on surface)
the flaw depth can be determined. Because the depth determination
by stereoscopy uses the acoustic images, the accuracy is governed
by image resolution.

Alternatively, flaw characteristics may be determined by holographic reconstruction techniques. The interferograms produced by
the SLAM are, in fact, in-focus-holograms which lend themselves adequately to direct optical reconstruction. With the aid of a digital
computer, of course, digital techniques may be employed. An example
of this process on SLAM data is presented in this volume by
C. H. Chou, B. T. Chieri-Yakub and G. S. Kino in the paper entitled,
Transmission Imaging: Forward Scattering and Scatter Reconstruction.

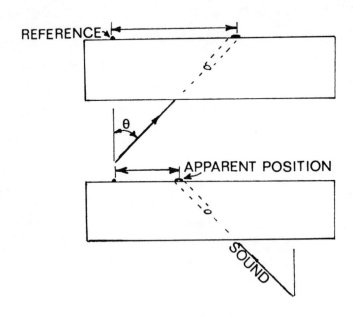

Fig. 10 - Schematic illustrating depth determination by stereoscopy.

SUMMARY

Examination of a variety of different flaws in ceramics has led
to the development of a series of unique image characteristics. These
acoustic signatures can be used to differentiate various flaw types.
Many of the interpretations rely on defect morphology. Thus, under-
scoring the utility of the imaging approach. This is particularly important, for example, in detecting solid inclusions in porous ceramics.
Acoustic images also provide a handle on the determination of flaw
sizes. The extension of cracks below the surface are directly obtainable from micrographs. Estimates of the size of solid inclusions are
also obtainable. However, in this case, it may be necessary to
correct recorded image size for beam spreading and diffraction effects.

REFERENCES

1. D. S. Kupperman, C. Sciammarella, N.P. Lapinski, A. Sather,
D. Yuhas, L. Kessler, and N. F. Fiore, "Preliminary Evaluation of
Several NDE Techniques for Silicon Nitride Gas-Turbine Rotors",
Argonne National Laboratory Report ANL-77-89, January 1978.

2. D. E. Yuhas, <u>Characterization of Surface Flaws by Means of</u>
<u>Acoustic Microscopy</u>, First International Symposium on Ultrasonic
Materials Characterization, June 7-9, 1978, National Bureau of
Standards, Gaithersburg, MD., (In Press) ed by: H. Berger (1979.

3. D. S. Kupperman, L. Pahis, D. Yuhas, and T. McGraw, <u>Acoustic</u>
<u>Microscopy for Structural Ceramics</u>, submitted Journal of American
Ceramic Society, 1979.

4. D. E. Yuhas and L. W. Kessler, <u>Scanning Laser Acoustic Micro-</u>
<u>scope Applied to Failure Analysis</u>; Proc. ATFA - 78 IEEE Inc., New York,
N. Y. Catalog No. 78CH1407-6 REG6., pp. 25-29 (1978).

5. <u>Acoustic Microscopy, SEM and Optical Microscopy: Correlative</u>
<u>Investigations in Ceramics</u>. Scanning Electron Microscopy 1979, 1,
SEM Inc., AMF O'Hare, Il. 60666, pp. 103-110.

6. D. E. Yuhas, T. E. McGraw, and L. W. Kessler, <u>Scanning Laser</u>
<u>Acoustic Microscope Visualization of Solid Inclusions in Silicon Ni-</u>
<u>tride</u>, Proc. ARPA/AFML Conf. on Quantitative NDE., LaJolla, CA. 1979.

7. Commercially available under trade name SONOMICROSCOPE^{TM} 100,
Sonoscan, Inc., Bensenville, Illinois 60106.

8. L. W. Kessler and D. E. Yuhas, <u>Acoustic Microscopy 1979</u>, <u>Proc.</u>
<u>IEEE</u>, 67, (4) pp. 526-536 (1979).

9. S. A. Goss and W. D. O'Brien, <u>Direct Ultrasonic Velocity Mea-</u>
<u>surements of Mammalian Collagen Threads</u>; T. Acoust. Soc. Amer. 65(2)
pp. 507-511 (1979)

10. G. S. Kino, <u>Nondestructive Evaluation</u>, Science, Vol. 206,
pp. 173-180, Oct. 1979.

11. Micrograph Courtesy of J.J. Schuldies, Airesearch Mfg. Co.,
Phoenix, AR.

IMAGING IN NDE

B. R. Tittmann

Rockwell International Science Center

Thousand Oaks, California 91360

ABSTRACT

The ultimate objective of most nondestructive evaluation (NDE) studies is to develop a capability for predetermining the inservice failure probabilities of a structural component with the best possible confidence. The role that ultrasonic imaging can be expected to play in the failure prediction process is reviewed. Included are discussions of the basic concepts of imaging in NDE, a survey of types of NDE imaging systems, and some key problem areas which need to be addressed in the future.

INTRODUCTION

In the past five years, the field of ultrasonic imaging for medical purposes has undergone a number of significant advances. Now the diversity of imaging systems that have become available for medical applications seems to be far larger than those available for NDE applications. At first glance one would expect that these advances in imaging techniques could be incorporated into instruments that are designed solely for NDE work. There are, however, some very fundamental differences between the two applications, some of which are outlined below.

One traditional way of forming ultrasonic images has been to display pulse-echoes in a two-dimensional pattern so that the spatial relationship between the interfaces and acoustic impedance discontinuities that give rise to the echoes are maintained. When the echo data are arranged in this way, usually referred to as B or C scan, it is possible for the eye and brain to serve as a very sophisticated pattern recognition system to detect flaws or defects within an object. This capability is particularly important when the object being examined normally has a considerable amount of internal structure. For medical imaging, the objective is to delineate the more or less known outlines of normal internal structures which typically show up by different shades and contours of grey. Abnormalities such as tumors or cysts are found more by recognition of abnormalities in patterns than by sharply distinctive features. In complete contrast, typical NDE images have most often been analogous to the night-sky, a featureless dark background with a few very bright, simply contoured shapes. The reason for this drastic difference in typical images lies in the acoustic impedance. The material encountered in medical imaging is typically tissue with high water content so that the patterns emerge as fine nuances in the acoustic impedance. In NDE imaging the flaws typically present in the metals or ceramics are voids, cracks, or inclusions with considerably different acoustic impedance.

The high water content in tissue has acted to confine medical imaging in specific directions, such as to limit the wave mode to longitudinal waves only, whereas all modes are possible in NDE, i.e., longitudinal, shear and Rayleigh. The presence of these modes raises the complexity of the situation for NDE, since mode conversion and re-radiation from creeping rays can produce degradations in the imaging process.

On the other hand, one of the restrictions imposed on many medical applications, not present in NDE, is the sensitivity of tissue to high amplitude sound waves. In NDE, high amplitudes are found to be useful not only in obtaining greater depth penetration into the part being examined, but also in the use of nonlinear effects, such as second-harmonic generation by nearly closed cracks.

The most important distinction between medical and NDE imaging lies in the objectives of the two schemes. While the medical diagnostician is trying to find out whether or not there is an abnormality in an organ or other internal structure, the NDE imager must obtain size, shape, and orientation in a sufficiently quantitative way to enable failure prediction, which takes into account the stress distribution of the part during operating conditions.

IMAGING IN NDE

Philosophy

To define what is needed in quantitative NDE, it is necessary to briefly discuss design philosophy.[1] For many years,

structures were designed according to the expected fatigue life
of defect-free components (safe life philosophy). It has been
found, however, that a significant number of failures can be
attributed to pre-existing flaws, such as occur by machining
surface damage etc., and hence a damage tolerant design philosophy is now being adopted. These, however, can be tolerated if
they are sufficiently small that they will not lead to failure
during the service life of the structure. The actual size of
these tolerable flaws will depend on the fracture toughness of
the material; their detectability will be given by the sensitivity of the nondestructive testing technique.

Prediction of remaining lifetimes of structural components
from a knowledge of defect structure, stress levels, and material properties is a rapidly emerging science essential to this
philosophy of design.[2] Figure 1 presents as an example an expression for the remaining life of a part under uniaxial, cyclic
loading tension.[3] Catastrophic failure is predicted when the
flaw has reached sufficient size that the expression in brackets
is equal to zero; this size varies dramatically from material to
material as shown[3] in Fig. 2 for a cyclic loading amplitude of
$\sigma_a = 0.5 \, \sigma_y$ (where σ_y is yield strength of the material). Also
included in Fig. 2 is the frequency at which the wavelength of a
longitudinal elastic wave is equal to the critical flaw diameter
for the case discussed. The great range of frequencies, coupled
with the variability of ultrasonic attenuation from material to
material, indicates that a variety of techniques may be required
to determine defect sizes. In some cases the regime $\lambda \ll a$ may be
accessible, in others it may be necessary to do the best that is
possible when $\lambda \tilde{>} a$.

REMAINING LIFETIME:

$$t_r = \frac{2}{v_o (n-2)} \left(\frac{K_o}{\sigma_a Y} \right)^n \left[\frac{1}{a^{(n-2)/2}} - \left(\frac{\sigma_a Y}{K_c} \right)^{n-2} \right]$$

Where σ_a is the tensile stress

2a is diameter of flaw in plane or propagation

K_o, K_c, v_o, n define material crack propagation
    resistance

Y is a flaw profile factor ($\sqrt{\pi} > Y > 2/\sqrt{\pi}$ typically)

Fig. 1

CRITICAL FLAW SIZE:

$$a_c \approx \left(\frac{K_c}{Y\sigma_a}\right)^2$$

Order of Magnitude Estimates of Critical Flaw
Sizes in Some Metal and Ceramic Systems

Materials		Flaw Size (mm)	Frequency for $\lambda_\ell = 2a_c$ (MHz)
Steels	4340	1.5	2.0
	D6AC	2.3	1.3
	Marage 250	5.0	0.59
	9N14Co 20C	27.5	0.10
Aluminum	2014-T651	8.0	0.4
Alloys	2024-T3511	25.0	0.26
Titanium	6Al-4V	8.0	0.375
Alloys	8Al-1Mo-IV(β)	14.5	0.21
Silicon	Hot Pressed	0.05	100
Nitrides	Reaction Sintered	0.02	250
Glasses	Soda Lime	0.001	2,500
	Silica	0.003	830

Fig. 2

Based on these considerations, the ultimate goal of quantitative NDE is defined in Fig. 3. For cases in which the remaining lifetime is controlled by the growth of a subcritical defect
as described, for example, by the expression in Fig. 1, ultrasonics must provide quantitative information about the size,
shape, and orientation of the defect. Figure 4 illustrates a
classical difficulty in accomplishing this. If an inspector
encounters a defect such as shown at the top, he will see a
large echo return. However, when this sample is loaded along

GOAL OF QUANTITATIVE NDE;

To predetermine the in-service failure
probability of structural component, with
the best possible confidence.

Fig. 3

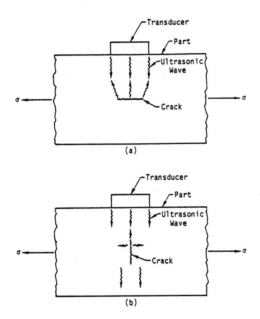

Fig. 4   Comparison of ultrasonic reflection from two
         crack orientations. (a) Crack favorably
         oriented, (b) crack unfavorably oriented.

its length, the defect may produce little reduction in the
strength of the material since it lies in the plane of applied
stress. On the other hand, the defect shown at the bottom can
produce premature failure if it is of sufficient size. Yet, in
general, it will produce a very weak ultrasonic echo.

    It is clearly necessary to gather, process, and interpret
more of the ultrasonic information than just the backscattered
amplitude if lifetime is to be predicted. Two general classes

of approaches may be defined: imaging and scattering.[4] The philosophy of the former is to process the scattered fields in a manner such that a visual outline of the object can be produced on a display. The latter attempts to infer geometric characteristics of the object from its far field scattering cross sections. As will be seen later, model based reconstruction using the scattered radiation can also be used to form images. The two philosophies are contrasted in Fig. 5. In order to compare these techniques, it is useful to view the interaction of ultrasound with a defect as a generalized scattering process in which an incident plane wave, varying as e^{+ikz}, is converted into a scattered wave, which (in the far field) varies as $r^{-1} S(\theta,\phi)$ e^{-ikr}. Here k is the ultrasonic wave vector and r, θ and ϕ are spherical coordinates centered at the defect. The function $S(\theta,\phi)$ contains information regarding the size, shape, and orientation of the defect, and a number of different techniques for processing and inerpreting this information have been proposed.

Image Formation

The most familiar processing of scattered wave fields is the formation of an image. From the Fourier optics[5] (acoustics) point of view, this is based on the recognition that for small values of θ, and in the far field, $S(\theta,\phi)$ is directly related to the spatial Fourier transform of the reflected field radiated by the defect. The purpose of a lens is to perform the inverse

COMPARISON OF IMAGING AND SCATTERING APPROACHES

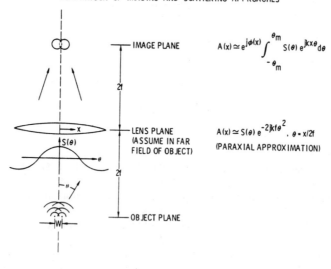

$$A(x) \simeq e^{j\phi(x)} \int_{-\theta_m}^{\theta_m} S(\theta) e^{jkx\theta} d\theta$$

$$A(x) \simeq S(\theta) e^{-2jkf\theta^2} , \quad \theta = x/2f$$

(PARAXIAL APPROXIMATION)

Fig. 5

transform and thereby recover the original geometric information. As an example, consider a lens (with unit magnification)
forming an image of a defect situated at a distance equal to
twice the focal length f as shown in Fig. 5. Then within the
paraxial approximation, and assuming the defect is sufficiently
small that the lens is in the far field, the amplitude of the
scattered field at the lens aperture is

$$A_L(x) \approx S(\theta)e^{-2jkf\theta^2} \qquad (1)$$

where, for simplicity, we have considered a two-dimensional case
where x = 2fθ. In the image plane, apart from phase factors,
the amplitude may be expressed as

$$A_I(x) = e^{i\phi(x)} \int_{-\theta_m}^{\theta_m} S(\theta)e^{+jkx\theta}d\theta, \qquad (2)$$

where $4f\theta_m$ is the diameter of the lens. This is the desired
inverse spatial Fourier transform of S. For a sufficiently
large wave vector k, the scattered field at the defect, and
hence its shape, is recovered with high resolution.

Limitations

     There are fundamental limitations to this technique.
First, both the aperture and wave vector are finite, and resolution is therefore limited. A commonly used criterion for the
lateral resolution[6] (which, although strictly applicable to two
point objects, a circular lens and incoherent illumination, is
useful as a guideline) is

lateral resolution = 0.61λ/NA   (3)

where λ is the wavelength and NA the numerical aperture sin θ_m.
The second is that, although imaging represents a useful way to
process data, it does not necessarily overcome the geometric
problem illustrated in Fig. 4. Here, however, some encouraging
results have been reported with systems operating at other than
normal incidence.[7-8] More complete discussions of the details of
various imaging approaches can be found in references (9-12).

     It is useful at this point to consider specific applications
in which an imaging system might be used to distinguish deleterious defects. The B-1 aircraft is one of the first major systems
in which nondestructive evaluation has been coupled with fracture
mechanics in the design process. After an extensive experimental
program to demonstrate capability, it was established that defects

with critical dimensions of 0.050 in (0.127 cm) could be reliably detected.[13] However, since the measurement of size is not absolute, a number of noncritical and nonexistent defects are also likely to be detected. In order to improve on this, let us say to achieve a resolution of 0.025 in (0.064 cm), an ideal imaging system with a numerical aperture of $\frac{1}{2}$ would have to operate in excess of 10 MHz. To the author's knowledge, systems having NA $> \frac{1}{2}$ and operating in the 10–50 MHz range are not yet commercially available.

Examination of Fig. 2 indicates that there are other materials in which the critical flaw sizes are much greater, and the resolution available today appears quite adequate. Conversely, in very brittle materials, an order of magnitude increase in frequency of operation is necessary.[14]

Figure 6 illustrates some of the differences in the imaging and scattering techniques, by comparing[4] the functions $|A_I|^2$ and $|S(\theta)|^2$ of Equations (1) and (2) for the case of two line objects separated by a distance W and coherently illuminated by a wavelength for the case of sin $\theta_m = \frac{1}{2}$. When W/λ = 2, the images of the objects are readily resolved. Also, their separation can be easily determined by noting the spatial frequency of the interference pattern of the far field scattering S, such as could be

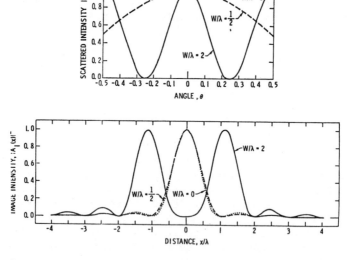

Fig. 6   Comparison of scattered fields and image for W/λ = 2, $\frac{1}{2}$, 0. The image intensity is normalized to a maximum value of unity, and an aperture, sin θ_m = $\frac{1}{2}$, is assumed for both cases. (Courtesy R.B.Thompson)

measured by moving a small pick-up transducer over the aperture. However, when $W/\lambda = \frac{1}{2}$, the image is nearly indistinguishable from that of a single point reflector (W = 0). The image amplitude is a convolution of the reflected amplitude distribution with an aperture function which, in this limit, spreads the large angle scattering information (which distinguishes the two cases in the scattering function) throughout image space. The two are mathematically equivalent, but in real situations, where measurement noise exists, recovery of the information may be more feasible by direct measurement of the scattered fields.

The above discussion has, for simplicity, treated the wave phenomenon as a scalar process. The existence of mode converted signals and other consequences of the true tensor nature of the problem are very real and should not be forgotten. In fact, these effects may contain vitally important information.

Thus, a wide variety of practical approaches have been developed for estimating the desired flaw parameters. These are ordered[15] in Fig. 7 in accordance with the ratio of flaw size to

TYPE	REGIME	ADVANTAGES
IMAGING	ka > 6.3 ($\lambda < a$)	HIGH INFORMATION CONTENT EASILY INTERPRETED DISPLAY RESOLVES MULTIPLE FLAWS
MODEL BASED RECONSTRUCTION	ka ⩾ 3	PHYSICAL PRINCIPLES USED TO IMPROVE RESOLUTION AND TREAT MODE CONVERSION
MODEL BASED ADAPTIVE LEARNING NETWORKS	0.4 < ka < 3	MULTIPLE SCATTERING TAKEN INTO ACCOUNT GAIN MORE INFORMATION IN DIFFICULT REGIME
LONG WAVELENGTH SCATTERING	ka < 0.5	FRACTURE RELATED PARAMETERS DEDUCED FROM A FEW MEASUREMENTS MAY BE USEFUL IN AUTOMATION

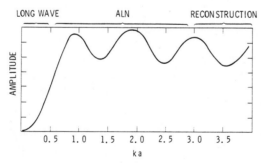

Fig. 7 Inversion techniques.

wavelength required to apply the technique. For reference, the
frequency dependence of the backscattering from a spherical cavity
is reproduced at the bottom of the figure.[16] At short wavelengths,
it is useful to reconstruct a detailed picture of the flaw.
Implicit in any such approach is a physical model of the elastic
wave–flaw interaction. Imaging systems are usually designed with
the implicit assumption that the flaw is a diffuse reflector.
That is, any incident ray is assumed to be converted into a divergent cone of rays, emanating from the point of illumination. For
cavities with rough surfaces, this model can accurately describe
the scattering process and high quality images can be formed.
However, if inclusions or flaws have surfaces that are smooth with
respect to a wavelength, then specular reflections, mode converted
signals, re–radiation from creeping rays, and internal reverbera-
tions can all produce degradations of the reconstruction to produce, for example, ghost images. To overcome these difficulties,
it is necessary to base the reconstruction on a more precise model
of the elastic wave–flaw interactions. The basic strategy is to
use the model to specify a relationship between measurement and
the Fourier transform of the object shape function. Inverse
Fourier transforms can then be used to reconstruct the object.
Several such approaches will be discussed.

At longer wavelengths, the information is not sufficient
information to develop a detailed reconstruction of a flaw
shape. Furthermore, because of nonlinearities caused by such
effects as strong multiple scattering, analytical solutions to the
inverse problem are difficult to develop. Consequently, adapta-
tive learning procedures have been utilized as tools to find
empirical solutions. At still lower frequencies when the wavelength is large with respect to the flaw size, recent work[17] has
shown that, if independent a priori information is available which
limits the flaw to membership in a certain class, the experimental
data are sufficient to determine flaw orientation and the reduced
stress intensity factor.

SELECTED SURVEY OF IMAGING SYSTEMS

Single and Double Transducer Systems

One of the most widely used imaging system is the C–scan
system. This system is reasonably simple in concept, in many
cases produce very good images, and most people active in the NDE
field are familiar with it.

A C–scan system is shown[18] in Fig. 8. Although it is shown
operating in the transmission made, it can be operated in the
reflection mode as well. The illustrated system utilizes two
focused transducers in a confocal geometry. The transmitted

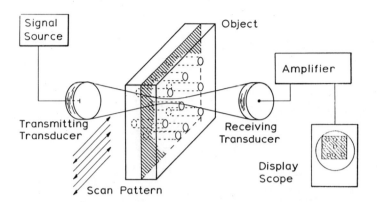

Fig. 8   Diagram of a focused C-scan system.
         (Courtesy R. C. Addison, Jr.)

signal is chiefly acted upon by the small region at the focal
point of the transducer; it is then received, detected and displayed as an intensity modulation of a scope or TV monitor. The
transducers are scanned synchronously and the image is built up
point by point. The image is an orthographic view of the part
being inspected.

       An interesting and important application of the mechanical
scanning technique (by Kino and his co-workers)[19] is the imaging
of inhomogeneous stress fields in externally loaded solids. This
method requires a measurement of transit time of a longitudinal
acoustic wave through a stressed metal specimen typically 10 mm
thick. A 3 mm diameter water-coupled 12.5 MHz transducer is
mechanically scanned over the surface of the sample by a computer
controlled system to take stress field contour plots. The system
for phase change measurements is described in detail in a forth-
coming paper by Kino and co-workers. It is capable of setting a
null at 12.5 MHz to within 60 cycles, or 5 parts in 10^6. Figure 9
shows stress and frequency contour plots for a 6061-T6 aluminum
double edge-notched panel near one of the notches under a tensile
load of 30,000 Newtons. The 1 mm spaced scan points are shown
compared with the numeric elastic solution (solid lines). In
addition to measuring stress fields, the technique also allows
determinations of stress intensity factors.

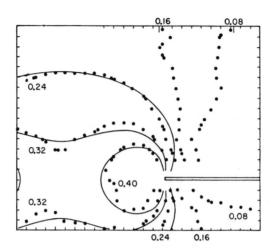

Fig. 9   Stress and frequency contour plots of results
         for 6061-T6 aluminum double edge-notched panel
         30,000 Newton tensile load.
         (Courtesy G. Herrmann and G.S. Kino)

     There are a considerable number of imaging systems that
currently exist in either a laboratory or prototype state that
seek to improve upon the performance parameters of the conventional C-scan systems. In order to describe them and point out
how they differ, it is helpful to group them into the various
categories listed in Table 1. The first group includes the C-scan
systems and for sake of completeness, the A and B system.

     All of the systems in group II use multiple transducers which
are arranged in different kinds of arrays. The first type is
known as a sequenced linear array. It was developed by Bom[20] and
consists of a row of transducers each one is sequentially used to
send out an ultrasonic beam and receive the returning echoes. The
electronically phased arrays make up the next group. In these
systems all of the transducers are pulsed simultaneously but the
relative phases of the r.f. signals sent to the transducers are
varied so that the direction and focus of the ultrasonic beam
emanating from the array can be changed electronically. In the
simplest of these systems,[21] the beam is left unfocused and is
scanned through a sector. In more sophisticated versions,[22] the
beam is focused while it is scanned and the focus can be dynamically varied while the array is receiving echoes.

Table 1

Types of Ultrasonic Imaging Techniques

I.   Single and Double Transducer Systems

        A.   Single Transducer (A,B,C, Scan; Reflection Mode)
        B.   Double Transducer (C Scan; Transmission Mode)

II.   Multiple Transducer Systems

        A.   Sequenced Linear Array
        B.   Electronically Phased Linear Array
        C.   Two-Dimensional Array (NXN)

III.   Systems Using Optical Diffraction or Reflection

        A.   Laser Illuminated Liquid-Air Interface
        B.   Bragg-Diffraction Imaging
        C.   Laser Illuminated Membrane
        D.   Laser Scanned Solid Surface

IV.   Model Based Reconstruction

        A.   Born Approximation
        B.   POFFIS

Several years ago, Kino and his colleagues developed an acoustic imaging system based on a phased array technique which used chirp-focusing.[23] This system demonstrated that various types of waves, including shear, Lamb, and Rayleigh waves could be excited by using mode conversion from a longitudinal wave array and that imaging could be performed by using these types of acoustic waves. All of these techniques are applicable to their new imaging system based on the synthetic aperture approach.[24]

A block diagram of the actual hardware system is shown in Fig. 10. This new system incorporates a number of design improvements over the previous hardware system and it contains 80 kilobytes of high-speed (50 nsec cycle time) memory as well as all of the electronics required for computer interfacing, display control, and array multiplexing. The memory is partitioned as thirty-two 1 kilo-byte blocks of signal memory and 48 kilo-bytes of focus memory.

Operation of this system consists of two phases: the signal acquisition phase and the image reconstruction phase. During the

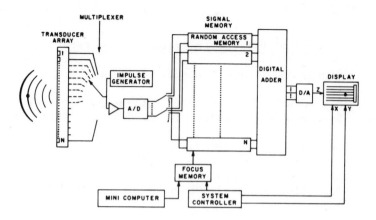

Fig. 10 Digital imaging system. (Courtesy G. S. Kino)

signal acquisition phase, an anlog multiplexer selects a single
element of the transducer array. The element is excited with an
impulse from the impulse generator, and an acoustic signal propagates out into the medium. The return echoes are received by that
same transducer element and, after passing through the multiplexer
and input amplifier, the signal is digitized by a high-speed
analog-to-digital converter and stored in a corresponding memory.
During the image reconstruction phase, the focus memory controls
the addressing of the signal memories. The outputs of the signal
memories are summed together in the digital adder and are converted to an analog voltage which modulates the intensity of a
raster scanned display; This produces a two-dimensional image.
This entire process is accomplished in approximately 30 msec, with
about 10 msec for signal acquisition and 20 msec for image reconstruction; it results in a frame rate of approximately 30 Hz, well
above the rate required for real-time imaging. This array is a
32-element transducer array, with a 1.6 cm aperture and a 3.3 MHz
center frequency. The results to date indicate a 3 dB resolution
of about 0.5 mm in the transverse direction and about 0.4 mm in
the range direction. The synthetic-aperture system was used with
an edge-bonded transducer array capable of exciting Rayleigh waves
in a test piece.[25] Figure 11 shows a Rayleigh wave image of an
EDM slot placed parallel to the transducer array (1 cm squares).
The clutter in the image is due mainly to spurious modes which are
excited by the EBT array.

The next category of systems listed in Table 1 are those
using two-dimensional arrays. There are several existing efforts
in this area[26-28] although only one will be described.

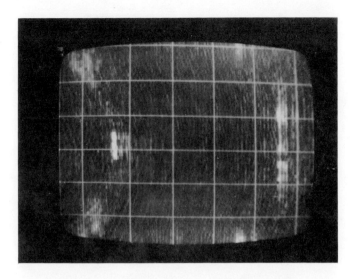

Fig. 11   Rayleigh wave image of an EDM slot placed parallel
to the transducer array (1 cm squares).
(Courtesy G. S. Kino)

Lakin[28] describes an imaging system involving the digital
reconstruction of the near fields of an object by first measuring
the amplitude and phase of the remove scattered field and then
operation on these fields with certain computational algorithms.
These algorithims involve a predistortion correction and phase
normalization, a fast Fourier transform of the two-dimensional
data, a post transform phase correction and allowance for parallel
or perpendicular plane reconstruction in a "side looking" configuration. In order to obtain a fast data through-put the scattered
fields are measured by a two-dimensional array which may be translated as well. The two-dimensional array is sequentially scanned
in a random access manner and may be used for either transmitter
or receiver. Each array element consists of a PVDF piezoelectric
film covering a high impedance surface electrode configuration.
In groups of eight the array elements are connected to a custom
packaged analog switch which is addressed by a microprocessor
controller.

Figure 12 shows the measured radiation pattern of the source
transducer. Clearly visible is the side lobe structure. Although
this array is now operational at 5 MHz, future designs at higher
frequencies (50 MHz) are feasible, in principal.

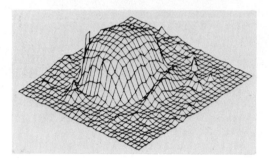

Fig. 12 Measured radiation pattern of source transducer.
         (Courtesy K. Lakin)

The next group of systems use either optical diffraction or
reflection to convert the ultrasonic image into a visible image.
For particular applications some of these systems have proven to
be very useful but, in general, they are not as sensitive as
systems that use a piezoelectric transducer. The first system
listed in Table 1 uses a liquid-air interface to form an ultrasonic ripple pattern. This system was developed a number of years
ago by Brenden;[29] it was one of the first ultrasonics imaging
systems to produce recognizable images. The laser beam that is
incident on the liquid surface is diffracted by the ripple pattern. The diffracted beam is modulated by the ripple pattern and
contains an image of the ripple pattern. This optical image can
be viewed directly or it can be focused onto a vidicon and displayed on a TV monitor. The chief drawback to this technique is
that it is a highly coherent imaging system. The numerous diffraction fringes and specularity appearing in the image prevent

any quantitative image analysis. The sensitivity and dynamic range of the system are also less than is desired.

Most recently, Schmitz and co-workers[30] have advanced the technique by substituting the optical with numerical reconstruction; this has several advantages versus optical reconstruction: objectivity, reproducibility, free scaling, only one image. Some shortcomings are: high costs and long reconstruction time with conventional data processing. The numerical reconsruction is accomplished by the calculation of the Fourier transform in the Fresnel approximation. This approach forms the basis for replacing the mechanically scanned detector with a linear or two-dimensional array. Figure 13 shows in three different views, a Fresnel reconstruction of experimental data taken from a series of holes of varying separation, with 3 mm diameter for the more numerous set. On the top is a C-scan without grey level; below that is a side projection of every line; and on the bottom an isometric view.

Bragg-diffraction imaging techniques[31,32] depend on an interaction between a laser beam and an ultrasonic beam in the bulk of a liquid. The acoustic wavefronts in the liquid cause

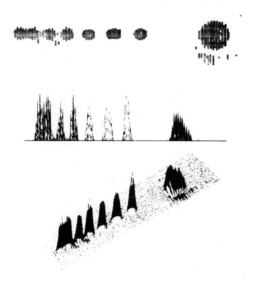

Fig. 13   Inconel reconstruction of experimental data from
          series of holes (3 mm in diameter for all except
          one). Top: C-scan. Middle: side projection.
          Bottom: isometric view. (Courtesy V. Schmitz)

compressions and rarefractions. The resulting changes in index of
refraction of the liquid cause the incident laser beam to be partially diffracted. The diffracted laser beam is modulated with
the acoustic image information. The drawbacks of this system are
its low sensitivity and the specularity of the images.

The next system which utilizes a laser and a thin membrane
or pellicle, was developed by Mezrich at RCA.[33] The pellicle is
placed in a water tank and couples to the ultrasonic image
focused onto it, as illustrated in Fig. 14; the amplitude of the
motion of the pellicle at each image point will be very nearly the
same as the amplitude of the molecular motion of the water. The
pellicle is made highly reflective by a gold coating, and serves
as one mirror of a Twyman-Green interferometer. The laser beam
enters the interferometer and the beam splitter sends part of the
beam to a reference mirror and part through a beam deflector to
the pellicle. The two beams are reflected and recombined in a
photodiode. The amplitude of the difference signal obtained will
be proportional to the image displacement amplitude of the pellicle. The beam defector scans the laser beam over the entire
pellicle in a raster pattern, and the ultrasonic image is reproduced in this way. The reference mirror is vibrated to stabilize
the system against ambient room vibrations. This system is capable of very accurately measuring the displacement amplitude of the
pellicle and, consequently, the absolute value of the ultrasonic
intensity at a point.

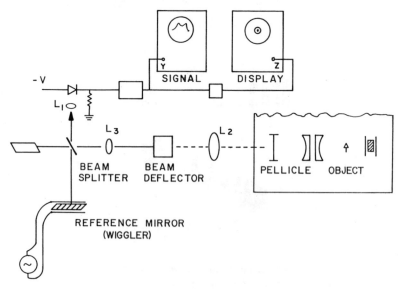

Fig. 14   Diagram of ultrasonic imaging system developed
          at RCA. (Courtesy of R. Mezrich)

In the past several years, the frequencies at which it is possible to obtain good images has been extended upward, and images can now be obtained at frequencies as high as a Gigahertz with a resolution approaching one micron. Several acoustic microscope systems, as these are termed, are currently being developed [34] that together span the frequency range from 100 MHz to 3 GHz. The unit that operates over the lower portion of the frequency range was first described by Korpel et al [35] and serves as an example of the last category of systems listed part III of Table I. A diagram of the system, which is available from Sonoscan, [36] is shown in Fig. 15. The 100 MHz acoustic signal is applied to a specimen which is placed upon the microscope stage. The energy transmitted through the sample creates a ripple pattern and as the ripple moves under the focused spot, the reflected beam wiggles and is partially intercepted by a knife edge. The beam reaching the photodetector is thus modulated by the acoustic signal. The output of the photodetector consists of two signals; one signal depends upon visual features of the specimen and the other signal depends on the acoustic features of the specimen. The laser beam is rapidly scanned over the field of interest in a raster pattern and the resulting signals are processed and displayed on the TV monitors as optical and acoustic images. One of the specimens that has been examined contained a crack 100 μ long and about 50 μ deep. This defect, introduced in a hot-pressed SiC bar via the Knoop indentor technique, is clearly visible in the acoustic micrograph of Fig. 16. The presence of the mode-converted surface-skimming bulk waves leads to the characteristic cone-shape ripple pattern observed in the vicinity of the flaw. An optical micrograph of the crack is shown in the lower photo.

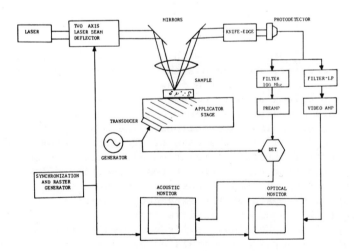

Fig. 15.   Diagram of Sonoscan acoustic microscope.
(Courtesy of L. W. Kessler)

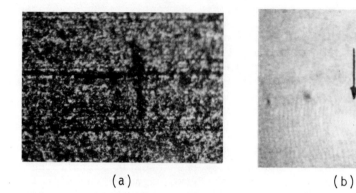

(a)   (b)

Fig. 16   A 100 × 50 μm dent (arrow) in hot-pressed SiC
          bar. (a) optical micrograph (.35 × 45 μm area);
          (b) acoustic micrograph (3 × 2 mm area) showing
          effect of interference by backscattered surface-
          skimming bulk wave. (Courtesy D. S. Kuperman,
          L. Pahis, D. Yuhas and T. McGraw).

          Part IV of Table 1 deals with model based reconstruction.
As discussed earlier, imaging systems are usually designed with
the implicit assumption that the flaw is a diffuse reflector. To
overcome the problems associated with smooth inclusions and voids,
it is necessary to base the reconstruction on a more precise model
of the elastic wave-flaw interactions. The basic strategy, as
illustrated in Fig. 17 for the case of the Born approximation[37] is
to use the model to specify a relationship between measurement and
the Fourier transform of the object shape function. Inverse
Fourier transforms can then be used to reconstruct the object.
One such approach, based on the extended quasi-static model, has
been developed by Rose[38] and is illustrated in Fig. 18. Here the
shape of a prolate spheroidal cavity has been reconstructed by
carrying out an inversion on simulated experimental data. A
second, by Bleistein and Cohen[39], is illustrated in Fig. 19. The
model used in this case was the physical optics approximation,
which has been shown[40,41] to asymptotically

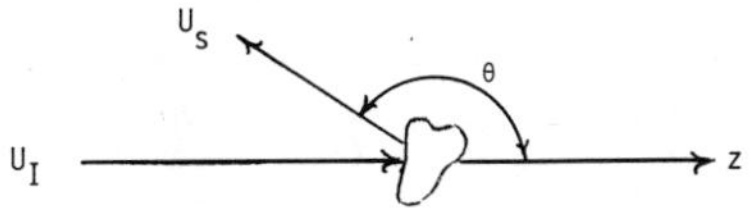

$$U_s = \frac{\vec{A}}{r} e^{iK_L r} + \frac{\vec{B}}{r} e^{iK_T r}$$

$$\vec{A} = \frac{K_L^2}{4\pi} \left[\frac{\delta\rho}{\rho} \cos\theta - \frac{\delta\lambda + 2\delta\mu\cos^2\theta}{\lambda + 2\mu} \right] F \left[\frac{f\sin\theta}{V_L}, 0, \frac{f(\cos\theta - 1)}{V_L} \right]$$

$$\vec{B} = \frac{K_T^2}{4\pi} \left[\frac{2\alpha\delta\mu\cos\theta\sin\theta}{\mu\rho} - \frac{\delta\rho\sin\theta}{\rho} \right] F \left[\frac{F\sin\theta}{V_T}, 0, f\left(\frac{\cos\theta}{V_T} - \frac{1}{V_L}\right) \right]$$

where

$$F(u,v,w) \triangleq \int_{-\infty}^{\infty} dx \int_{-\infty}^{\infty} dy \int_{-\infty}^{\infty} dz\ f(x,y,z)e^{-i2\pi(ux+vy+wz)}$$

$$f = \begin{cases} 1 \text{ INSIDE OBJECT} \\ 0 \text{ OUTSIDE OBJECT} \end{cases}$$

Fig. 17   Use of Born approximation to relate measurements
to Fourier transform of flow shape function.

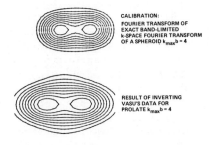

CALIBRATION:
FOURIER TRANSFORM OF
EXACT BAND-LIMITED
k-SPACE FOURIER TRANSFORM
OF A SPHEROID $k_{max}b = 4$

RESULT OF INVERTING
VASU'S DATA FOR
PROLATE $k_{max}b = 4$

Fig. 18   Model based reconstruction of synthetic data
for prolate spheroidal cavity. (Courtesy J. Rose)

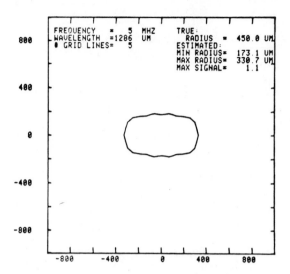

Fig. 19   Reconstruction of oblate spheroid data from
           synthetic data using POFFIS algorithm.
           (Courtesy N. Bleistein and J. K. Cohen)

approach the exact result, for the case of elastic wave back-
scattering, in the high frequency limit. Here the shape of an
oblate spheroidal cavity has been reconstructed by applying the
POFFIS (physical optics far field inverse scattering) algorithm to
simulated experimental data.[42]

A review of reconstruction procedures has recently been
prepared by Lee.[41]

CONCLUSION - DISCUSSION OF FUTURE NEEDS AND PROBLEM AREAS

There are several problem areas for imaging in NDE which
must be overcome before this technique is reduced to practical
application in the field. One important problem is the geometry
of the NDE sample. Whether it is a curved airplane wing or a
cooling tube or a turbine, the reconstruction must be carried out
through the surface in spite of the complexity of the specimen
shape. Speckle is a problem for the coherent systems, particularly for finding surface flaw in the presence of surface irregu-
larities such as are due to machining damage. If the flaw is
interior to the part then time-gating is helpful, but cannot
eliminate contributions from, for example, grain scattering such
as in weld joints.

For phased arrays, there are other more specific problems. For example the problem of <u>grating lobes</u> can be tackled by spacing the elements closer together, so that they occur at rather large angles from the main lobe, or by using sufficiently short pulses such that the information from the grating lobes occurs at substantially different times. <u>Sidelobes</u> have been a source of problems, but much progress has been made in this area, by astute choices for the amplitude distribution of the array. Many of the imaging schemes, particularly the one and two-dimensional arrays, need to be developed to operate at a higher frequency; inability to do so produces sampling errors and <u>aliasing</u>.

A more general problem is the need to investigate <u>error tolerance</u>. In contrast to optical transmission media, ultrasonic media, particularly in NDE applications, are typically strongly scattering (isolated large grains) inhomogeneous (as in weldments), anisotropic (due to texture), and attenuating (Rayleigh grain scattering). The use of a highly developed and sophisticated imaging system may be severely limited by the hostile transmission medium. For the sake of cost effectiveness, the development of a system should be limited to achieve a compromise. To illustrate the point further, the use of an imaging system on a ceramic test piece in a water bath involves about a seven-fold change in sound velocity. The refraction at the water-ceramic interface produces a spherical aberration which considerably lowers the effective NA of the higher NA systems. Clearly, the intended use of an imaging system should be kept in mind during the development and, if possible, the error tolerance should be examined quantitatively in order to achieve a meaningful compromise.

ACKNOWLEDGEMENT

The author is indebted to R. C. Addison, Jr. for his critical review of the manuscript.

REFERENCES

1. M. D. Coffin and C. F. Tiffany, How the Air Force Assures Safe, Durable Progress, "Metal Progress" p. 26 (1976).
2. S. T. Rolfe, "Use of Fracture Mechanics in Design," International Metallurgical Reviews, 19: 183 (1974)
3. B. R. Tittmann, O. Buck, L. Ahlberg, M. deBilly, F. Cohen-Tenoudji, A. Jungmen, and G. Quentin, Surface Wave Scattering from Elliptical Cracks for Failure Prediction, Journ. Appl. Physics, in press.
4. R. B. Thompson and A. G. Evans, Goals and Objectives of Quantitativ NDE, IEEE Trans. on Sonics and Ultrasonics SU23:292 (1976).

5. J. W. Goodman "Introduction to Fourier Optics," McGraw-Hill, New York (1968).

6. M. Born and E. Wolf, "Principles of Optics," Pergamon Press, New York (1964).

7. H. D. Collins and B. B. Brendon, Acoustical Holography Transverse Wave Scanning Technique for Imaging Flaws in Thick-Walled Pressure Vessels, in "Acoustical Holography", V. 5, P. S. Green, ed., Plenum Press, N.Y. (1974), p. 175.

8. H. D. Collins, Acoustical Focussed Holographic Scanning Technique for Imaging Inner Bore Radial Defects in Minuteman Miussle Sections, in "Proceedings of 10th Symposium on NDE," Southwest Research Institute, San Antonio (1975), p. 77.

9. B. P. Hildebrand and B. B. Brenden, "An Introduction to Acoustical Holography," Plenum Press, N.Y. (1972).

10. R. K. Mueller, Acoustical Holography, Proc. IEEE 59:1319 (1971).

11. N. Both, ed., "Acoustical Holography", Vol. 6, Plenum Press, N.Y. (1975), and previous volumes of the series.

12. T. Waugh, G. S. Kino, C. Desilets and J. Fraser, Acoustic Imaging Techniques for Nondestructive Testing, IEEE Trans. on Sonics and Ultrasonics, SU23:313 (1976).

13. E. L. Caustin, State-Of-The-Art NDE in Quantitative Inspection, in "Proceedings of the Interdisciplinary Workshop for Quantitative Flaw Definition," D. O. Thompson, ed., AFML-TR-74-238, p. 123 (1974).

14. G. S. Kino, and A. G. Evans, Prospects for Non-Destructive Flaw Detection in Ceramics, Report of the Materials Research Council (1975), ARPA 2341/2.

15. R. K. Elsley, J. M. Richardson, R. B. Thompson, and B. R. Tittmann, Comparison Between Experimental and Computational Results for Elastic Wave Scattering in "Recent Developments in Classical Wave Scattering, Focus on the T-Matrix Approach," Ohio State Univ., Columbus (1979).

16. B. R. Tittmann, E. R. Cohen, and J. M. Richardson, J. of Acoust. Soc. of America, 63:68 (1978).

17. B. R. Tittmann, W. L. Morris, and J. M. Richardson, Elastic Wave Scattering at Long Wavelengths, App. Phys. Lett., in press.

18. R. C. Addison, Recent Advances in Imaging, in "Proc. of ARPA/AFML Rev. of Progr. in Quant. NDE," held Jan. 1976 at Science Center, Rockwell Int. Air Force Report AFML-TR-75-212, p. 273.

19. G. Herrmann and G. S. Kino, Ultrasonic Measurements of Inhomogeneous Stress Fields, in "Proc. of ARPA/AFML Rev. of Progr. in Quant. NDE," held July 1978 at La Jolla, Calif, Air Force Report AFML-TR-78-205, p. 447.

20. N. Bom, C. T. Lancee, G. Zwieten, F. E. Kloster, and J. Roelendt, Multiscan Echocardiography. I. Technical Description, Circulation, 48:1066 (1973).

21.   J. C. Somer, Electronic Sector Scanning for Ultrasonic
      Diagnosis, Ultrasonics 153 (1968).
22.   F. L. Thrustone and O. T. von Ramm, A New Ultrasound Imaging
      Technique Employing Two-Dimensional Electronic Beam
      Steering,"Acoustical Holography," Vol. V, P. S. Green, ed.
      Plenum Press, New York (1974).
23.   G. S. Kino, Acoustic Imaging for Nondestructive Evaluation,
      Proc. IEEE, 67:510 (1979), also T. M. Waugh and G. S. Kino,
      Real Time Imaging with Shear Waves and Surface Waves, in
      "Acoustical Holography," Vol. 7, Plenum Press, 103 (1977).
24.   G. S. Kino, P. M. Grant, P. D. Corl, and C. S. DeSilets,
      Digital Synthetic Aperture Acoustic Imaging for NDE, in
      "Proc. ARPA/AFML Rev. of Progr. in Quant. NDE," held in
      July 1978 at La Jolla, Calif, Air Force Report AFML-TR-78-
      205, p. 459.
25.   G. S. Kino, B. T. Khun-Yacub, A. Selfridge, and H. Tuan,
      Development of Transducers for NDE, in "Proc. ARPA/AFML Rev.
      of Progress in Quant. NDE held in July 1979 at La Jolla.
      Calif, Air Force Report, in press.
26.   P. Alais, Real Time Acoustical Imaging with a 256   256
      Matrix of Electrostatic Transducers, "Acoustical
      Holography," Vol. V, P.S. Green, ed., Plenum Press, New York
      (1974).
27.   M. G. Maginness, J. D. Plummer, and J. D. Meindl, An
      Acoustic Image Sensor Using a Transmit-Receive Array,
      "Acoustical Holography," Vol. V, P. S. Green, ed., Plenum
      Press, New York (1974).
28.   K. Lakin, Acoustic Imaging and Image Processing by Wavefront
      Reconstruction Techniques, in "Proc. ARPA/AFML Rev. of
      Progr. in Quant. NDE," held in July 1979 at La Jolla,
      Calif., Air Force Report, in press, also Proc. IEEE
      Ultrasonic Symposium, New Orleans, 1979, in press.
29.   B. B. Brenden, A Comparison of Acoustical Holography
      Methods, "Acoustical Holography," Vol. I, A. Metherell,
      H. M. A. El-Sum, L. Larmore, eds., Plenum Press, New York
      (1969).
30.   V. Schmitz, Real Time Ultraschall Holographie, Report
      780141-TW of the Traunhofer-Institut fur Zerstorungs freie
      Prufverfahren, Saarbricken, Germany, p. 38 (1978), also K.
      J. Langenberg, R. Kiefer, M. Wosnitza, and V. Schmitz,
      Recent Advances and Techniques in Ultrasonic Holography with
      Numerical Reconstruction for Improved NDT, unpublished.
31.   A. Korpel, Visualization of the Cross Section of a Sound
      Beam by Bragg Diffraction of Light, Appl. Phys. Lett. 9:425,
      (1976).
32.   H. Keyani, J. Landry, and G. Wade, Bragg Diffraction
      Imaging: A Potential Technique for Medical Diagnosis and
      Material Inspection, Part II. "Acoustical Holography,"
      Vol. V, P.S. Green, ed., Plenum Press, New York (1974).

33. R. Mezrich, K. F. Etzold, and D. H. R. Vilkomerson, System
     for Visualizing and Measuring Ultrasonic Wavefronts, R.C.A.
     Review, 35:483 (1974).
34. R. A. Lemons, and C. F. Quate, Acoustic Microscope-Scanning
     Version, App. Phys. Lett. 24:163 (1974).
35. A. Korpel, L. W. Kessler, and P. R. Palermo, Acoustic
     Microscope Operating at 100 MHz, Nature 232:110 (1971).
36. L. W. Kessler and A. Madeyski, Acoustic Microscopy of Steel,
     Proc. IEEE, 75 CHO 994-4SU:57 (1975).
37. R. B. Thompson, Introduction to Defect Characterization by
     Quantitative Ultrasonics, "Proceedings of the ARPA/AFML
     Review of Progr. in Quant. NDE," AFML-TR-78-55, p. 15.
38. J. H. Rose and J. A. Krumhansl, A. Technique for Determining
     Flaw Characteristics from Ultrasonic Scattering Amplitudes,
     in Proc. ARPA/AFML Rev. of Prog. in Quant. NDE," held in
     July 1979 at La Jolla, Calif. Air Force Report, in press.
39. N. Bleistein and J. K. Cohen, Application of a New Inverse
     Method to Nondestructive Evaluation, Denv. Research Report,
     MS-R-7716 (1977).
40. N. Bleistein and J. K. Cohen, Application of Inverse Method
     to Non-Destructive Evaluation of Flaws, Interdisciplinary
     Program for Quantitative Flaw Definition, Semi-Annual Report
     (Rockwell International Science Center, Thousand Oaks,
     Calif. 1979) p. 112.
41. D. A. Lee, Mathematical Principles of Data Inversion," (in
     preparation).
42. J. K. Cohen, N. Bleistein and R. K. Elsley, Nondestructive
     Detection of Voids by a High Frequency Inversion Technique,
     Interdisciplinary Program for Quantitative Flaw Definition,
     Special Report Fourth Year Effort (Rockwell International
     Science Center, Thousand Oaks, Calif. 1978) p. 81.

A REAL-TIME SYNTHETIC-APERTURE IMAGING SYSTEM

P. D. Corl
G. S. Kino
Ginzton Laboratory
Stanford University
Stanford, CA. 94305

ABSTRACT

A synthetic-aperture acoustic imaging system has been developed which employs digital electronics to perform real-time imaging with a 32-element acoustic transducer array. This system has also been implemented on a computer (not real-time) so that new concepts can be tested before they are implemented in hardware. A number of real-time and computer-reconstructed images are presented to illustrate the performance of this system, including images obtained using longitudinal waves in a water tank and Rayleigh (surface) waves on aluminum samples. Computer-generated contour plots are shown to illustrate the point response of this system and to demonstrate the resolution and sidelobe levels obtained with this synthetic-aperture technique.

INTRODUCTION

The aim of this work is the development of a real-time synthetic-aperture acoustic imaging system. In previous papers,[1,2] the theory of synthetic-aperture imaging was discussed and some preliminary experimental results were presented. These early results included images obtained using computer-reconstruction techniques with both 8- and 32-element acoustic transducer arrays as well as an image obtained using a prototype real-time acoustic imaging system designed to operate with an 8-element acoustic transducer array. Based on the success of this prototype system and the encouraging results obtained with computer reconstruction, a new real-time acoustic imaging system, designed to operate with a 32-element transducer array, was constructed. Preliminary measurements have now been made on this new system and several acoustic images have been obtained.

Prior to this time, most of these images have been made using a longitudinal wave array in a water tank to look at wires and metal blocks. However, a new type of array, the edge-bonded transducer (EBT) array, has recently been developed for the excitation of Rayleigh (surface) waves on solid substrates. Using this new EBT array as well as the wedge-coupling technique which has been used with previous systems, images of both real and simulated flaws in aluminum samples have been obtained with both real-time and computer-reconstruction techniques.

REAL-TIME IMAGING SYSTEM

Fig. 1 shows a photograph of the new real-time synthetic-aperture imaging system. This new system incorporates a number of improvements over the prototype system and it is designed to utilize a 32-element acoustic transducer array in order to produce real-time images. A block diagram of the hardware system is shown in Fig. 2. The hardware consists of the following major components: 80 kilo-bytes of high-speed (50 nsec cycle time) memory, partitioned as 32 one kilo-byte blocks of signal memory and 12 four kilo-byte blocks of focus memory; a fast (33 nsec conversion time) eight-bit analog-to-digital (A-to-D) converter; a 32-element acoustic transducer array with an analog multiplexer and a pulser-amplifier; a digital adder tree and digital-to-analog converter; and the miscellaneous electronics required for timing, display control, and computer interfacing.

Fig. 1.   Real-time acoustic imaging system.

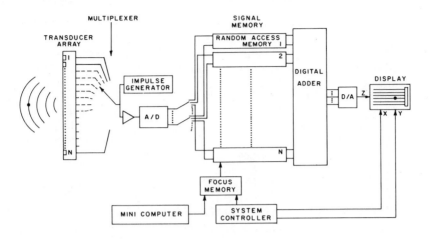

Fig. 2. Digital synthetic-aperture acoustic imaging system.

The operation of this system consists of two phases: the
signal acquisition phase and the image reconstruction phase. During
the signal acquisition phase, an analog multiplexer is used to
select a single element of the acoustic transducer array. This
element is excited with an impulse from the impulse generator and
an acoustic signal propagates out into the medium. The return
echoes are received by that same transducer element and after
passing through the multiplexer and input amplifier, the signal is
digitized by a high-speed A-to-D converter and stored in a digital
memory. This process is repeated for successive elements of the
transducer array, with each of the signals being stored in a
corresponding memory. During the image reconstruction phase, the
focus memory is used to control the addressing of the signal memories.   The outputs of the signal memories are summed together in
the digital adder and converted to an analog voltage which modulates
the intensity of a 256-line raster-scanned display in order to

produce a two-dimensional image. This entire image reconstruction process is accomplished in 30 msec , resulting in a frame rate of 30 Hz .

This system is normally operated with an acoustic transducer array having a center frequency in the 2-5 MHz range. Since the signal is transmitted and received by only a single element at any given time, the array is required to have a very high efficiency. In addition, to give good range resolution, the array is required to have a compact impulse response, or equivalently, broad bandwidth. A significant effort has gone into the design of acoustic transducer arrays[3,4] to satisfy these conflicting requirements and this work will be carried on in the future.

There are a number of advantages to synthetic-aperture imaging in general and to this digital implementation in particular. First of all, there are the inherent advantages of using digital elec- tronics: reliability, low cost, adjustment-free operation, and the ready availability of a wide range of complex circuit elements. In addition, since this system has only a single input amplifier, a number of signal-processing techniques such as matched filtering or nonlinear gain compression can be considered which would be prohibitively complex, or at least very difficult if the electronics had to be duplicated for every element of the transducer array. Matched filtering of relatively long transmitter codes can be used to improve the signal-to-noise ratio if that becomes necessary. Also, nonlinear gain compression, which has been described in detail previously,[1,2] can be used to improve the sidelobe levels obtained with this imaging technique. Finally, this synthetic aperture technique gives a factor of two improvement in transverse resolution[1,2] over that of a conventional imaging system which transmits a plane wave and focuses on reception.

LONGITUDINAL WAVE IMAGING

In order to illustrate the performance of this synthetic- aperture imaging technique, a test object was made which consisted of three 1 mm diameter wires placed in front of a metal block in a water tank. Using this test object, a set of images was produced using real-time and computer-reconstruction techniques, 8 and 32-element acoustic transducer arrays, and both linear and nonlinear processing. This particular object was chosen in order to demonstrate the capability of this synthetic-aperture technique to image both specular reflectors (the surface of the metal block) and non-specular reflectors (the wires). Images have also been made of a single small wire so that quantitative measurements of the resolution and sidelobe levels could be taken.

One such image is shown in Fig. 3, where the 32-element real- time imaging system was used to look at a .25 mm diameter wire,

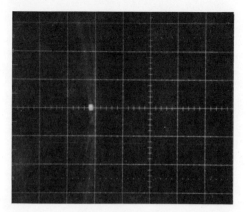

Fig. 3.   32-element real-time image of a .25 mm wire (5 mm
          per division).

placed at a range of 6 cm from the acoustic transducer array
which has a center frequency of 3.3 MHz and an aperture width
of 1.6 cm . In this, as in all previous experiments, the
measured transverse resolution was in agreement with theoretical
(diffraction-limited) prediction of 0.9 mm , obtained from the
equation[1,2]

$$\Delta x \; = \; \frac{z\lambda}{2D} \; . \tag{1}$$

The range resolution, which is somewhat better than 1 mm , is
determined principally by the duration of the transducer impulse
response or equivalently by the transducer bandwidth. The measured
sidelobe level of -16 dB , far away from the mainlobe, is encouraging, but it is still somewhat less than the -22 dB sidelobe
level which is expected based on the computer-reconstruction
results. However, there is no reason to doubt that the performance
of this real-time imaging system will eventually equal that already demonstrated with computer reconstruction of images.

     After these initial quantitative measurements had been made,
a more complicated test object was chosen in order to more graphically demonstrate the capabilities of this synthetic-aperture
imaging technique. The test object and the imaging geometry are
shown schematically in Fig. 4. The metal block was placed about
7 cm from the acoustic transducer array which has a 1.6 cm
aperture. The scales are given by the grid lines in each of the
images so that the spacing of the wires and block can be determined, and the resolution can be observed.

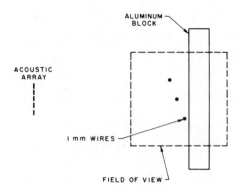

Fig. 4.   Schematic view of test object in the water tank.

Fig. 5 shows a set of images obtained using computer reconstruction with both 8 and 32-element acoustic transducer arrays
and using linear and non-linear processing to look at the test
object described previously. Fig. 6 shows the real-time equivalents
of the images in Fig. 5, although the 32-element, real-time nonlinear result was not available at the time this paper was written.
In all of these images, the three wires and the front surface of
the metal block are visible, although in some of the 8-element
images, one of the wires is nearly obscured by the sidelobes from
the block. In comparing these images to one another, it is clearly
evident that there is a significant reduction in sidelobe levels
as more transducer elements are used and as nonlinear processing
techniques are employed.

There are a number of significant details apparent in the
images of Figs. 5 and 6 which had not been observed with the earlier
chirp-focused imaging system.[5] Since the metal block acts as a
specular reflector, its image is only seen over a length which is
approximately equal to the width of the array aperture, 1.6 cm .
In addition, the surface of the metal block acts as an acoustic
mirror and the wires are seen both as true images in their correct
spatial positions to the left of the block and as virtual (mirror)
images to the right of the reflecting surface. There is also a
secondary image of each of the wires which appears just to the
right of the main image. This secondary image results from the
converted surface wave which travels around the circumference of
the wire and returns to the array somewhat delayed in time from
the main echo.[6] This result is in agreement with direct measurements of pulse echoes from a wire. These secondary images may
be useful in the interpretation of an acoustic image for determining the size and nature of a flaw.

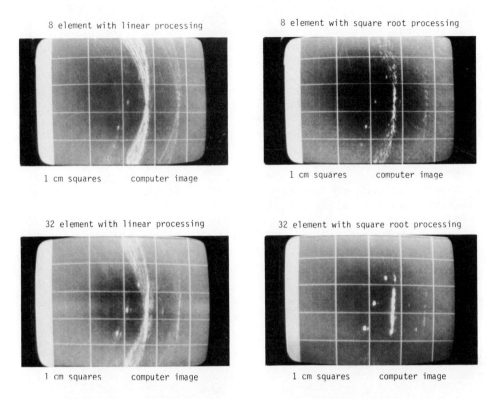

Fig. 5.   Computer-reconstructed images.

RAYLEIGH WAVE IMAGING

A number of techniques for the excitation of Rayleigh (surface) waves in solids have been investigated. Looking at wires in a water tank, as has been done in the past, is very convenient during the early stages of development of an imaging system, but for non-destructive evaluation (NDE) applications, bulk (longitudinal and shear) waves in solids, Rayleigh (surface) waves and other types of Lamb waves are more likely to be of interest. Recently, a new type of array, the edge-bonded transducer (EBT) array, has been developed. This EBT array is a major breakthrough for the excitation of Rayleigh waves and several computer-reconstructed images obtained with this new type of array will be shown.

Several years ago, Kino and his colleagues developed an acoustic imaging system based on a phased array technique which used chirp-focusing.[5,7] With this system, it was demonstrated that various types of waves, including shear, Lamb, and Rayleigh waves could be excited using mode conversion from a longitudinal wave array and that imaging could be performed using these types of

acoustic waves. All of these techniques are applicable to the
new synthetic-aperture imaging system and the wedge-coupling technique has been used with this system in order to obtain Rayleigh [8]
wave images of both real and simulated flaws in aluminum samples.

Fig. 7 shows an image obtained using this wedge coupling technique with the 8-element real-time imaging system to examine a
7 mm long fatigue crack in an aluminum sample. This crack was
tightly closed so that it was just barely visible to the unaided
eye, but it shows up quite clearly in this acoustic image. One
drawback to this technique is that it introduces aberrations due
to the compound angles involved, which in turn lead to a degradation
in the performance of our imaging system. In the past, this degradation was not noticeable, but as the performance of these
imaging systems is improved from a resolution of 2 mm to a resolution of approximately 0.5 mm , this effect becomes significant.

8 element Real-time with square root compression

5 mm per division

8 element Real-time with linear processing

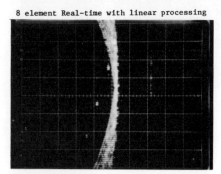

5 mm per division

32 element Real-time with linear processing

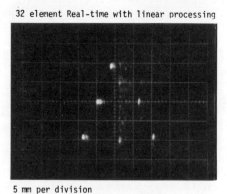

5 mm per division

Fig. 6. Real-time images.

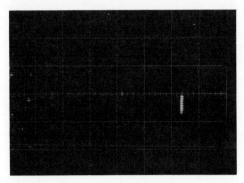

Fig. 7.   Real-time image of a   7 mm   long fatigue crack (5 mm
          per division).

        The edge-bonded transducer array has been developed to over-
come these problems. This array has a number of important advantages
over other techniques used for the excitation of Rayleigh waves.
First of all, since the transducer elements are unslotted and do
not require extra matching layers, the EBT array is very simple to
construct and it gives both high efficiency and broad bandwidth.
Secondly, as it is used in a configuration where a Rayleigh wave
on a solid substrate excites a Rayleigh wave on a test object of
the same material, where there is no difference in acoustic velocity,
the EBT array is free from the aberrations introduced by the wedge
coupler formerly employed.

        The array is shown schematically in Fig. 8, and it is described
in detail elsewhere.[4] The array is constructed by epoxy-bonding
a slab of shear polarized ceramic onto an aluminum substrate.
Thin film electrodes are deposited on the back of the ceramic which
is one-half of a Rayleigh wavelength in thickness. The electrodes
are approximately one wavelength long in the direction perpendicular
to the surface of the substrate; therefore they efficiently excite
a Rayleigh wave whose penetration depth is on the order of a
wavelength. Each element is approximately one wavelength wide and
the individual elements are shielded from one another by ground
strips. Our first edge-bonded transducer array has exhibited
high efficiency(14 dB round trip insertion loss), broad bandwidth
(about an octave), and wide angle of acceptance ($\pm 35°$) . The
spurious modes excited by this first array are quite high, but we
have already developed a number of techniques for reducing
these spurious signals to acceptable levels in our future arrays.

        The edge-bonded transducer array can be used to excite Ray-
leigh waves in a test sample by placing the test sample on top of
the array substrate with a thin strip of coupling material (typically plastic or rubber) placed between them as shown in Fig. 8.

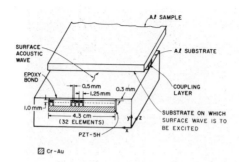

Fig. 8.   Edge-bonded transducer array.

        The loss in going from one substrate to the other has been
measured to be approximately 2 dB , which adds another 4 dB
to the round trip insertion loss. Computer-reconstructed images of
several test objects were obtained using this EBT array. Fig. 9
shows an image of four center-punch marks and one drilled hole on
the surface of the array substrate. This image was obtained using
square root gain   compression to reduce the sidelobes from
the end of the block. Three of the defects show up quite clearly,
while the last two defects are difficult to see amidst the clutter
from the end of the substrate.

        Figs. 10 and 11 show images obtained using the EBT array to
look at an electron discharge machined (EDM) slot in a separate
test sample. In Fig. 10 the slot was placed parallel to the
transducer array, and it shows up quite clearly over its entire
length. In Fig. 11, the slot was placed at a 25° angle to the
array, and once again the slot shows up quite clearly over its
entire length, since the array was able to receive the specular
reflection from the slot. This illustrates the width of the
aperture and large angle of acceptance which can be used with this
imaging system. In the future, the slot will be placed at an even
steeper angle to the array so that only the ends of the slot will
be observed. The clutter in this image is due mainly to the
spurious modes which are excited by the EBT array, which will be
eliminated in a later version of the array.

        The success of the edge-bonded transducer array has been
encouraging and the development of this new type of array is continuing. In the near future, the EBT array will be coupled to the
new 32-element real-time imaging system and it is expected that
this will lead to a much more rapid development of Rayleigh wave
imaging techniques.

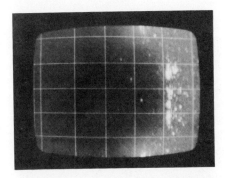

Fig. 9. Rayleigh wave image of center-punch marks in aluminum
          substrate (1 cm squares).

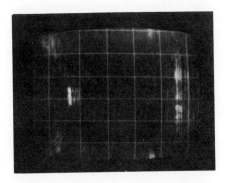

Fig. 10. Rayleigh wave image of an EDM slot placed parallel to the
           transducer array (1 cm squares).

COMPUTER PROCESSING

     Real-time and computer-reconstruction techniques have been
employed to produce images of wires in a water tank and it has been
demonstrated that good resolution and low sidelobe levels can be
obtained. As this work has progressed, it has become evident that
there are serious limitations to the use of intensity display of
images obtained from a high quality imaging system. One problem
is with the limited grey scale range of many display monitors which
makes it difficult to observe a weak reflector in the vicinity of
a strong reflector. This method of display also makes it hard to
obtain quantitative information about the image and it is expected
that as the nondestructive evaluation (NDE) field progresses, more
quantitative information will be required from an image in order to
make a better determination of the shape, size, and nature of a flaw.

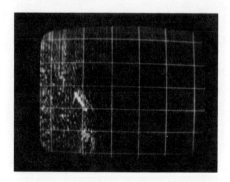

Fig. 11.   Rayleigh wave image of an EDM slot placed at   25°   to
the transducer array (1 cm squares).

There are a number of simple examples which can be used to
illustrate how quantitative measurements can help in the interpretation of an image obtained from a flaw. First, consider a
point defect which gives rise to a scattered field that falls off
inversely as the distance from the flaw. If the defect has a higher
impedance than the surrounding medium, the reflected signal will
have the same sign as the incident signal, while in the case of a
lower impedance defect, such as a void, the reflected signal
will be opposite in sign to the incident signal. Now, consider
the case of a wave scattered from a crack which is aligned at an
angle to the wave so that the specular reflection cannot return to
the transducer array. In this case, only the two ends of the
crack are observed and they appear in the image as two isolated
points.[7,9] However, the fields associated with these two radiating points fall off inversely as the square root of the distance
from the end of the crack, and furthermore, the scattered signals
from the two ends of the crack are of opposite sign. It can
be seen that there is quantitative information available from this
imaging technique, which if adequately displayed, could be utilized
to aid in the interpretation of the images produced. Accordingly,
a contour plotting program, which at the present time gives
magnitude plots of the image, but which at a later time might
give plots of both amplitude and phase, has been written. This
contour plotting program has proven to be quite useful in making
quantitative measurements of the performance of this imaging
system.

In order to measure the point-spread function of this imaging
system, a small (.25 mm diameter) wire was placed in a water tank
at a range of 3.2 cm from a 32-element transducer array having

a 1.6 cm aperture and a 3.3 MHz center frequency. By using
computer reconstruction techniques, an image of this wire was produced. From this image, a contour plot was obtained which shows
lines of constant amplitude around the wire. The result shown in
Fig. 12 was obtained using 32 transducer elements with linear
processing. Here, 3 dB resolutions of .5 mm in the transverse
direction and .4 mm in the range direction are seen. These results
are essentially in agreement with the theoretical predictions of
approximately .5 mm resolution in both the range and transverse
directions. This contour plot shows the sharp fall-off from the
main lobe down to a background sidelobe level of about -24 dB .
It is important to note the smooth contours around the main lobe.
Others have predicted that there might be higher sidelobe levels
at a 45° angle to the axis of the array,[10] but it is apparent
from the results shown that there is no such problem with this
system. Fig. 13 shows a similar contour plot obtained using the
nonlinear processing technique which has been discussed in the
past.[1,2] This technique, using square root gain compression, is
expected to give a significant reduction in the sidelobe levels.
This can be seen quite clearly in the contour plot, with the
background sidelobe level reduced to about -36 dB . Again, there
is a sharp fall-off from the main lobe, and with this nonlinear
processing technique, the resolution in both the range and transverse directions is actually somewhat improved over that obtained
with linear processing.

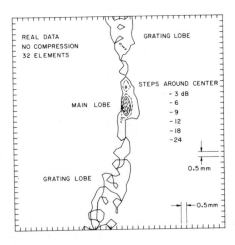

Fig. 12. Contour plot of the image of a .25 mm wire obtained
          with linear processing.

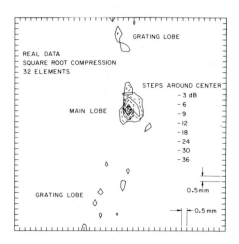

Fig. 13.   Contour plot of the image of a   .25 mm   wire obtained
           with nonlinear processing.

CONCLUSIONS

     A 32-element real-time (30 msec frame time) synthetic-aper-
ture acoustic imaging system has been developed and acoustic images
have been obtained. These initial results, comparable to those
obtained using computer reconstruction, have been very encouraging.
Both real-time and computer-reconstruction techniques have been
applied to longitudinal wave imaging in water and these techniques
are now being applied to Rayleigh wave imaging of solids. A new
edge-bonded transducer array has been developed for the excitation
of Rayleigh waves on a solid substrate and both this array as well
as the wedge-coupled array have been used to obtain Rayleigh wave
images of surface cracks and holes in aluminum samples. Accurate
measurements of the performance of this imaging system have been
made using computer-generated contour plots. This synthetic-
aperture imaging technique has demonstrated a diffraction-limited
resolution which is better than that of any comparable system
described in the literature, and at the same time, low sidelobe
levels have been obtained, particularly with nonlinear processing.

ACKNOWLEDGMENTS

     This work was sponsored by the Center for Advanced NDE
operated by the Rockwell International Science Center for the
Advanced Research Projects Agency and the Air Force Materials
Laboratory under Contract RI74-20773 and through the Air Force
Office of Scientific Research under Contract F49620-79-C-0217.

REFERENCES

1. P. D. Corl, G. S. Kino, C. S. DeSilets and P. M. Grant, "A Digital Synthetic Focus Acoustic Imaging System," to be published in Acoustical Imaging and Holography, Vol. 8, A. F. Metherell (ed.), Plenum Press, New York, (1978).

2. P. D. Corl, P. M. Grant, and G. S. Kino, "An Imaging System for NDE," Proc. of the 1978 IEEE Ultrasonics Symposium, Cherry Hill, New Jersey.

3. C. S. DeSilets, A. Selfridge, and G. S. Kino, "Highly Efficient Transducer Arrays Useful in Nondestructive Testing Applications," Proc. of the 1978 IEEE Ultrasonics Symposium, Cherry Hill, New Jersey.

4. G. S. Kino, B. T. Khuri-Yakub, A. Selfridge, and H. Tuan, "Development of Transducers for NDE," to be published in Proc. ARPA/AFML Review of Progress in Quantitative NDE, La Jolla, California, July, 1979.

5. T. M. Waugh, and G. S. Kino, "Real Time Imaging with Shear Waves and Surface Waves," Acoustical Holography, Vol. 7, L.W. Kessler (ed.), Plenum Press,New York, 103-115, (1977).

6. H. Uberall, L. R. Dragonette, and L. Flax, "Relation Between Creeping Waves and Normal Modes of Vibration of a Curved Body," J. Acoust. Soc. Am., Vol. 61, No. 3, March, 1977.

7. G. S. Kino, "Acoustic Imaging for Nondestructive Evaluation," Proc. IEEE, Vol. 67, 510-527 (1979).

8. G. S. Kino, "New Techniques for Acoustic Non-Destructive Testing," ARPA-AFML Interdisciplinary Program for Quantitative Flaw Detection, Contract F33615-74-C-5180, Annual Progress Report, January 1979.

9. J. D. Achenbach, A. K. Gantesen, and H. C. McMaken, "Diffraction of Elastic Waves by Cracks - Analytic Results," Elastic Waves and Nondestructive Testing of Materials, AMD-Vol. 29, Y. H. Pao (ed.), ASME (1978).

10. A. Macovski, "Ultrasonic Imaging Using Arrays," Proc. IEEE, Vol. 67, 484-495 (1979).

TRANSMISSION IMAGING: FORWARD SCATTERING AND SCATTER

RECONSTRUCTION

C. H. Chou, B. T. Khuri-Yakub,
and G. S. Kino
Ginzton Laboratory
Stanford University
Stanford, CA. 94305

ABSTRACT

We give here a theory for the nature of the image if produced
by the scanning laser acoustic microscope (SLAM) at a surface in
the far field of the flaw. A physical model and an exact calculation of scattering of longitudinal acoustic waves by spherical
defects are presented. The total displacement field (incident plus
scattered) is evaluated at planes at a distance from the defect by
using the Ying and Truell scattering theory. It is found that
evaluation of the exact type and size of a defect is not possible
from direct observation of the total field. A holographic type
reconstruction technique using either 2-dimensional Fourier transform, or Rayleigh-Sommerfeld diffraction formulae, was used to
reconstruct the field at the location of the defect from a knowledge of the field at an observation plane, a distance z away
from the defect. It is found that accurate defect characterization
can be made using this reconstruction technique.

INTRODUCTION

A discussion was given at the 1979 Conference on the Review
of Progress in Quantitative NDE by Yuhas [1a] of the results obtained with the Scanning Laser Acoustic Microscope (SLAM). He
described how this device had been employed to observe various
types of defects in NC132 Si_3N_4 ceramics. It is the purpose of
this paper to discuss his observations and their interpretation
and to show how computer-type holographic reconstruction techniques could be used to determine this size and nature of the flaw.

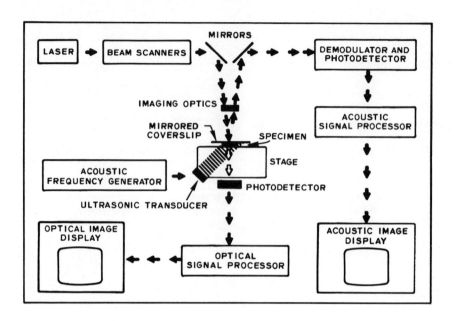

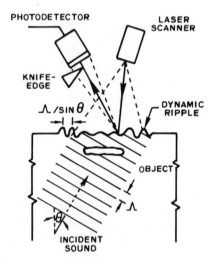

Fig. 1.   (a) Schematic diagram of scanning laser acoustic micro-
          scope; (b) illustration of laser detection of acoustic
          energy at surface.   Ref. 1b.

The SLAM device uses a plane shear or longitudinal wave to insonify the flaws as illustrated in Fig. 1.[1b] The laser beam detects the dynamic ripple of the ceramic surface due to the sum of the directly transmitted acoustic waves and the waves scattered from the flaw. The definition of the system is that of the laser beam itself, which is focused by an optical lens. So a void very near the surface casts a shadow and can be observed with good definition.

It is important to realize that except when the flaw is very close to the surface of the ceramic, the scattered signal observed is in the far field or the Fraunhöfer region of the wave scattered from the flaw. Consider for instance, a 100 MHz shear wave scattered from a flaw in the shape of a flat disk of radius a . For Si_3N_4 the shear wave velocity is 6.25 mm/μsec and the wavelength $\lambda = 60$ μm . So the distance z_F to the boundary of the Fresnel zone is $z_F = a^2/\lambda$ or $z_F = 0.7$ mm for a 400 μm diameter flaw. For smaller diameter flaws or longitudinal waves z_F is still smaller than this value.

In this paper we first consider some simple examples of scattering theory. We then employ an exact calculation, from the work of Ying and Truell[2] and Johnson and Truell[3], of the scattering from spherical voids or inclusions. This allows us to determine the total field at a plane a distance z from the flaw. This is equivalent to determining the displacement u_z at the surface of the ceramic which is detected by the Kessler microscope.

In these examples, a reference beam u_i and a scattered beam u_s are present. Thus by using holographic reconstruction techniques it should be possible to determine the size and nature of the flaw. We have demonstrated the use of such computer-type reconstruction techniques on this theoretical problem. So for the first time we have been able to show, theoretically, how acoustic imaging methods should perform on different types of flaws. In particular, we show how the reconstructed image differs for voids and inclusions. The examples which we have chosen to illustrate involve the scattering of a longitudinal wave incident normally on the surface of the substrate. But the techniques employed are equally applicable to shear waves incident at an oblique angle; the results would be similar to those given in this paper.

SIMPLE PHYSICAL PICTURE OF SCATTERING FROM VOIDS AND INCLUSIONS

Voids

We regard a void in the form of a disc, to be cutting off a portion of the beam of radius a ; so it is as if there is a piston source of radius a emitting a wave with an amplitude displacement $-u_i$ equal and opposite in amplitude to the incident wave on the disc. Here, for simplicity, we also consider the disc to be placed with its plane parallel to the surface of the ceramic and normal to the incident plane wave. Within the Fresnel region, where $z < a^2/\lambda$, this piston source emits a beam which has both axial and radial variations in amplitude and has a varying phase with respect to the primary transmitted beam. For instance, on axis, the incident beam has an amplitude $u_i e^{-jkz}$, whereas it can be shown by using the analogy for compressional waves, that a piston transducer emits a beam which has an amplitude on its axis $u_s(z)$ given by the relation[4]:

$$u_s(z) = -u_i e^{-jkz}(1 - e^{-jka^2/2z})$$ (1)

where z is the distance from the disc, $k = 2\pi/\lambda$ the propagation constant of the wave, and λ its wavelength.

It follows that on axis the total amplitude of the wave due to the primary and scattered beams is

$$u(z) = u_i e^{-jkz} + u_s(z) = u_i e^{-jkz} e^{-jka^2/2z}$$ (2)

So there is a bright spot on the axis of the same intensity as the incident beam, which only will exist if the disc is perfectly circular. The same results would be expected to hold for a spherical void or axially symmetric elliptical void, for it casts essentially the same shadow as a disc of the same diameter.

In the Fraunhöfer region $(z \gg a^2/\lambda)$, it can be shown that the scattered field has the form[4,5]:

$$u_s(r,z) = -\left(ju_i \frac{2\pi a^2}{z\lambda}\right)\left(\frac{J_1(2\pi ar/\lambda z)}{2\pi ar/\lambda z} e^{-jkz} e^{-j\pi r^2/z\lambda}\right)$$ (3)

where $J_1(x)$ is a Bessel function of the first kind and first order, and r is the radial distance from the z-axis.

This scattered beam has as its first zeroes at r_0, where

$$r_0 = 0.61\lambda z/a$$ (4)

We note that for a 350 μm diameter flaw situated at 5 mm from the surface of the ceramic, i.e., roughly in the middle of a 1 cm thick sample, a probing 100 MHz beam $(\lambda = 110 \ \mu m, ka = 10)$ gives $r_0 = 1.9$ mm. In this case $u_s(0,z) = .175|u_i|$ and

$u_s(0,z)$ is $\pi/2$ out of phase with $u_i(z)$. But at
$r = (z\lambda/2)^{1/2} = 524$ μm , $u_s(z)$ has the opposite sign to $u_i(z)$.
So the total amplitude of the wave drops to $.86|u_i|$. Thus the
amplitude variation of the total signal at the surface of the
ceramic is dependent to a large extent on the phase variation term
$\exp -j\pi r^2/z\lambda$ and the flaw size. We conclude that the flaw should
appear in the image, because even at distances far into the Fraunhöfer region, the amplitude change due to the scattered field can
be quite large. However, the correlation of the results with the
flaw diameter is likely to be quite difficult.

Inclusions

Consider now tne effect of an inclusion on the transmitted
beam. If the inclusion is made of material with a lower velocity
of propagation than the surrounding matrix, it will act as a
convex lens of focal length f . As an example, it can be shown
from the paraxial focusing theory that a sphere of radius a
would have a focal point a distance f from the center of the
sphere, where

$$f = \frac{an}{2(n - 1)} \tag{5}$$

and n is the relative refractive index of the sphere in the
surrounding medium.

Consider, for instance, a silicon inclusion of 350 μm
diameter in a silicon nitride matrix. In this case, the longitudinal wave velocities are respectively $v = 8.98$ mm/μsec and
11.3 mm/μsec ; so $n = 1.26$ and $f = 2.42a$ or 420 μm . Thus
there is a true focus outside the silicon. For an iron inclusion,
$v = 5.95$ mm/μsec , $n = 1.90$, $f = 1.06a$ or 185 μm . So the
calculated focal point is just outside the sphere. We conclude
that if the inclusion is located near the surface of the ceramic
there will be an intense focal spot on the axis. At distances
far from the inclusion, the beam passing through it spreads and
its intensity is very small. So the image observed should be
very like that of a void, and a void or inclusion would be indistinguishable from each other.

SCATTERING CALCULATION

Symmetrical System

The scattering calculation is carried out for the case of
plane longitudinal wave incident on a spherical inclusion of
radius a , as illustrated in Fig. 2. We restricted our calculation to the case of $ka = 10$ in Si_3N_4 when the longitudinal

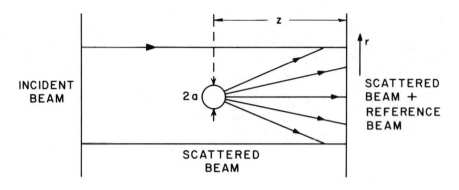

Fig. 2.   Scattering geometry from a spherical defect.

acoustic wave frequency is 100 MHz, corresponding to a defect of
radius 175 μm . We use the theory developed by Ying and Truell[2]
and Johnson and Truell[3] for scattering from spherical inclusions.
The calculation was carried out using a PDP 11-34 minicomputer.
We have calculated u_z the component of displacement in the direction of the incident wave on a plane of constant z . The value
of the scattered field and the total field $u_{zi} + u_{zs}$ is evaluated
at different planes of constant z in the near field and far field
of the defect.

    Fig. 3a shows the total field after scattering from a spheri-
cal void at the plane kz = 10 in a host material of Si$_3$N$_4$. This
would correspond to the observation plane being tangent to the
defect. It will be noted that a shadowing effect is observed
indicating that, indeed, no acoustic energy is transmitted through
the void. The top part of Fig. 3a shows the total field along a
diametrical line in the observation plane while the lower part of
Fig. 3a shows the intensity modulation of the field as would be
displayed in an intensity image. This ring structure is typical
of the scattering from spherical inclusions observed in actual
results from the SLAM. Fig. 3b shows the total field behind a
void at the plane kz = 50 (z = 875 μm) . It will be seen that
the shadowing effect is now completely absent, but a bright spot
is present in the center of the defect, as indicated by the
simplified theory. The loss of shadowing is further apparent in
Fig. 3c showing the field at the plane kz = 100π (z = 5.5 mm) .
We note that the first minimum of intensity occurs roughly where
predicted by the approximate theory and the amplitude is approximately that predicted.

    Fig. 4a shows the total field after scattering from a spherical
silicon inclusion at the plane kz = 10 . Notice the strong transmission through the inclusion and the sharp focusing observed due

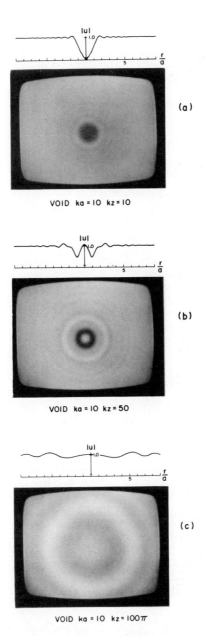

Fig. 3. Total field cross-section and intensity modulation from
a spherical void at (a) kz = 10 ; (b) kz = 50 ;
(c) kz = 100π .

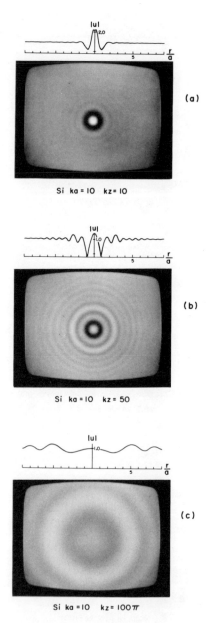

Fig. 4. Total field cross-section and intensity modulation from
         a spherical silicon inclusion at (a) kz = 10 ;
         (b) kz = 50 ; (c) kz = 100π .

to the spherical lens formed by the silicon. The result obtained
is quite different from the equivalent one for a void.

The field on the axis of the silicon defect is plotted in
Fig. 5. It will be seen that the axial field amplitude passes
through a maximum, as predicted by the simple focusing theory.
However, the maximum is nearer to the defect than the simple calculation indicates due to the fact that the paraxial theory is not
very accurate, and because there are multiple reflections of the
waves inside the inclusion. The period of the "wiggles" in the
total field inside the defect correspond very well to a half wavelength of a longitudinal wave in silicon.

Figs. 4b and 4c show the total field after scattering from the
Si inclusion at $kz = 50$ $(z/a = 5)$ and $kz = 100\pi$ $(z/a = 10\pi)$,
respectively. Notice that the transmission through the sample is
still observed and that the total field at $kz = 100\pi$, at
least to $r/a = 10$, is indistinguishable from that of a void.

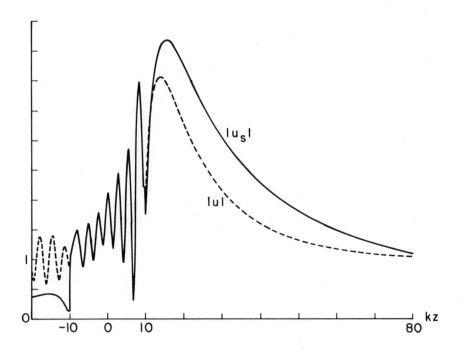

Fig. 5.   On axis total and scattered fields from a spherical
          silicon inclusion versus distance.

Figs. 6a, 6b, and 6c show the total fields for the case of an Fe inclusion for the same distances from the defect as in Figs. 3 and 4. Notice the same type of focusing effect is present as with a Si inclusion and that, again, in the far field, the total field is very similar to that of a void or Si .

Thus, we conclude from Figs. 3, 4, 5, and 6 that it is difficult to estimate the size or type of defect from direct observation of the diffraction pattern of the display at distances well into the Fraunhöfer region of the defect.

IMAGE RECONSTRUCTION OF THE FLAW

Hankel Transform Technique

We have used two techniques for image reconstruction. The first makes use of the symmetry of the problem and employs a Hankel transform technique. The second method employs the more general Rayleigh-Sommerfeld[5] approach and does not require using the symmetry of the problem. This latter technique, or a 2D Fourier transform technique, are obviously the more suitable ones to employ for image reconstruction from experimental data.

We consider only the scattered longitudinal wave components by assuming that in an experiment, a pulsed rf signal can be employed so the shear wave signals can be separated from the longitudinal wave signal. In this case we can treat the problem by using the longitudinal wave equation. A cylindrically symmetric solution of this equation for u_{zs} can be written in the form

$$u_{zs}(r,z) = \int_{\alpha=0}^{\infty} F_{\alpha} e^{-j(k^2 - \alpha^2)^{\frac{1}{2}}z} \, J_0(\alpha r) \, \alpha d\alpha \qquad (6)$$

where α is the radial propagation constant and F_{α} is the amplitude of the spatial frequency component of the spectrum with radial wave number α .

If $u_{zs}(r,z_0)$ at the plane z_0 is known, it follows from Hankel transform theory that

$$F_{\alpha} = e^{j[(k^2 - \alpha^2)^{\frac{1}{2}}z_0]} \int_0^b u_{zs}(r,z_0) \, J_0(\alpha r) \, r dr \qquad (7)$$

where b is the radius of the image region being investigated, and we have assumed that $u_{zs} = 0$ for $r > b$.

It follows that the displacement at the plane z can be determined by first employing Eq. 7 to determine F_{α} and then carrying out a further Hankel transform using Eq. (6) to find $u_{zs}(r,z)$ at any plane z .

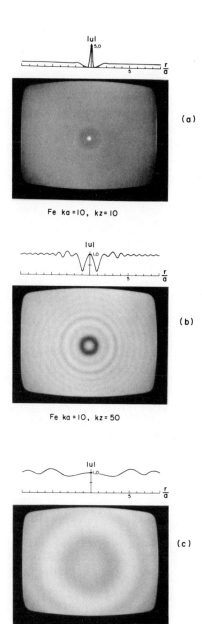

Fig. 6. Total field cross-section and intensity modulation from a
spherical iron inclusion at: (a) kz = 10; (b) kz = 50;
and (c) kz = 100π .

By carrying out this analysis for a point scatterer, it is possible to find the form of the reconstructed field. It can be shown most easily by use of the equivalent Rayleigh-Sommerfeld scattering theory that if we use the paraxial approximations $r^2 \ll z_0^2$, the reconstruction of a point source is of the form

$$u_{zs}(r,z) \;=\; AJ_1(krb/z_0)/(krb/z_0) \tag{8}$$

where A is a constant.

In this case, it follows that the radius r_{3dB} to the 3 dB points is

$$r_{3dB} \;\approx\; 0.25 \; \lambda z_0/b \tag{9}$$

and to the zero amplitude points (r_0) is

$$r_0 \;=\; 0.61 \; \lambda z_0/b \tag{10}$$

This gives some measure of the radius of the imaging region required to give good definition of the flaw. As an example, with ka = 10 , $z_0 = b$, $kr_0 = 3.83$. Thus, the definition is adequate for imaging the flaw.

We assume that the available information at plane z is the magnitude of u_z . Thus, our problem is that of holographic reconstruction. It is therefore convenient to write

$$|u_z|^2 \;=\; |u_{zi} + u_{zs}|^2 \;=\; |u_{zi}|^2 + |u_{zs}|^2 + u_{zs}u_{zi}^* + u_{zs}^* u_{zi} \tag{11}$$

where u_{zi} and u_{zs} are the incident and scattered fields, respectively. Our interest is in reconstructing u_{zs} at the defect.

We define the quantity $\phi(r,z)$ given by the relation

$$\phi(r,z) \;=\; |u_z|^2 - |u_{zi}|^2 \;=\; |u_{zs}|^2 + u_{zs}u_{zi}^* + u_{zs}^* u_{zi} \tag{12}$$

If the observation plane z is in the far field of the defect, $|u_{zs}|^2$ can be neglected in comparison to $u_{zs}u_{zi}^*$. When we are reconstructing u_{zs} at the plane of the defect (z = 0) , the contribution due to u_{zs}^* can be neglected because it would correspond to an unfocused contribution from the mirror image to the defect that is located at a distance of 2z . It is thus possible to write

$$\phi(r,z) \approx u_{zs}u_{zi}^{*} \tag{13}$$

Thus, we can employ the quantity $\phi(r,z_0)$ in the reconstruction process. If this is done, we actually reconstruct $u_{zs}(r,z)$. We can always, however, determine the total field $u_z \stackrel{\geq}{=} u_{zs} + u_{zi}$ by a simple addition. Thus, the procedure is:

1. Find $|u_z|$;

2. calculate $|u_z|^2$;

3. subtract $|u_{zi}|^2$ to find ϕ (this is equivalent to removing the $\alpha = 0$ component in the spatial frequency spectrum);

4. carry out the Hankel transforms to find $u_{zs}(r,z)$;

5. find $u_z(r,z)$ if required.

If both amplitude and phase information is available, the procedure can be simplified to find $u_{zs}(r,z)$. In this case, the artifact due to the mirror image does not occur. In all cases, we are reconstructing the displacement field due to the defect, not the shape of the defect itself.

Figure 7 illustrates the reconstruction process for a spherical void located with its center at a normalized distance $kz = 100\pi$ from the surface where the image is formed. $|u|^2 - 1$ (set $|u_i| = 1$) is plotted at tht top of Fig. 7 from the results given in Fig. 3. F_{α} is shown in the next curve followed by the reconstruction of the scattered and total displacement fields at $kz = 0$, the center plane of the flaw. It will be seen that the total reconstructed field exhibits a deep shadow, whose width is exactly $2a$. Thus, the diameter of the void can be very clearly determined by this reconstruction technique. Furthermore, when reconstruction is carried out along the axis, the total field in the neighborhood of the flaw is close to zero and gradually increases with z to the amplitude of the incident field.

Fig. 8 shows a comparison of the field at $kz = 10$ from the exact calculation and the reconstruction. It is clearly seen that for a region outside the flaw, the reconstruction result compares well with that of the exact calculation.

Figs. 9 and 10 show the results of reconstruction for spherical Si and Fe inclusions. In both cases, when $kz = 0$, a zero in the total field is observed at $r/a = \pm1$, indicating a defect with a diameter of $2a$. The zero in the total field at $r/a = \pm1$ is present because very little energy is transmitted through the defect at that location. The lack of transmission is due to the

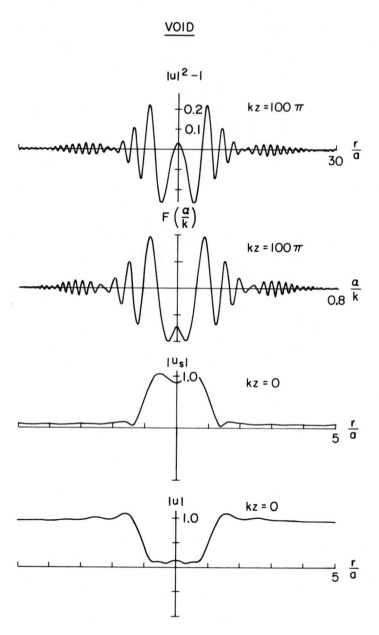

Fig. 7.   Scattered field at kz = 100π , and the reconstructed
scattered and total fields at kz = 0 for a spherical
void.

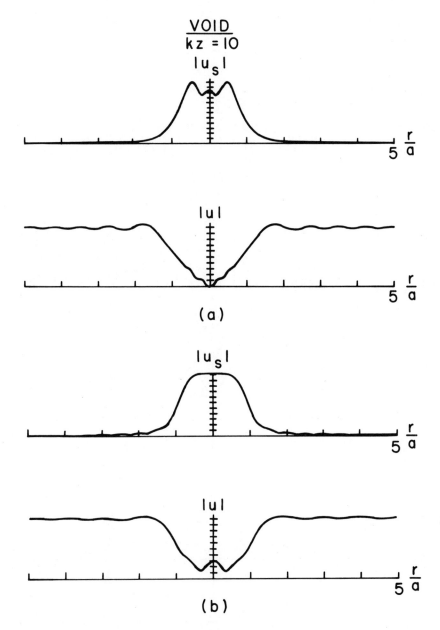

Fig. 8. Comparison of (a) exact total and scattered fields,
         and (b) reconstructed total and scattered fields,
         at kz = 10 for a spherical void.

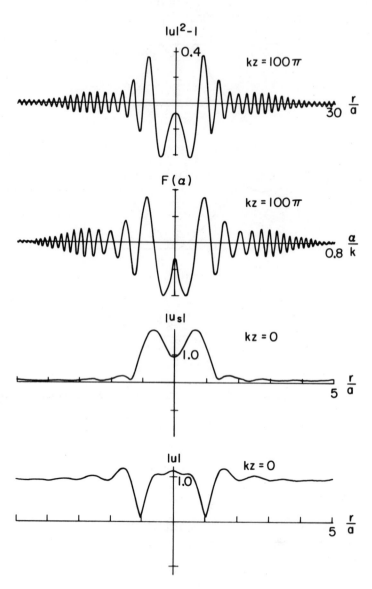

Fig. 9. Scattered field at $kz = 100\pi$, and the reconstructed scattered and total fields at $kz = 0$ for a spherical silicon inclusion.

IRON

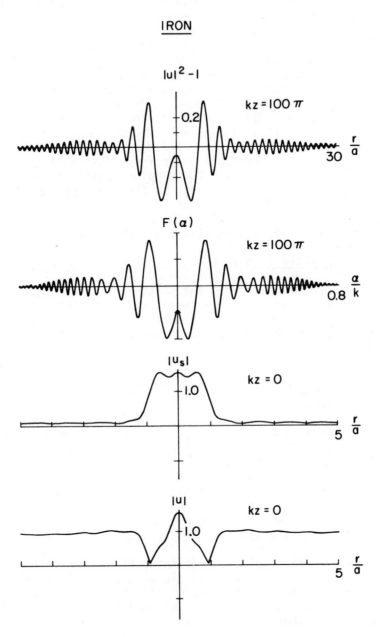

Fig. 10. Scattered field at $kz = 100\pi$, and the reconstructed scattered and total fields at $kz = 0$ for a spherical iron inclusion.

fact that the velocity of the wave in the defect material is much lower than that in silicon nitride; hence, the component of the wave incident at the edge of the defect is refracted into the material. The component of displacement in the z direction near this point is thus very small. From this argument, it seems plausible to expect the edges of an inclusion to be delineated by zero total field even if the defect is not spherical. Figs. 7, 9, and 10 show that there is enough difference between the reconstructed fields to distinguish between a void and a Si or Fe inclusion. With a void, the total field on the axis is very small near the sphere. With the Fe inclusion, a strong field on the axis is observed at a vertical image front just inside the sphere in approximate agreement with the simple paraxial focusing theory, while with Si a less strong peak in field is observed further from the center of the sphere than for Fe . This increased field strength is due to the focusing obtained. The effect is stronger in an Fe inclusion because it has a lower velocity than Si .

Rayleigh-Sommerfeld Technique

In the general case when the scatterer is not symmetrical, it is possible to reconstruct the defect either by using the Rayleigh-Sommerfeld formula or to use two-dimensional Fourier transforms. The Rayleigh-Sommerfeld diffraction formula for monochromatic scalar fields is[5]

$$u_z(r,z) = \frac{1}{j\lambda} \iint_\Sigma u_z(z_0) \frac{e^{jkR}}{R} \cos{(\overline{n},\overline{R})}\, ds$$

where, as illustrated in Fig. 11, R is the distance between two points $\overline{R}_0$ and $\overline{R}_1$, and $\cos{(\overline{n},\overline{R})}$ is the cosine of the angle between the normal to the aperture and the line joining R_0 and R_1 . This formula can be integrated directly and is economical to use when the number of image points and scatter points employed is limited. In this case, we have employed a square array of points enclosing the circular region already described and make no assumptions about the symmetry of the scatterer.

We have carried out such reconstructions using the same scatterers as before. When a sufficient number of image points (8 x 8) is employed in a circular field, the results obtained are indistinguishable from those obtained with the Hankel transform method, as shown in Fig. 12a.

As the number of points is reduced, the quality of the reconstruction deteriorates, as can be seen in Fig. 12b. However, the results appear to be adequate and show very little evidence of problems with grating lobes with as few as 4 x 8 image points.

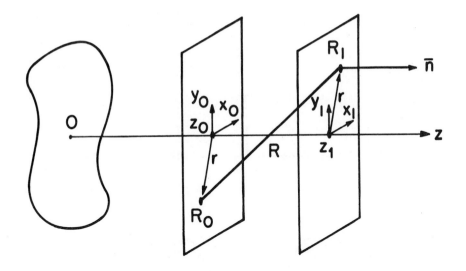

Fig. 11.   Diffraction geometry.

CONCLUSIONS

We conclude that the SLAM microscope should yield information on the nature and size of burried flaws when either laser
or computer type holographic reconstruction techniques are employed.

Similar techniques are, of course, applicable when purely
acoustic detection methods are employed. Thus, it should be
possible to employ a mechanically-scanned focused transducer,
illuminating and receiving from a point on the surface of the
object, to carry out imaging by this technique.

The results obtained show that, as normally expected, it
should be possible to employ holographic imaging techniques to
determine the size of flaws that are present, particularly their
transverse dimensions. In addition, the results indicate that it
should be possible to distinguish between different types of
flaws. It is particularly simple to distinguish between a void
and an inclusion, and the method also yields some information on
the type of inclusion. Furthermore, we have been developing simple
ray tracing techniques which provide information comparable to
that obtained from the exact calculations. This has the advantage
that it should be possible to generalize the approach to deal
with a more general shape of flaw, and carry out the reconstruction
with a limited number of imaging points.

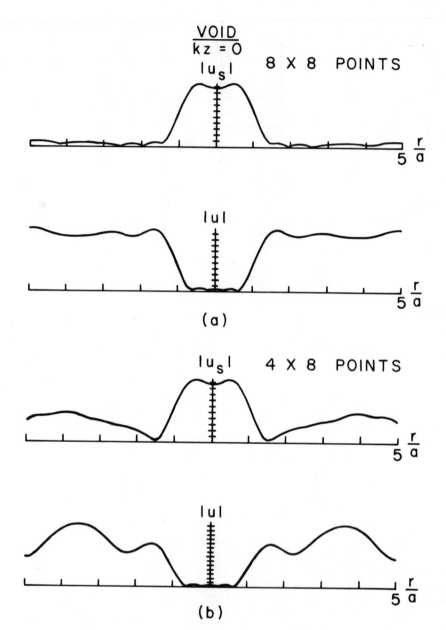

Fig. 12.   Reconstructed total and scattered fields at   kz = 0
           using: (a) 8 x 8 points; and   (b) 4 x 8 points in
           the plane   kz = 100π .

ACKNOWLEDGMENTS

This work was sponsored by the Center for Advanced NDE, operated by the Rockwell International Science Center for the Advanced
Research Projects Agency under Contract R174-20773 and by the
Defense Advanced Research Projects Agency under Contract MDA903-
76C-0250 with the University of Michigan.

REFERENCES

1a. T.E. McGraw, D.E. Yuhas, and L.W. Kessler, "Flaw Charac-
     terization in Silicon Nitride at 100 MHz." ARPA/AF Review
     of Progress in Quantitative NDE, July 8-31, 1979, Scripps
     Institute of Oceanography.

1b. Sonomicroscope ^{TM}, Sonoscan Inc., Bensenville, IL 60106.

2.   C.F. Ying, R. Truell, "Scattering of a Plane Longitudinal
     Wave by a Spherical Obstacle in an Isotropically Elastic
     Solid," J. Appl. Phys., 27, 1086 (1956).

3.   G. Johnson, R. Truell, "Numerical Computations of Elastic
     Scattering Cross-Section," J. Appl. Phys., 36, 3466 (1965).

4.   G.S. Kino, "The Application of Reciprocity Theory to
     Scattering of Acoustic Waves by Flaws," J. Appl. Phys.,
     49, 6 (1978).

5.   J.E. Goodman, "Introduction to Fourier Optics, McGraw-
     Hill, 1968.

ULTRASONIC IMAGING BY RECONSTRUCTIVE TOMOGRAPHY

Glen Wade

Department of Electrical & Computer Engineering

University of California at Santa Barbara

ABSTRACT

The recent development of X-ray computer reconstructive tomography has brought about a revolution in radiology. It has given the practicing physician a new degree of access to what is going on inside a patient's body. The computer has become an important factor for implementing diagnostic techniques in the major hospitals of the world. In recognition of this fact, G.N. Hounsfield and A.M. Cormack were made co-recipients of the 1979 Nobel Award in physiology and medicine.

In conventional radiology, X rays diverge from a single source to project onto film a shadowgraph of the structure along the paths of the rays. But structural elements, cleanly separated in the three-dimensional object, often overlap in the final two-dimensional image in such a way as to make them hard to distinguish. This is particularly true of structural elements of similar density.

In reconstructive tomography there is no overlap. An image is computed from a large number of projections. The image has the form of a two-dimensional mapping of the discrete non-overlapping structural elements in a single plane of the body. Ordinary X-ray technology is combined with sophisticated computer processing to make this possible.

The X-ray source and detector move around the body, and in effect, hundreds of X-ray pictures are made. Instead of being recorded on film, the information is sent to a computer to be

I. INTRODUCTION

It is appropriate for one of the assigned tutorial papers in this volume to be on the subject of reconstructive tomography. The 1979 Nobel Prize for physiology and medicine was awarded to G.N. Hounsfield, a major inventor in the field of X-ray computer tomography, and A.M. Cormack, a major contributor to the solution of its mathematical problems. Their work and that of others in this area has brought about a revolution in radiology. The practicing physician has been given a new degree of access to what is going on inside a patient's body. Today there are an estimated 2,000 computer-assisted tomographic scanners in operation throughout the world. According to the Royal Caroline Medico-Chirurgical Institute in Stockholm, which selected the Nobel Prize winners, "It is no exaggeration to state that no other method within X-ray diagnostics has led within such a short period of time to such remarkable advances in research and in a multitude of applications." The computer has thus become an important factor for implementing new diagnostic techniques in the major hospitals of the world.

In conventional radiology, X rays, coming from a single source, pass through the patient's body and cast on film a shadowgraph of the structure along the paths of the rays. The picture produced is a record of the attenuation of the rays as they are transmitted through the body. This generally results in an overlapping of anatomical features so that some information may be hidden from view. The film itself detects less than 1% of the projected X rays and this fact produces a limitation on how well differences among the various soft tissues of the body can be resolved by this technique. Structures of similar density tend to get in each other's way in the final picture, with the result that they are not well distinguished. In order to overcome these problems with conventional X-ray technology, radiologists often make a number of exposures from different angles in an effort to project the internal parts in different relative positions.

Reconstructive tomography is essentially a systematic extension of this procedure. The patient is viewed by means of X-ray imaging from numerous angles, though the X rays are confined to the vicinity of a single plane of the body. A large number of edge-wide projections of that plane are made and a cross-sectional image of the structure in the plane is calculated by a computer. The resulting image is a two-dimensional mapping of that structure. It is in this way that ordinary X-ray technology is combined with sophisticated computer processing to make reconstructive X-ray tomography possible.

processed by it in making a tomogram. With this approach it is possible in principle to obtain the image of any cross section within the body. The technique has proven to be invaluable for the diagnosis of brain tumors and many other pathologies.

However, it has one disadvantage: it is invasive. It depends on ionizing radiation. Too much exposure to X rays will harm the patient.

But X rays are not the only kind of radiation for which reconstructive tomography is feasible. Microwaves, electron beams, ultrasound, fast subatomic particles from accelerators, gamma rays from such sources are positron annihilation, and even magnetic fields can be used. Ultrasound is particularly attractive in reconstructive tomography, and that is what this paper is all about.

Acoustic energy can often give a view of a cross section not available with X rays or other types of radiation. A mapping of acoustic and elastic variations can be expected to provide a basically different pattern than a mapping of variations in X-ray absorption and scattering coefficients. In addition, a mapping of one kind of acoustic parameter will yield quite a different picture than that of another. So far, two acoustic parameters have received the most attention in research: acoustic attenuation and acoustic refractive index. Mappings of variations in either parameter require transmitting the ultrasound through the object. But ultrasound can also be reflected from objects. Inhomogeneities within an object provide echoes and research is going on to produce mappings of variations in ultrasonic reflection. Thus, computer-assisted acoustic-echo tomography is being examined and also tomographic extensions of Doppler processing.

In reconstructive tomography it is essential to know the paths taken by the rays in going from source to detector. With X rays the paths are all essentially straight and therefore easy to take into account. This is not the case with ultrasound. When an ultrasonic beam propagates through an object, it undergoes deflection (refraction and reflection) at almost every interface between regions of different refractive index. Diffraction of the acoustic waves also occurs. Because of the wavelength differences between X rays and ultrasound, diffraction is far more important for ultrasound than for X rays. Thus refraction, reflection and diffraction are special problems peculiar to ultrasonic reconstructive tomography.

With this approach we can, in principle, obtain the image of any cross section within the body. The technique has proven to be invaluable for the diagnosis of brain tumors and many other pathologies. In some of the more modern systems, the scan time to produce all the edge-wise projections is less than 10 seconds and the computer time to reconstruct an image from the projections is less than 30 seconds.

Figure 1 shows two X-ray tomograms, one displaying a cross section of the head and the other, a cross section of the chest.

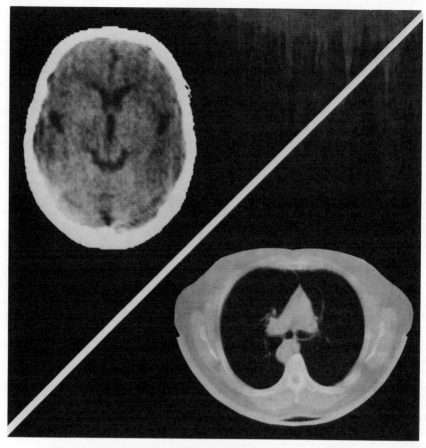

Fig. 1   Reconstructive tomograms of cross sections of the head and of the chest. (Taken from a 1976 General Electric pamphlet entitled "Introduction to Computed Tomography.")

In a good tomogram of the head, it is possible to distinguish between normal blood in the brain, clotted blood, brain tissue, fatty tissue, bone structure and tumors. Figure 2 shows another chest tomogram where we see heart tissue, bone in the spinal column and elsewhere, blood vessels and bronchi. The original picture was in pseudocolor where the lungs were presented in black, muscular tissue in red, fat in lavender, and bones in white. Pseudocolor of course is not real color. An arbitrary color is assigned to the tissue that shows up in the tomogram having a particular shade of gray. Thus red can be specified for

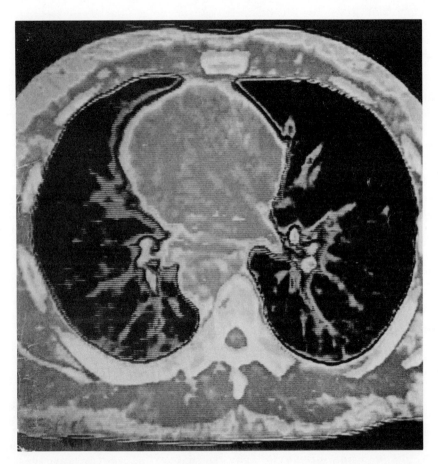

Fig. 2   Reconstructive tomogram of a cross section through the chest. (This picture originally appeared in pseudo-color on the cover of the October 1975 issue of Scientific American. It was produced by a Delta Scanner from Technicare and is used by permission of the Technicare Corporation.)

the shade of gray representing muscle, white for bone, pink for fat, and black for airspaces in the lungs.

The technique of computer reconstructive tomography has also been successfully applied to imaging the emission centers of nuclear medicine. The clinical usefulness of emission tomography, as it is called, is well established. X-ray tomograms are useful because they can show the anatomical morphology of various cross sections of the human body with good sensitivity and resolution. Emission tomograms are useful because, after a gamma-ray emitting isotope is administered to a patient, the tomograms can accurately show the distribution of the isotope in the body and give important information to a physician regarding the fundamental state of the various organs. Dynamic studies may be made of the organs simply by examining the temporal dependence of this distribution.

Both of these tomographic approaches have at least one disadvantage: they are invasive. They depend on ionizing radiation and exposure to such radiation can be harmful to the patient.

But X rays and gamma rays are not the only kind of radiation for which reconstructive tomography is feasible. Microwaves, electron beams, ultrasound, fast subatomic particles from accelerators, and even magnetic fields can also be used [G75b], [W78a], [M79]. Ultrasound is particularly attractive in reconstructive tomography and is currently undergoing much research and even clinical trials. The basic goal with ultrasound is the same as with X rays: to obtain quantitative cross-sectional images depicting the morphological detail of the human body.

Acoustical energy can give a view of a cross section not available with X rays or other types of radiation. A mapping of acoustic and elastic variation can be expected to provide a basically different pattern than a mapping of variation in X ray absorption and scattering coefficients or a mapping of the distribution of gamma-ray emitting isotopes after the isotope has been administered to a patient. In addition, a mapping of one kind of acoustic parameter will yield quite a different picture than that of another kind. The two acoustic parameters that have received the greatest attention in research laboratories are the acoustic attenuation coefficient and the acoustic refractive index. A mapping of variations in either parameter involves a transmission mode of operation; that is, the ultra-sound is transmitted through the object whose cross section is to be imaged. But reflection modes may also be of value and research is also going on to achieve computer reconstructions with reflected ultrasound. Computer-assisted acoustic-echo tomography is being examined as well as tomographic extensions of Doppler processing [N79], [M79], [W80a], [M80], [W80b].

In all reconstructive tomography it is essential to know the path that a ray traverses in going from source to detector. With X rays, the paths are essentially straight lines. There is little reflection, refraction or diffraction. However, this is not the case with ultrasound. When an ultrasonic beam propagates through tissue, it undergoes reflection and refraction at almost every interface between tissues of different refractive index. Especially serious is the refraction that takes place when an acoustic beam passes between soft and hard tissues. Because of this, applications of ultrasonic reconstructive tomography appear to be limited, at least in the foreseeable future, to those parts of the body that are free of hard tissues such as bone. Most of the current effort in this area is therefore focused on the clinically important problem of detecting tumors in the female breast [K79b].

But even in this case, diffraction of the acoustic waves needs to be taken into account. Because of the wavelength differences between X rays and ultrasound, diffraction is a far more important effect for ultrasound than for X rays.

This paper will continue with four major sections. In section II, certain basic principles of reconstructive tomography will be briefly considered to provide an elementary understanding of what is involved. Section III will describe the early ultrasonic work on reconstructive tomography and what it accomplished. Section IV will survey what has transpired since that early work and will discuss some of the lastest progress and results. Section V will give an overall perspective of the effort and where it is going, and will state conclusions that can be drawn from the results to date.

II. BASIC PRINCIPLES

A tomogram is simply a picture of a slice. In its most ideal manifestation, tomography can be described as a diagnostic technique, employed most commonly with X rays, in which a cross section of a body is imaged, the structural elements on either side of the cross section being completely absent in the image. In diagnostic medicine, a tomogram displays a slice of the body at a desired location and from a desired viewpoint.

There are two major tasks in reconstructive tomography (RT). One is to acquire information about the slice (data acquisition) and the other is to use the information to compute an image of

the slice (image reconstruction). As already indicated, an ideal
tomogram involves only a single plane and contains no interference from image elements in unwanted planes. Data can be obtained to do this if we are careful to irradiate only a thin layer of
the body in complete isolation from all other layers. This can
be done by using a narrow beam of the irradiating energy (X rays,
electrons, ultrasound, microwaves or other radiation) having the
shape of a thin pencil, a sheet or a fan. Such radiation can be
produced by collimation and directed so that it irradiates only
the layer in question. In practice the radiation is restricted
by the collimator to a thickness of one centimeter or less. The
emitted radiation can therefore pass through the desired anatomical plane without entering into other areas. It is then detected as is indicated in Fig. 3.

The simplest case uses a pencil beam and a single detector.
Only a thin pencil-like region of the object lying in the desired
plane is irradiated. The detector will measure the total radiation passing through the object along the path defined by the
thin region. After such a measurement is made for a single pulse
of radiation, both the source and the detector are moved as a
unit a short distance in the tomographic plane, that is in the
direction perpendicular to the plane of the paper in Fig. 3. An-

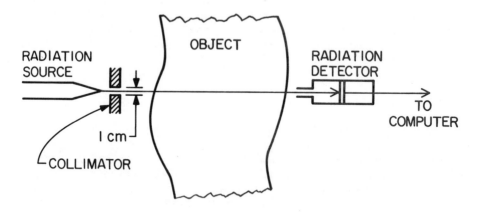

Fig. 3   Data acquisition in reconstructive tomography showing
          collimated radiation passing through the plane of the
          image. The radiation traverses only the layer under
          examination so that unwanted planes are completely
          omitted.

other pulse is then emitted and detected. This procedure is repeated over and over again until the entire tomographic plane has been traversed. The detector readings when plotted as a function of path position provide a profile of transmissions made during the scan of the positions. These profiles are generally called "projections." A complete projection in the tomographic plane is measured by sampling the output of the detector for each pulse of radiation at a large number of equally-spaced positions (for example, 160 such positions) for a single scan direction. The sampled values are stored in the memory of a computer.

As soon as the first scan is completed, the entire source-detector unit is rotated a small angle (for example, 1°) while remaining in the plane of the tomogram and the process recurs. This is done repeatedly, with data being collected for all the scan directions, until the sum of the rotations of the unit adds up to 180°. A diagram to illustrate a typical scanning pattern is shown in Fig. 4. For clarity, only a small percentage of the source-detector positions have been drawn for the two projections shown.

When the projections have all been made, the resulting information is processed by a computer to give an image of the cross section at the tomogram plane. This is the second of the two major tasks, the image reconstruction. The radiation intensity data from the projections is converted from an analog signal to digital impulses in an analog-to-digital converter and fed into the computer which reconstructs the image. The resulting RT picture is a cross-sectional image of the object.

Reconstruction from a large number of projections is a mathematical process of substantial complexity and, in practice, necessitates the use of a digital computer. However, in principle, digital-computer use is neither necessary nor unique to this method. That is why I prefer the term "reconstructive tomography" (RT) to "computed tomography" (CT) or "computer-assisted tomography" (CAT), although the latter two terms are in more general usage.

The mathematical basis for a solution to the RT problem was solved more than 60 years ago by the Austrian mathematician Johann Radon [R17]. He proved that any two-dimensional object can be reconstructed uniquely from an infinite set of its projections. This result was independently rediscovered a number of times after Radon did his work, by other mathematicians, radio astronomers, electron microscopists, workers in optics and medical radiologists. The first practical reconstructions of this kind were accomplished by the radio astronomer Ronald Bracewell in 1956 [B56].

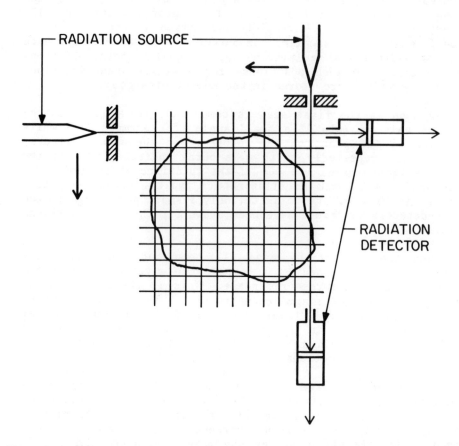

Fig. 4   Diagram to illustrate a typical scanning pattern for
            the radiation source-detector unit of Fig. 3 as it
            moves in the tomographic plane. The detector samples
            the radiation for a large number of paths through the
            object along each projection. Successive projections
            are taken at small-angle intervals around the object.
            Eventually projections around a 180° arc are taken.
            Only two projections are illustrated here, one at right
            angles to the other.

To speed up the process of data acquisition, the modern RT systems use fan beams with an array of detectors operating simultaneously instead of the single source-single detector unit I have described for illustrative purposes. No translational motion of the fan-beam unit is necessary, only rotation. However, the basic process of the tomographic technique is adequately illustrated and more easily understood by using the single source-and-detector combination.

When the radiative beam passes through the body it is attenuated by absorption and scattering as it goes from volume element to volume element. The detector determines the total radiation and hence provides a measure of the total attenuation for each fixed-beam position. The job of the computer is to unscramble this attenuation information and present it quantitatively for each volume element. In practice, the computer can do this with a precision far greater than that of any other radiographic technique. From the measured projections the computer calculates the individual attenuation coefficient for each volume element. The values may be arranged in a two-dimensional array of numbers called a matrix. The computer can then display the matrix numbers on a cathode ray tube as an array of picture cells (pixels) by using various shades of gray to represent the computed RT numbers. This display constitutes the tomogram.

Assume that we divide the tomographic plane into an N×N array of volume elements as shown in Fig. 5. At each position of the source and the detector the sum of the attenuation coefficients is obtained by the detector. An equation representing that sum can then be written. If, in moving the source-detector unit about in the tomographic plane, a minimum of N^2 such equations are obtained, the attenuation coefficient for each volume element can be solved for, at least in principle, providing that the equations are independent.

As an illustration, consider the simple problem of reconstructing the four values in a 2×2 array from a few projections. To get four equations we need at least two projections such as the two in Fig. 6. At first glance we might think that the solutions could be found unambiguously by solving the system of equations

$$A + B = E$$

$$C + D = F$$

$$A + C = G$$

$$B + D = H$$

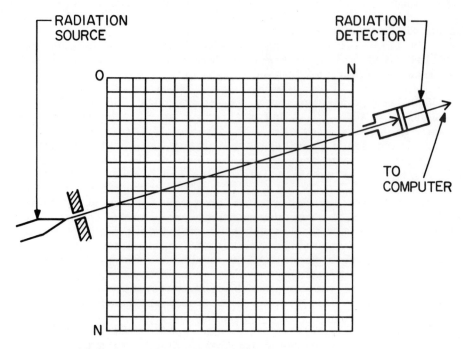

Fig. 5   Geometry illustrating how the various sums of attenua-
          tion coefficients are obtained for an N×N array of
          volume elements.

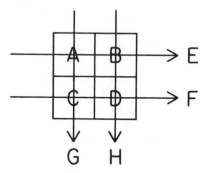

Fig. 6   A 2×2 array of volume elements with a horizontal
          projection and a vertical projection.

In matrix notation we have for the equations:

$$\underline{J} \times \underline{K} = \underline{L}$$

$$
\begin{bmatrix}
1 & 1 & 0 & 0 \\
0 & 0 & 1 & 1 \\
1 & 0 & 1 & 0 \\
0 & 1 & 0 & 1
\end{bmatrix}
\times
\begin{bmatrix}
A \\
B \\
C \\
D
\end{bmatrix}
=
\begin{bmatrix}
E \\
F \\
G \\
H
\end{bmatrix}
$$

Since the determinant for $\underline{J}$ is zero, we see that the above four equations are not independent and we can not obtain a unique solution with the two projections we have used. That the equations are not independent is also apparent if we note that for each projection each cell is passed through exactly once so that

$$E + F = G + H$$

Any one of the four measurements is equal to a combination of the other three and therefore is not independent of them. To obtain at least four independent equations we must take one other measurement such as that from a diagonal projection. Thus $A + D = I$ would provide a fourth independent equation in this case.

This simple example suggests that it may indeed be difficult to insure the existence of N^2 independent equations when imaging an N×N array. That problem and the computational burden of inverting large matrices have led to the development of alternative methods of reconstruction such as back projection, filtered back projection, two-dimensional Fourier reconstruction (in which the Fast Fourier Transform is used) and convolution filtering. These are organized and systematic approaches called algorithms. It is the algorithm for convolution filtering that is believed to be the one most used in commercial X-ray RT scanners at the present time.

III. EARLY WORK IN ACOUSTIC TOMOGRAPHY

The early work in acoustic tomography was based on algorithms developed for X-ray tomography and the objective was very similar — to make cross-sectional images corresponding to a parameter of the tissue involved. The actual parameters employed in the two types of tomography were different of course. In fact in the case of ultrasound, two parameters were used: the acoustic attenuation coefficient and the acoustic refractive index. Such workers as Greenleaf et al. [G74], Carson et al. [C76a], and Jakowatz and Kak [J76] were the first to produce tomograms involving these parameters.

Long before this work was accomplished, however, cross-sectional images and hence "tomograms" of quite a different kind had been produced by conventional pulse-echo B-scan systems. These systems are simple in concept and stem from the development of pulse-echo techniques first used in the sonar systems of World War I [W76]. Just as in a reconstructed tomogram, the tissue structures of a B-scan image are not superimposed upon each other. Although excellent images of this kind can readily be produced and are of major significance in diagnostic medicine, the pulse-echo systems can make visible only the structure that causes the echoes, that is the tissue interfaces. It is not presently possible to quantatively correlate the returning echoes with the local properties of the tissues. The problem of making quantitative maps of acoustical parameters by means of B-scan systems is very severe. Research in an area called impediography [L80a], [L80b] is an attempt to quantify pulse-echo data, but so far this work has not been very successful. Because of this difficulty, more and more workers over the last few years have been attracted to ultrasonic computer tomography for which quantification is much easier.

The fundamental approach in all the early work was to use transmission of the ultrasound. For simplicity, it was assumed that the transmitted wave traveled in a straight line between source and receiver. The detected wave at each point on the observation plane was related to some characteristic of the medium only along the straight path connecting the receiver and the point. Such an assumption is a good one indeed for X-ray tomography where the wavelength is very much less than the size of the scatterers or the absorption centers. With X rays, diffraction, refraction and reflection can be regarded as negligible. But this is not true for ultrasound where the wavelength is of the order of the size of the scattering and absorption centers. The assumption of straight-path propagation was obviously severely restrictive, but reasonably good tomograms were obtained by using it [G74], [G75c], [C76a], [J76]. Figure 7 shows two such tomograms obtained by Greenleaf and colleagues of acoustic velocity variation within an excised canine heart [G75c]. Figure 8 shows four reconstructed tomographic sections of the breast, again displaying velocity variation, along with a pulse-echo tomogram (mammogram) indicating the position of the sections.

The effect of refraction at the interfaces between soft and hard tissues was more serious than the straight-path assumption.

(Water – Filled, 22° C)

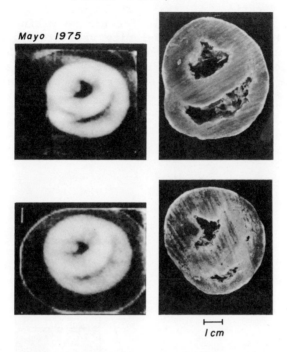

Fig. 7   Tomographic reconstructions of acoustic velocity variation within a canine heart (left) compared to photographs of cross-sections through corresponding levels (right). (Taken from Greenleaf, et al. [G75c].)

In fact, the effect is still so serious that applications of ultrasonic RT in diagnostic medicine appear to be limited, at least in the foreseeable future, to those parts of the human body that are free of hard tissue such as bone. Most of the current effort is focused on the clinically important problem of tumor detection in the female breast.

Nevertheless workers have attempted to solve the problem of deflection in various ways. One suggestion was to calculate corrections to the straight-path approximation using geometrical optics and ray-tracing methods. This approach is iterative in character. The first step is to construct a tomogram of velocity variation by ignoring refraction and using the straight-line assumption. Since velocity variation is caused by variation in the refractive index, the tomogram can then be used to digitally trace the rays through the various regions. This provides a first-order indication of the actual propagation path. The corrected path is then available for use in making a new tomographic reconstruction of the velocity variation. This process is repeated over and over for greater and greater accuracy. The method is obviously time consuming and may never be suitable for near real-time applications. Furthermore, only refraction is taken into account; diffraction is not.

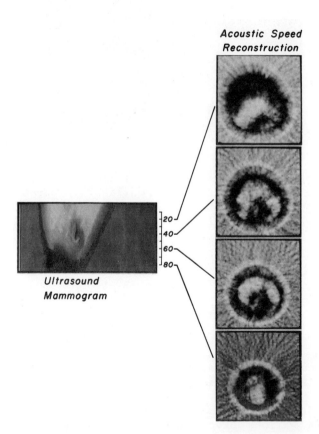

Fig. 8   Tomographic reconstruction of acoustic velocity variation in four sections of the breast (right) along with an ultrasonic mammogram (left) showing the position of the sections. (Courtesy of J.F. Greenleaf.)

Johnson, Greenleaf, and colleagues performed some initial experiments using digital ray tracing as described above to correct for refraction [J75a], [J75b]. They found that the approach did improve the reconstruction of gross shapes, but not of fine detail. This was disappointing. The experimental results are illustrated in Figure 9. The top picture on the left is of a tomogram of velocity variations using the straight-line assumption. The picture just below shows the same tomogram after the iterative correction procedure has been followed. The pictures on the right show the same two tomograms except that low-pass filtering has been applied to the images. From pictures like these it can be seen that overall configurations are improved, but not resolution.

The question of diffraction was also examined. Some early papers were published in Acoustical Imaging, Vol. 8, and elsewhere [M80b], [K79c], [W78a], [M79], [J78b]. The problem was found to be tractable, at least in principle. The work reported in the above references was purely analytical. Computer simulations were made but no actual experiments were performed. The approach will be described in some detail shortly.

Thus many inaccuracies were inherent in the early experimental systems; inaccuracies caused primarily by diffraction, refraction and reflection. From an historical standpoint it is interesting to note that early workers in acoustical tomography switched from attenuation tomograms to velocity tomograms to alleviate inaccuracies due to refraction. Most of the researchers in those days nevertheless continued to use the algorithms and assumptions developed for X-ray tomograms and the straight-line assumption remained a major part of the picture.

It gradually became increasingly evident that more accurate models of the physical mechanisms by which acoustic waves generate images were crucial for high resolution and correct reconstruction. Johnson, Greenleaf, and colleagues were among the first to propose a solution to the diffraction problem involving a perturbation method known as Rytov's approximation [J78b]. But their approach required parallel incident wave fronts which precluded its adaption to the fan-beam ultrasound scanner employed at the Mayo Clinic where they worked. They did not experimentally test their method and published only preliminary work.

Mueller and Kaveh, et al., devised two methods that included diffraction effects [K79c], [M80b], [M79], [W78a]. In these methods, the wave equation is applied to sound propagating through a medium with spatially varying parameters. The equation is solved approximately. The solution is then used to develop an

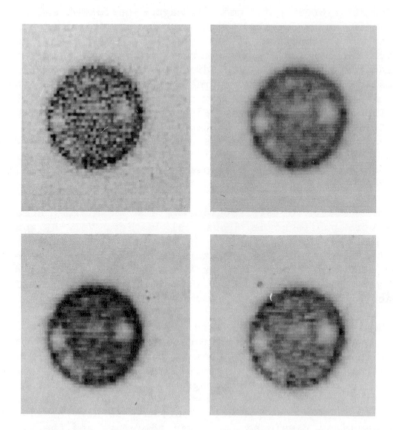

Fig. 9   Velocity tomograms illustrating the effect of cor-
           recting for ray-path variations from straight-line
           geometry. For those at the top, straight-line paths
           were assumed; for those at the bottom, the iterative
           correction procedure was followed. For those at the
           left, there was no filtering; for those at the right,
           low-pass filtering was applied. (Courtesy of S.A.
           Johnson.)

algorithm for reconstructing one or more of the parameters from
observed two-dimensional complex amplitude distributions of sound
waves that have traveled through the medium. The idea was a spe-
cialization for ultrasound of an approach previously proposed by
Iwata, Nagata [I70], [I75] and Wolfe for light [W69].

     In order to solve the wave equation, a perturbation method
is used. The solution for the wave function is developed into a
power series containing the unperturbed portion of the wave and
higher-order perturbation terms. It is the unperturbed portion
that is assumed to insonify the object.

A formal solution to the wave equation is made by Fourier transformation. The procedure is relatively simple and computationally efficient. In the solution, either of two approximations can be employed. If it is assumed that the perturbations are small, Born's approximation is used. If it is assumed instead that the perturbations are smooth, Rytov's approximation is the appropriate one.

As previously stated these approaches were not verified experimentally in the early work. However two-dimensional reconstructions were made from computer-simulated data. The physical setup assumed for simulation involved an object imbedded in an infinite, homogeneous acoustic medium. The object, constituting a nonhomogeneity with a spacially-varying velocity distribution, was assumed to be rotated to successive discrete angular positions as a planar, ultrasonic wave impinged upon it. A receiver array was assumed to coherently detect all of the waves reaching the array. These waves consisted of incident and scattered components. Detection was made for each angular position. Figure 10 shows one of the computer-calculated perspective reconstructions made from simulated data. For this particular reconstruction, Born's approximation was used, but a similar reconstruction based on Rytov's approximation looks almost exactly the same [K79c], [W78a], [M79], [M80b].

It could be seen from the simulations that the Born and Rytov approximations gave excellent results which were almost identical for a disturbance of circular shape with a radius approximately equal to the acoustical wavelength. Although still valid, neither approximation was as good for disturbances of large radius. These and other results emphasized the importance of both the magnitude and the geometry of the disturbance on the validity of the approximation used. It was apparent from a close comparison of the results, that small differences in the reconstructions cropped out depending upon whether Born's approximation or Rytov's approximation was used. But the differences were usually very small indeed.

The wave equation in this early work did not include the possibility of attenuation. This omission would certainly lead to error in practical applications. The work was therefore only a first step toward feasible diffraction tomography. Only velocity variations were considered. Not only attenuation, but also variations in density and all nonadiabatic processes were neglected. All of these complexities must be included in a realistic model.

So far I have discussed in detail only transmission modes of operation. As we have seen, the transmission tomograms gave the

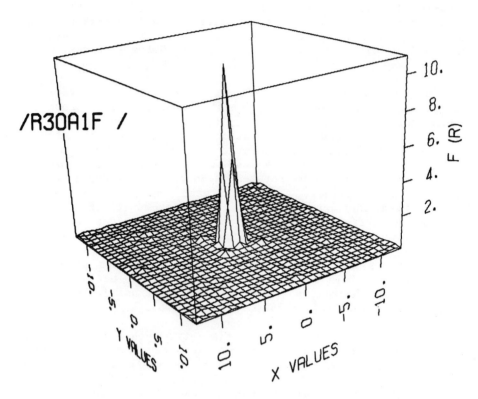

Fig. 10   Perspective reconstruction of velocity variation for a
           diffraction tomogram [W78a], [M79].

spacial distribution of variation in acoustic absorption coefficients or variation in acoustic velocity within the object
under scan.   Another technique is to use a reflection mode of
operation.   Workers at the National Bureau of Standards (NBS) and
the University of California at Santa Barbara (UCSB) were among
the first to delve into this approach [N79], [W80a].

    In Acoustical Imaging, Vol. 8, the UCSB group described an
experiment with a reflection mode of operation in which a beam of
ultrasound was reflected from the scattering centers of an object
[W80a].   The beam was sufficiently wide to encompass the object's
largest lateral dimension.   The beam was pulsed and the scattering centers produced pulsed echoes that were detected by the
receiver.   The set-up is illustrated in Figure 11a.   The single
transducer shown acted both as source and receiver.   Echoes from

scatterers at the same range arrived at the same time at the receiver and their sum appeared in the output voltage.

Such a sum can be viewed as the integral of the echoes from
that range. Because echoes from successively greater ranges
arrive at successively later times, the received echo train is a
projection of these integrals, where range has been replaced by
time.

In this fashion, one transmitted pulse generates data for
one complete projection. In the experiment, the transducer was
rotated in small-angle steps around the object and successive
projections were recorded. An artist's view of the experimental
set-up is shown in Figure 12. This procedure was repeated for
the desired angular span, a full 360°.

Using the projections so obtained, the computer was able to
reconstruct an image of the distribution of the reflection
centers of the object. A relatively good tomogram was produced
of an object with the simple geometrical elements shown in Figure
12 [W80a]. An ideal perspective reconstruction for this object

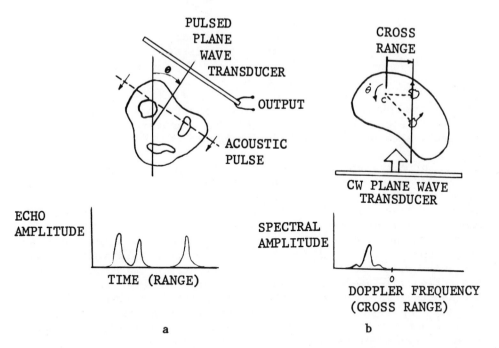

Fig. 11   Data acquisition in a) reconstructive reflection tomography using pulses; b) reconstructive reflection tomography using Doppler shift.

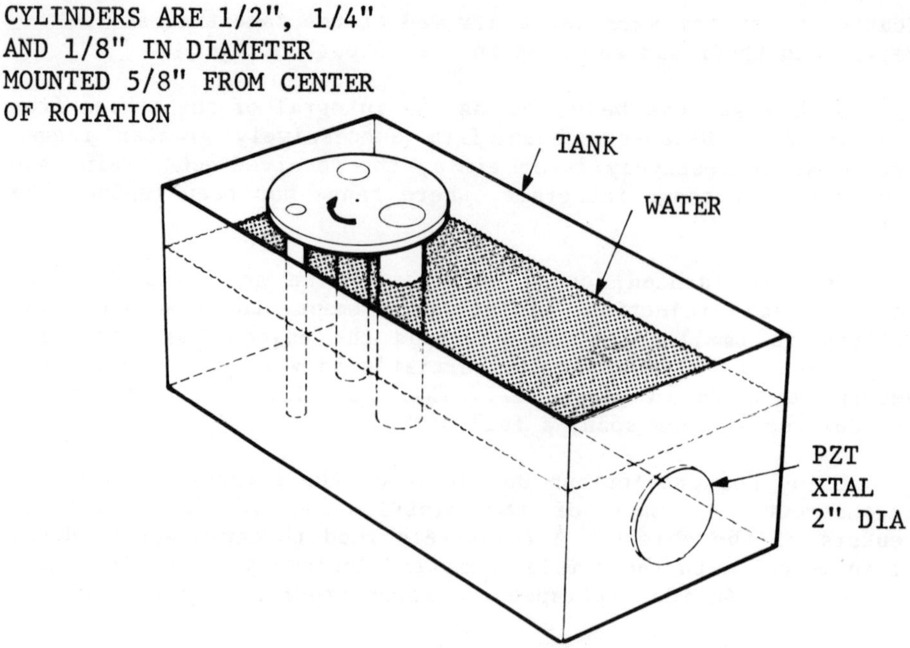

CYLINDERS ARE 1/2", 1/4"
AND 1/8" IN DIAMETER
MOUNTED 5/8" FROM CENTER
OF ROTATION

Fig. 12   Experimental arrangement to demonstrate reconstructive
           reflection tomography using pulses.

is displayed in Figure 13. Using the experimental data and some
unsophisticated image processing, the computer calculated the
perspective reconstruction shown in Figure 14.

     A primary advantage of this technique is that a single
collimated beam of ultrasound in the plane of the tomogram is
used in lieu of slow mechanical scanning or of multitransistor
arrays.   Line integrals for successively larger ranges are
produced as a time sequence in the transducer output voltage and,
as stated, the plot of output voltage versus time constitutes a
projection. It is only necessary to sample this voltage and
store the data for the computer. A hundred or more samples per
projection are needed to produce an image of good quality. If
the sampling is done as described, the electronic data collection
can be very fast, a couple of orders of magnitude faster than
with a point-by-point mechanical scan. Only a single transmit-
receive transducer is required.

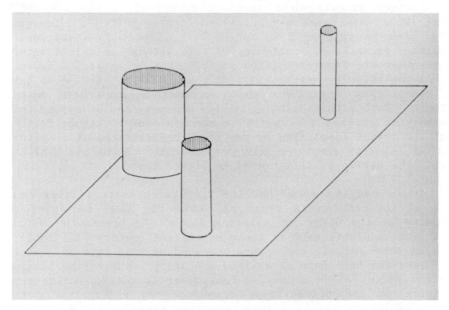

Fig. 13   Ideal perspective reconstruction of the reflection
           tomogram. The vertical coordinate represents the
           reflectivity distribution [W80a].

Fig. 14   Computer calculated perspective reconstruction of the
           experimental reflection tomogram after simple image
           processing [W80a].

Another acoustic-echo approach described in the same UCSB paper [W80a] uses what may be called Doppler-reflection tomography. When an object is in motion with respect to a stationary acoustic transmitter, scattering centers within the object return echoes that are Doppler shifted in frequency by an amount depending on the velocity of the individual scatterers. Assume we have a transmitting transducer that produces a sheet-beam output pattern. Rather than emitting pulses however, let us assume this transmitter emits a steady (CW) sinusoidal wave. Assume further that the object insonified by the sheet beam rotates at a uniform angular velocity about an axis perpendicular to the plane of the beam. The set-up is illustrated in Figure 10b.

It is easy to show that the scattering centers lying on a line of constant cross range all have the same component of velocity in the direction pointing toward the transducer. Thus at a given instant, each of these scattering centers will reflect back an echo and all the echoes produced by this line of centers will have the same Doppler-shifted frequency in traveling towards the transducer. The receiving transducer is placed at the same position as the transmitting transducer or alternatively the system employs a single transmit-receive transducer. By measuring the receiver output at any particular frequency we obtain the sum or the line integral of the scattered radiation at the cross range corresponding to that frequency. The amplitude of the returned signals at other frequencies in the receiver output give the line integrals for the scatterers at other cross ranges. An instantaneous plot of amplitude as a function of frequency can be interpreted as a projection. With the passage of time, the object will rotate to new positions and new projections will be produced. In this fashion, a continuum of one projection after another is generated. Many individual projections can then be recorded until the body has rotated through 360°. An RT algorithm using the data from these projections can produce an image of the distribution of scattering centers in the insonified slice.

I have described in substantial detail some of the early work on diffraction and reflection tomography. Because of time and space limitations, I cannot treat other early work with the same fastidiousness. This is unfortunate because much of it was outstanding and of great significance. Most of what I have covered in detail concerns work I am personally familiar with -- some of it from my own laboratory. I am anxious to reiterate that much highly meritorious research, accomplished in many different places, cannot be adequately treated here. It deserves detailed coverage also, and I feel badly that it is not possible for me to do so in this article.

I will conclude this section by listing names of laboratories and workers involved in the early work. These are as follows:

(1) Battelle Pacific Northwest Laboratories, where B. P. Hildebrand, D. E. Hufford, T. P. Harrington, and associates reconstructed velocity fields for mapping residual stress.

(2) National Bureau of Standards where M. Linzer, S. J. Norton and colleagues worked on ultrasonic echo tomography.

(3) Mayo Clinic, where J. F. Greenleaf, S. A. Johnson, C. R. Hanson, W. F. Samayoa, M. Tanaka, and associates followed a number of approaches in the pursuit of clinical applications for ultrasonic reconstructive tomography.

(4) Purdue University, where C. A. Kak and colleagues worked on ultrasonic tomography using attenuation data.

(5) Washington University, where J. G. Miller and associates worked on single-frequency tomography employing attenuation data.

(6) University of Colorado where P. L. Carson and colleagues examined clinical applications of ultrasonic tomography.

(7) General Electrical Company where G. H. Glover, J. C. Sharp and associates experimented with acoustic velocity-field reconstruction.

(8) University of Minnesota where R.K. Mueller, M. Kaveh and colleagues worked on ultrasonic diffraction tomography.

(9) University of California at Santa Barbara where G. Heidbreder, D. Mensa, G. Wade and colleagues experimented with acoustic reflection and Doppler-shift tomography.

The above list is certainly not complete, but it does indicate the wide variety of research that went on in the early effort. Most notably the list does not include activity outside the United States. Although most of the early published work in ultrasonic reconstructive tomography did take place in the U.S., some very excellent work was done elsewhere. For example,

Professor Bates and colleagues at the University of Canterbury
(New Zealand) have, over the years, carried out striking work in
both X-ray and ultrasonic RT. A 1971 paper by Bates and Peters
[B71a] was perhaps the first publication specifically suggesting
clinical X-ray computed tomography. Dr. Rolf Koch of the
Queensland Institute of Technology (Australia) has had an equally
ambitious program in the ultrasonic area.

IV.   RECENT WORK

The theoretical work on the diffraction tomography of
Mueller and Kaveh has recently been followed by experiment.
Objects were placed in a water tank and insonified by an acoustic
planar wave at 3 Mhz. A small receiving transducer was moved
along a path parallel to the face of the large transmitting
transducer with objects located on a turntable between the two
transducers. The objects were made of gelatin with a sound
velocity 2% greater than that in the surrounding water. The
gelatin was molded into cylinders of diameter 60 wavelengths and
120 wavelengths. The cylinders were perforated with holes of
various sizes and orientations. The reconstructed images all
showed very good fidelity to the original objects displaying
correct size and orientation. Figure 15 shows one such image.
The continuing work includes investigating the sensitivity of
image fidelity as a function of various system parameters and
refining the reconstruction algorithms [C80].

Work is also proceeding on the theoretical aspects of the
above diffraction tomography. The early theoretical work, as
described in the last section, did not include loss. The most
recent work does take loss into account. The theory has been
expanded to cover the general case of acoustic-wave propagation
in an isotropic visco-elastic medium. Calculations show that the
process is very sensitive to unavoidable noise in the experimental data. In order to evaluate the range of applicability of
diffraction tomography in lossy media, the effect of the noise on
the reconstruction process must be a major consideration.

A relationship between the image constrast obtainable and
the resolution has been derived. Image contrast is limited by
system noise. The derived relationship includes as parameters
the size of the object and the acoustic power propagating through
it. By means of this relationship Mueller can quantify the
trade-off between contrast and resolution. A computer modelling
effort to demonstrate these theoretical results is currently in
process.

An interesting and important potential application of
ultrasonic tomography is that of mapping residual stress in such

Fig. 15   Reconstructed diffraction tomogram showing a gelatin
           cylinder, 60 wavelengths in diameter, in water. The
           gelatin contained various perforations including two
           large holes, each 8 wavelengths in diameter, separated
           by a distance of 30 wavelengths [C80].

large structures as pressure vessels. One of the outstanding
problems of nondestructive testing is how to locate and measure
areas of residual stress. It is well known that internal stress
concentrations give rise to microcracks which grow when the
structure is subjected to external forces. Frequently in the
manufacture of large structures a number of very large welds are
required. The heat-affected zone surrounding the welds will

always contain residual stress due to uneven cooling. The velocity of sound is altered as it propagates through a region of such stress. Workers at the Battelle Pacific Northwest Laboratories, including B. P. Hildebrand, D. E. Hufferd, and T. P. Harrington, have examined the application of ultrasonic reconstructive tomography for locating and mapping regions of this residual stress [H77b], [H79].

In various computer simulations and in actual experiments they have demonstrated that velocity abnormalities as little as 2% can be quite easily resolved. In fact in extreme cases they have found it possible to experimentally map velocity variations as low as 0.2% and they feel that 0.05% is technically feasible.

Standard nondestructive testing examinations involve the use of radiography and ultrasonic pulse-echo techniques. But neither of these approaches can reveal the presence of residual stress. Simulation and experimental results indicate that residual stress can be mapped by means of ultrasonic reconstructive tomography. The Battelle workers have concluded that the tomographic approach is unique in its potential for the important task of locating and mapping residual stress in metals.

The UCSB work on Doppler tomography is also continuing. Two possible approaches to data processing have been examined. In one, the ensemble of the received Doppler-shifted signals is passed through a suitable frequency analyzer during a time period short compared to the time required for rotating the object through a small angle. As has been pointed out, the spectral distribution of the signal is linearly related to the spatial distribution of the scattering centers within the object under scan. Data concerning the spectral distribution at an instant of time are suitably digitized and the process is repeated for a large number of angular positions of the transmitter-receiver around the object. In this approach, the primary data fed to the computer consists of the spectral values representative of the spatial distribution of the scattering centers. Phase information, although present, is not employed.

In the second approach to data processing, phase and amplitude information are employed, but not the frequency spectra per se. The system, as in the first approach, uses a stationary source and sensor of continuous wave (CW) radiation, co-located in the plane being imaged. By expressing the signal as a function of the angular position of the object, an equation can be derived that has the form of a two-dimensional Fourier transformation. When inverted, this equation will give the spatial distribution of the scattering centers within the object. The analysis shows that if we rotate the object and process the data in this way, we can synthesize an aperture which is a segment of a circle centered on the center of rotation of the object. If the rotation is over a full 360°, the aperture completely encloses the object. Under these conditions high resolution may be obtained for all directions in the object plane without the use of a wideband signal. Two similar point objects, separated by a quarter of a wavelength, may then be resolved [M80a]. The process provides high resolution for sparse arrays of objects of similar strength where the objects are small compared to a wavelength. However, the imaging capability is severely limited in the case of dense object arrays or large objects. It also has a somewhat restricted dynamic range.

The significant feature of the above imaging process is the ability to produce highly resolved, two-dimensional images from measurements conducted with CW signals. This eliminates the requirement for large bandwidths, a requirement associated with conventional means of obtaining high resolution in range. The procedure of rotating the object through 360° and processing coherent signals collected from a stationary source-receiver is, as stated above, equivalent to synthesizing a circular aperture which completely surrounds the object. The high resolution is achieved because the aperture is focused to each point in the object space. This imaging process bears considerable similarity to conventional tomographic techniques but the processing algorithms are different.

Although Doppler-shift ultrasonic tomography has yet to be tested for ultrasound, experiments have been performed on a precisely analogous system using microwaves. The theoretically predicted behavior was shown to be readily achievable. Figures 16, 17, 18, and 19 show various perspective reconstructions produced with the microwave system [M80a].

Benefits that may be expected to accrue from applying Doppler-shift principles to ultrasonic reconstructive tomography include:

1.   The use of a CW, coherent source of ultrasonic energy, eliminating problems associated with obtaining rapid response of piezoelectric transducers to fast transients.

2.   The use of narrow-band receivers (made possible because of the coherent source) with attendant noise reduction and high sensitivity.

3.   The fast acquisition of the scanned data, limited only by the rotation speed of the object and the speed of response of the analyzer used in processing the reflected energy.

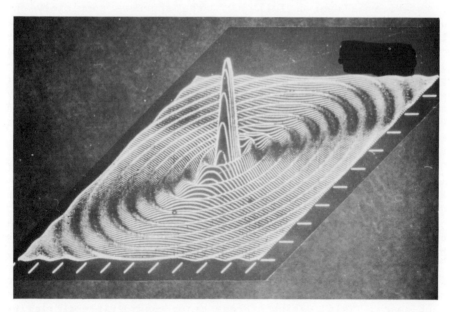

Fig. 16   The point-spread intensity function shown in perspec-
           tive for the microwave system [M80a].

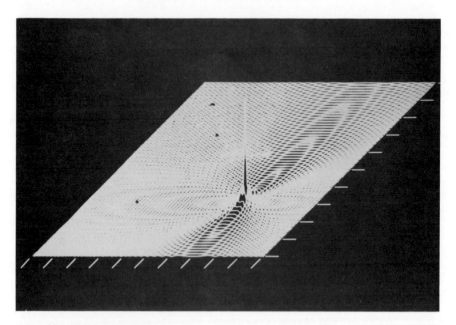

Fig. 17   Perspective plot of image intensity of a line object
           computed from simulated data for a microwave system
           operating with a wavelength of 3 cm [M80a].

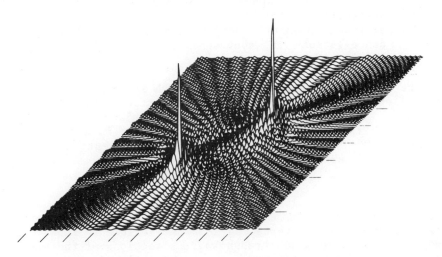

Fig. 18   Perspective plot of image intensity from measured data
           in a microwave system. The object consisted of a pair
           of identical rods 0.64 cm in diameter and 17 cm apart.
           The system operated with a wavelength of 6 cm [M80a].

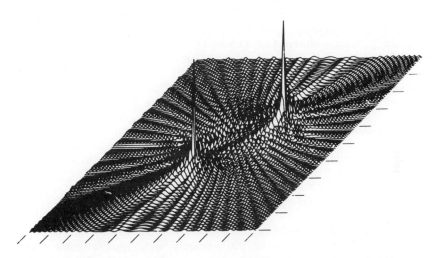

Fig. 19   Perspective plot of image intensity from simulated data
           for the microwave system and object of Figure 17
           [M80a]. Note how remarkably similar are the plots from
           simulated data (in this figure) and from measured data
           (in Figure 18).

4. The inherent capability of the system to image a
    scattering center which is surrounded by a matrix of
    different acoustic contrast that would tend to produce
    cancellation in conventional reflection tomography
    [J79].

The disadvantage of the echo approach in general is that it
uses reflection and hence only tissue interfaces are imaged. As
previously stated, it is difficult to correlate the scattered
returns with the local properties of the tissue in any quantitative way. One of the major advantages of conventional tomography, that of quantitative tissue characterization, is difficult
to achieve by this type of imaging.

Greenleaf, Rajagolalan, and colleagues at the Mayo Clinic
have continued their clinical testing of ultrasonic reconstructive tomography with interesting results. Figure 20 shows eight
tomograms they have made through various sections of the breast.
These are attenuation tomograms and were displayed originally in
blue color. An introductal papiloma is shown. The tumor was not
malignant. As is apparent, the tumor is not well seen in the
tomograms, although a mapping of attenuation variation such as
this does have a tendency to permit discrimination between
tumorous tissue and non-tumorous tissue. These tomograms have a
typically granular appearance. The bright spots are due to
reflections from water.

Figure 21 displays velocity variations through the same
tomographic planes. Here the tumor is seen a little better. The
original displays for these tomograms were in green color.

Greenleaf and his colleagues were able to superimpose the
two sets of tomograms in exact register using the original colors
and obtain the pictures shown in Figure 22. These combination
tomograms contain features from both sets. The original display
was multicolored with certain elements in blue due to attenuation
variations and other elements in green due to velocity variations.

Clinical work is also continuing at the University of
Colorado Medical Center by Paul Carson and colleagues. Figure 23
displays experimental results obtained by that group in correlat-
ing ultrasonic reconstructive tomography with pulse-echo images
[C79a]. At the left, vertical cross sections of the patient's
breast are shown (A and B). Two coronal planes in each cross
section are indicated by the straight, white parallel sets of
lines. The reconstructed tomographic images are shown to the
right corresponding to these planes and are presented with the
viewer facing the patient, the patient's head being above the

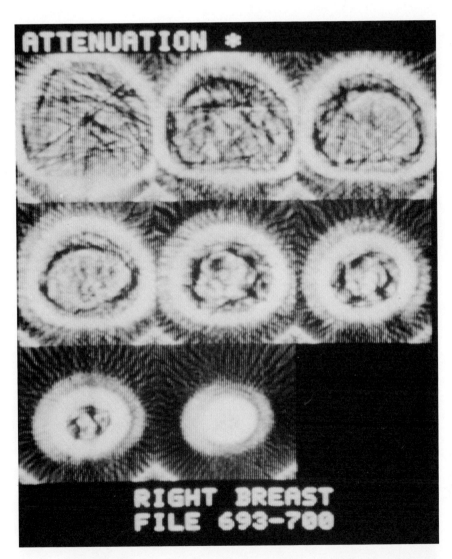

Fig. 20 Tomograms of eight sections of the breast produced in clinical experiments by Greenleaf, Rajagolalan, and colleagues at the Mayo Clinic. The tomograms display an introductal papiloma by mapping the attenuation variations. (Courtesy of Greenleaf and Rajagolalan.)

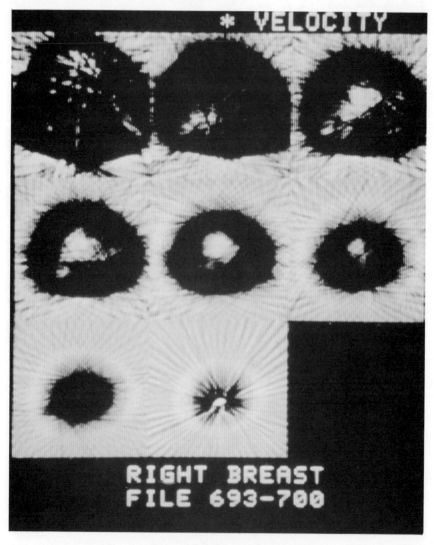

Fig. 21   Tomographic mappings of velocity variation for the same
          sections of the breast as in Figure 20.   (Courtesy of
          Greenleaf and Rajagolalan.)

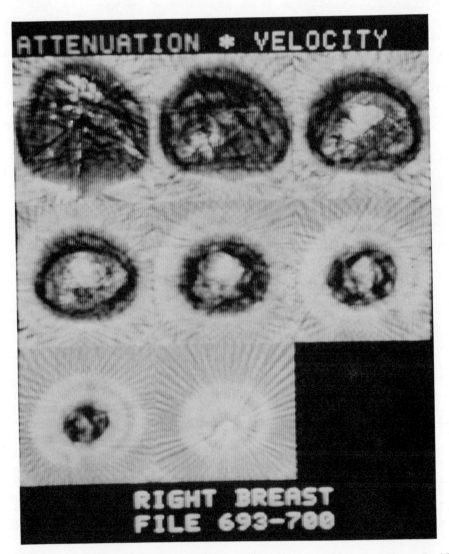

Fig. 22   Composite tomograms of those in Figures 20 and 21. (Courtesy of Greenleaf and Rajagolalan).

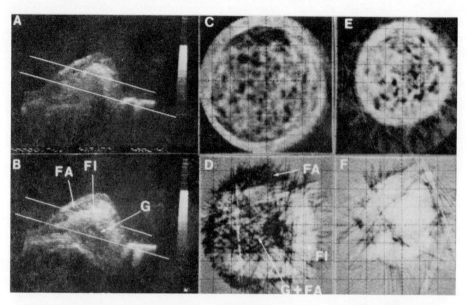

Fig. 23   Pulse-echo images and attenuation and velocity tomo-
           grams of a female breast. A and B are pulse-echo
           images, C and E are attenuation tomograms, and D and F
           are velocity tomograms. The white parallel lines in A
           show the tomographic planes for C and E; those in B,
           for D and F. (This figure was provided by Paul Carson
           and appeared in reference [C79a].)

images. The two tomograms at the top (C and E) are mappings of
attenuation variations and the two at the bottom (D and F) are
mappings of velocity variations. The patient was 40 years old
and exhibited a classic pattern for that age in the distribution
of breast tissues. The tomograms of velocity variations seem to
give a reasonably active quantitative image of the breast architecture. Fat and fibrous tissues in particular can be distinguished from other breast components such as glandular tissue
and fluids. Carson and his colleagues report that, in many
cases, the tomograms of velocity variations proved to be quite
complementary to the images provided by pulse-echo modalities,
but the latter have higher spatial resolution.

V. PERSPECTIVE AND CONCLUSIONS.

     As reported above, much research is being carried out in the
United States and elsewhere on ultrasonic reconstructive tomog-

raphy. A great deal of progress has been realized in various laboratories. In the case of diagnostic imaging, the work has progressed beyond the strictly laboratory stage. Certain instruments and systems are now in the clinic undergoing operational testing and evaluation.

Applications of widely diverse nature are envisioned for ultrasonic RT. The applications range from mapping residual stress in metallic components to mapping thermal, attenuation, velocity, and reflective variations in the human body.

Severe problems have been encountered due to diffraction, refraction, and reflection and also due to artifact generation, a topic not previously discussed in this paper. But in spite of such problems, a few systems are sufficiently attractive for medical diagnostics that they have, as mentioned above, been taken out of from the research laboratory and put into the clinic where their usefulness can be evaluated under practical working conditions.

So far, the most important application seems to be tumor detection in the female breast. The basic goal here is the same as with X-ray tomography, namely to obtain quantitative cross-sectional images depicting morphological detail. But this is easier to do with X rays than with sound because of the accuracy of the straight-line assumption for X-ray paths. As previously stated, rays of sound, with their long wavelengths, do not travel in straight lines through human tissue. This fact is particularly serious in regions of interface between hard and soft tissue, and because of it, applications of ultrasonic RT appear at present to be limited to parts of the human body not containing bone. It is for this reason that most of the current effort in clinical testing is focused on detecting tumors in the breast.

From those tests, it appears probable that ultrasonic reconstructive tomography will contribute significantly to future diagnosis of breast diseases. When this tomography is combined with pulse-echo techniques the information obtainable appears to be especially valuable. However, there still remain problems with artifact generation and variable image quality. In addition, the present ultrasonic systems are limited by slow imaging speed. Certain clinical tests, but not all, have shown that edge enhancement in attenuation tomograms has substantial potential for detecting tumors.

Nevertheless, the diagnostic utility of the use of ultrasonic reconstructive tomography in medicine is not defined well enough at present to justify great optimism as to its long-range utility. This is true in other areas of application as well.

More study, more development and more evaluation are needed before we can say for certain that the ultrasonic modality will eventually be useful in the applications presently being contemplated.

VI. ACKNOWLEDGMENT

I would like to state my indebtedness to many persons for supplying information and comments concerning the topics dealt with in this paper. Their inputs have been vital in providing me with historical facts and overall perspective on the research that has taken place. Among these people are R.H.T. Bates, Paul L. Carson, Gary H. Glover, James F. Greenleaf, T.P. Harrington, B.P. Hildebrand, Steven A. Johnson, Avinash C. Kak, Mostafa Kaveh, Melvin Linzer, James G. Miller, Rolf K. Mueller and Balu Rajagolalan.

I wish also to express great appreciation to Tracy Hamilton-Ricketts and Suzanne M. Garcia for their patience and skill in typing the many drafts that were necessary for producing the paper.

VII. REFERENCES & BIBLIOGRAPHY

A68 H.O. Anger, "Tomographic gamma-ray scanner with simultaneous readout of several planes," in Fundamental Problems in Scanning, Eds. A. Gottschalk, R.N. Beck, Springfield, Illinois: Thomas, 1968, pp. 195-211.

A76a R.E. Alvarez and A. Macovski, "Energy-selective reconstructions in X-ray computerized tomography," Phys. Med. Biol., vol. 21, pp. 733-744, 1976.

A76b R.E. Alvarez and A. Macovski, "Noise and dose in energy dependence computerized tomography," Proc. SPIE, vol. 96, pp. 131-137, 1976.

A77 H.C. Andrews and B.R. Hunt, Digital Image Restoration, Englewood Cliffs, New Jersey: Prentice Hall, 1977.

B56 R.N. Bracewell, "Strip integration in radioastronomy," Aust. J. Phys., vol. 9, pp. 198-217, 1956.

B61 R.N. Bracewell and G. Swarup, "The Stanford microwave spectroheliograph antenna, a microsteradian pencil-beam interferometer," Trans. I.R.E., vol. AP-9, pp. 22-30, January 1961.

B67   R.N. Bracewell and A.C. Riddle, "Inversion of fan-beam scans in radio astronomy," Astrophys. J., vol. 150, pp. 427-434, November 1967.

B70a R. Bender, et.al., "ART and the ribosome: A preliminary report on the three-dimensional structure of individual ribosomes determined by an algebraic reconstruction technique," J. Theor. Biol. vol. 29, pp. 483-487, 1970.

B70b M.V. Berry and D.F. Gibbs, "The interpretation of optical projections," Proc. Roy. Soc. Lond. A., vol. 314, pp. 143-152, January 6, 1970.

B71a R.H.T. Bates and T.M. Peters, "Towards improvements in tomography," New Zealand J. Sci., vol. 14, pp. 883-896, 1971.

B71b S.H. Bellman, R. Bender, R. Gordon, and J.E. Rowe, "ART is science, being a defense of algebraic reconstruction techniques for three-dimensional electron microscopy," J. Theor. Biol., vol. 32, pp. 205-216, 1971.

B73   G.L. Brownell and C.A. Burnham, "MGH positron camera," in Tomographic Imaging in Nuclear Medicine, G.S. Freedman, Ed., Soc. Nucl. Med., 1973, p. 154.

B74   T.F. Budinger and G.T. Gullberg, "Three dimensional reconstruction of isotope distributions," Phys. Med. Biol., vol. 19, pp. 387-389, 1974.

B76a N.A. Baily and R.A. Keller, "A physical comparison of a fluoroscopic CAT system and the EMI head scanner," Proc. SPIE, vol. 96, pp. 210-215, 1976.

B76b N.A. Baily, R.A. Keller, C.V. Jakowatz, and A.C. Kak, "The capability of fluoroscopic systems for the production of computerized axial tomograms," Investigative Radiology, vol. 11, pp. 434-439, Sept.-Oct. 1976.

B76c R.A. Brooks and G. DiChiro, "Beam hardening in X-ray reconstructive tomography," Phys. Med. Biol., vol. 21, pp. 390-398, 1976.

B76d R.A. Brooks and G. DiChiro, "Principles of computer assisted tomography (CAT) in radiographic and radioisotopic imaging," Phys. Med. Biol., vol. 21, pp. 689-732, 1976.

B76e R.A. Brooks and G. DiChiro, "Statistical limitations in X-ray reconstructive tomography," Med. Phys., vol. 3, pp. 237-270, 1976.

B76f R.A. Brooks and G.H. Weiss, "Interpolation problems in image reconstruction," Proc. SPIE, vol. 96, pp. 313-319, 1976.

B76g T.F. Budinger and G.T. Gullberg, "Transverse section reconstruction of gamma-ray emitting radionuclides in patients," in Reconstruction Tomography in Diagnostic Radiology and Nuclear Medicine, TerPogossian et al., Eds. Baltimore, MD: University Park Press, 1976.

B77a N.A. Baily, "Computerized tomography using video techniques," Opt. Eng., vol. 16, pp. 23-27, Jan.-Feb. 1977.

B77b R.N. Bracewell, "Correction for collimator width (restoration) in reconstructive X-ray tomography, " J. Comput. Assist. Tomog., vol. 1, pp. 6-15, Jan. 1977.

B77c R.A. Brooks, "A quantitative theory of the Hounsfield unit and its application to dual energy scanning," J. Compt. Assist. Tomog., vol. 1, pp. 487-493, 1977.

B77d R.A. Brooks and G. DiChiro, "Slice geometry in computer assisted tomography," J. Compt. Assist. Tomog., vol. 1, pp. 191-199, 1977.

B77e T.F. Budinger, S.E. Derenzo, G.T. Gullber, W.L. Greenber, and R.H. Huesman, "Emission computer assisted tomography with single photon and positron annihilation photon emitters," J. Compt. Assist. Tomog., vol. 1, pp. 131-145, 1977.

B78a C. Bohm, L. Eriksson, M. Bergstrom, J. Litton, R. Sundman, and M. Singh, "A computer assisted ringdetector positron camera system for reconstruction tomography of the brain," IEEE Trans. Nucl. Sci., vol. NS-25, pp. 624-637, 1978.

B78b R.A. Brooks and G. DiChiro, "Split-detector computer tomography: A preliminary report," Radiology, vol. 126, pp. 255-257, Jan. 1978.

B78c R.A. Brooks, G.H. Weiss, and A.J. Talbert, "A new approach to interpolation in computed tomography," J. Compt. Assist. Tomog., vol. 2, pp. 577-585, Nov. 1978.

B78d G.L. Brownell and S. Cochavi, "Transverse section imaging with carbon-11 labeled carbon monoxide, " J. Compt. Assist. Tomog., vol. 2, pp. 533-538, Nov. 1978.

B80 J.S. Ball, S.A. Johnson, and F. Stanger, "Explicit inversion of the Helmholtz equation for ultrasound insonification and spherical detection," this volume pp. ___ - ___, 1980.

C63 A.M. Cormack, "Representation of a function by its line integrals, with some radiological applications," J. Appl. Phys., vol. 34, no. 9, pp. 2722-2727, September 1963.

C64 A.M. Cormack, "Representation of a function by its line integrals with some radiological applications, II," J. Appl. Phys., vol. 35, pp. 2908-2913, October 1964.

C69 L.A. Chernov, Wave Propagation in a Random Medium, New York: McGraw-Hill, 1969.

C76a P.L. Carson, T.V. Oughton and W.R. Hendee, "Ultrasonic transaxial tomography by reconstruction," Ultrasound in Medicine, vol. 2, Eds. D. White and R. Barns, New York: Plenum Press, pp. 341-350, 1976.

C76b P.L. Carson, T.V. Oughton, and W.R. Hendee, "Ultrasound transaxial tomography by reconstruction," in Ultrasound in Medicine, vol. 2, D.N. White and R.W. Barnes, Eds. New York: Plenum Press, pp. 391-400, 1976.

C76c Z.H. Cho, L. Eriksson, And J. Chan, "A circular ring transverse axial positron camera," in Reconstruction Tomography in Diagnostic Radiology and Medicine, M.M. TerPogossian et al., Ed. Baltimore, MD: University Park Press, pp. 393-421, 1976.

C77a P.L. Carson, T.V. Oughton, W.R. Hendee, and A.S. Ahuja, "Imaging soft tissue through bone with ultrasound transmission tomography by reconstruction," Med. Phys., vol. 4, pp. 302-309, July/Aug. 1977.

C77b D.A. Chesler, S.J. Riederer, and N.J. Pelc, "Noise due to photon counting statistics in computed X-ray tomography," J. Compt. Assist. Tomog., pp. 64-74, vol. 1, Jan. 1977.

C77c Z.H. Cho, M.B. Cohen, M. Singh, L. Eriksson, J. Chan, N. MacDonald, and L. Spotter, "Performance and evaluation

of the circular ring transverse axial positron camera,"
IEEE Trans. Nucl. Sci., vol. NS-24, pp. 532-543, 1977.

C77d G. Chu and K.C. Tam, "Three dimensional imaging in the positron camera using Fourier techniques," Phy. Med. Biol.,
vol. 22, pp. 245-265, 1977.

C78a L.R. Carroll, "Design and performance characteristics of a
production model positron imaging system," IEEE Trans.
Nucl. Sci., vol. NS-25, pp. 606-614, Feb. 1978.

C78b P.L. Carson, D.E. Dick, G.A. Thieme, M.L. Dick, E.J. Bayly,
T.V. Oughton, G.L. Dubuque, and H.P. Bay, "Initial
investigation of computed tomography for breast imaging
with focused ultrasound beams," Ultrasound in Medicine,
vol. 4, D.N. White and E.A. Lyons, Eds. New York:
Plenum Press, pp. 319-322, 1978.

C78c L.T. Chang, "A method for attenuation correction in radionuclide computed tomography," IEEE Trans. Nucl. Sci.,
vol. NS-25, pp. 638-643, Feb. 1978.

C78d R.C. Chase and J.A. Stein, "An improved image algorithm for
CT scanners," Med. Phy., vol. 5, pp. 497-499, Dec.
1978.

C78e Z.H. Cho, O. Nalcioglu, and M.R. Furukhi, "Analysis of a
cylindrical hybrid positron camera with Bismuth
Germanate (BGO) scintillation crystals," IEEE Trans.
Nucl. Sci., vol. NS-25, pp. 952-963, Apr. 1978.

C79a P.L. Carson, A.L. Scherzinger, T.V. Oughton, J.E.
Kubitschek, P.A. Lambert, G.E. Moore, M.G. Dunn, and
D.E. Dick, "Progress in ultrasonic computed tomography
(CT) of the breast," SPIE Vol. 173, Application of
Optical Instrumentation in Medicine VII, pp. 372-381,
1979.

C79b L.T. Chang, "Attenuation correction and incomplete projection in single photon emission computed tomography,"
IEEE Trans. Nucl. Sci., vol. NS-25, Apr. 1979.

C79c C.R. Crawford and A.C. Kak, "Aliasing artifacts in CT
images," School Elec. Eng., Purdue Univ., West
Lafayette, IN, Res. Rep. TR-EE 79-25, 1979.

C80 T.R. Coulter, M. Kaveh, R.K. Mueller, and R.L. Rylander,
"Experimental results with diffraction tomography," to
be published in the Proceedings of the 1979 Ultrasonics
Symposium, 1980.

D68   D.J. De Rosier and A. Klug, "Reconstruction of three-dimensional structures from electron micrographs," _Nature_, vol. 217, pp. 130-134, January 13, 1968.

D76a K.A. Dines and A.C. Kak, "Measurement and reconstruction of ultrasonic parameters for diagnostic imaging," School Elec. Eng., Purdue Univ., West Lafayette, IN, Res. Rep. TR-EE 77-4, Dec. 1976.

D76b P. Dreike and D.P. Boyd, "Convolution reconstruction of fan-beam reconstructions," _Comp. Graph. Image Process._, pp. 459-469, vol. 5, 1976.

D77a S.E. Derenzo, "Positron ring cameras for emission computed tomography," _IEEE Trans. Nucl. Sci._, vol. NS-24, pp. 881-885, Apr. 1977.

D77b S.E. Derenzo, T.F. Budinger, J.L. Cahoon, R.H. Huesman, and H.G. Jackson, "High resolution computed tomography of positron emitter," _IEEE Trans. Nucl. Sci._, col. NS-24, pp. 544-558, Feb. 1977.

D78a G. DiChiro, R.A. Brooks, L. Dubal, and E. Chew, "The apical artifact: Elevated attenuation values toward the apex of the skull," _J. Compt. Assist. Tomog._, vol. 2, pp. 65-79, Jan. 1978.

D78b A.J. Duerinckx and A. Macovski, "Polychromatic streak artifacts in computed tomography images," _J. Compt. Assist. Tomog._, vol. 2, pp. 481-487, Sept. 1978.

D79   K.A. Dines and A.C. Kak, "Ultrasonic attenuation tomography of soft biological tissues," _Ultrasonic Imaging_, vol. 1, pp. 16-33, Jan. 1979.

E66   E.R. Epp and H. Weiss, "Experimental study of the photon energy spectrum of primary diagnostic X-rays," _Phys. Med. Biol._, vol. 11 pp. 225-238, 1966.

E76   L. Eriksson and Z.H. Cho, "A simple absortpion correction in positron (annihilation gamma coincidence detection) transverse axial tomography," _Phys. Med. Biol._ vol. 21, pp. 429-433, 1976.

F78   E.J. Farrell, "Processing limitations of ultrasonic image reconstruction," in _Proc. 1978 Conf. Pattern Recognition and Image Processing_, May 1978.

G70 R. Gordon, R. Bender and G.T. Herman, "Algebraic reconstruc-
     tion techniques (ART) for three-dimensional electron
     microscopy and X-ray photography," J. Theor. Biol.,
     vol. 29, pp. 471-481, December 1970.

G72 M. Goiten, "Three dimensional density reconstruction from a
     series of two dimensional projections," Nucl. Instrum.
     Methods, vol. 101, pp. 509-518, 1972.

G74 J.F. Greenleaf, S.A. Johnson, S.L. Lee, G.T. Herman, and
     E.H. Wood, "Algebraic reconstruction of spatial distri-
     bu tions of acoustic absorption within tissue from
     their two-dimensional acoustic projections," in
     Acoustical Holography, vol. 5, Ed. P.S. Green, New
     York: Plenum Press, 1974, pp. 591-603.

G75a M. Gado and M. Phelps, "The peripheral zone of increased
     density in cranial computed tomography," Radiology,
     vol. 117, pp. 71-74, 1975.

G75b R. Gordon, G.T. Herman and S.A. Johnson, "Image reconstruc-
     tion from projections," Scientific American, vol. 233,
     no. 4, pp. 56-68, October 1975.

G75c J.F. Greenleaf, S.A. Johnson, W.F. Samayoa and F.A. Duck,
     (1975a) "Algebraic reconstruction of spatial distribu-
     tions of acoustic velocities in tissue from their
     time-of-flight profiles," in Acoustical Holography,
     vol. 6, N. Booth, Ed., New York: Plenum Press, pp.
     71-90, 1975.

G77 G.H. Glover and J.C. Sharp, "Reconstruction of ultrasound
     propagation speed distribution in soft tissue: time-of-
     flight tomography," IEEE Trans. on Sonics and Ultra-
     sonics, vol. SU-24, no. 4, July 1977.

G78a J.F. Greenleaf, S.K. Kenue, B. Rajagopalan, R.C. Bahn, and
     S.A. Johnson, "Breast imaging by ultrasonic computer-
     assisted tomography," in Acoustical Imaging, vol. 8, A.
     Metherell, Ed. New York: Plenum Press, 1978.

G78b D.E. Gustafson, M.J. Berggren, M. Singh, and M.K. Dewanjee,
     "Computed transaxial imaging using single gamma
     emitters," Radiology, vol. 129, pp. 187-194, Oct. 1978.

H71 G.T. Herman and S. Rowland, "Resolution in ART: An experi-
     mental investigation of the resolving power of an
     algebraic picture reconstruction technique," J. Theor.
     Biol., vol. 33, pp. 213-223, 1971.

H72 G.N. Hounsfield, "A method of and apparatus for examination of a body by radiation such as X ray or gamma radiation," The Patent Office, London, Patent Specification 1283915, 1972.

H73a T. Hagfors and D.B. Campbell, "Mapping of planetary surfaces by radar," Proc. IEEE, vol. 61, no. 9, pp. 1219-1225, September 1973.

H73b G.T. Herman and S.W. Rowland, "Three methods for reconstructing objects from X-rays: a comparative study," Computer Graphics and Image Processing, vol. 2, pp. 151-178, 1973.

H73c G.N. Hounsfield, "Computerized transverse axial scanning (tomography): Part 1. Description of system," Br. J. Radiol., vol. 46, pp. 1016-1022, November 1973.

H75a T.S. Huang, M. Kaveh and S. Berger, "Some further results in iterative image restoration," presented at the Annual Meeting of the Optical Society of America, Boston, Mass., October 1975.

H75b H. Hurwitz, Jr., "Entropy Reduction in Bayesian analysis of measurements," Phys. Rev., vol. 12, pp. 698-704, 1975.

H76a G.T. Herman, A.V. Lakshminarayanan, A. Naparstek, E.L. Ritman, R.A. Robb, and E.H. Wood, "Rapid computerized tomography," Med. Data Process., pp. 582-598, 1976.

H76b E.J. Hoffman, M.E. Phelps, N.A. Mullani, C.S. Higgins, and M.M. TerPogossian, "Design and performance characteristics of a whole body transaxial tomography," J. Nucl. Med., vol. 17, pp. 493-502, 1976.

H76c R.C. Hsieh and W.G. Wee, "On methods of three-dimensional reconstruction from a set of radioisotope scintigrams," IEEE Trans. Syst. Man Cybern., vol. SMC-6, pp. 854-862, Dec. 1976.

H77a G.T. Herman and A. Naparstek, "Fast image reconstruction based on a Radon inversion formula appropriate for rapidly collected data," SIAM J. Appl. Math., vol. 33, pp. 511-533, Nov. 1977.

H77b B.P. Hildebrand and D.E. Hufferd, "Computerized reconstruction of ultrasonic velocity fields for mapping of residual stress," in Acoustical Holography, Vol. 7, L.W. Kessler, Ed., New York: Plenum Press, pp. 245-262, 1977.

H78a P. Haque, D. Pisano, W. Cullen, and L. Meyer, "Initial performance evaluation of the CT 7000 scanner," paper presented at the 20th Meeting of AAPM, San Francisco, CA, Aug. 1978.

H78b B.K.P. Horn, "Density reconstruction using arbitrary ray sampling scheme," Proc. IEEE, vol. 66, pp. 551-562, May 1978.

H79 B.P. Hildebrand and T.P. Harrington, "Maping residual stress by ultrasonic tomography," Proc. Inst. Mech. E., pp. 105-116, May 1979.

I70 K. Iwata and R. Nagata, "Calculation of three-dimensional refractive index distribution from interferograms," J. Opt. Soc. of Am., vol. 60, pp. 133-135, 1970.

I75 K. Iwata and R. Nagata, "Calculation of refractive index distribution from interferograms using the Born and Rytov's approximations," Jap. J. Appl. Phys., vol. 14, pp. 1921-1927, 1975.

I78 R.D. Iverson, M.S. Thesis, Department of Electrical Engineering, University of Minnesota, 1978.

J75a S.A. Johnson, J.F. Greenleaf, A. Chu, J.D. Sjostrand, B.K. Gilbert and E.H. Wood, "Reconstruction of material characteristics from highly refraction distorted projections by ray tracing," Image Processing for 2-D and 3-D Reconstruction from Projections: Theory and Practice in Medicine and the Physical Sciences. A Digest of Technical Papers, Stanford, California, pp. TRB2-1-TUB2-4, August 4-7, 1975.

J75b S.A. Johnson, J.F. Greenleaf, W.F. Samayoa, F.A. Duck and J.D. Sjostrand, "Reconstruction of three-dimensional velocity fields and other parameters by acoustic ray tracing," 1975 Ultrasonic Symposium Proceedings, IEEE Cat. No. 75, pp. 46-51, CHP994-1SU, 1975.

J76 C.V. Jakowatz, Jr. and A.C. Kak, "Computerized tomography using X-rays and ultrasound," School Elec. Eng., Purdue Univ., West Lafayette, IN, Res. Rep. TR-EE 76-25, 1976.

J78a P.M. Joseph and R.D. Spital, "A method for correcting bone induced artifacts in computed tomography scanners," J. Compt. Assist. Tomog., vol. 2, pp. 100-108, Jan. 1978.

J78b S.A. Johnson, J.F. Greenleaf, M. Tanaka, B. Rajagopalan, and R.C. Bahn, "Quantitative synthetic aperture reflection imaging with correction for refraction and attenuation: application of seismic techniques in medicine," Proc. of the San Diego Biomedical Symposium, Vol. 17, Western Periodicals Co., North Hollywood, CA, pp. 337-349, 1978.

J79 S.A. Johnson and J.F. Greenleaf, "New ultrasound and related imaging techniques," IEEE Trans. on Nuclear Science, Vol. NS-26, No. 2, pp. 2812-2816, April 1979.

K63 D.E. Kuhl and R.Q. Edwards, "Image separation radioisotope scanning," Radiology, vol. 80, pp. 653-661, 1963.

K77 A.C. Kak, C.V. Jakowatz, N.A. Baily, and R.A. Keller, "Computerized tomography using video recorded fluoroscopic images," IEEE Trans. Biomed. Eng., vol. BME-24, pp. 157-169, Mar. 1977.

K78a A.C. Kak and K.A. Dines, "Signal processing of broadband pulsed ultrasound: Measurement of attenuation of soft biological tissues," IEEE Trans. Biomed. Eng., vol. BME-25, pp. 321-344, July 1978.

K78b P.N. Keating, "More accurate interpolation using discrete Fourier transforms," IEEE Trans. Acoust. Speech Signal Processing, vol. ASSP-26, pp. 368-369, 1978.

K78c D.K. Kijewski and B.E. Bjarngard, "Correction for beam hardening in computed tomography," Med. Phys., vol. 5, pp. 209-214, 1978.

K79a A.C. Kak, "Image reconstruction from projections," School Elec. Eng., Purdue Univ., West Lafayette, IN, Res. Rep. TR-EE-79-26, 1979.

K79b A.C. Kak, "Computerized tomography with X-ray emission and ultrasound sources," PROC. IEEE, Vol. 67, No. 9, pp. 1245-1272, September 1979.

K79c M. Kaveh, R.K. Mueller and R.D. Iverson, "Ultrasonic tomography based on perturbation solutions of the wave equation," Computer Graphics and Image Processing, February 1979.

K80 M. Kaveh, R.K. Mueller, R. Rylander, T.R. Coulter, and M Soumekh, "Experimental results in ultrasonic diffraction tomography," this volume, pp. 433 - 450, 1980.

L64 C. Lanczos, <u>Linear Differential Operators</u>, London: Van Nostrand, 1964.

L75 A.V. Lakshminarayanan, "Reconstruction from divergent ray data," Dep. Computer Science, State Univ. New York, Buffalo, Tech. Rep. 92, 1975.

L80á C.Q. Lee, "Tissue characterization by ultrasonic impediography," this volume, pp. 521 - 532, 1980.

L80b S. Leeman, "Impediography Revisited," this volume, pp. 513 - 520, 1980.

M73 R.M. Mersereau, "Recovering multidimensional signals from their projections," <u>Computer Graphics and Image Processing</u>, vol. 1, pp. 179-195, 1973.

M74a E.C. McCullough, H.L. Baker, Jr., O.W. Houser, and D.F. Reese, "An evaluation of the quantitative and radiation features of a scanning X-ray transverse axial tomograph: The EMI scanner," <u>Radiation Phys.</u> vol. 3, pp. 709-715, June 1974.

M74b R.M. Mersereau and A.V. Oppenheim, "Digital reconstruction of multidimensional Signals from Their Projections," <u>Proceedings of the IEEE</u>, vol. 62, October 1974.

M75a E.C. McCullough, "Photon attenuation in computed tomography," <u>Med. Phys.</u>, vol. 2, pp. 307-320, 1975.

M75b W.D. McDavid, R.G. Waggener, W.H. Payne, and M.J. Dennis, "Spectral effects on three-dimensional reconstruction from X-rays," <u>Med. Phys.</u>, vol. 2, pp. 321-324, 1975.

M77 J.G. Miller, M. O'Donnell, J.W. Mimbs, and B.E. Sobel, "Ultrasonic attenuation in normal and ischemic myocardium," in <u>Proc. 2nd Int. Symp. Ultrasonic Tissue Characterization</u> (National Bureau of Standards), 1977.

M78a M. Millner, W.H. Payne, R.G. Waggener, W.D. McDavid, M.J. Dennis, and V.J. Sank, "Determination of effective energies in CT calibration," <u>Med. Phys.</u>, vol. 5, pp. 543-545, 1978.

M78b N.A. Mullani, C.S. HIggins, J.T. Hood, and C.M. Curie, "PETT IV: Design analysis and performance characteristics," <u>IEEE Trans. Nucl. Sci.</u>, vol. NS-25, pp. 180-183, Feb. 1978.

M79   R.K. Mueller, M. Kaveh and G. Wade, "Reconstructive tomog-
      raphy and applications to ultrasonics," PROC. IEEE,
      Vol. 67, No. 4, pp. 567-587, April 1979.

M80a D. Mensa, G. Heidbreder, and G. Wade, "Aperture synthesis by
      object rotation in coherent imaging," to be published
      in IEEE Transactions on Nuclear Science, April 1980.

M80b R.K. Mueller, M. Kaveh and R.D. Iverson, "A new approach to
      acoustic tomography using diffraction techniques," in
      Acoustical Imaging, Vol. 8, Ed. A. Metherell, New York:
      Plenum Press, pp. 615-628, 1980.

M80c R.K. Mueller and M. Kaveh, "Ultrasonic Diffraction Tomog-
      raphy," Internal Report, Department of Electrical
      Engineering, University of Minnesota, 1980.

N79   S.J. Norton and M. Linzer, "Ultrasonic reflectivity tomog-
      raphy: Reconstruction with circular transducer ar-
      rays," Ultrasonic Imaging, vol. 1, no. 2, Apr. 1979.

O61   W.H. Oldendorf, "Isolated flying spot detection of radioden-
      sity discontinuities displaying the internal structural
      pattern of a complex object," IRE Trans. Biomed. Elec.,
      vol. BME-8, pp. 68-72, 1961.

P72   J.E.B. Ponsonby, I. Morison, A.R. Birks, and J.K. Landon,
      "Radar images of the moon at 75 and 185 cm wave-
      lengths," The Moon, vol. 5, pp. 286-294, Nov./Dec.
      1972.

P75   M.E. Phelps, E.J. Hoffman, and M.M. TerPogossian, "Attenua-
      tion coefficients of various body tissues, fluids and
      lesions at photon energies of 18 to 136 deV,"
      Radiology, vol. 117, pp. 574-583, 1975.

P77a T.M. Peters and R.M. Lewitt, "Computed tomography with
      fan-beam geometry," J. Comput. Assist. Tomog., vol. 1,
      pp. 429-436, 1977.

P77b M.E. Phelps, "Emission computed tomography," Sem. Nucl.
      Med., vol. 7, pp. 334-365, Oct. 1977.

P78   M.E. Phelps, E.J. Hoffman, S.C. Huang, and D.E. Kuhl, "ECAT:
      A new computerized tomographic imaging system for
      positron-emitting radiopharmaceuticals," J. Nucl. Med.,
      vol. 19, pp. 635-647, 1978.

R17   J. Radon, "Uber die bestimmung von funktionen durch ihre in-
      tergralwerte langs gewisser mannigfaltigkeiten," ("On
      the determination of functions from their integrals
      along   certain   manifolds"),   Berichte   Saechsische
      Akademie der Wissenschaften, vol. 69, pp. 262-277,
      1917.

R69   P.D. Rowley, "Quantitative interpretation of three-dimen-
      sional weakly refractive phase objects using holo-
      graphic interferometry," J. Opt. Soc. Am., vol. 59, pp.
      1496-1498, November 1969.

R71a G.N. Ramachandran and A.V. Lakshminarayanan, "Three-dimen-
      sional reconstruction from radiographs and electron
      micrographs: II. Application of convolutions instead of
      Fourier transforms," Proc. Nat. Acad. Sci., vol. 68,
      no. 9, pp. 2236-2240, 1971.

R71b G.N. Ramachandran and A.V. Lakshminarayanan, "Three-dimen-
      sional reconstruction from radiographs and electron
      micrographs: III. Description and application of the
      convolution method," Indian J. Pure Appl. Phys., vol.
      9, pp. 997-1003, 1971.

R76   S.W. Rowland, "The effect of noise in the projection data on
      the reconstruction produced by computerized tomog-
      raphy," Proc. soc. Photo-Opt. Instrum. Eng., vol. 96,
      pp. 124-130, 1976.

R78   S.J. Riederer, N.J. Pelc, and D.A. Chesler, "The noise power
      spectrum in computer X-ray tomography," Phys. Med.
      Biol., vol. 23, pp. 446-454, 1978.

S67   D. Slepian, "Linear least squares filtering of distorted
      images," J. of Opt. Soc. of America, vol. 57, pp.
      918-922, July 1967.

S68   J.W. Strohbehn, "Line of sight wave propagation through tur-
      bulent atmosphere," IEEE Proceedings, vol. 56, pp.
      1301-1318, August 1968.

S73   D.W. Sweeney and G.M. Vest, "Reconstruction of three-dimen-
      sional refractive index fields from multidimensional
      interferometric data," Applied Optics, vol. 2, 1973.

S77a L.A. Shepp and J.A. Stein, "Simulated reconstruction arti-
      facts in computerized X-ray tomography," in Reconstruc-
      tion Tomography in Diagnostic Radiology and Nuclear
      Medicine, M.M. TerPogossian et al., Eds. Baltimore, MD:
      University Park Press, 1977.

S77b R.A. Shulz, E.C. Olson, and K.S. Han, "A comparison of the number of rays vs. the number of views in reconstruction tomography," Proc. Soc. Photo-Opt. Instrum. Eng., vol. 127, pp. 313-320, 1977.

S78 H.J. Scudder, "Introduction to computer aided tomography," Proc. IEEE Trans. Nucl. Sci., vol. NS-21, pp. 21-43, 1974.

S79 F. Stanger and S.A. Johnson, "Ultrasonic transmission tomography based on the inversion of the Helmholtz wave equation for plane and spherical wave insonification," Applied Mathematics Notes, Canadian Mathematical Soc., Ottawa, Ontario, Canada, Vol. 4, No. 3-4, pp. 102-127, December 1979.

T61 V.I. Tatarski, Wave Propagation in a Turbulent Medium, New York: McGraw-Hill, 1961.

T65 J.H. Thomson, Talk presented at the Symposium on Planetary Atmospheres and Surfaces, Dorado, Puerto Rico, 1965.

T67 M.M. TerPogossian, The Physical Aspects of Diagnostic Radiology. New York: Harper and Row, 1967.

T68 J.H. Thomson and J.E.B. Ponsonby, "Two-dimensional aperture synthesis in lunar radar astronomy," Proc. Roy. Soc. A., vol. 303, pp. 477-491, March 1968.

T69 O. Tretiak, M. Eden, and M. Simon, "Internal structures for three dimensional images," in Proc. 8th Int. Conf. Med. Biol. Eng. (Chicago, IL), 1969.

T73 O. Tretiak, "Recovery of multi-dimensional signals from their projections," Computer Graphics and Image Processing, vol. 1, pp. 179-195, October 1973.

T75 E. Tanaka and Y.A. Iinuma, "Correction functions for optimizing the reconstructed image in transverse section scan," Phys. Med. Biol., vol. 20, p. 789, 1975.

T77 M.M. TerPogossian, "Basic principles of computer axial tomography," Sem. Nucl. Med., vol. VII, pp. 109-127, Apr. 1977.

T78a K.C. Tam, G. Chu, V. Perez-Mendez, and C.B. Lin, "Three dimensional reconstruction in planar positron cameras using Fourier deconvolution of generalized tomograms," IEEE Trans. Nucl. Sci., vol. NS-25, pp. 152-159, Feb. 1978.

T78b M.M. TerPogossian, N.A. Mullani, J. Hood, C.S. Higgins, and C.M. Curie, "A multistate positron emission computed tomography (PETT-IV) yielding transverse and longitudinal images," Radiology, vol. 128, pp. 477-484, Aug. 1978.

T78c M.M. TerPogossian, N.A. Mullani, and J.J. Hood, C.S. Higgins, and D.C. Ficke, "Design cnsideration for a positron emission transverse tomography (PETT-IV) for the imaging of the brain," J. Comput. Assist. Tomog., vol. 2, pp. 439-444, Nov. 1978.

T78d O.J. Tretiak, "Noise limitations in X-ray computed tomography," J. Comput. Assist. Tomog., vol. 2, pp. 477-480, Sept. 1978.

W51 F.R. Wrenn, Jr. M.L. Good, and P. Handler, "The use of positron-emitting radioisotope for the localization of brain tumors," Nature, vol. 113, pp. 525-527, 1951.

W69 E. Wolf, "Three-dimensional structure determination of semi-transparent objects from holographic data," Optics Communications, vol. 1, pp. 153-156, 1969.

W76 G. Wade, "Historical perspectives," in Acoustic Imaging, Ed. G. Wade, New York: Plenum Press, pp. 21-42, 1976.

W77a P.N.T. Wells, "Ultrasonics in medicine and biology," Phys. Med. Biol., vol. 22, pp. 629-669, 1977.

W77b L. Wang, "Cross-section reconstruction with a fan-beam scanning geometry," IEEE Trans. Comput., vol. C-26, pp. 264-268, Mar. 1977.

W78a G. Wade, R.K. Mueller, and M. Kaveh, "A survey of techniques for ultrasonic tomography," in Proceedings of IFIP TC-4 Working Conference on Computer-Aided Tomography and Ultrasonics in Medicine, Ed. J. Raviv, Amsterdam: North-Holland, pp. 165-215, 1978.

W78b G.H. Williams, "The design of a rotational X-ray CT scanner," Medita, (Proc. MEDEX 78), vols. 6 and 7, pp. 47-53, June 1978.

W80a G. Wade, S. Elliott, I. Khogeer, G. Flesher, J. Eisler, D. Mensa, N.S. Ramesh, and G. Heidbreder, "Acoustic echo computer tomography," in Acoustic Holography, vol. 8, Ed. A. Metherell, New York: Plenum Press, pp. 565-576, 1980.

W80b G. Wade and D. Mensa, "Pulse echo and Doppler reconstructive tomography," to be published in Proceedings of the 1980 International Optical Computing Conference, 1980.

Y77   M. Yaffe, A. Fenster, and H.E. Johns, "Xenon ionization detectors for fan-beam computed tomography scanners," J. Comput. Assist. Tomog., vol. 1, pp. 419-428, 1977.

Y77   Y. Yamamoto, C.J. Thompson, E. Meyer, J.S. Robertson, and W. Feindel, "Dynamic positron emission tomography for study of cerebral hemodynamics in a cross-section of the head using positron-emitting ^{68}Ga-EDTA and ^{77}Kr," J. Comput. Assist. Tomog., vol. 1, pp. 43-56, Jan. 1977.

Z35   F. Zernike, "Das Phasenkontrastverfahren bei der mikroskopischen beobacktung," ("The phase contrast method of microscopic observation"), Z. Tech. Phys., vol. 16, pp. 454, 1935.

EXPERIMENTAL RESULTS IN ULTRASONIC DIFFRACTION TOMOGRAPHY

M. Kaveh, R. K. Mueller, R. Rylander, T. R. Coulter,
M. Soumekh
Department of Electrical Engineering
University of Minnesota
Minneapolis, MN 55455

ABSTRACT

This paper presents the measurement system and experimental
results in ultrasonic diffraction tomography. The tomographic reconstruction is based on the measurements of profiles of the complex
amplitude of the waves scattered from the object. The results are
tomograms of various phantoms in water and show a resolution of a few
wavelengths. Signal processing issues such as errors due to the
measurement and quantization noises are discussed. Simulation of
these effects are given and related to the scattered characteristics,
number of profiles, and frequency domain resolution.

INTRODUCTION

Ultrasonic diffraction tomography is an attempt at imaging
the internal structure of an object, such as its density, compressibility or loss factors, from the measurement of the waves
scattered from the object at different object orientations. The
general form of diffraction tomography has recently been formulated[1]. This scheme, however, has not yet been validated in simulations, due to the complexity of calculating the scattered field
and will be reported on in the future. Diffraction tomography for
objects with negligible loss and no density variation has been attempted and detailed simulations of such reconstruction has been
given [2,3,4].   In this paper, we present examples of the actual velocity tomograms of phantoms using the simplified diffraction technique. We believe that these are the first experimental ultrasonic
diffraction tomograms. Because of this, the quality of the images do
not represent accurately what may be obtainable with refinements in

433

the algorithms and experimental techniques. In the following, the reconstruction method is briefly reviewed. The experimental system and some results are then presented and finally some signal processing issues are discussed.

The Reconstruction Method

The assumption in deriving this reconstruction method is that the disturbance due to the object is reflected through the introduction of an inhomogeneous velocity of ultrasound propagation inside the surrounding medium. The problem is then posed as solving the inverse problem of the Helmholtz equation that describes the propagation of the sound field in the object and measurement medium.

Assuming single frequency insonifaction, the wave equation may be written as

$$c^2(r)\nabla^2\psi(\vec{r}) + \omega^2\psi(\vec{r}) = 0 \text{ or } \nabla^2\psi(\vec{r}) + k^2\psi(\vec{r}) = 0 \qquad (1)$$

where $c(\vec{r})$ is the velocity at position $\vec{r}$, $\psi(\vec{r})$ is the wave function, $\vec{r}$ is the position vector, ω is the angular frequency and $k = \omega/c$ is the wave number. It is now assumed that $c(\vec{r})$ in the object can be approximated by a small perturbation to the velocity in the surrounding (measurement) medium, i.e.,

$$c(\vec{r}) = (1-f(\vec{r}))c_o \qquad (2)$$

where c_o is the velocity in the medium surrounding the object and $f(\vec{r})$ is the perturbation to the velocity inside the object, with the assumption that $|f(r)|<<1$. This is a reasonable assumption for media and objects of interest.

Consider an object located at the origin of an x-y coordinate system insonified by a plane wave incident along the -x axis. A receiver scans along a line parallel to the y axis at x_o coherently measuring the wave comprised of the incident and scattered components

$$\psi(\vec{r}) = \psi_o(\vec{r}) + \psi_s(\vec{r}) \qquad (3)$$

A first order perturbation approximation is used with $|f(\vec{r})|<<1$ and $|\psi_s(r)|/|\psi_o(\vec{r})| = |\psi_s(\vec{r})|<<1$. Substituting $\psi(\vec{r})$ and $c(\vec{r})$ in Eq. (1) and retaining only the first-order terms in $\psi_s(r)$ and $f(r)$ one obtains[2,3,4]

$$\nabla^2\psi_s(\vec{r}) + k^2\psi_s(\vec{r}) = -2k^2f(\vec{r})\psi_o(\vec{r}) \qquad (4)$$

The calculation of $f(\vec{r})$ from the measurements of $\psi_s(\vec{r})$ is done by rewriting Eq. (4) in the Fourier domain. Let $S(\vec{\Lambda})$ and $F(\vec{\Lambda})$ be the Fourier transforms of $\psi_s(r)$ and $f(r)$ respectively. Equation (4) may now be written as

$$-|\vec{\Lambda}|^2 S(\vec{\Lambda}) + k^2 S(\vec{\Lambda}) = -2k^2 F(\vec{\Lambda}) \tag{5}$$

This equation may now be used to solve for the transform of the object function $F(\vec{\Lambda})$ in terms of the transform of the observable $S(\vec{\Lambda})$. The accessible spatial frequencies $\vec{\Lambda}$ lie on a circle of radius k passing through the origin in the $\vec{\Lambda}$ plane with the center of the circle determined by the angular position of the object. Thus by repeating the measurements of $\psi(\vec{r})$ for a number of angular orientations of the object, the Fourier transform of the object is filled in slice by slice. In practice interpolation is used to obtain values of $F(\vec{\Lambda})$ at a predetermined frequency grid from these calculated values. The function $F(\vec{\Lambda})$ is then inverse transformed to reconstruct the object function $f(r)$[2,3,4]. Figure 1 shows the tomogram of an inhomogeneous torid obtained from simulated scattered fields. The tubular diameter is 6λ, the center hole has a radius of λ, and the maximum velocity perturbation is 2% • 32 profiles, with 32 samples at $\lambda/2$ spacing were used in the reconstruction.

Fig. 1. Tomogram of a toroid

THE EXPERIMENTAL SYSTEM

Data Acquisition

Figure 2 shows the block diagram of the experimental data acquisition system. The information required for reconstruction is a set of samples of the complex (amplitude and phase) acoustic field scattered by the object for multiple angular orientations. To

collect this data, test objects are placed on the object turntable
which is immersed in a water tank and are insonified with a 3 MHz
acoustic plane wave (λ=.5mm in water). A small (1λ diameter) receiving transducer mechanically scans a line parallel to the face
of the large (400λ wide by 50λ high) transmitting transducer with
the object turntable located between the two. The receiver line to
turntable center distance is 130λ and the receiver line to transmitter distance is 1000λ. The walls of the water tank are lined
with sound absorbing material to eliminate reflections and resonances. Sample locations are registered by optically reading an
encoded disc attached to a shaft which drives the receiver element.
Hardwired logic controls: 1) The number of samples taken per scan,
2) the receiver scan motor, 3) the turntable stepper motor,
4) the A/D converters and 5) microcomputer data storage.

Signal Detection and Data Storage

    Figure 3 is a block diagram representation of the method used
in the derivation of the normalized scattered wave. The top branch
corresponds to the measurement of the unperturbed incident plane
wave ψ_0. This figure shows the receiver signal detection hardware
and the computer algorithm used to determine the normalized
scattered wave ψ_1.

    A crystal-controlled oscillator provides a continuous 3 MHz
tone to a power amplifier which drives the transmitting transducer.
This signal is also passed through a 90° phase shift network providing the signal a need for coherent demodulation via in phase and
quadrature component detections. The sine and cosine waves are
mixed with the amplified received signal. The mixer outputs are low-
pass filtered, amplified, and then these baseband quadrature signal
components are fed to sample-and-hold modules. The motor speed and
gear ratios for the scanning receiver were chosen to give a sampling
period of 250 msec for the finest sample spacing (1/2λ). This
allows a cut-off frequency of 20 Hz to be used in the low-pass
filters to reduce noise while tracking the signals accurately for
the highest sample frequency.

    When the optical sensor detects a sample location, the analog
quadrature signal components are held constant while an 8-bit A/D
converter digitizes the data. The "data valid" signal from the A/D
converter initiates a "store data" cycle in the microcomputer. The
microcomputer codes each sample value into a four digit hexadecimal
word. The two most significant digits correspond to the in-phase
signal component and the two least significant digits correspond to
the quadrature signal computer. Each coded sample value is stored
on magnetic disc. Since the data storage operation is completed in
less than 20 msec and the minimum time between samples is 250 msec,
there is no synchronization problem in performing these operations.

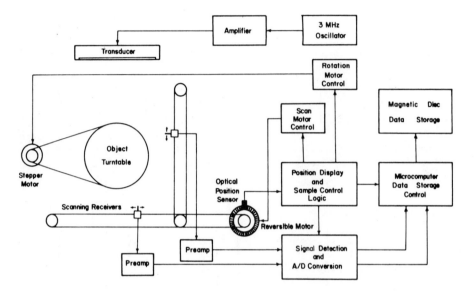

Fig. 2   Experimental System

The large transducer used to generate the incident plane wave is made of four individual crystal sections driven in parallel. The phase of the incident wave is fairly constant along the receiver scan line but the amplitude varies due to differences in the four sections. To correct for this imperfect plane wave, down and back reference scans are made (with no object on the turntable) at the start of each data collection run. These records of the incident reference wave are used to determine the normalized scattered wave when an object is present. After the reference scans are made the object is placed on the turntable and data is collected for the

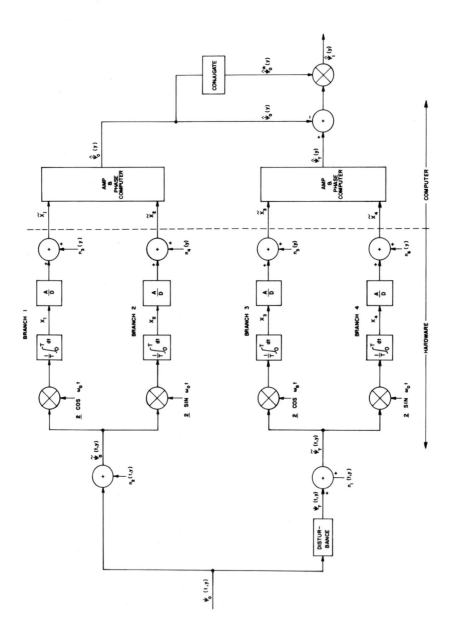

Fig. 3 Signal Detection System

measurement of the total wave for various angular orientations of the object.

After the object completes one revolution, the system automatically shuts off and signals the completion of data collection. The data is transmitted via a telephone link to a central computer facility for processing.

The preprocessing involved before reconstruction is shown schematically in the computer portion of figure 3. First, all sample values are assigned a complex representation by decoding each hexodecimal word into two real floating point words and then calculating their amplitude and phase. Since the reconstruction algorithm utilizes both forward and back scans, the forward and back reference scans are averaged to minimize mechanical backlash effects. This averaged reference scan is used to calculate the normalized scattered field for each object profile.

EXPERIMENTAL RESULTS

The reconstruction algorithm assumes the object to produce a small perturbation in the incident wave. To satisfy this requirement, objects were made of gelatin with a sound velocity 2% greater than the surrounding water medium. This gelatin was molded into 60λ and 120λ diameter cylinders which were perforated with holes of various sizes and orientations. Small plastic and glass rods were also used to estimate the impulse response and resolving power of the system. Although the material of the rods was significantly different from water, their small size (3λ) kept the total scattered wave small.

Sixty-four samples were taken spaced two wave-lengths apart for each of 64 angular orientations of the objects. The reconstruction program produced from this data an approximate velocity profile of the object with a complex value to each element of a 64x64 array. The real part, imaginary part, or complex magnitude of the grid values could then be used to produce a contour plot, as z values in perspective view of the grid, or to assign gray-scale values to the elements of a black and white photograph.

The reconstructions of the gelatin objects all showed good fidelity to the original bojects. The overall sizes of the cylinders were correct as were the sizes and orientations of the water-filled cavities wihtin them. Some ringing and edge enhancement was present due to, we believe, the finite measurement window used. The higher spatial frequencies were intentionally attenuated in the reconstruction algorithm to suppress much of the background noise without sacrificing too much resolution. Even with this high spatial frequency filtering, the image of a glass rod 3λ in diameter was roughly 5λ.

The object center to scan line and sample spacing distances used in the reconstruction program were varied by 15% using the same measurement data to see the effect of errors in these parameters on the final image. A change in either distance produced a lower signal-to-noise ratio in the image but the geometry remained constant. This imaging method seems to be relatively tollerant of small errors in defining the system parameters.

The pictures shown in figure 4 are examples of the reconstructions of various gelatine phantoms. The inaccuracies in modeling the disturbances as only a perturbation in the velocity of propagation, as well as signal acquisition limitations are evident in the form of edge effects at the discontinuities. Figure 5 depicts the image of two small glass rods. This is essentially a shadow tomogram and does not carry any quantitative information. It does, however, indicate the resolution of the experimental system. As a final example of our experimental results, figure 6 shows the tomogram of a 50λx30λ rectangular block of natural sponge in water. Because of the coarse and random nature of the sponge, no well defined structure can be distinguished. The general outline of the phantom however, is evident and indicates an example of diffraction tomography of a random medium.

SIGNAL PROCESSING CONSIDERATIONS

An important problem in practical image restoration and reconstruction, is the nature and effect of variuos forms of errors, e.g. round off error and measurement noise. In the process of solving the inverse problem using a deterministic approach, noise is generally amplified. In this section we summarize some results obtained in analyzing such errors in the type of diffraction tomography described in this paper.

Three types of errors are considered: 1) additive receiver noise, 2) quantization noise resulting from the digitization of the measured wave and 3) interpolation errors resulting from the frequency plane filling procedure used. In the following the effect of uniform quantization noise is first considered and related to the scatterer strength. Receiver noise and interpolation errors are then discussed and trade-offs on their effects are indicated.

Receiver and Quantization Noise

Figure 3 shows the manner in which several types of noise are introduced in the determination of the normalized scattered wave.

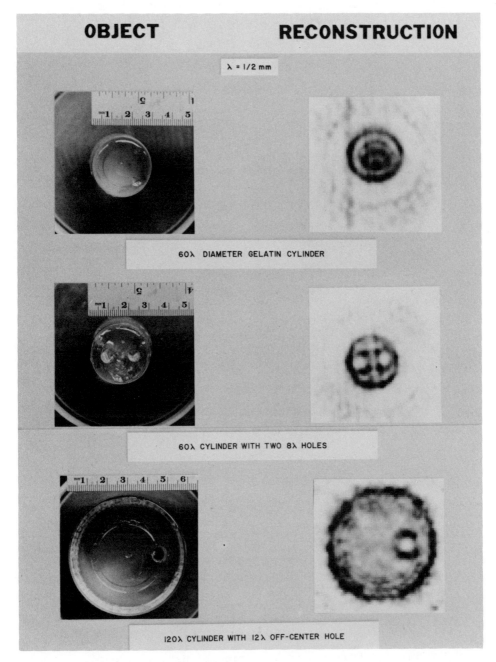

OBJECT   RECONSTRUCTION

$\lambda = 1/2$ mm

60λ DIAMETER GELATIN CYLINDER

60λ CYLINDER WITH TWO 8λ HOLES

120λ CYLINDER WITH 12λ OFF-CENTER HOLE

Fig. 4 Tomograms of gelatine phantoms in water.

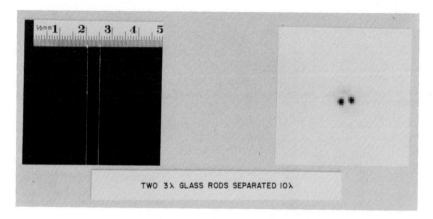

Fig. 5   Tomogram of two glass rods.

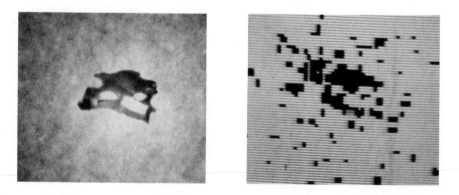

Fig. 6   Tomogram of a section of natural sponge.

We use the in-phase and quadrature representation of the reference wave ψ_o and the total wave ψ_T. These signal are then given by

$$\psi_o(x_o,y,t) = S_c(x_o,y)\cos \omega_o t + S_q(x_o,y)\sin \omega_o t$$

$$\psi_T(x_o,y,t) = [S_c(x_o,y) + S_c^{\,s}(x_o,y)]\cos \omega_o t \qquad (6)$$

$$+ [S_q(x_o,y) + S_q^{\,s}(x_o,y)]\sin \omega_o t$$

where S_c is the in-phase signal component and S_q is the quadrature signal component. The s superscript denotes the signal components due to the scattered wave. The additive noise which is introduced into the measurement may be characterized by

$$n_i(y,t) = n_c(y,t)\cos \omega_o t + n_q(y,t)\sin \omega_o t \qquad (7)$$

where the $n_i(y,t)$ is assumed to be a white indpendent zero mean Gaussian random process in both variables y and t. Spatial noise will be ignored. The power spectral density with respect to the time variable has height $N_o/2$ over a bandwidth of W which is the noise equivalent bandwidth of the receiving transducer preamplifier. The outputs of the integrate-and-dump circuits are Gaussian random variables with

$$x_1 = S_c$$

$$x_2 = S_q$$

$$x_3 = S_c + S_c^{\,s} + n_c$$

$$x_4 = S_q + S_q^{\,s} + n_q$$

$$VAR(x_i) = \sigma_x^2 = \frac{N_o W}{T}$$

where $<\ >$ is the statistical expectation operator.

Next we consider the noise due to the digitization of the signal components. Assume that the quantization error components are independent identically distributed uniform random variables having probability density function of height Q/2 over the range [-1/Q, 1/Q] for Q levels of quantization. In vector notation the normalized scattered wave is given by

$$\hat{\psi}_1 = \frac{\hat{\psi}_T - \hat{\psi}_o}{\hat{\psi}_o} \tag{8}$$

It should be noted that the small perturbation assumption implies a small value for $|\psi_R/\psi_o|$. If this normalized difference is on the order of a quantization step, significant phase errors may result due to round off. The resultant vector ψ_R may be written as

$$\hat{\psi}_R = \psi_R + \varepsilon_R$$

where ψ_R is the actual resultant vector and ε_R is the complex quantization error vector. The joint density of the components of $\varepsilon_R = \varepsilon_1 + i\varepsilon_2$ is[5]

$$p(\varepsilon_1, \varepsilon_2) = \frac{Q^2}{4} [1 - |\varepsilon_1| \frac{Q}{2}] [1 - |\varepsilon_2| \frac{Q}{2}]$$

$$\text{for } |\varepsilon_1|, |\varepsilon_2| \le 2/Q \tag{9}$$

$$0 \qquad \text{otherwise} \qquad .$$

The mean-square-error resulting from quantization is $4/3Q^2$. For analytical tractability, $P(\varepsilon_1, \varepsilon_2)$ is represented by a conic surface, namely,[5]

$$p(\varepsilon_1, \varepsilon_2) = \frac{3Q^2}{4\pi} [1 - \sqrt{\varepsilon_1^2 + \varepsilon_2^2} \frac{Q}{2}] \tag{10}$$

the density of the amplitude of the complex error $|\varepsilon_R|$ is then given by[5]

$$p(|\varepsilon_R|) = \frac{3Q^2}{2} |\varepsilon_R| [1 - |\varepsilon_R| \frac{Q}{2}], |\varepsilon_R| \le 2/Q \tag{11}$$

For the above density the expected value of the amplitude of the complex quantization error is

$$|\varepsilon_R| = 1/Q$$

which is the midpoint in the range of $p(|\varepsilon_R|)$.

The complex error ε_R may be expressed in aplitude and phase form as

$$\varepsilon_R = |\varepsilon_R| e^{i\phi_{\varepsilon_R}}$$

where ϕ_{ε_R} is assumed to be uniformity distributed over $[-\pi, \pi]$.

Defining θ_ϵ as the angle between the actual resultant vector ψ_R and its quantized version $\hat{\psi}_R$, the phase error due to quantization can be shown to be[5]

$$\theta_\epsilon = \tan^{-1} \left(\frac{\sin \beta}{\frac{|\psi_R|}{|\epsilon_R|} + \cos \beta} \right) \tag{12}$$

where β is the angle between ψ_R and ϵ_R and is assumed to be uniformly distributed (See figure 7).   As the ratio $|\psi_R|/|\epsilon_R|$ tends to zero, θ_ϵ approaches a uniformly distributed random variable.   As $|\psi_R|/|\epsilon_R|$ becomes large, θ_ϵ is approximately

$$\theta_\epsilon \sim \frac{|\epsilon_R|}{|\psi_R|} \sin \beta. \tag{13}$$

under this assumption the density of the phase error is given by

$$p(\theta_\epsilon) = \frac{1}{\pi} \frac{1}{\sqrt{|\epsilon_R|^2/|\psi_R|^2 - \theta_\epsilon^2}} \qquad |\theta_\epsilon| < \frac{|\epsilon_R|}{|\psi_R|} \tag{14}$$

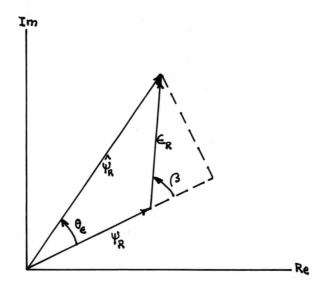

Fig. 7.   Phase error of the quantized resultant vector

     An important conclusion to be drawn from the previous analysis is that the effect of uniform quantization is dependent on the size of the disturbance.   Reconstructions of smaller disturbances will

be more severely degraded. Figure 8 shows this effect for two
disturbances.

Interpolation and Effects of Noise

We have used, for computational efficiency, an averaged zero-
order interpolation scheme[3] to fill the frequency plane. The interpolated values of the Fourier transform $F[u(j), v(k)]$ is then
given by

$$F[u(j), v(k)] = \frac{1}{[N_1]} \sum_{i=1} F_i[u_R(j), v_R(k)] \tag{15}$$

where $[N_1]$ is the integer number of times the rotated frequencies
$(u_R(j), v_R(k))$ fall within the neighborhood of $(u(j), v(k))$. Given
that we are in an interpolation cell and ignoring interpolation
errors we may write

$$F(u(j), v(k)) = \tilde{F}[u(j), v(k)] + \frac{1}{[N_1]} \sum_{i=1}^{[N_1]} Q_i[u(j), v(k)] \tag{16}$$

where $Q_i(u(j), v(k))$ are complex random variables due to additive
and quantization noise and where $\tilde{F}[u(j), v(k)]$ is the actual signal
component.

To obtain an expression which approximates the interpolation
error, a first order Taylor expansion is used yielding

$$\varepsilon_i(u,v) = \nabla F(u_{Ri}, v_{Ri}) \cdot \vec{R}_i(u,v) \tag{17}$$

with

$$R_i(u,v) = \begin{pmatrix} u - u_{Ri} \\ v - v_{Ri} \end{pmatrix}$$

where $\varepsilon_i(u,v)$ is the error in interpolating from the rotated frequencies (u_R, v_R) to the grid frequencies (u,v). The average
interpolation error is defined as

$$\varepsilon_a(u, v) = \begin{cases} \dfrac{1}{[N_1]} \sum_{i=1}^{[N_1]} \nabla F(u_{Ri}, v_{Ri}) \cdot \vec{R}_i(u, v) & [N_1] > 0 \\[4ex] F(u,v) & [N_1] = 0 \end{cases} \tag{18}$$

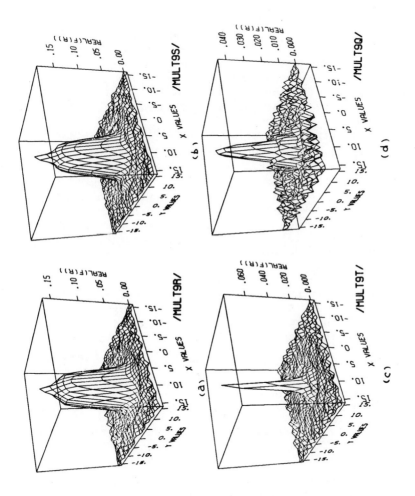

Fig. 8   Effect of quantization errors on reconstruction.   181 quantization levels and 1% disturbance used.   a) Non-quantized, R=5λ, b) quantized, R=5λ, c) Non-quantized, R=1λ, d) quantized, R=1λ.

The model used to describe the interpolation pattern is given by

$$N_1(u,v) \sim k \exp\left(-\frac{u^2 + v^2}{2\alpha^2}\right)$$ (19)

where k is the number of profiles. The parameter α is dependent on the number of profiles k, the profile vector frequency spacing Δvp, and the reconstruction grid frequency spacing Δv. For a particular k as the ratio $\Delta v/\Delta vp$ becomes large (i.e. a coarser grid spacing, $N_1(u,v)$ broadens and hence more interpolations occur at a point than would have occurred for a finer grid spacing. Since $N_1(u,v)$ is large at low spatial frequenices, the measurement noise power is significantly reduced at these frequencies due to the averaging performed in the interpolation scheme. This reduction is even more pronounced for a coarser grid spacing.

The average interpolation error is object dependent and is dependent on $N_1(u,v)$. The object dependence is twofold: 1) if F(u,v) is fairly flat over a grid cell, minimum errors occur given that the interpolation has occurred, and 2) for those grid locations where no interpolation occurs and F(u,v) is negligible, minimum errors occur. If the reverse of 1) is true, on the average, the errors for the lower frequencies will be less due to the nature of $N_1(u,v)$. However, if F(u,v) is not small at those grid locations where interpolation does not occur, large errors result.

The trade-off between measurement noise and interpolation noise reduction is apparent. The measurement noise power has $[N_1]$ dependence for $[N_1] \neq 0$. A coarser grid spacing further reduces the low frequency noise power. However, interpolation errors increase due to the increase in the distance vectors and due to the possible wider variation of F(u,v) over a grid cell. Figure 9 shows a profile and the reconstruction of two cylindrical disturbances in the presence white Gaussian noise. The smoothing effect on the low frequency noise, due to averaging is apparent. Higher frequency noise is filtered with a post reconstruction filter.

CONCLUSIONS

We have demonstrated experimentally, that ultrasonic diffraction tomography is a viable imaging technique. A measurement technique was described and examples of successful diffraction tomograms for simple phantoms were presented. The effects of noise, quantization and interpolation errors were then discussed. These erros were then related to such parameters as scatterer strength and bandwidth.

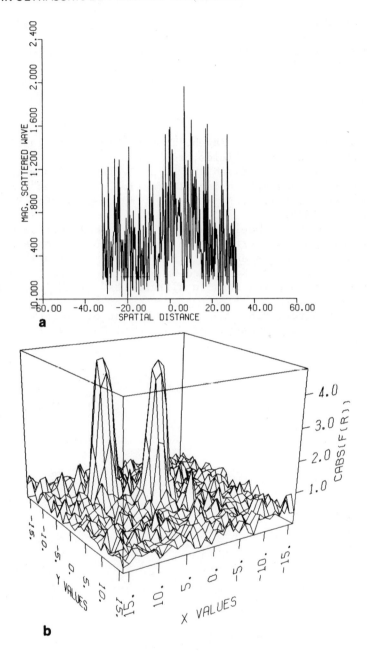

Fig. 9   Diffraction tomogram of two cylindrical disturbances.
$R_1=R_2=2\lambda$, 5% disturbance each, signal-to-noise ratio
=5dB.   (a)   A profile, (b)   tomogram with 36 views 256
samples/profile.

ACKNOWLEDGMENT

This work was supported by the National Science Foundation under Grant EiG 76-84521.

REFERENCES

[1] R. K. Mueller and M. Kaveh, Tomographic reconstruction from scattered acoustic fields, to be published.

[2] R. K. Mueller, M. Kaveh and R. D. Iverson, "A new approach to acoustic tomography using diffraction techniques", Acoustic Holography, Vol. 8, A. Metherell, Ed., New York: Plenum Press, (1978).

[3] M. Kaveh, R. K. Mueller and R. D. Iverson, Ultrasonic tomography based on perturbation solutions of the wave equation, Computer Graphics and Image Processing, 37: (1974).

[4] R. K. Mueller, M. Kaveh and G. Wade, "Reconstructive tomography and applications to ultrasonics", IEEE Proceedings, 67: (1979).

[5] T. R. Coulter, M.S. Thesis, Department of Electrical Engineering, University of Minnesota, December 1979.

EXPLICIT INVERSION OF THE HELMHOLTZ EQUATION FOR ULTRA-SOUND INSONIFICATION AND SPHERICAL DETECTION

[o] James Ball
[*] Steven A. Johnson[1]
[+] Frank Stenger[2]

[o] Department of Physics
University of Utah
Salt Lake City, UT 84112

[*] Department of Bioengineering
University of Utah
Salt Lake City, UT 84112

[+] Department of Mathematics
University of Utah
Salt Lake City, UT 84112

INTRODUCTION

In this paper we wish to describe a method by which scattering data measured on a closed surface can be used to reconstruct the sound speed as a function of position in the enclosed volume. When the variation of the sound speed from its average value is small, the ray paths become straight lines, and the inversion can then be easily performed. For largervariations of the sound velocity, the paths become curved, and the three dimensional nature of the problem becomes essential. In our analysis we will employ the Rytov approximation, which has the following desirable properties:

1. When variations are small, it reduces to the straight line case.

[1]Supported by NIH Grant Numbers HL-006 87 and CA-234 30.
[2]Supported by U. S. Army Research Contract Number DAAG-29-07-G-0139 and by NIH Grant Number CA-234 30.

2. It reduces to the Born approximation when the scatterer is small compared to the wavelength of the sound.

3. It is a three dimensional treatment and includes contributions from none-straight line paths.

In Sec. 2 the Rytov approximation for the scattering data will be obtained for both the case of a plane wave ultrasound source and that of a spherical wave source. These expressions are then inverted, resulting in an exact spherical harmonic expansion for the spacial Fourier transform of sound velocity. This Fourier transform is obtained within a sphere having center at the origin and radius $2k$ in the three dimensional space of the Fourier transform variable, where $k = \omega/c_0$ and where c_0 is the average of the sound velocity.

In Sec. 3 we obtain approximate integral expressions for the sound velocity within a sphere as a function of $\bar{r}$. The sound velocity is calculated directly from data measured on the surface of the sphere.

THE RYTOV APPROXIMATION AND ITS INVERSION

Sound excitation in a stationary medium with spacial variation of the local density ρ and compressibility κ is described by the equation

$$(\nabla^2 - \frac{1}{c^2(\bar{r})} \frac{\partial^2}{\partial t^2})p = 0 \quad . \tag{2.1}$$

Here $p = p(\bar{r},t)$ is the pressure, and $c(\bar{r}) = (\kappa(\bar{r})\rho(\bar{r}))^{-\frac{1}{2}}$ denotes the local sound velocity in the medium. In deriving this equation it has been assumed that the amplitude of sound waves is small enough so that the second order quantities such as $|\Delta\rho|^2$ and $|\Delta\rho|^2$ can be ignored relative to $\Delta\rho$ and $\Delta\rho$ [1].

Assuming sound of a single frequency, i.e.,

$$p(\bar{r},t) = u(\bar{r})e^{i\omega t} \quad , \tag{2.2}$$

we see from Eq. (2.1) that u will satisfy the Helmholtz equation

$$\nabla^2 u + \frac{\omega^2}{c^2(\bar{r})} u = 0 \quad . \tag{2.3}$$

This equation will be solved approximately for various sound sources by means of the Rytov approximation.

We begin by defining

$$f(\bar{r}) = \omega^2 \left\{ \frac{1}{c^2(\bar{r})} - \frac{1}{c_0^2} \right\} , \quad k = \frac{\omega}{c_0} \tag{2.4}$$

where c_0 is the average speed of sound in the medium, and will also be assumed to be the speed of sound of the medium surrounding the body.

(a)   The Case of a Spherical Wave Source

For the case of a unit point source located at the point $\bar{r}_s$, we may rewrite Eq. (2.3) in the form

$$\nabla^2 u + k^2 u = -fu + \delta^3(\bar{r} - \bar{r}_s) \quad . \tag{2.5}$$

The outgoing wave Green's function for the Helmholtz equation is

$$G(\bar{r}, \bar{r}_s) = \frac{e^{ik|\bar{r} - \bar{r}_s|}}{4\pi |\bar{r} - \bar{r}_s|} \quad ; \tag{2.6}$$

this is the $f = 0$ solution to (2.5) and will be useful in obtaining our results.

Setting

$$u = Ge^V \tag{2.7}$$

in (2.3), we get

$$\nabla^2 V + 2(\bar{\nabla}V) \cdot \frac{\bar{\nabla}G}{G} + (\bar{\nabla}V) \cdot (\bar{\nabla}V) + f = (e^{-V} - 1)\delta^3(\bar{r} - \bar{r}_s) \tag{2.8}$$

where the source term is added on the right hand side.

Assuming that V is slowly varying so that $(\bar{\nabla}V) \cdot (\bar{\nabla}V)$ is negligible relative to the other quantities in (2.7), we may drop this term. This is the assumption of validity of the Rytov approximaton. We may also drop the term on the right hand side of equation (2.8), by assuming that $V(\bar{r}_s) = 0$. Setting

$$V = \frac{w}{G} \tag{2.9}$$

in the resulting equation, we get

$$\nabla^2 w + k_0^2 w = -fG \; ; \tag{2.10}$$

the solution to this equation by the usual Green's function method is

$$w(\bar{r},\bar{r}_s) = - \int_{\mathbb{R}^3} d^3\bar{r}' G(\bar{r},\bar{r}') f(\bar{r}') G(\bar{r}',\bar{r}_s) . \tag{2.11}$$

Since $f(\bar{r}) = 0$ if $|\bar{r}| \geq a$, $w(\bar{r},\bar{r}_s)$ is finite at $\bar{r} = \bar{r}_s$, and from equation (2.9) we do indeed have $V(\bar{r}_s) = 0$.

(b) The Case of a Plane Wave Source

Similarly, for the case of a plane wave source

$$u_0(\bar{r},\bar{k}) = e^{i\bar{k}\cdot\bar{r}}; \; |\bar{k}| = k , \tag{2.12}$$

we set

$$u = u_0 e^w \tag{2.13}$$

in (2.3) and drop the $\bar{\nabla}w \cdot \bar{\nabla}w$ term (the Rytov approximation) to get

$$w(\bar{r},\bar{k}) = \int_{\mathbb{R}^3} d^3\bar{r}' f(\bar{r}') G(\bar{r},\bar{r}') e^{i\bar{k}\cdot\bar{r}'} \tag{2.14}$$

(c) Solution of Equation (2.11) via Spherical Harmonics

Let $\bar{x}$ denote an arbitrary vector in $\mathbb{R}^3$. We use the notation $x = |\bar{x}|$ for the magnitude of $\bar{x}$, and $\hat{x} = \bar{x}/|\bar{x}|$ for a unit vector in $\mathbb{R}^3$ having the direction of $\bar{x}$.

We shall assume that both the source and the detector are located on sphere of radius a containing the body. We shall thus measure the solution w of equations (2.11) and (2.14) at $\bar{r} = \bar{r}_d = a\hat{r}_d$. In that way, the solution w of equation (2.11) becomes a function of $\hat{r}_s$ and $\hat{r}_d$ only, i.e., we define W by

$$W(\hat{r}_s,\hat{r}_d) = w(a\hat{r}_d,a\hat{r}_s) . \tag{2.15}$$

Each of the following two identities is instrumental in the solution of equations (2.11) and (2.14) [2 equations 8.533 and 8.511].

If $\bar{r}$ and $\bar{k}$ denote arbitrary vectors in $\mathbb{R}^3$, then

$$e^{i\bar{k}\cdot\bar{r}} = e^{ikr\hat{k}\cdot\hat{r}} = \sqrt{\frac{\pi}{2kr}} \sum_{\ell=0}^{\infty} i^{\ell}(2\ell+1)J_{\ell+\frac{1}{2}}(kr)P_{\ell}(\hat{k}\cdot\hat{r}). \quad (2.16)$$

The Green's function can also be expanded as follows, if $r'<r$, then

$$\frac{e^{ik|\bar{r}-\bar{r}'|}}{4\pi|\bar{r}-\bar{r}'|} = \frac{i}{8\sqrt{rr'}} \sum_{\ell=0}^{\infty} (2\ell+1)J_{\ell+\frac{1}{2}}(kr')H^{(1)}_{\ell+\frac{1}{2}}(kr)P_{\ell}(\hat{r}\cdot\hat{r}') \quad (2.17)$$

In the expansions (2.16) and (2.17) $J_{\ell+\frac{1}{2}}$, $H^{(1)}_{\ell+\frac{1}{2}}$ and P_{ℓ} denote the usual Bessel functions, Hankel functions and Legendre polynomials respectively.

Substituting the expansion (2.17) into Eq. (2.11) and using the definition (2.15), we get

$$W(\hat{r}_s,\hat{r}_d) = \frac{1}{64a} \int_{\mathbb{R}^3} d^3r \frac{f(\bar{r})}{r} \cdot$$

$$\cdot [\sum_{\ell=0}^{\infty} (2\ell+1)J_{\ell+\frac{1}{2}}(kr)H^{(1)}_{\ell+\frac{1}{2}}(ka)P_{\ell}(\hat{r}_s\cdot\hat{r})] \quad (2.18)$$

$$\cdot [\sum_{m=0}^{\infty} (2m+1)J_{m+\frac{1}{2}}(kr)H^{(1)}_{m+\frac{1}{2}}(ka)P_m(\hat{r}_d\cdot\hat{r})].$$

Now if m and M are non-negative integers, then (see [2 Eqs. 8.814, 8.815, 7.112])

$$\int_{\Omega_s} d\Omega_s P_m(\hat{r}_s\cdot\hat{r}')P_M(\hat{r}_s\cdot\hat{r}) = \frac{4\pi}{2M+1} \delta_{Mm}P_M(\hat{r}'\cdot\hat{r}). \quad (2.19)$$

We therefore multiply Eq. (2.18) by

$$\frac{(2L+1)i^{L}P_L(\hat{r}_s\cdot\hat{r}')}{H^{(1)}_{L+\frac{1}{2}}(ka)} \quad \frac{(2M+1)i^{M}P_M(\hat{r}_d\cdot\hat{r}'')}{H^{(1)}_{M+\frac{1}{2}}(ka)}$$

where L and M are non-negative integers and where $\hat{r}'$ and $\hat{r}''$ are arbitrary unit vectors in $\mathbb{R}^3$, integrate over $\Omega_s \times \Omega_d$

use Eq. (2.19), sum over L and M and then use Eq. (2.16), to get

$$\int_{\Omega_s} d\Omega_s \int_{\Omega_d} d\Omega_d W(\hat{r}_s, \hat{r}_d) g(\hat{r}_s \cdot \hat{r}') g(\hat{r}_d \cdot \hat{r}'')$$

$$= \frac{k\pi}{2a} \int_{\mathbb{R}^3} d^3\overline{r} \; e^{ik\overline{r} \cdot (\hat{r}' + \hat{r}'')} f(\overline{r}) \tag{2.20}$$

$$= \frac{k\pi}{2a} \hat{f}(k(\hat{r}' + \hat{r}''))$$

where $\hat{f}$ denotes the Fourier transform of f, and where g is defined by

$$g(z) = \sum_{L=0}^{\infty} \frac{(2L+1)i^L P_L(z)}{H^{(1)}_{L+\frac{1}{2}}(ka)}. \tag{2.21}$$

If we vary $\hat{r}'$ and $\hat{r}''$ in (2.20) we can thus evaluate $\hat{f}$ at all points of a sphere of radius $2k$.

(d) Solution of Eq. (2.14) via Spherical Harmonics.

Let us set

$$W(\hat{r}_s, \hat{r}_d) = w(a\hat{r}_d, k\hat{r}_s) \tag{2.22}$$

in Eq. (2.14) and let us make the substitution (2.17) for G, Then, after multiplying each side by

$$\frac{i^L (2L+1) P_L(\hat{r}_d \cdot \hat{r})}{H^{(1)}_{L+\frac{1}{2}}(ka)} ,$$

using (2.19), summing over L from 0 to ∞ and then using (2.16), we get

$$\int_{\Omega_d} d\Omega_d W(\hat{r}_s, \hat{r}_d) g(\hat{r}_d, \hat{r})$$

$$= \sqrt{\frac{k}{32\pi a}} \int_{\mathbb{R}^3} d^3 r' \, e^{i k \overline{r}' \cdot (\hat{r}_s + \hat{r})} \, f(\overline{r}') \qquad (2.23)$$

$$= \sqrt{\frac{k}{32\pi a}} \, \hat{f}(k(\hat{r}_s + \hat{r}))$$

where g is defined as in (2.18) and where $\hat{f}$ denotes the Fourier transform of f.

We note again that by varying $\hat{r}_s$ and $\hat{r}$ we can use (2.23) to evaluate $\hat{f}$ at any point in the interior of a sphere of radius 2k. In practice it may be sufficient to know $\hat{f}$ only within a sphere of radius k. In this case it is necessary to vary $\hat{r}_s$ only along two orthogonal geodesics on the sphere.

INVERSION OF THE FOURIER TRANSFORMS

Clearly, if f denotes the Fourier transform of f, then

$$f(\overline{r}') = \frac{1}{(2\pi)^3} \int_{\mathbb{R}^3} d^3 \overline{R} \, e^{-i\overline{r}' \cdot \overline{R}} \, \hat{f}(\overline{R}). \qquad (3.1)$$

However, in the formulas of Sec. 2, $\hat{f}(\overline{R})$ is known only for $|\overline{R}| \leq 2k$. We therefore define F by the inverse transform of $\hat{f}$ over $|\overline{R}| \leq 2k$, that is

$$F(\overline{r}') = \frac{1}{(2\pi)^3} \int_{|\overline{R}| \leq 2k} d^3 \overline{R} \, e^{-i\overline{r}' \cdot \overline{R}} \, \hat{f}(\overline{R}). \qquad (3.2)$$

This means that we can detect objects larger than a wavelength. In applications with frequencies greater than 1 megahertz we may therefore expect the approximation

$$F(\overline{r}') \cong f(\overline{r}') \qquad (3.3)$$

to be quite accurate.

(a) Inversion of Transform in the Plane Wave Source Case

Let us return to (2.20) in the form

$$\hat{f}(k(\hat{r}_s + \hat{r})) = \sqrt{\frac{32\pi a}{k}} \int_{\Omega_d} d\Omega_d W(\hat{r}_s, \hat{r}_d) g(\hat{r}_d \cdot \hat{r}). \qquad (3.4)$$

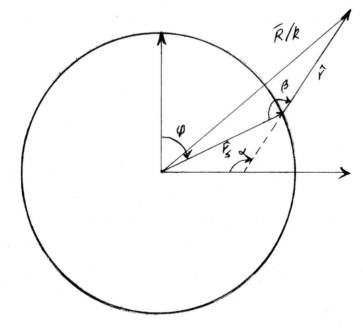

FIGURE 3.1

For purposes of combining Eqs. (3.2) and (3.4), we consider the variables in Figure 3.1, and set

$$\begin{cases} \overline{R} = k(\hat{r}_s + \hat{r}) \\[2mm] \hat{r}_s = (\sin\phi\cos\theta, \ \sin\phi\sin\theta, \ \cos\phi) \\[2mm] \hat{r} = (\sin(\phi+\beta-\pi)\cos\theta, \ \sin(\phi+\beta-\pi)\sin\theta, \ \cos(\phi+\beta-\pi)) \\[2mm] \quad = -(\sin(\phi+\beta)\cos\theta, \ \sin(\phi+\beta)\sin\theta, \ \cos(\phi+\beta)). \end{cases} \qquad (3.5)$$

Clearly, the region

$$\{(\phi,\theta,\beta) : 0 \leq \phi \leq \pi, \ 0 \leq \theta \leq 2\pi, \ 0 \leq \beta \leq \pi\} \qquad (3.6)$$

combined with (3.5) is the region $\{\overline{R} \in \mathbb{R}^3 : |\overline{R}| \leq 2k\}$, where, by (3.5)

$$\bar{R} = 2k\sin(\beta/2)(-\cos(\phi+\beta/2)\cos\theta, \ -\cos(\phi+\beta/2)\sin\theta, \ \sin(\phi+\beta/2)).$$
$$(3.7)$$

By considering an element of volume under the transformation (3.5)-(3.7), we have

$$d^3\bar{R} = |\bar{R}|^2|\cos(\phi+\beta/2)| \ d|\bar{R}|d\phi \ d\theta$$

$$= 4k^3\sin^2(\beta/2)\cos(\beta/2)|\cos(\phi+\beta/2)|d\phi d\theta d\beta.$$
$$(3.8)$$

Combining Eqs. (3.2), (3.4), (3.5), (3.6), (3.7) and (3.8) we obtain

$$\begin{cases} F(\bar{r}') = \dfrac{2k^3}{\pi^3}\sqrt{\dfrac{2\pi a}{k}} \displaystyle\int_{\Omega_d} d\Omega_d \int_{\Omega_s} d\Omega_s \int_0^\pi d\beta \ \cdot \\[2ex] \qquad \cdot \ W(\hat{r}_s,\hat{r}_d)g(\hat{r}_d \cdot \hat{r})e^{-ik\bar{r}' \cdot (\hat{r}_s + \hat{r})} \qquad \cdot \\[2ex] \qquad \cdot \ \sin\phi_d \ \sin^2(\beta/2)\cos(\beta/2)|\cos(\phi+\beta/2)|. \end{cases} \qquad (3.9)$$

Finally, setting

$$K(\hat{r}_d,\hat{r}_s,\bar{r}') = \dfrac{2k^3}{\pi^3}\sqrt{\dfrac{2\pi a}{k}} \ \sin\phi_d \ e^{-ik\bar{r}' \cdot \hat{r}_s} \cdot$$

$$\cdot \int_0^\pi d\beta \ \sin^2(\beta/2)\cos(\beta/2)|\cos(\phi+\beta/2)|g(\hat{r}_d \cdot \hat{r})e^{-ik\bar{r}' \cdot \hat{r}} \qquad (3.10)$$

we have

$$F(\bar{r}') = \int_{\Omega_d} d\Omega_d \int_{\Omega_s} d\Omega_s \ K(\hat{r}_d,\hat{r}_s,\bar{r}')W(\hat{r}_s,\hat{r}_d).$$

(b)   Inversion of Transform in the Spherical Wave Source Case

In this case, Eq. (2.17) yields

$$\hat{f}(k(\hat{r}'+\hat{r}'')) = \dfrac{2a}{k\pi} \int_{\Omega_s} d\Omega_s \int_{\Omega_d} d\Omega_d \ g(\hat{r}_s \cdot \hat{r}')g(\hat{r}_d \cdot \hat{r}'')W(\hat{r}_s,\hat{r}_d).$$
$$(3.12)$$

Proceeding as for (3.4), we set

$$\begin{cases} \overline{R} = k(\hat{r}' + \hat{r}'') \\ \hat{r}' = (\sin\phi\cos\theta, \ \sin\phi\sin\theta, \ \cos\theta) \\ \hat{r}'' = -(\sin(\phi+\beta)\cos\theta, \ \sin(\phi+\beta)\sin\theta, \ \cos(\phi+\beta)) \end{cases} \quad (3.13)$$

so that

$$\overline{R} = 2k\sin(\beta/2)(-\cos(\phi+\beta/2)\cos\theta, \ -\cos(\phi+\beta/2)\sin\theta, \ \sin(\phi+\beta/2)). \quad (3.14)$$

Then, as in (3.8),

$$d^3\overline{R} = 4k^3 \sin^2(\beta/2)\cos(\beta/2)|\cos(\phi+\beta/2)|d\phi d\theta d\beta. \quad (3.15)$$

Substituting Eqs. (3.12), (3.13) and (3.15) into (3.2) and considering (3.6), we get

$$F(\overline{r}) = \frac{ak^2}{\pi^4} \int_{\Omega_s} d\Omega_s \int_{\Omega_d} d\Omega_d \int_0^{2\pi} d\theta \int_0^\pi d\phi \int_0^\pi d\beta \ \cdot$$

$$\cdot \ W(\hat{r}_s,\hat{r}_d)g(\hat{r}_s\cdot\hat{r}')g(\hat{r}_d\cdot\hat{r}'')e^{ik\overline{r}\cdot(\hat{r}'+\hat{r}'')} \quad (3.16)$$

$$\cdot \ \sin\phi_s\sin\phi_d\sin^2(\beta/2)\cos(\beta/2)|\cos(\phi+\beta/2)| \ .$$

Proceeding similarly as in Eq. (3.10), we now set

$$K(\hat{r}_s,\hat{r}_d,\overline{r}) = \frac{ak^2}{\pi^4} \int_0^{2\pi} d\theta \int_0^\pi d\phi \int_0^\pi d\beta \ \cdot$$

$$\cdot \ g(\hat{r}_s\cdot\hat{r}')g(\hat{r}_d\cdot\hat{r}'')e^{ikr\cdot(\hat{r}'+\hat{r}'')}$$

$$\cdot \ \sin\phi_s\sin\phi_d\sin^2(\beta/2)\cos(\beta/2)|\cos(\phi+\beta/2)|$$

in (3.16), to get

$$F(\overline{r}) = \int_{\Omega_s} d\Omega_s \int_{\Omega_d} d\Omega_d \ K(\hat{r}_s,\hat{r}_d,\overline{r})W(\hat{r}_s,\hat{r}_d). \quad (3.18)$$

Detailed algorithms for computing the kernels K and K will be published elsewhere.

REFERENCES

[1] P. M. Morse and K. U. Ingard, "Theoretical Acoustics," McGraw-
     Hill, N.Y. (1965).
[2] I. S. Gradshteyn and I. M. Ryzhik, "Table of Integrals, Series
     and Products," Academic Press, N.Y. (1965).

AN APPROXIMATE PROPAGATION SPEED INVERSION OVER A PRESCRIBED SLAB

S. Raz
The Technion-Israel
Institute of Technology
Department of Electrical Engineering

Abstract

An algorithm is proposed which facilitates an explicit velocity profile inversion in the presence of a presumedly known, background slab. The formulation is outlined within the confine of the so called, "distorted wave" Born approximation in which the background effects are properly accounted for. A plane wave excitation is considered.

The presence of a non-negligible background slab presents the following, previously unencountered, difficulties. Firstly, the slab's multiple reflections impose a sequential reconstruction procedure, in which each iterative step yields the reconstruction of a finite spatial slab. This contrasts the one-step reconstruction, achieved by Bleistein and Cohen over a homogeneous background.

Secondly, difficulties associated with the reality of limited band measurements may be magnified. Finite bandwidth data may render the inversion procedure nonunique.

1. Introduction

The feasibility of an explicit velocity profile inversion from scattered field data, presuming the validity of the single-scatter (Born) model was reported by Bleistein, Cohen and Gray.[1-4] Here, an algorithm is proposed which facilitates the inversion of a stratified profile in the presence of a, presumedly known, background slab (Fig. 1). The procedure is formulated within the confine of the, so called, "distorted wave Born" approximation, in which the background effects are properly accounted for.

Although the sequence may be formulated for arbitrary, finite aperture, excitations,[5] detailed attention is paid to the specifics of an arbitrarily oriented incident plane wave.

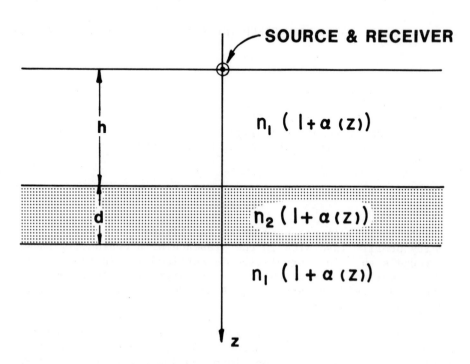

Fig. 1   The Physical Configuration

The slab's multiple reflections impose a sequential, piece-meal reconstruction sequence. Each iterative step yields the reconstruction of a finite slab of thickness $(n_2/n_1)d$ where n_2 and n_1 are the refractive indexes in and out of the slab's region, and d is its thickness (Fig. 1). At each step the successful completion of all the preceeding iterations is presumed. This contrasts the complete, one-step reconstruction achievable in the absence of multiple reflections. Furthermore, difficulties associated with the reality of limited band measurements may be magnified. The finite bandwidth will not only limit the resolu-tion but may render the inversion procedure non-unique whenever the resolution range exceeds $2(n_2/n_1)d$.

We end this section on the following note. An outstanding deficiency of the procedures suggested in references (1-5) is that, given the physical realities of acoustical scattering, the resultant reconstruction does not, in fact, represent a true velocity profile. Even within the single scatter regime, the scattered acoustical field is generated by two separate mechanisms, the medium's spatial velocity variations and the corresponding density gradients. The formulations in (1-5) constitute a true velocity inversion only subject to the, generally unrealistic, assumption of ignorable density gradients. This problem, is addressed to elsewhere[6] and is presently ignored.

II.   The Scattering Model

Let $U(\underline{r},t)$ denote a scalar field obeying the wave equation

$$[\nabla^2 - c^{-2} n^2(z) \frac{\partial}{\partial t^2}] U(\underline{r},t) = -s(\underline{r}) \delta(t) \tag{1}$$

and let $U(\underline{r},\omega)$ be its Fourier transform

$$U(\underline{r},t) = (2\pi)^{-1} \int_{-\infty}^{+\infty} U(\underline{r},\omega) e^{-i\omega t} \, d\omega \tag{2}$$

satisfying the reduced wave equation

$$[\nabla^2 + k^2 n^2(z)] U(\underline{r},\omega) = -s(\underline{r}), \quad k = \omega/c \tag{3}$$

where $U(\underline{r},t)$ and $U(\underline{r},\omega)$ are subject to the requirements of causality and the radiation condition, respectively, as well as to the appropriate continuity (or finite jump discontinuity) conditions at z=h and z=h+d.

The physical interpretation of U (e.g. as the velocity potential or the pressure field in acoustics or any of the cartesian components of the electric field in electromagnetics) is of secondary importance. It will effect the specific forms of the reflection and transmission coefficients at a medium discontinuity but not the eventual inversion scheme. The medium, characterized by its index of refraction may be decomposed into a presumedly known background $(n_0(z))$ on which a perturbation

$$\alpha(z) = n(z)/n_0(z)^2 - 1 \tag{4}$$

is superimposed.

Correspondingly, a field decomposition into a background $(U_0 \equiv U(\alpha=0))$ and a scattered

$$U_s = U - U_0 \tag{5}$$

wave constituents is presumed. This field decomposition is physically meaningful as we shall assume $U_0(\underline{r},\omega)$ to be known everywhere and U_s to be measurable, as a function of either time or frequency, and hence known at least at one point on the surface $z=0$.

The following analytical sequence will be carried out in the frequency domain, however, pertinent time-domain results are also given. Equations (3-5) may be rewritten for a plane wave excitation as

$$L_0 U_0(\underline{r},\omega) = 0 \tag{6}$$

$$L_0 U_s(\underline{r},\omega) = -k^2 n_0^2(z)\alpha(z)U(\underline{r}) \ . \tag{7}$$

The operator $L_0 \equiv \nabla^2 + k^2 n_0^2(z)$ may be inverted via Green's theorem[7] yielding

$$U_s(\underline{r},\omega) = k^2 \int_{-\infty}^{+\infty} d^2\rho' \int_0^\infty dz' n_0^2(z')\alpha(z')G(\underline{r},\underline{r}',\omega)U(\underline{r}',\omega) \tag{8}$$

where $\alpha(z<0) = 0$ is presumed, $\int_{-\infty}^{\infty} d^2\rho'$ denotes integration over the entire transverse spatial domain, and $G(\underline{r},\underline{r}')$ is the Green function satisfying

$$L_0 G(\underline{r},\underline{r}',\omega) = -\delta(\underline{r}-\underline{r}') \tag{9}$$

and subject to the radiation condition.

We would, of course, like to view (8), say for a known $U_s(z=0)$, as an integral equation for $\alpha(z)$. Unfortunately, as it now stands, this is not the case since $U(\underline{r}',\omega)$ itself is an unknown function of $\alpha(z)$. This dilemma is standardly resolved by resorting to the approximate single scatter model. $U(\underline{r}',\omega)$ is replaced by $U_0(\underline{r}',\omega)$ in the integrand of (8). Since U_0 is not only known but, by definition, independent of $\alpha(z')$, (8) may now be viewed as a linear integral equation for $\alpha(z)$ (Fredholm of the first type). In this connection, the following comments are in order. Firstly, the implications of the Born approximation on the direct scatter has been investigated thoroughly in conjunction with random media scattering.[8] Secondly, better approximate forms for $U(\underline{r},\omega)$ and, for that matter, for G can be envisioned (e.g. replacing $U(\underline{r})$ by a renormalized "coherent" wave, or alternatively, by a Rytov approximated form).[8] However, these α-dependent forms render (8) non-linear. Iterative inversion schemes might be considered.

The three-dimensional, slab Green function is not known in closed form, however it is readily representable in terms of the following plane-wave (Fourier) superposition:

$$G(\underline{r},\underline{r}',\omega) = (2\pi)^{-2} \int_{-\infty}^{+\infty} d\xi \int_{-\infty}^{\infty} d\eta g(\underline{k}_t,z,z') e^{i\underline{k}_t \cdot (\rho-\rho')}$$

(10)

where $\underline{k}_t = \hat{\underline{x}}\xi + \hat{\underline{y}}\eta$ is the transverse wavevector, $\rho = \underline{r} - \hat{\underline{z}}z$, $\rho' = \underline{r}' - \hat{\underline{z}}z'$ and $g(\underline{k}_t,z,z')$ is the one-dimensional slab Green's function, given in Appendix A. The background field (U_0) is given

$$U_0(\underline{r}',\omega) = (2\pi)^{-2}(-2i\kappa_1)g(\underline{k}_{t0},0,z') e^{i\underline{k}_{t0} \cdot \rho'} .$$

(11)

where $\underline{k}_{t0} = \hat{\underline{x}}\xi_0 + \hat{\underline{y}}\eta_0$ is the incident transverse wavevector.

The substitution of (10,11) into (8) yields for observation points on the plane $z=0$

$$U_s(\rho,\omega) = (2\pi)^{-2}k^2(-2i\kappa_1) e^{i\underline{k}_{t0} \cdot \rho} \int_0^{\infty} dz' n_0^2(z')\alpha(z')$$

$$\cdot g^2(\underline{k}_{t0},0,z') .$$

(12)

We end this section by pointing out that (12) may be cast into the explicit format

$$U_s(\rho,\omega) = \sum_{j=1}^{10} F_j^{m\ell}(\rho,\omega)$$

(13)

where $F_j^{m\ell}$ are defined in Appendix B. This format, despite its cumbersome appearance, constitutes a convenient starting point for establishing the desired inversion scheme.

III. An Inversion Scheme

In this section we seek solutions for Equation (12) (or (B1)). The sequence, pursued below, leads to an explicit and thus efficient inversion algorithm.

We direct our attention to (B1). Having carried out, rather trivially, the z"-integration one may proceed as follows. Multiply the resultant equation by the factor $-2i\kappa_1 k_1^{-2}$, Fourier invert with respect to ω, and subsequently, integrate over z'. This straight-forward algebraic sequence leads to

$$4\pi v_1^{-1} \int_{-\infty}^{\infty} (-2i\kappa_1 k_1^{-2} U_s(0,\omega)) e^{-i\omega\tau} d\omega$$

$$\equiv -16\pi^2 \cos^2\theta_1 \int_0^{2z/v_1} dt U_s(0,t)$$

$$= \alpha(z) P(0<z<h) + r_1^2 \alpha(2h-z) P(h<z<2h)$$

$$+ t_1^2 t_2^2 r_2^{2(1+\ell+m)} \alpha(2h-z+(2+\ell+m)\tilde{d}) P(0<2h-z+(2+\ell+m)\tilde{d}<h)$$

$$+ 2[r_1\delta(h-z)+t_1 t_2 r_2^{1+2m}\delta(h-z+(1+m)\tilde{d})] \int_0^h dz'\alpha(z')$$

$$+ 2t_1 t_2 r_1 r_2^{1+2m} \alpha(2h-z+(1+m)\tilde{d}) P(0<2h-z+(1+m)\tilde{d}<h)$$

$$+ \cos^2\theta_1 \sec^2\theta_2 [t_1 t_2 r_2^{2(\ell+m)} \alpha(h-(\ell+m)d-(v_2/v_1)(h-z))$$

$$\cdot P(h<h-(\ell+m)d-(v_2/v_1)(h-z)<h+d)$$

$$+ t_1 t_2 r_2^{2(\ell+m+1)} \alpha(h+(\ell+m+2)d+(v_2/v_1)(h-z))$$

$$\cdot P(h<h+(\ell+m+2)d+(v_2/v_1)(h-z)<h+d)$$

$$+ 2t_1 t_2 r_2^{2(\ell+m)+1}\delta((\ell+m+1)d+(v_2/v_1)(h-z)) \int_h^{h+d} dz'\alpha(z')]$$

$$+ t_1^2 t_2^2 r_2^{2(\ell+m)} \alpha(z+d-(\ell+m+1)\tilde{d}) P(h+d<z+d-(\ell+m+1)\tilde{d}<\infty) \quad (14)$$

where the appearance of $\ell, m = 0,1,2\ldots$, implies summation over these indexes; without loss of generality, we set $\underline{\rho=0}$ and

$$P(z_1<z<z_2) = 1, \quad \text{if } z_1<z<z_2$$

$$= 0, \quad \text{otherwise.} \quad (15)$$

Also defined are the quantities: $v_{1,2} = c_{1,2}\sec\theta_{1,2}$, $z = v_1\tau/2$, $\tilde{d} = (v_1/v_2)d$ and $\theta_{1,2}$ denote the directions of propagation in the respective media. For clarity, we shall refer to consecutive terms in (14) by $F_j^{m\ell}(z)$, $j = 1,2,\ldots,10$ in accord with their order of appearance. We observe that each of the terms $F_j^{m\ell}$ is readily interpretable in terms of the slab's multireflections (depicted symbolically in Fig. 2).

At first glance (14) seems hopelessly tangled; however, once its structure is clearly understood, it may be unraveled. The left-hand-side (LHS) of (14) is presumed known from scattered field measurements and the subsequent Fourier processing or temporal integration. The right-hand-side (RHS) of (14) contains an infinite number of terms, all of which, except $F_{10}(z)$, are finite extent functions.

The extraction of $\alpha(z)$ may proceed as follows. We first direct our attention to the terms F_4, F_5 and F_9. These may be ignored as they vanish everywhere except at the discrete set of points $z = h + \ell d$. These impulsive contributions are readily recognizable and may be subtracted from the LHS of (14). The appearance of $F_{4,5,9}$ in (14) is attributable, physically, to forward scattering events by the perturbation $\alpha(z)$ and thus yields only integral information ($\int_0^h \alpha(z)dz$ and $\int_h^{h+d} \alpha(z)dz$). These integrals may always be constrained to zero by an appropriate redefinition of the background.

By omitting $F_{4,5,9}$, denoting the known LHS by $F(z)$ and recognizing

$$\sum_{\ell=m}^{\infty} \sum_{m=0}^{\infty} f(\ell+m) = \sum_{p=0}^{\infty} (p+1)f(p) , \tag{16}$$

Equation (14) reduces to

$$F(z) = \alpha(z)P(0<z<h) + r_1^2\alpha(2h-z)P(h<z<2h)$$

$$+ (p+1)t_1^2 t_2^2 r_2^{2(1+p)}\alpha(2h-z+(2+p)\tilde{d})P(0<2h-z+(2+p)\tilde{d}<h)$$

$$+ 2t_1 t_2 r_1 r_2^{1+2p}\alpha(2h-z+(1+p)\tilde{d})P(0<2h-z+(1+p)\tilde{d}<h)$$

$$+ \cos^2\theta_1 \sec^2\theta_2[(p+1)t_1 t_2 r_2^{2p}\alpha(h-pd-(v_2/v_1)(h-z)) .$$

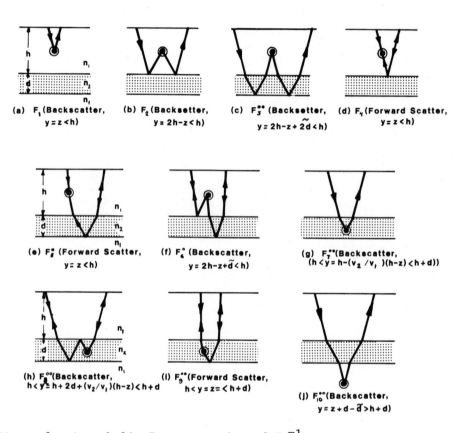

Figure 2.   A symbolic Interpretation of F_j^{m1} (y-the scattering point)

$$P(h<h-pd-(v_2/v_1)(h-z)<h+d)+(p+1)t_1t_2r_2^{2(p+1)}$$

$$\alpha(h+(p+2)d+(v_2/v_1)(h-z))P(h<h+(p+2)d+(v_2/v_1)(h-z)<h+d)]$$

$$+ (p+1)t_1^2t_2^2r_2^{2p}\alpha(z+d-(p+1)\tilde{d})P(h+d<z+d-(p+1)\tilde{d}<\infty) \quad . \quad (17)$$

We observe that $F_1(z)$ is the sole non-vanishing contribution in the range $0<z<h$. Hence,

$$\alpha(z) = F(z), \quad 0<z<h \tag{18}$$

where $F(z)$ denotes the known LHS of (17). Having determined $\alpha(z)$ in $0<z<h$, the terms $F_2(z)$, $F_3(z)$ and $F_6(z)$ may be computed as they require information of $\alpha(z)$ within the, now known, region.

Next, we consider the region $h<z<h+\tilde{d}$ in which (17) takes the form

$$F(z) = r_1^2\alpha(2h-z)P(h<z<2h)+\cos^2\theta_1\sec^2\theta_2t_1t_2\alpha(h-(v_2/v_1)$$

$$\cdot (h-z))P(h<h-(v_2/v_1)(h-z)<h+d) \tag{19}$$

with the remaining terms vanishing identically. (19) may be re-written as

$$\alpha(y) = (t_1t_2)^{-1}\sec^2\theta_1\cos^2\theta_2[F(h+(v_1/v_2)(y-h))$$

$$- r_1^2F(h-(v_1/v_2)(y-h))], \quad h<y<h+d \tag{20}$$

where the coordinate transformation $y = h+(v_2/v_1)(z-h)$ has been introduced and (18) utilized. (20) constitutes an explicit solution for $\alpha(y)$ in $h<y<h+d$ as its RHS is known. With α reconstructed in the slab region the terms $F_{7,8}(z)$ in (17) may be determined.

Similarly, by reconsidering (17) in the region $h+\tilde{d}<z<h+2\tilde{d}$ one obtains

$$\alpha(y) = (t_1 t_2)^{-2}\{F(y-d+\tilde{d})-r_1^2\alpha(2h-y+d-\tilde{d})P(h+\tilde{d}<z<2h)$$

$$- 2r_1 r_2 t_1 t_2 d(2h-y+d)P(h+\tilde{d} \ z \ 2h+\tilde{d})$$

$$- \cos^2\theta_1 \sec^2\theta_2 [2t_1 t_2 r_2^2 \ \alpha(h-(v_2/v_1)(h-y+d))$$

$$+ t_1 t_2 r_2^2 \ \alpha(h+d+(v_2/v_1)(h-y+d))]\} \tag{21}$$

where

$$h+\tilde{d}<y = z+d-\tilde{d}<h+d+\tilde{d} . \tag{22}$$

Equation (21) facilitates the explicit reconstruction of $\alpha(y)$ in the region $h+d<y<h+d+\tilde{d}$ since its RHS is known either from scattered field measurements or from the previously accomplished reconstruction of $\alpha(y)$ in $0<y<h+d$.

The procedure may be repeated indefinitely, with each additional iterative step yielding the reconstruction of the next slab of width $\tilde{d}$. The general term is,

$$\alpha(y) = (t_1 t_2)^{-2}\{F(y-d+\tilde{d})-r_1^2\alpha(2h-y+d-\tilde{d})P(h<z<2h)$$

$$- 2r_1 t_1 t_2 r_2^{2p+1}\alpha(2h+p\tilde{d}+d-y)P(h+(p+1)\tilde{d}<z<2h+(p+1)\tilde{d})$$

$$- pt_1^2 t_2^2 r_2^{2P}\alpha(2h+p\tilde{d}+d-y)P(h+(p+1)\tilde{d}<z<2h+(p+1)\tilde{d})$$

$$- \cos^2\theta_1 \sec^2\theta_2 [(p+2)t_1 t_2 r_2^{2(p+1)}\alpha(h-pd-(v_2/v_1)(h-y+d))$$

$$+ (p+1)t_1 t_2 r_2^{2(p+1)}\alpha(h+(p+1)d+(v_2/v_1)(h-y+d))]$$

$$- \sum_{p'=0}^{p-1} (p'+2)t_1^2 t_2^2 r_2^{2(p'+1)}\alpha(y-(p'+1)\tilde{d})P(0<y-h-d-$$

$$- (p'+1)\tilde{d}<\infty)\} \tag{23}$$

Here,

$$h+d+p\tilde{d} < y = z+d-\tilde{d} < h+d+(p+1)\tilde{d} \ , \tag{24}$$

$p = 0,1,2...$, and the RHS is known as in (21).

The inversion scheme based on (14) or (24) implies that $U_s(\omega)$ is measurable over an infinite frequency band. This, of course, contradicts the physical and system realities which, invariably, confine us to limited band observations.

Let us assume that the acquired data is confined to the band $\omega_{min} < |\omega| < \omega_{max}$. We may define the aperture function as $P(\omega_{min} < \omega < \omega_{max})$. Its Fourier transform, the band-limited δ-function is given by

$$\bar{\delta}(z) = [(\pi\tau)^{-1}(\sin\omega_{max}\tau - \sin\omega_{min}\tau)]_{\tau=2z/v_1}$$

$$= 2(\pi\tau)^{-1}\sin(\Delta\omega)(\tau/2)\cos\Omega\tau \tag{25}$$

where $\Omega = (\omega_{max}+\omega_{min})/2$, $\Delta\omega = \omega_{max}-\omega_{min}$, $\bar{\delta}(z)$ is characterized by a main lobe peak $(\Delta\omega)/\pi$ at $z=0$ and a characteristic width $\bar{z} = \pi v_1(2\Omega)^{-1}$.

Equation (14) (and similarly (25), will be modified as follows. The ω-integration on the LHS will now range over $\omega_{min} < |\omega| < \omega_{max}$, and the αs on the RHS are replaced by their respective convolutions with $\bar{\delta}(z)$. The z-resolution of α is obviously limited to an interval of order $\bar{z}$. Furthermore, insufficient resolution renders the reconstruction algorithm, non-unique. This can be anticipated since the proposed inversion scheme depends on our capability to isolate and subsequently subtract consecutive, multireflected constituents. This task, readily accomplished with infinite band information becomes impossible if consecutive multiple reflections are non-resolvable, giving rise to the requirement $2\Omega < \pi v_2/d$ which sets a low bound on the frequency range.

ACKNOWLEDGEMENTS

This work was partly supported by the UH Seismic-Acoustic Laboratory. The author wishes to thank Professors Fred Hilterman and Keith Wang for the fruitful discussions and continued interest.

REFERENCES

1. J. K. Cohen and N. Bleistein, "An Inversion Method for De-
    termining Small Variations in Propagation Speed", SIAM J.
    App. Math., Vol. 32, No. 4, pp. 784-799, June 1977.
2. N. Bleistein and J. K. Cohen, "A Survey of Recent Progress
    on Inverse Problems", in Underwater Acoustics, Ed. De Santo
    J. A., Springer Verlag, New York, 1979.
3. J. K. Cohen and N. Bleistein, "A Velocity Inversion Proce-
    dure for Acoustic Waves", Geophysics, Vol. 44, No. 6, pp.
    1034-1040, June 1979.
4. S. H. Gray, N. Bleistein and J. K. Cohen, "Direct Inversion
    for Strongly Depth Dependent Velocity Profiles", Research
    Report MS-R-7902, Department of Mathematics, University of
    Denver.
5. S. Raz, "An Algorithm for Profile Inversion in the Presence
    of a Known Background Slab (submitted to Radio Science).
6. S. Raz, "Direct Reconstruction of Velocity and Density Pro-
    files from Scattered Field Data", (submitted to Geophysics).
7. P. M. Morse and H. Feshbach, "Methods of Theoretical Physics,
    I", McGraw Hill, N.Y., 1953.
8. A. Ishimaru, "Wave Propagation and Scattering in Random
    Media, I", Academic Press, 1978.

APPENDIX A

The Explicit Form of $g(k_t, z, z')$

All the results contained in this appendix are well known
and readily derivable by elementary means. Therefore, the pertinent forms are listed below without derivations:

$$g = (-2i\kappa_1)^{-1}[e^{i\kappa_1|z-z'|} + Re^{-i\kappa_1(z+z'-2h)}] \ , \tag{A1}$$

in the region: $0<z'<h, \quad 0<z<h.$

$$g = (-2i\kappa_2)^{-1}t_1 Se^{i\kappa_1(h-z)}[e^{i\kappa_2(z'-h)} + r_2 e^{i\kappa_2(h+2d-z')}] \tag{A2}$$

,in the region: $h<z'<h+d, \quad 0<z<h.$

$$g = (-2i\kappa_1)^{-1}t_1 t_2 Se^{i\kappa_1(z'-z-d)+i\kappa_2 d} \ , \tag{A3}$$

in the region: $h+d<z'<\infty, \quad 0<z<h.$

Here,

$$S \equiv \sum_{m=0}^{\infty} (r_2^2 e^{2i\kappa_2 d})^m = (1-r_2^2 e^{2i\kappa_2 d})^{-1} ; \qquad (A4)$$

$$\kappa_{1,2} = [k_{1,2}^2 - \xi^2 - \eta^2]^{1/2}, \quad \text{Re}(\kappa_{1,2}) > 0, \quad \text{Im}(\kappa_{1,2}) > 0 \qquad (A5)$$

are the wavenumbers along z in the respective media (1,2) and

$$k_{1,2} = kn_{1,2} . \qquad (A6)$$

Furthermore,

$$R = r_1 + t_1 t_2 r_2 Se^{2i\kappa_2 d} \qquad (A7)$$

and

$$T = t_1 t_2 Se^{i\kappa_2 d} \qquad (A8)$$

denote the slab's reflection and transmission coefficients, respectively. r_j and t_j are the reflection and transmission coefficients at a two-media interface. The subscript (j = 1,2)
denotes the medium supporting the incident field.

If interface continuity constraints are imposed on the field
as well as on its normal derivative the following expression results:

$$t_j = 1 + r_j = 2\kappa_j (\kappa_1 + \kappa_2)^{-1}, \quad j = 1,2. \qquad (A9)$$

APPENDIX B

An Explicit Form for $U_s(\rho,\omega)$

The substitution of (A1–A8) into (12) yields,

$$(2\pi)^2 e^{-ik_1 0 \cdot \rho} U_s(\underline{\rho},\omega) = k_1^2 \int_0^h dz' \alpha(z') [\int_{z'}^\infty dz'' e^{2i\kappa_1 z''}$$

$$+ r_1^2 \int_{-\infty}^{z'} dz'' e^{2i\kappa_1(2h-z'')}$$

$$+ t_1^2 t_2^2 r_2^2 (1+\ell+m) e^{2i\kappa_2(2+\ell+m)d} \int_{-\infty}^{z'} dz'' e^{2i\kappa_1(2h-z'')}$$

$$+ 2r_1 \int_h^\infty dz'' e^{2i\kappa_1 z''} + 2t_1 t_2 r_2^{1+2m} e^{2i\kappa_2(1+m)d} \int_h^\infty dz'' e^{2i\kappa_1 z''}$$

$$+ 2t_1 t_2 r_1 r_2^{1+2m} e^{2i\kappa_2(1+m)d} \int_{-\infty}^{z'} dz'' e^{2i\kappa_1(2h-z'')}]$$

$$+ k_2^2 \int_h^{h+d} dz' \; \alpha(z') t_1 t_2 r_2^{2(\ell+m)} e^{2i\kappa_2(\ell+m)d+2i\kappa_1 h}$$

$$\cdot [\int_{z'}^\infty dz'' e^{2i\kappa_2(z''-h)} + r_2^2 \int_{-\infty}^{z'} dz'' e^{2i\kappa_2(2d+h-z'')}$$

$$+ 2r_2 \int_d^\infty dz'' \; e^{2i\kappa_2 z''}] + k_1^2 \int_{h+d}^\infty dz' \alpha(z') t_1^2 t_2^2 r_2^{2(\ell+m)}$$

$$\cdot e^{2i\kappa_2(1+\ell+m)d} \int_{z'}^\infty dz'' \; e^{2i\kappa_1(z''-d)} \tag{B1}$$

where the appearance of $\ell, m = 0,1,2,\ldots$ implies summations over the entire range of these indexes.

For clarity of referencing we denote consecutive terms in (B1) by $F_1(\underline{\rho},\omega)$, $F_2(\underline{\rho},\omega)$ etc. We also observe that each $F_j^{m\ell}(j = 1,2\ldots,10)$ in (B1) is readily interpretable in terms of the slab's multi-reflected contributions (Fig. 2).

THEORY AND MEASUREMENTS OF ULTRASONIC SCATTERING

FOR TISSUE CHARACTERIZATION

Robert C. Waag

Departments of Electrical Engineering and Radiology
University of Rochester, Rochester, New York

INTRODUCTION

It is known among scientists and clinicians working in medical ultrasound that more information than time of arrival and amplitude is contained in the ultrasonic echoes produced by tissue. This has led to a variety of investigations (1-5) with the common objective of extracting additional information to characterize tissue more completely from its ultrasonic properties. Among the various studies are those concerned with sound speed, absorption, and a general class of research that falls under the category of acoustic scattering. The scattering investigations are the subject of this review.

Broadly stated, the objective of scattering-based studies is to infer tissue structure or architectural properties from measurement and analysis of data obtained as a function of the scattering vector which is defined in terms of the incident wave direction and wavelength as well as the direction of observation relative to the scattering volume. Since the scattering vector is a function of wavelength or frequency, information as a function of scattering vector can be obtained by sending a short or broadband pulse and determining the wavenumber spectrum of the backscattered signal or of those signals scattered at some other angle. Since the scattering vector is also a function of geometry, information can be obtained by changing beam-reflector orientation using angle scanning.

This tutorial paper describes acoustic scattering characterization of tissue by reviewing a mathematical model useful for interpreting tissue characterization results, reporting selected examples from the work of others to show the nature of existing studies, and summarizing investigations under way in our laboratory which has dealt primarily with models but has also obtained some results from human tissue in vitro.

THEORY

The simplest form of the mathematical model commonly employed by those studying ultrasonic scattering by tissue is developed in terms of a wave equation for a lossless medium having a spatially varying sound speed, $c(\underline{r})$ (6,7). The equation is readily derived from conservation of mass and momemtum relations and the thermodynamic equation of state under the assumptions of small acoustic perturbations, spatially varying compressibility or velocity, and time harmonic dependence. The result, expressed in terms of a harmonic amplitude velocity potential, ϕ, which is proportional to pressure and whose negative gradient gives acoustic particle velocity, is

$$\nabla^2 \phi + \frac{\omega^2}{c^2(\underline{r})} \phi = 0$$

In this equation,

$$\nabla^2 = \frac{d^2}{dx^2} + \frac{d^2}{dy^2} + \frac{d^2}{dz^2} \qquad = \text{Laplacian operator}$$

ω = radian harmonic frequency

and

$$c(\underline{r}) = \frac{1}{(\kappa\rho)^{1/2}}$$

where ρ is the medium ambient density and κ is the medium compressibility that is assumed to dominate the speed of sound variations. Dividing the velocity potential into a component ϕ_i representing the incident wave and a component ϕ_s representing the scattered wave and writing the sound speed in terms of a steady value c_0 and a small fluctuation c_1

results in an equation explicitly containing the scattered wave and the medium variations. The equation can be rewritten in an analytically tractable form by using a binomial expansion for the quantity $\{c(\underline{r})\}^{-2} = (c_0 + c_1)^{-2}$ and retaining only the first two terms. Rearranging the equation yields

$$\nabla^2\phi_s + k^2\phi_s = 2k^2\gamma_c(\phi_i + \phi_s)$$

where

$$k = \frac{\omega}{c_0} = \text{wavenumber}$$

$$\gamma_c = \frac{c_1}{c_0} = \text{normalized velocity fluctuation}$$

The incident wave velocity potential does not appear on the left hand side of this equation because it satisfies the homogeneous equation

$$\nabla^2\phi_i + k^2\phi_i = 0$$

The equation for the scattered wave may be solved by using the Green's function for unbounded space and assuming that the variations in sound speed produce a very small scattered wave so that

$$|\phi_i| >> |\phi_s|$$

The result is

$$\phi_s = \frac{k^2}{2\pi} \int_V \phi_i(\underline{r}')\gamma_c(\underline{r}') \frac{e^{jk|\underline{r}-\underline{r}'|}}{|\underline{r}-\underline{r}'|} d^3r'$$

This expression gives the velocity potential at an observation point $\underline{r}$ due to variations in sound speed within a scattering volume $\overline{V}$.

If the incident wave, ϕ_i, is taken as plane with amplitude A and propagating in direction $\underline{n}$, and if the observation point $\underline{r}$ is far from the scattering volume so that the expression

$$|\underline{r} - \underline{r}'|$$

may be approximated by

$$r - \underline{r}' \cdot \underline{m} , \quad \underline{m} = \frac{\underline{r}}{r}$$

in the exponent of the equation and by

$$r$$

in the denominator, the scattered wave velocity potential can be written as

$$\phi_s = \frac{Ak^2 e^{jkr}}{2\pi r} \int_V e^{jk(\underline{n}-\underline{m})\cdot\underline{r}'} \gamma_c(\underline{r}')d^3r'$$

Defining the scattering vector, $\underline{K}$, by the relation

$$\underline{K} = k(\underline{n}-\underline{m})$$

yields a vector quantity whose magnitude, K, is given by

$$K = 2k \sin(\nu/2)$$

where

$$\nu = \text{angle between } \underline{n} \text{ and } \underline{m}$$

and is known as the scattering angle. The direction of the scattering vector is along the unit vector $\underline{a}_k$ given by

$$\underline{a}_k = \frac{\underline{n}-\underline{m}}{|\underline{n}-\underline{m}|}$$

The scattered wave can be expressed compactly in terms of the scattering vector as

$$\phi_s(\underline{K}) = \frac{Ak^2 e^{jkr}}{2\pi r} \int_V e^{j\underline{K}\cdot\underline{r}'} \gamma_c(\underline{r}')d^3r'$$

Thus, the scattered pressure is defined by the product of a frequency-dependent factor and the 3-dimensional Fourier transform of the variations in sound speed with respect to the scattering vector $\underline{K}$. The implication of this relationship is that, under certain conditions, scattered wave measurements spanning all values of the scattering vector can be inverted via a Fourier transform to obtain the fluctuations in sound speed.

A similar relation can be written in terms of compressibility variations. If density variations are also important, another term is added that produces an angular dependence like that of a dipole source instead of the simpler monopole form of the velocity or compressibility variations.

Although density variations may be negligible in some situations, more data is needed to identify the role of density variations in scattering by tissue.

A dipole source term is also produced by medium motion in the direction of observation. However, aside from Doppler studies which deal with mean values of medium motion, little has been done using this term to characterize scattering medium movement such as may occur in turbulent blood flow.

Since in many cases of biological interest the sound speed is a random function of space, it is appropriate to consider the mean square value of the velocity potential for the scattered wave. Under the additional assumptions that the second-order statistics of the fluctuations are independent of origin within the sample volume and the correlation length of the scattering process is much less than the size of the scattering volume, the mean square value of the scattered wave velocity potential can be related to the 3-dimensional Fourier transform of the correlation function of the medium fluctuations. The relation is

$$< |\phi_s (\underline{K})|^2 > = \frac{A^2 k^4 V}{4\pi^2 r^2} \int_V e^{j\underline{K}\cdot\underline{r}'} B_{\gamma_c} (\underline{r}') d^3 r'$$

where

$$B_{\gamma_c} (\underline{r}') = <\gamma_c (\underline{r}_{-1} + \underline{r}') \gamma_c (\underline{r}_{-1})>$$

and

$$< > = \text{ensemble average}$$

The Fourier transform relationship between the mean square velocity potential of the scattered wave and the correlation of speed fluctuations implies that, under certain restrictions, determinations of mean square scattered pressure encompassing all values of scattering vector can be inverted to obtain the correlation function of the source of scattering. Alternatively, a correlation function can be assumed and the scattered pressure can be computed for comparison with measurements.

If the correlation function depends only on the magnitude of the distance between points and not on their relative orientation, the 3-dimensional transform relationship collapses to a 1-dimensional relationship. The result can be expressed as

$$<|\phi_s(K)|^2> = \frac{A^2k^4V}{\pi r^2 K} \int_0^\infty B_{\gamma_c}(r')r'\sin(Kr')dr'$$

or

$$<|\phi_s(K)|^2> = - \frac{A^2k^4V}{r^2} \frac{1}{2\pi K} \frac{d}{dK}\Phi_1(K)$$

where

$$\Phi_1(K) = \frac{1}{2\pi} \int_{-\infty}^\infty B_{\gamma_c}(r')e^{jkr'}dr'$$

This permits inference of the correlation function from backscatter measurements alone either by employing broadband pulses and averaging the results obtained from statistically similar sample volumes or by averaging backscatter power spectra obtained at different angles.

More general expressions for scattered signals can be developed without the plane incident wave and point receiver assumptions. In addition, wavefront distortion arising from inhomogeneous intervening media can be included and the assumption of observation point in the far-field of the scattering volume can be relaxed. However, the more complex results obscure the insight provided by the basic Fourier transform relationships developed here.

Comparison of different measurements with theoretical predictions is facilitated by a normalization which eliminates dependencies on incident wave intensity, receiver area, and scattering volume size. The normalization follows from determining the power scattered into the solid angle subtended by a receiver per unit intensity and unit scattering volume. Averaging and expressing the result as a limit per unit solid angle yields a quantity known as average differential scattering cross-section per unit volume, $< \frac{d\sigma}{d\Omega} >$, which has the integral form

$$\frac{d\sigma}{d\Omega} = \frac{k^4}{4\pi^2} \int e^{j\underline{K}\cdot\underline{r}'}B_{\gamma_c}(\underline{r}')d^3r'$$

where

> $d\sigma$ = power scattered into a differential aperture per unit incident intensity and unit scattering volume.

> $d\Omega$ = solid angle subtended by the differential aperture.

One common differential cross-section is for a scattering angle of $180°$ and is known as the backscatter cross-section. The total scattering cross-section or energy removed from a beam by scattering is just the integral of the differential cross-section over all angles.

The foregoing expressions provide an analytic foundation for experimental studies. However, correlation with the existing base of optical data that physicians have collected is required before results can be interpreted for tissue characterization. The correlation may be accomplished by simply looking at stained sections of tissues examined acoustically and empirically relating the acoustic pattern with the optical pattern. This has been the most widespread approach. A quantitative way involves serial sectioning of tissue volumes studied acoustically, staining for tissue components thought to be responsible for acoustic scattering, and computing transmitted light power spectra on a scale comparable to the acoustic measurement (8). Then the optical power spectra can be compared to the acoustic spectrum of scattered power for correlation of structure.

MEASUREMENTS

The interpretation of scattered pressure as the Fourier transform of a variation function and the similar interpretration of the mean square scattered pressure as a Fourier transform of the correlation function of variations establishes a frame of reference for the review of experiments. The Fourier transform relations provide information in 3-dimensional wavespace about scattering region structure in a way that is analogous to the way frequency domain measurements describe the time response of a filter.

Experiments have been performed to determine backscatter as a function of frequency or angle, measure the frequency dependence of scattering at various other angles, or show the angular dependence of scattering at a fixed frequency.

Determinations of backscatter as a function of frequency give a
portion of a radial line in wavespace. The length of the line
depends on the bandwidth over which the backscatter measurements
are made. The position of the line or its direction in
wavespace depends on the orientation of the emitter and detector
relative to the scattering region. Wavenumber space is spanned
by frequency analysis of backscatter for all directions of the
incident wave relative to the scattering volume. As already
noted, one radial line is sufficient to characterize the
correlation function if the basic structure is radially
symmetric or has statistics with radial symmetry.

Examples of frequency-dependent backscatter measurements of
biological media are found in work of investigators who have
studied the eye and also in the work of investigators who have
studied blood. In vivo data (9) obtained from an eye containing
a detached retina shows backscatter peaks and nulls resulting
from front and back surface echoes that interfere (Fig. 1).
The spacing of peaks implies a retinal thickness of 190 microns.
The eye, because of its relatively simple structure and
uniformity among individuals is one of the few organs where
useful in vivo scattering measurements have been made.

Blood is a randomly organized medium that backscatter
measurements (10) indicate follows a 4th-power Rayleigh
scattering dependence at low red cell concentrations (Fig. 2).
Other measurements (11) of scattering by blood have been
obtained and the analysis of the data indicates that the
scattering phenomena can be predicted analytically although the
predictions (12) employ a more complicated model than that
presented here.

Backscatter measurements at a single frequency as a
function of angle with respect to the scattering region give a
line segment on a spherical surface in wavenumber space.
Wavenumber space is spanned in this type of measurement by
varying the frequency and, thus, the radius of the sphere over
which lines are traced during the angle variations. If the
scattering region is anisotropic, the measurement of angular-
dependent backscatter at a suitably chosen frequency can show
the nature of the anisotropy.

Backscatter measurements as a function of the angle have
been made by investigators for liver, brain and spleen (13), and
for arterial tissue (14). The results indicate differences
between scattering from different tissues. In addition, data
from liver containing metastatic disease shows differences from
data produced by normal liver (15) (Fig. 3). Although the
liver data is from a single determination of backscatter

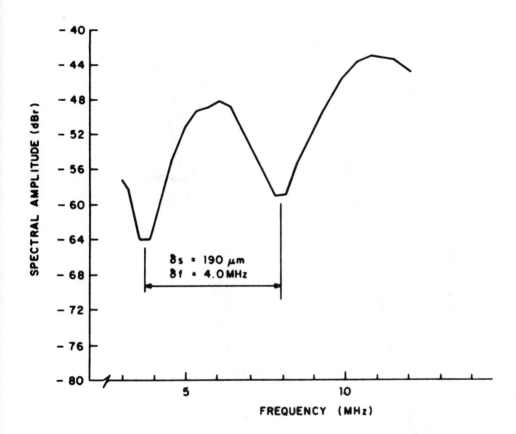

Fig. 1. Spectrum of detached retina. The scalloping is the result of interference between the front and back surfaces of the thin retinal membrane. From Lizzi et al. (9).

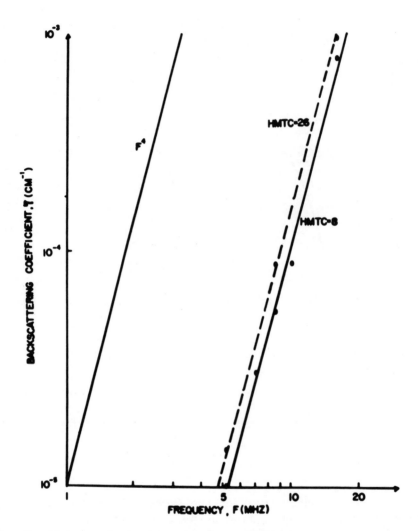

Fig. 2. Volumetric backscattering coefficient for red blood
cells versus frequency. Low concentrations of red cells (HMTC
expressed in percent) are shown to follow fourth-power frequency
law. From Reid, et al. (10).

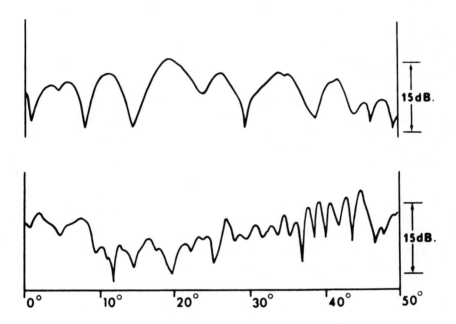

Fig. 3. Observed backscattered Bragg diffraction patterns,
taken at 1 MHz from normal liver parenchyma (upper) and
secondary liver tumor (lower).   The backscatter from the
cancerous liver shows a higher rate of change with angle.   From
Hill et al. (15).

amplitude recorded as the specimen was rotated in front of a transducer fixed in a water tank, it is a result from a medium usually considered random.

The frequency-dependence of scattering at angles other than 180° also provides information along a radial line in wavespace. The direction of the line is established by the scattering vector which is oriented along the line defined by the difference between the incident wave vector and the vector in the direction of the observer. Wavespace is scanned with this technique by selecting emitter and detector positions relative to the scattering volume so that all directions of the scattering vector are spanned. Determination of frequency-dependent scattering at angles other than 180° provides the same information as frequency analysis of backscatter when variations in medium density and medium motion are unimportant.

Frequency scanning measurements have been carried out on monofilament cylinders arranged in regular arrays with spacings in the millimeter range (16). Fourier transforms of the results indicate that the spacings of elements in the arrays are accurately predicted with a scattering model like that described here (Fig. 4).

The measurement of scattering dependence on angle at a fixed frequency is another way to obtain a portion of a radial line in wave space. The basic measurement technique maintains the scattering vector along the same line in wave space by equal and opposite angular increments of the emitter and detector. The magnitude of the scattering vector which is proportional to $\sin(\nu/2)$ changes because the scattering angle ν changes. Wavespace is spanned by making repeated angular scattering measurements with the initial positions of the emitter and detector defining all possible K̲-vector directions relative to the scattering volume. This type of measurement eliminates the need for an isotropy assumption in the plane of rotation and also affords the possibility of sorting out contributions of scattering due to compressibility and density variations should the density variations be significant.

Angular-dependent scattering measurements · made in our laboratory have employed a special apparatus to provide precise control of transducer positions relative to the scattering medium. The medium was contained in a cylindrical column by a thin polyethylene tubing to minimize reflection of energy at the perimeter of the sample. A typical angle scanning experiment consisted of 28 angle scans taken at 5 mm increments along the axis of the cylindrical sample. In each angle scan, data was collected at 160 equally spaced increments of 1.96° from a scattering angle of -160° to $+152^{\circ}$. Mean square values of scattering were obtained by averaging angle scans taken at different levels along the axis of the cylinder. At each angular position, data was acquired at 256 frequency points equally spaced between 2 and 8 MHz so that the frequency dependence of angle scattering could be determined.

Results (8) obtained at 6 MHz from collections of closely packed small spheres show more omnidirectional scattering for an average particle radius of 84 microns than for an average radius of 93 microns (Fig. 5). Similar results (8) were obtained for particles having an average radius of 128 micrometers when the frequency was reduced to 3.8 MHz from 6.0 MHz. The particle sizes in these experiments were determined by a spatial frequency analysis of optical data in a manner like that described for correlation of optical and acoustic data.

Measurements in our laboratory of angular dependent scattering by tissues have been confined to liver. Using the same angle scanning and averaging techniques as described for the spherical particles, data was obtained from pig and human liver (17,18,19). The pig liver data shows little angular dependence of scattering at 3.0 MHz but an increased angular dependence at 7.0 MHz (Fig. 6). This implies scatterer spaces small compared to the wavelength (.43 mm) at 3.5 MHz but more comparable to the wavelength (.21 mm) at 7.0 MHz.

Optical and acoustic microscopy carried out on selected 1 mm^2 fields of sections cut from pig liver specimens after they were fixed shows the normal structure of pig liver which is composed of lobules separated by sheets of collagen (Fig. 7). Acoustic interferograms obtained during these studies reveal that sound speed was higher in the collagen sheets than in the pig liver lobules. The sound speed changes produce ultrasonic scattering.

Optical studies were conducted on 4.4 x 4.4 mm^2 fields by first computing average 2-dimensional power spectra and then finding the average radial distribution of energy to demonstrate the influence of various sizes and spacings of tissue components on scattering (Fig. 8). The results for optical fields containing all the spatial frequency information in the original images show a monotonic decrease of the scattered light with angle. However, the same optical field when edge-enhanced to remove the contribution of spectral power from large structures

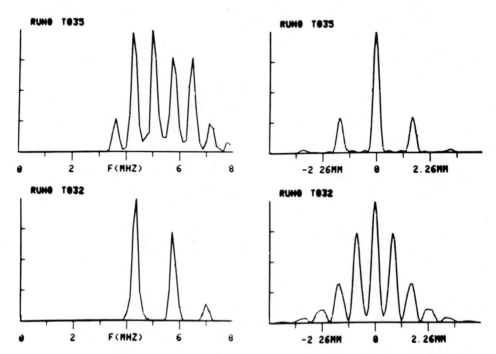

Fig. 4. Swept-frequency diffraction by arrays. Measured values
of scattered energy as a function of frequency are shown (left)
on a linear scale with their corresponding Fourier transforms
(right) also on a linear scale for array spacings of 1.52 mm
(top) and 0.76 mm (bottom) at a transmitter range of 11.0 cm.
The ultrasonically determined spacings were 1.55 mm and 0.78 mm
respectively. Adapted from Lee et al. (16).

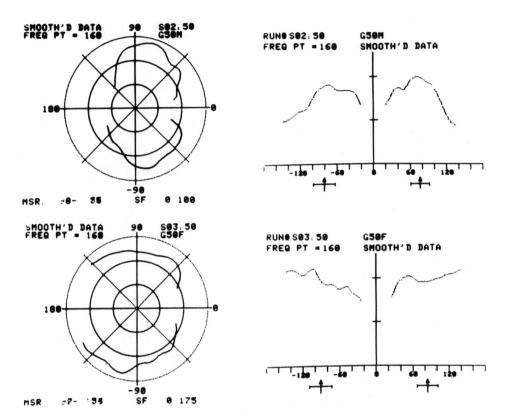

Fig. 5. Size-dependent angular scattering by a random medium model. The polar plots and corresponding Cartesian displays show average intensity on a linear scale as a function of scattering angle at 5.9 MHz for distributions of scatterers with an average particle radius of 93μ (top) and with an average radius of 84μ (bottom). The increase in omnidirectional scattering for smaller scatterers is evident in both plots and also demonstrated by the mean scattering angle defined by arrows crossing standard deviation bars below the Cartesian plots. From Waag et al. (8).

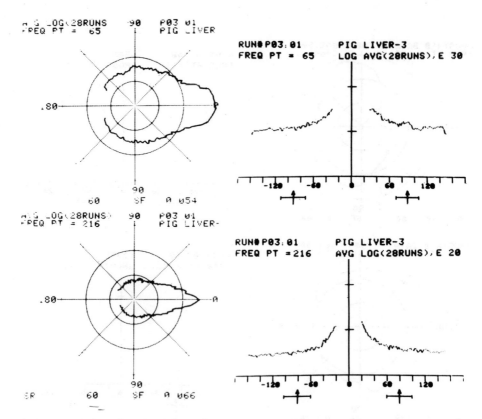

Fig. 6. Pig liver scattering. Average scattered energy at 3.5 MHz (upper) and 7.0 MHz from pig liver is plotted as a function of scattering angle in polar and Cartesian coordinates on a log scale (26 db/division left and 35 db/division right) for scattering angle ranging from −160° to +152°. Data corresponding to the direct beam has been blanked in the Cartesian representations and the average angle (arrows with variance shown by bars) of the remaining energy has been computed separately for positive and negative scattering angles. The data shows a greater directional dependence at 7.0 MHz than 3.5 MHz, indicating that at 7.0 MHz the wavelength (.21 mm) is more comparable to the scatterer spacing. From Waag et al. (19).

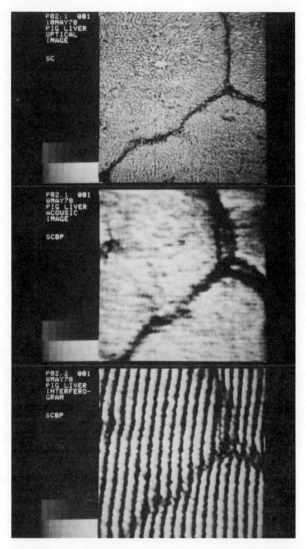

Fig. 7. Pig liver. Optical micrograph (top) of a thin 1x1 mm^2
section stained to emphasize collagen displays collagenous
septae as dark bands between liver lobules. A transmission
acoustic micrograph (center) of approximately the same field in
a thicker section also demonstrates the septae as dark bands.
An acoustic interferogram (bottom) of the same section contains
interference lines bending to the right as they cross the
septae. The acoustic micrographs provide a relative
determination of sound attenuation and speed implying that the
septae cause ultrasonic scattering in normal pig liver. From
Waag et al. (19).

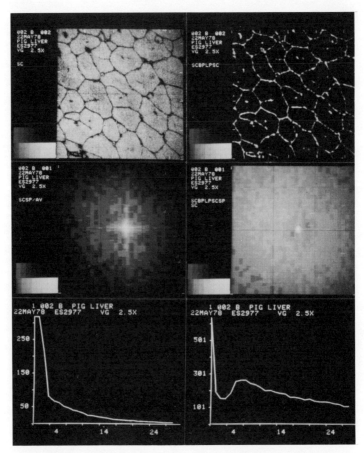

Fig. 8. Pig liver. Optical image analysis. Typical 4.4x4.4 mm^2
image (upper left) of normal pig liver with collagen stained
dark is shown across from an edge-enhanced version. The
2-dimensional power spectrum (center left) found from typical
images decreases monotonically with angle while the comparable
spectrum (center right) computed from the corresponding
edge-enhanced images shows a ring pattern. The average radial
distribution (lower left) of power from the typical images is
qualitatively the same as the scattered ultrasound intensity
pattern while the average radial distribution (lower right)
from the edge-enhanced images contains a peak arising from the
septal pattern regularity. The 2-dimensional power spectra are
displayed on a log amplitude scale with the horizontal distance
from the center point to the edge of the field equal to 3.64
cycles/mm and the 1-dimensional data is plotted on a linear
amplitude scale with each major division equal to 3 1/3
harmonics or .76 cycles/mm. From Waag et al. (19).

and low spatial frequency aberrations from uneven staining shows a ring structure associated with the regular lobular pattern. Since the acoustic data does not show scattering peaks which can be associated with the regular lobular pattern, it appears that the observed angular scattering in the low megahertz range was due to a combination of scattering from a variety of structures or spacings which include the lobular pattern.

Measurements were made of acoustic scattering from a fixed human liver containing large focal lesions which were sometimes completely within the beam and sometimes absent from the beam (Fig. 9). In general, there was a greater angular dependence of the scattering at both 3.5 MHz and 7.0 MHz than in the pig liver scattering, indicating an average scatterer spacing or size that is larger.

Optical and acoustic microscopy performed on this human liver specimen shows that portal tracts containing collagen attenuated sound more than the surrounding tissue and that collagen produced velocity changes just as in the pig liver studies. The average power spectrum and average radial distribution of energy are both concentrated at small angles (Fig. 10) and agree qualitatively with the acoustic studies.

A fixed human liver containing small diffusely distributed lesions has also been studied. The average acoustic scattering data (Fig. 11) shows a slight angular dependence of scattering at both the low and the high frequencies much like that from pig liver. The scattering from this liver was more omnidirectional than that from the other human liver, implying scatterer spacings that are smaller than the other human liver and more comparable to the pig liver.

Optical and acoustic studies show that the specimen consisted of glandular nests of tumor cells surrounding by connective tissue (Fig. 12). The central regions of the tumor nests were necrotic and, like the connective tissue, attenuated sound. Acoustic interferograms indicate that the scattering was associated with connective tissue and necrotic cells. The optical studies of the serial sections prepared from this specimen indicate that no part of the specimen was uninvolved (Fig. 13). The average power spectrum has a greater spread than that for the large focal lesions described earlier, again indicating a qualitative agreement between acoustic and optical scattering data.

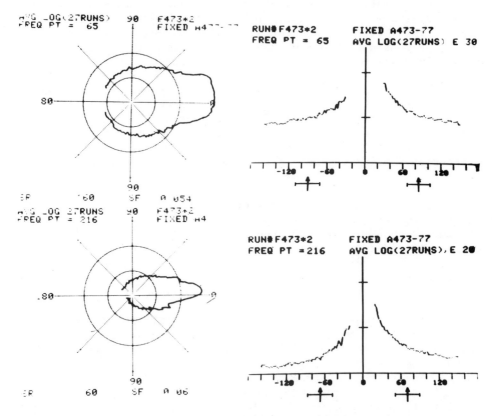

Fig. 9. Acoustic scattering from a fixed human liver containing large, widely separated, metastatic tumor nodules. Average scattered energy at 3.5 MHz (upper) and 7.0 MHz is plotted as a function of scattering angle on a log amplitude scale in polar (26 db/division) and Cartesian (35 db/division) coordinates for scattering angle ranging from $-153°$ to $+165°$ Data corresponding to transmission of the direct beam has been removed from the Cartesian plots and the average (arrows with variance shown by bars) of the remaining energy has been computed separately for positive and negative scattering angles. The data shows the same trend of increasing directional dependence at higher frequencies as the pig liver (Fig. 6) but exhibits a greater directional dependence at both the low and high frequencies than the pig liver. From Waag et al. (19).

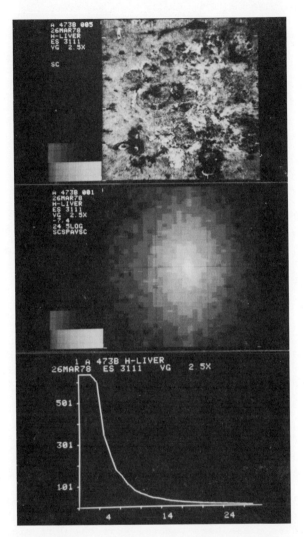

Fig. 10. Optical image analysis of human liver containing large, widely separated metastatic tumor. A typical 4.4x4.4 mm^2 image (top) with connective tissue stained dark shows part of a tumor nodule surrounded by normal tissue. The 2-dimensional power spectrum (center) is displayed on a log amplitude scale with the horizontal axis from the center point to the edge of the field equal to 3.64 cycles/mm. The average radial distribution of power (bottom) is plotted on a linear amplitude scale with each major division equal to 3 1/3 harmonics or 0.76 cycles/mm and shows a concentration of intensity at small angles. From Waag et al. (19).

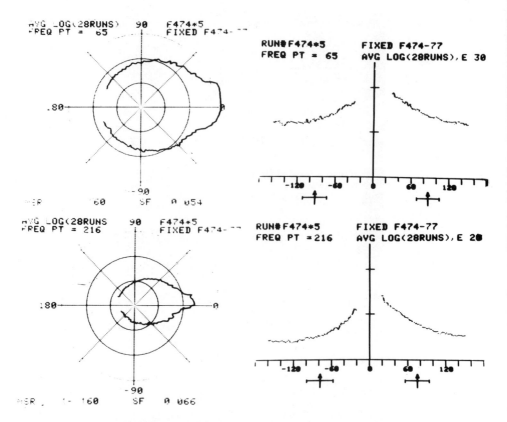

Fig. 11. Ultrasound scattering from a human liver containing small diffusely distributed metastatic nodules. Average scattered energy at 3.5 MHz (upper) and 7.0 MHz is plotted as a function of scattering angle in polar (26 db/division) and Cartesian (35 db/division) coordinates for scattering angles ranging from -153° to $+165^{\circ}$. Data corresponding to transmission of the direct beam has been removed from the Cartesian plots and the average angle (arrows with variance shown by bars) of the remaining energy has been computed separately for positive and negative scattering angles. The data from this specimen is relatively more omnidirectional than the data from the other liver. This indicates that the scattering objects and spacings are smaller in this specimen than in the other human liver (Fig. 9). From Waag et al. (19).

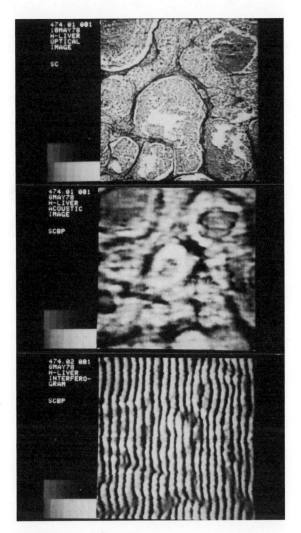

Fig. 12. Human liver containing metastatic adenocarcinoma. An optical micrograph (top) of a thin 1x1 mm^2 section stained to emphasize collagen demonstrates collagenous septae as dark bands between glandular nests of tumor cells. Some of these tumor cell nests have necrotic centers which appear darker than the surrounding tumor cells. The septae and necrotic areas also appear dark in transmission acoustic micrograph (center) of approximately the same field in a thicker section. An acoustic interferogram (bottom) with lines bending as they traverse the collagenous septae and necrotic tissue indicates that the speed of sound is different than in the adjacent tissue. From Waag et al. (19).

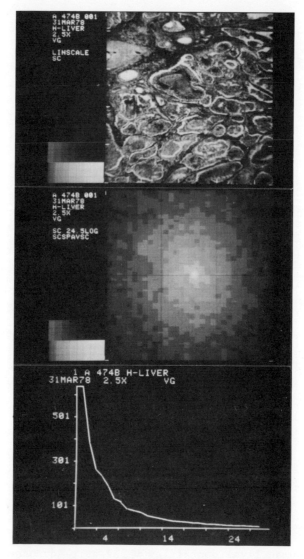

Fig. 13. Optical image analysis of human liver containing
metastatic tumor nodules diffusely distributed. A typical
4.4x4.4 mm^2 image (upper) is shown with collagen and necrotic
tissue stained dark. The 2-dimensional power spectrum (center)
is displayed on a log amplitude scale with the horizontal
distance from the center point to the edge of the field equal to
3.64 cycles/mm. The average radial distribution of power
(lower) is plotted on a linear amplitude scale with each major
division equal to 3 1/3 harmonics or 0.76 cycles/mm and shows
relatively more energy at high angles than the other human liver
(Fig. 10). From Waag et al. (19).

REFERENCES

(1) Proc. Ultrasonic Tissue Characterization, M.   Linzer
     (editor), National Bureau of Standards, Gaithersburg, Md.,
     May 28-30, 1975, Spec. Publ. 453 (U.S.   Govt.   Printing
     Office, Washington, D. C.).
(2) Proc. Ultrasonic Tissue Characterization II, M.   Linzer
     (editor), National Bureau of Standards, Gaithersburg, Md.,
     June 13-15, 1977, Spec. Publ. 525 (U. S.   Govt.
     Printing Office, Washington, D. C.).
(3) S.   A.   Goss,   R.   L.   Johnston,   and   F.   Dunn,
     "Comprehensive   Compilation   of   Empirical   Ultrasonic
     Properties of Mammalian Tissues," J. Acoust. Soc. Am 64,
     423-457 (1978).
(4) R.   C.   Chivers and R.   J.   Parry, "Ultrasonic Velocity and
     Attenuation in Mammalian Tissues," J. Acoust. Soc. Am.
     63, 940-953 (1978).
(5) P.   N.   T. Wells, "Absorption and Dispersion of Ultrasound
     in Biological Tissue," Ultrasound in Med.   and Biol.,
     Vol.1, 369-376 (1975).
(6) R.   C.   Waag,   R.   M.   Lerner,   P.   P.   K.   Lee,   and
     R. Gramiak, "Ultrasonic Diffraction Characterization of
     Tissue," Recent Advances in Ultrasound in Biomedicine, Vol.
     1, D.N.   White (editor), Research Studies Press, Forest
     Grove, Oregon, 1977, p 87.
(7) D.   Nicholas, "An Introduction to the Theory of Acoustic
     Scattering by Biological Tissues," Recent Advances in
     Ultrasound in Biomedicine, Vol. 1, D. N.   White (editor),
     Research Studies Press, Forest Grove, Oregon, 1977, p 1.
(8) R. C. Waag, P. P. K. Lee, R. M. Lerner, L. P. Hunter,
     R. Gramiak, and E. A. Schenk, "Angle Scan and Frequency-
     Swept Ultrasonic Scattering Characterization of Tissue,"
     Proc. Ultrasonic Tissue Characterization II, M. Linzer
     (editor), National Bureau of Standards, Gaithersburg, Md.,
     June 13-15, 1977, Spec. Publ. 525 (U. S. Govt. Printing
     Office, Washington, D. C.).
(9) F. L. Lizzi and M. A. Laviola, "Tissue Signature
     Characterization Utilizing Frequency Domain Analysis,"
     Proc. Ultrasonics Symposium, J. deKlerk and B. McAvoy
     (editors), 29 Sept-1 Oct 1976, Annapolis, Md., IEEE Cat.
     No. 76 CH1120-5SU, p 714.
(10) K. K. Shung, R. A. Sigelmann, and J. M. Reid, "The
     Scattering of Ultrasound by Blood," IEEE Trans. on
     Biomedical Engineering, BME-23, 6, 1976, p 460.
(11) K. K. Shung, R. A. Sigelmann, and J. M. Reid, "Angular
     Dependence of Scattering of Ultrasound from Blood," IEEE
     Trans. on Biomedical Engineering, BME-24, 4, 1977, p 325.
(12) A. S. Ahuja, "Effect of Particle Viscosity on Propagation
     of Sound in Suspensions and Emulsions," J. of Acoust. Soc.

Am. 51, p 182 (1972).

(13) C. R. Hill, "Interactions of Ultrasound with Tissues," Ultrasonics in Medicine, M. de Vlieger et al. (editors), Excerpta Medica, Amsterdam, 1974, p 14.

(14) F. E. Barber, III, "Ultrasonic Microprobe: For Modeling and Measuring the Angle Distribution of Echos From Diseased Arterial Tissues," University of Washington, Ph. D. Thesis, 1976, Xerox University Microfilms, Ann Arbor, Michigan 48106.

(15) C. R. Hill, R. C. Chivers, R. W. Huggins, and D. Nicholas, "Scattering of Ultrasound by Human Tissue," Ultrasound: Its Applications in Medicine and Biology, F. J. Fry (editor), Elsevier, 1979, Chapter 9.

(16) P. P. K. Lee, R. C. Waag, and L. P. Hunter, "Swept-Frequency Diffraction of Ultrasound by Cylinders and Arrays," J. Acoust. Soc. Am. 63, 600-606 (1978).

(17) R. C. Waag, P. P. K. Lee, H. W. Persson, E. A. Schenk, and R. Gramiak, "Frequency-Dependent Angle Scattering of Ultrasound by Liver," Proc. 8th International Symposium Acoustical Imaging, 29 May-2 June 1978, Key Biscayne, Florida, p 38. (abstract)

(18) R. C. Waag, P. P. K. Lee, H. W. Persson, E. A. Schenk, and R. Gramiak, "Frequency-Dependent Angle Scattering of Ultrasound by Liver," Proc. Annual Meeting Am Inst of Ultrasound in Medicine, 27-31 August 1979, Montreal, Canada, p 39. (summary)

(19) R. C. Waag, P. P. K. Lee, H. W. Persson, E. A. Schenk, and R. Gramiak, "Frequency-Dependent Angle Scattering of Ultrasound by Liver," Report DURL 79-1, University of Rochester, Rochester, New York, Nov 1979.

A COMPUTERIZED DATA ANALYSIS SYSTEM FOR ULTRASONIC TISSUE

CHARACTERIZATION

Joie Pierce Jones and Roger Kovack

Department of Radiological Sciences
University of California Irvine
Irvine, California 92717

INTRODUCTION

We have implemented a computerized data analysis system for ultrasonic tissue characterization at the University of California Irvine. One important feature of this system is its ability (via a fast analog to digital converter) to record both the phase and the amplitude of A-mode waveforms selected through regions of interest on a conventional B-mode ultrasonogram. The measuring system is well calibrated and characterized so that data obtained by it are essentially independent of the instrumentation used for acquisition. One section of the system, designed specifically for one-dimensional or A-line analysis, is built around PDP 11/40 and PDP 11/60 computers. A parallel section of the system, designed specifically for two-dimensional signal processing and display techniques, is built around an Interdata (Perkin-Elmer) 8/32 computer. This paper will describe our data analysis system and will present some representative results taken on tissue phantoms and on in vivo human liver as well as some typical computer simulations performed by the system.

DATA ACQUISITION AND ANALYSIS SYSTEM

A block diagram of the major components of our ultrasound system is presented in Figure 1. The details of this measurement system are shown in Figure 2. It should be noted that this system, which represents a major extension and improvement of one previously described,[1] is based on equipment, instrumentation, and facilities which were made available to us and therefore suffers from restrictive design constraints. Given sufficient initial

503

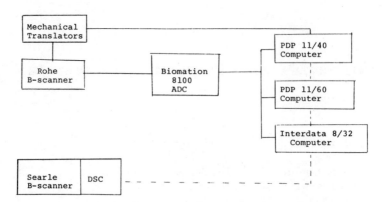

Fig. 1.   Major systems components for the data analysis system.

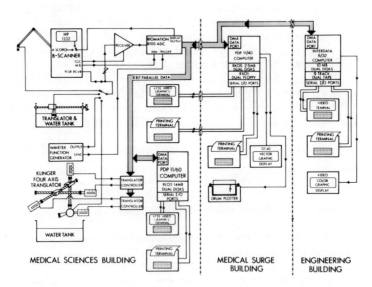

Fig. 2.   Details of the ultrasound data acquisition and analysis system.

resources we would not have chosen to spread our laboratory over
several buildings via an interconnected computer network. A future
paper[2] will explore in some detail the design considerations of a
system for the digital analysis of ultrasound waveforms.

The functional requirements of our system basically involve
the provision of a signal path from some echogenic medium to the
computer and methods for controlling all elements of this signal
path. The components of the signal path from sonic medium to
computer are shown in Figure 2. An ultrasound B-scanner has been
set up in such a manner as to output the ultrasound waveforms
directly from the receiver without the intervention of normal B-
scan processing. This has been done by modifying a Rohnar prototype
B-scan system. Such modifications provide for pre-processed receiver
output, time gain control (TGC) voltage output, main bang trigger
control and instrumentation of several operator and interally controlled functions that effect the receiver output.

To convert these signals from the receiver into a format useable
by a computer, we have set up a high speed analog-to-digital converter (ADC) in such a manner as to accept the ultrasound signal from
the B-scan system and transmit the digital equivalent of the instantaneous signal amplitudes to the computer. The ADC we use is the
Biomation 8100 featuring 8 bit resolution and sample rates to 100
megasamples/second. For ultrasound waveforms with center frequencies
at 2.25 MHz, we sample at a 20 MHz rate.

The computers serve the dual purpose of supplying a recording
and storage system for the digitized waveforms and providing a tool
for the implementation of signal processing algorithms. We use
Digital Equipment Corporation's (DEC) PDP 11/40 with 28K words of
memory and 5 megabytes on-line storage, DEC's PDP 11/60 with 128K
bytes of memory and 28 megabytes of storage, and Interdata's (Perkin-
Elmer) 8/32 with 383K bytes of memory and 40 megabytes of storage.
The PDP 11/60 is located in a room adjoining our ultrasound lab while
the PDD 11/40 and the Interdata 8/32 are located in other buildings
on campus. To facilitate data communications between our lab and
the computers, we have installed high speed data links with custom
driver/receiver circuits to interface the direct memory access (DMA)
data ports added to the computer. This allows data transfer at a
one megaword/second rate. For convenience and rapid experimental
execution, a data terminal with interactive software and graphics
display, as well as a separate printing terminal, are located in the
lab.

This facility presently functions in the following manner.
Ultrasound phantoms and test objects are placed in a water tank
equipped with a linear motion mechanism to carry the transducer
(a standard B-scan mechanical arm or a precision Klinger four-axis
translator are also available). The transducer is aligned with the

target of interest (this could be a line on a B-mode scan) and a command is issued to the computer through the data terminal keyboard. This initiates the digitization of the ultrasound waveform along with the TGC voltage. The digitized waveform is transferred to the computer main memory and the system awaits further commands. The experimenter may then command the display of the waveform in part or whole, the recording of the waveform on digital disc storage, or the processing of the data by any algorithm presently implemented. The actual processing techniques may be altered at the data terminal and immediately used on previously recorded waveforms or new waveforms. The results of modifying the processing techniques and the ultrasound subjects may also be compared on the graphics display screen and photographed for documentation. A drum plotter is also available to generate hard copy. The current system provides a tight and responsive feedback loop around the experimenter, the ultrasound subjects, the waveform acquisition hardware, and the signal processing system.

Space limitations require that the system characterization and calibration procedures, which are important to insure accurate and repeatable measurements, be discussed elsewhere.[2]

COMPUTER SIMULATIONS

Using our data analysis system we have conducted extensive computer simulations of the reflection of A-mode waveforms. Work now in progress is seeking to describe the reflection process for various distributions of reflectors and scatterers. By way of illustration we present some results obtained for reflection from simple periodic structures. Figure 3 is the simulated signal of reflected waveforms from a periodic structure with a 64 point spacing. Here the incident waveform is an actual ultrasonic pulse (recorded from the reflection of the incident pulse from a glass block in a water tank) made by a typical medical ultrasound transducer (center frequency: 2.25 MHz). In Figure 3 this actual incident pulse has been laid down at periodic intervals with alternating phase reversal to simulate reflection from a periodic structure. Various signal processing techniques can be applied to this simulated reflected waveform to extract the periodicity of the model structure. For example Figure 4 shows the cross correlation of the incident with the reflected waveform, Figure 5 shows the power spectrum of the reflected waveform, and Figure 6 shows the ECS of the reflected waveform. This latter process, termed envelope correlation spectrum (ECS), requires that we take a cross correlation of the incident with the reflected waveform, envelope detect, and display the power spectrum of the result. All three of these signal processing procedures clearly indicate the periodic nature of the model. Of course this is a trivial example since the periodic nature of the model is readily apparent from only a casual

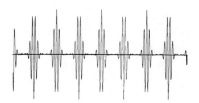

Fig. 3. Simulation of reflection
         from a simple periodic
         structure (64 point
         spacing) with alternat-
         ing phase reversal.
         Sample rate: 20 MHz.

Fig. 4. Cross correlation of
         incident wave (first
         wave packet in Fig. 3)
         with reflected wave
         (Fig. 3).

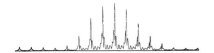

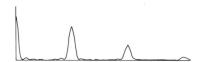

Fig. 5. Power spectrum of
         reflected waveform. The
         envelope of the display
         is the spectrum of the
         transducer. The spacing
         of the spectral lines is
         a measure of the period-
         icity of the model
         structure.

Fig. 6. ECS (envelope correla-
         tion spectrum) of the
         reflected waveform. The
         major peak is at a fre-
         quency corresponding to
         the periodicity of the
         model structure.

Fig. 7.   Simulation of reflection   Fig. 8.   Cross-correlation of
          from two interleafed   incident waveform with
          periodic structures   reflected waveform of
          (64 and 50 point spac-   Figure 7.
          ings).

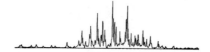
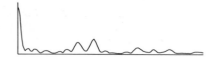

Fig. 9.   Power spectrum of   Fig. 10. ECS of reflected wave-
          reflected waveform of   form of Figure 7.
          Figure 7.

examination of the reflected waveform (Figure 3). However, this
example does illustrate the nature of our processing and gives
confidence that our algorithms are correct. A more realistic example
is shown in Figure 7. Here simulated reflections at two different
periodicities (64 and 50 point spacings) have been superimposed.
The periodic nature of each of the two structures is no longer
apparent in the reflected waveform from the interleafed structure.
Neither the cross correlation function (Figure 8) nor the power
spectrum (Figure 9) offer any insight as to the periodic nature of
the model structure. Even in this simple example conventional
analysis fails to extract details of the interleafed periodicities.
However, the two periodicities are readily apparent in the ECS of
the reflected waveform shown in Figure 10. These results are
encouraging and have led us to explore methods of describing a distribution of reflectors from the analysis of the reflected A-mode
waveform. Results from such a study should appear in the literature
in the near future.

A-MODE STUDIES

     Computer simulations of various scattering models have been
compared to A-mode data from various phantoms and from human tissue
taken in vivo. As an example of such work, we present a comparison
between data from a particular phantom (termed a gray-scale test
object) and normal human liver. Figure 11 is a B-scan of a phantom
made by Dr. John Ophir of the University of Kansas. This phantom
is designed to simulate the full range of echoes observed in a B-
mode ultrasonogram (hence the name gray-scale test object) and, in
particular, to simulate scattering from normal human liver. Figure
12 is a B-scan of a normal human liver taken under identical conditions. Even a detailed comparison between Figures 11 and 12 fails
to describe differences in texture between the two objects. Thus,
the phantom of Figure 11 correctly replicates the acoustical properties (here, distribution of echo intensities) of human liver as
measured on a B-mode ultrasonogram. However, quantitative differences between the phantom and liver are observed when various signal
processing procedures are applied to the A-mode ultrasound waveforms
from these objects. Figure 13 presents the ECS taken at two different sites in normal liver and the ECS taken at two different sites
in the phantom. Figure 14 gives a comparison of the power spectrum
of liver with that of the phantom. In both cases, major differences
are noted.

     The dip in the power spectrum from the phantom is easily ex-
plained by the fact that the phantom is encased in plastic and is
scanned through a 1/16 inch plexiglass window. The A-mode reflected
waveform from this window is shown in Figure 15. Although pulses
from the front and back face of the window interfere with each
other, the layer thickness can still be measured. Using handbook

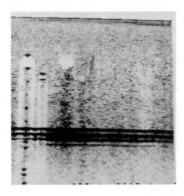

Fig. 11. B-scan of phantom (gray-
          scale test object) de-
          signed to simulate
          normal liver. The
          circular objects at the
          left simulate various
          vessels and lesions.

Fig. 12. B-scan of normal
          lver.

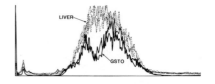

Fig. 13. ECS of A-mode waveforms
          at two different sites
          in the phantom and in
          normal liver. Window
          size is 1024 points.
          Data taken with a 1 cm
          delay.

Fig. 14. Power spectra of A-
          mode waveforms in the
          phantom (GSTO) and in
          normal in vivo liver.
          Window size is 1024
          points. Data taken
          with a 1 cm delay.

values for the velocity of sound in plexiglass, we find that the
window is 1.7 mm or about 1/16 inch as required. Figure 16 shows
the power spectrum of the reflected waveform (Fig. 15) from the
plexiglass window. The oscillations in the spectrum correspond to
a spatial periodicity of about 1/16 inch as required. Note that
the peak in this spectrum from the window occurs exactly at the dip
in the spectrum from the phantom material as would be expected.
Figures 11-16 demonstrate that information lost or not apparent in
a B-mode ultrasonogram can be extracted or made available through
analysis of the A-mode waveforms. The effects of overlying materials on quantitative measurements and the need to correct for their
effects is also clearly demonstrated.

B-MODE STUDIES

     Finally we present an example of two-dimensional signal pro-
cessing of B-mode ultrasonograms which we can perform on our
system. Figure 17 is a B-scan of a normal liver recorded on the
digital scan converter of a conventional ultrasound unit and transferred to our measuring system via magnetic tape. We are currently
exploring various two-dimensional filters which could possibly
enhance conventional B-mode ultrasonograms and are seeking to
develop image improvement techniques which are uniquely suited to
ultrasound data. Figure 18 shows the results of applying a particular filter to Figure 17. On the right half of Figure 18, we have
enhanced those edges perpendicular to the interrogating acoustic
beam, while on the left half, we have enhanced those edges parallel
to the beam. It should be stressed that Figure 18 is not indicative
of any proposed characterization scheme but is merely an illustration of the games we can play in a two-dimensional format.

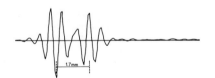

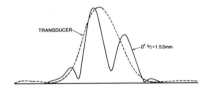

Fig. 15. A-mode waveform   Fig. 16. Power spectrum of
          reflected from 1/16   Figure 15 waveform.
          inch plexiglass window
          on phantom.

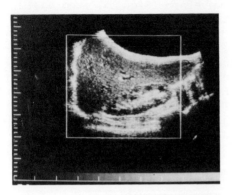
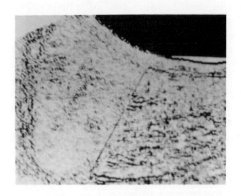

Fig. 17. B-scan of normal liver   Fig. 18. Two different edge
          as recorded by digital   enhancement filters
          scan converted and   applied to Figure 17.
          transferred to the
          Interdata 8/32.

ACKNOWLEDGEMENTS

    We gratefully acknowledge the kind and generous support of
Rohe Scientific, Searle Ultrasound, and the University of California
Cancer Research Coordinating Committee.

REFERENCES

1. J. P. Jones and C. Cole-Beuglet, In-vivo Characterization of
    Several Lesions in the Eye Using Ultrasonic Impediography,
    in "Acoustical Imaging", Vol. 8, A. F. Metherell (Ed.),
    Plenum (1980).
2. R. Kovack and J. P. Jones, Design Considerations of a System
    for Digital Analysis of A-mode and B-mode Ultrasound Signals,
    IEEE Trans. on Biomedical Engineering, to be published.

IMPEDIOGRAPHY REVISITED

S. Leeman

Department of Medical Physics
Royal Postgraduate Medical School and Hammersmith
Hospital
London, W12 OHS, U.K.

INTRODUCTION

Ultrasound pulse-echo scanning relies on the interaction of a propagating pulse with scattering centers within tissue. The measurement of echo return-time (back-scattering) may give a quite accurate picture of the location of scattering centers, but no further information. Grey-scale imaging extends the technique and provides, in addition, some knowledge of the strength, but little about the type, of scattering center. B-mode imaging may thus be seen as a crude "reconstruction" technique and, because of its impressive diagnostic usefulness, the natural question arises as to whether other techniques may provide further, and possibly even more valuable, information about the scattering interaction sites.

What are the tissue properties that give rise to the scattering of ultrasound pulses? It is well-known that tissue density and elasticity fluctuations may generate scattered waves, and it is demonstrated below that absorption fluctuations may do likewise. Hence, as far as interaction with ultrasound is concerned, tissue is a three-dimensional distribution of (at least) three parameters: density, $\rho(\underline{r})$, elasticity, $k(\underline{r})$, and absorption, $\alpha(\underline{r})$, where r is a vector denoting location in the object. Assume static tissue.

An inverse scattering technique (see, for example, Mueller et al, 1979) attempts to unambiguously reconstruct one (or more) of either these parameter distributions, or of associated tissue properties, the most important being ultrasound velocity, $C(\underline{r})$, attenuation, $a(\underline{r})$, and specific impedance, $Z(\underline{r})$. All these parameters are tissue characteristics, and it is easy to perceive the close relationship between tissue characterisation and inverse scattering methods. The latter should be (near-) exact, but the former are content with merely characterising or typifying the parameter distributions (or combinations of them) so that values typical of a certain disease state may result. The mere knowledge that a structure loses its regularity may be sufficient for tissue characterisation, even if it is not clear which parameter is changing. Nevertheless, the importance of inverse scattering methods as a preliminary to a tissue characterisation exercise seems apparent. In this context it should be remembered that computerised reconstruction imaging, and even doppler imaging (parameter : tissue velocity) are examples of inverse scattering methods. Impediography is a particular example of an inverse scattering technique (reconstruction of impedance from back-scattering) that is often seen as tissue characterisation ("parameter mapping"), and some of the problems besetting its successful implementation in practice are considered here.

UNIQUENESS OF INVERSE SCATTERING METHODS

An inverse scattering imaging method is of little value unless it is capable of unambiguously reconstructing the desired interaction parameter. Clearly, since at least three independent scattering parameter distributions exist in tissue, more than one <u>independent</u> scattering experiment needs to be performed in order to establish a unique reconstruction. If $K(r)$ and $\alpha(r)$ are not scalar functions, the number of experiments may be embarrassingly large; moreover, any experiments will not do — they must be independent (i.e. not provide equivalent information). The point is illustrated by the following simple example.

Consider the continuous-wave scattering from a $K(r)$-distribution. Assume that the scattering is weak enough for the first Born approximation to be valid, and that the scattering amplitude, Φ, is exactly measurable. It may be shown that (Morse and Ingard, 1968)

$$\dot{\Phi}(\underline{\mu}) = \frac{k^2}{4\pi} \int_R \gamma(\underline{r}) \, e^{i\underline{\mu}\cdot\underline{r}} \, d^3\underline{r}$$

Here, γ denotes the elasticity fluctuations within the scattering region R, k is the magnitude of the incident wave vector, $\underline{k}_i$, and $\underline{\mu} = \underline{k}_s - \underline{k}_i$, with $\underline{k}_s$ the scattered wave vector. Clearly, $\gamma(\underline{r})$ may be reconstructed by Fourier transformation of $\Phi(\underline{\mu})$.

$$\gamma(\underline{r}) = \frac{1}{2\pi^2} \int \frac{\Phi(\underline{\mu})}{k^2} \, e^{-i\underline{\mu}\cdot\underline{r}} \, d^3\underline{\mu}$$

In principle, the integral, and hence the measured values of $\Phi(\underline{\mu})$, needs to be taken over all $\underline{\mu}$ values. These values may be accessed by performing scattering experiments, at all incident frequencies, and measuring Φ at all scattering angles for each frequency (with the direction of the incident beam fixed in space, and then reversed)

However, if the possibility of density fluctuations, $\lambda(\underline{r})$, is also allowed, then (Morse and Ingard, 1968)

$$\Phi(\underline{\mu}) = \frac{k^2}{4\pi} \int_R \{\gamma(\underline{r}) + \lambda(\underline{r}) \; (\frac{\underline{k}_i \cdot \underline{k}_s}{k^2})\} \; e^{i\underline{\mu} \cdot \underline{r}} \; d^3\underline{r}$$

Different types of experiment are called for in order to unscramble the γ and λ distributions. If scattering is always measured at right angles to the incident beam, so that $\underline{k}_i \cdot \underline{k}_s = 0$, $\gamma(\underline{r})$ may be obtained by Fourier transformation of the corresponding $\Phi_\perp(\underline{\mu})$. Now, however, $\underline{\mu}$ values are accessed by performing experiments for all incident frequencies, and for all directions of $\underline{k}_i$ (keeping $\underline{k}_s \perp \underline{k}_i$ throughout). A similar back-scattering experiment ($\underline{k}_i \cdot \underline{k}_s = -k^2$ would give $\{\gamma(\underline{r}) - \lambda(\underline{r})\}$, and hence these two independent experiments suffice to measure the two desired distributions. The procedure is as unique, as the Fourier transformation involved, providing there are no ambiguities in any of the measured quantities.

For weak scattering, the fluctuations in the impedance are

$$\zeta = -\tfrac{1}{2}\{\gamma - \lambda\}$$

so that back-scattering data are seen, in this approximation, to be generated by a single (derived) parameter, Z. This, of course, is a restatement of the well-known result (Jones, 1977) that, in this approximation, the line-of-sight impedance profile may be reconstructed from the tissue impulse response ($\equiv$ experiments at all frequencies). In order to construct $Z(\underline{r})$, line-of-sight reconstructions for all possible incident directions would have to be made.

It has been pointed out that, in the next order of approximation, there is some ambiguity in the standard impediography reconstruction method (Kak and Fry, 1976). This certainly does not imply that impedance reconstructions are, per se, ambiguous, as Lefebvre has shown (in: "Lectures on the Applied Inverse Problem", Springer,Verlag, in press) that the problem may be transformed to an equation of the Gelfand-Levitan type, which permits unique inversion. The dilemma is partially resolved by examining the impediography approximation.

$$f(t) = \xi - \frac{4}{3}\xi^3$$

where f(t) denotes the (measured) data values at time t, and

$$\xi = \tfrac{1}{2} \ln \frac{Z}{Z_o}$$

with Z_o the initial impedance value. While it is true that each f(t) value gives three possibilities, for ξ, we would expect that these lie on distinct branches of values - i.e. at f(t + Δt) we would expect each of the three roots to differ only slightly from their previous values. If this is the case, then the knowledge of ξ at a particular location (t $\neq$ 0) will determine the particular branch on which the desired root falls, for all t. Thus, even in this approximation, uniqueness may not be a problem provided there is some additional information about the impedance values. Where large impedance changes appear discontinuously, as across a major boundary, the above continuity argument breaks down, but there may yet be sufficient information available to be able to apply the method.

ABSORPTION PROCESSES

Human tissues are strongly, and dispersively, absorbing. This complicates the impediography method in two ways: it may severely modify the form of the propagating pulse, and absorption fluctuation may generate back-scattering even where the impedance (ρc) remains constant.

It is easily demonstrated, in a one-dimensional model (Lindsay, 1960), that the ratio of back-reflected to incident wave amplitude, R, from a boundary across which the impedance jumps from Z_1 to Z_2, the absorption from α_1 to α_2, and the acoustic velocity from C_1 to C_2, is given by

$$R = \frac{Z_2(\alpha_1 C_1 + iw) - Z_1(\alpha_2 C_2 + iw)}{Z_1(\alpha_2 C_2 + iw) - Z_2(\alpha_1 C_1 - iw)}$$

where w is the circular frequency of the incident (plane) wave. [The dependence on w is interesting: it implies that the pulse shape is changed on reflection from such a boundary] Thus, absorption compromises the basic impediography assumption, that only impedance fluctuations give rise to the back-scattering. It may be noted that it is the ratio $\alpha C/w$ that determines to what degree R is influenced by the presence of absorption. Since this parameter is quite small ($\lesssim$ 1%) for cases of practical interest, with frequencies in the diagnostic range, it follows that

$$R \approx \frac{Z_2 - Z_1}{Z_1 + Z_2}$$

which is its usual (non-absorptive) value. We may thus suppose that
absorption-modified reflections may be neglected in impediography, to
good approximation. The effect of absorption on pulse propagation is
rather more difficult to quantitate in general. It is convenient to
consider the following dissipative one-dimensional wave equation
model for ultrasound transmission on homogeneous (i.e. non-scattering)
tissue:

$$\frac{\partial^2 p}{\partial x^2} - \frac{1}{c^2}\frac{\partial^2 p}{\partial t^2} - 2A\frac{\partial p}{\partial t} + 2B\frac{\partial^2 p}{\partial x \partial t} = 0$$

where $p(x,t)$ is the acoustic pressure at time t and distance x, and
A, B and C are constants. The absorption is essentially due to the
parameter A. If this is allowed to vary, $A = A_o(1 + a(x))$ say, then
it is easy to see that the fluctuation $a(x)$ generates scattered waves.
The expectation is that, to first approximation, these are weak enough
to neglect! It may be shown (Leeman, S., "Pulse Propagation in Dispersive Media", to be published in Phys. Med. Biol.) that the dispersion predicted by this model is in reasonable accord with measured
values for human tissues in the frequency range 0.5 ⌐ 7 MHz. The
advantage of this approach is that exact solutions for a propagating
impulse may be derived. An impulse, $\delta(t)$, incident at time $t = 0$ on
such a lossy medium filling the half-space $x > 0$, will be deformed to

$$H(x,t) = e^{-\alpha_o x}\,\delta(t \frown x/C_o)$$

$$+ \alpha_o x e^{-\alpha_o(x+b(t-x/C_o))}\,\frac{I_1(\alpha_o b\sqrt{(t-Bx)^2 - x^2/b^2}\,)}{\sqrt{(t-Bx)^2 - x^2/b^2}}\,u(t-x/C_o) \qquad (1)$$

where u is the unit step function, I_1 the modified Bessel function of
order 1,

$$\alpha_o = Ab$$

$$C_o = (B + 1/b)^{-1}$$

with $b = (B^2 + 1/C^2)^{-\frac{1}{2}}$

The propagator-reflector method (Leeman, 1979) may now be invoked to
show that the impulse response, $h(t)$ for a slab of material of (traveltime) "thickness" T, and with an impedance profile $Z(t)$, is given by

$$h(t) = \int_0^T d\tau\,\frac{d\ell nZ(\tau)}{d\tau}\,H(2\tau,t) \qquad (2)$$

This result holds only in the Born approximation (weak) back-scattering, such that the propagating pulse is unaffected by scattering) and provided that the reflection process itself is unaffected by the absorption.

It would be rather difficult, if not well nigh impossible, to invert (2) to give Z as a function of h. However, it should be remembered that h(t) is derived from a pulse-echo experiment, by deconvoluting the back-scattering with respect to the <u>incident</u> pulse, $p_{in}(t)$. When no absorption occurs, the measured back-scattering, $B(t)$, is given by

$$B_o(t) = h_o(t) * p_{in}(t)$$

where * denotes convolution and where $h_o(t)$ is related to $Z(t)$ in the usual way:

$$\int_o^t h_o(2\tau)d\tau = \frac{1}{2} \ln(Z(t)/Z_o)$$

When absorption of the above type is present, the propagated pulse in the tissue has the form

$$P(x,t) = p_{in}(t) * H(x,t)$$

(To avoid confusion, the distance variable, x, rather than the travel-time, is used). Then,

$$B(t) = h(t) * p_{in}(t)$$

$$= \int h_o(s) . P(s,t)ds$$

Hence, if we can manage to 'deconvolute' the measured back-scattering with respect to $P(x,t)$, rather than $p_{in}(t)$, we will be left with a tractable problem. Such deconvolution is difficult, but not impossible, if recourse is made to the cross-correlation technique developed by Herment et al (1979).

VELOCITY DISTORTIONS

The above dispersive absorption model predicts and automatically takes account of velocity dispersion. Thus pulse distortions due to this effect are already incorporated. An examination of (1) shows that the pulse actually propagates with a velocity, C_o, the "signal" velocity. This is the velocity that must be used in the determination of the travel time variable, which is so important in impediography.

On the other hand, impedance is defined with respect to the phase velocity, and, in a dispersive situation, there may be difficulties associated with the implied frequency dependence of Z. Experiment (Cartensen and Schwan, 1959) suggests that velocity dispersion is small, so this complication is probably negligible.

When a one-dimensional treatment is not valid, fluctuations in velocity will distort the surfaces of constant travel time into non-planar ones, and will complicate the notion of "effective impedance" necessary in that situation (Leeman, 1979). However, it may be deduced that this effect is quite small, as such velocity distortions would tend to degrade conventional B-scan images to a comparable degree, and this does not appear to be seen in practice.

CONCLUSIONS

Some of the problems besetting an effective implementation of the impediography technique have been reviewed. Evidence, and opinion (!), has been adduced to show that the major obstacle is probably the handling of the pulse deformations occasioned by the absorption of the medium. This persists even in the case of weak scattering. The problem may be tractable, however, and a further investigation of this matter is being made.

ACKNOWLEDGEMENTS

The funding of the Department of Health (U.K.) for a tissue characterisation project, which prompted this research, is gratefully acknowledged. Miss Ann Freemantle very capably generated the final copy.

REFERENCES

Carstensen, E. L. and Schwan, H. P., 1959, Acoustic Properties of Hemoglobin Solutions, J. Acoust. Soc. Am., 31(3);305.
Herment, A., Peronneau, P. and Tabbara, W., 1979, Contribution of Correlation Techniques to Impediography, in: "Abstracts of the Fourth International Symposium on Ultrasonic Imaging and Tissue Characterization", National Bureau of Standards, Gaithersburg, Maryland.
Jones, J. P., 1977, Ultrasonic Impediography and its Applications to Tissue Characterization, in; "Recent Advances in Ultrasound in Biomedicine", D. N. White, ed., Research Studies Press, Forest Grove.
Kak, A. C. and Fry, F. J., 1976, "Acoustic Impedance Profiling; An Analytical and Physical Model Study", in: "Ultrasonic Tissue Characterization", M. Linzer, ed., NBS Special Publication 453, National Bureau of Standards, Gaithersburg.

Leeman, S., 1979, The Impediography Equations, in: "Acoustical
    Imaging", A. F. Metherell, ed., Plenum Press, New York.
Lindsay, R. B., 1960, "Mechanical Radiation", McGraw-Hill, New York.
Morse, P. M. and Ingard, K. V., 1968, "Theoretical Acoustics",
    McGraw-Hill, New York.
Mueller, R. K., Kaveh, M. and Wade, G., 1979, Reconstructive Tomo-
    graphy and Applications to Ultrasonics, Proc. IEEE, 67(4);567.

TISSUE CHARACTERIZATION BY ULTRASONIC IMPEDIOGRAPHY

C.Q. Lee

University of Illinois at Chicago Circle
Department of Information Engineering
Box 4348, Chicago, IL 60680

ABSTRACT

An analytical method is presented which can be utilized for tissue characterization. This method permits the simultaneous determination of the attenuation and the impedance profiles. Higher-order reflections are included in the formulation. Computer simulated results for a hypothetical case are included.

INTRODUCTION

Acoustic characterization of living tissue is of increasing diagnostic value since many types of normal and abnormal tissues may be classified according to their acoustical properties. These include propagation velocity, tissue density, specific acoustical impedance, absorption and scattering of sound. A variety of acoustic techniques are available such that these characteristic parameters can be measured to provide information revealing the pathological or morphological states of specimens under examination. However, most of these techniques give only qualitative information, which is inadequate for precise tissue identification.

This paper deals with one of the more recent acoustic techniques which has been termed impediography[1,2]. Unlike the conventional A scan or B scan techniques which are based only on the amplitude response of the return wave, this approach utilizes both the amplitude and phase information available in the acoustic waveform and it provides quantitative characterization of tissue by means of impedance profiles. The subject matter is closely related to that of the inverse scattering problem in electromagnetic waves, in which the unknown region is an inhomogeneous medium whose refractive index varies in one spatial dimension[3]. The existing analytical methods used in acoustic impediography are essentially based on those of the propagation of electromagnetic waves in a stratified medium. The fundamental limitation of these methods, when they are applied in tissue characterization, is the failure of taking the medium loss into consideration. In any tissue or organ system, however, the medium is always lossy. The impedance profile calculated from an ideal lossless model can be much different from that of a real tissue. Furthermore, high-order reflections between different impedance layers in tissues are often neglected in the final analysis. This approximation would be valid only in regions having sufficient absorptions to highly attenuate such reflections.

In this paper, we present a different method for acoustic impediography. Using the Lagrangian approach in the formulation, we are able to include the medium loss as well as multiple reflections in our results. Thus, our method will permit simultansous determination of both the attenuation and the impedance profiles. Some computer simulated results for a hypothetical case will also be presented in this paper.

PROFILE RECONSTRUCTION

Consider a tissue medium having its constitutive parameters vary along the path of the interrogating ultrasonic pulse but remaining constant in directions perpendicular to this disturbance. To facilitate the mathematical analysis, we assume that the acoustic properties of this medium can be approximated by a finite number of piecewise uniform layers. Taking advantage of the similarities between the governing equations of acoustic waves and electromagnetic waves, we model the tissue medium with finite sections of piecewise uniform lossy transmission lines with each section having its own constitutive parameters R_k, G_k, L_k and C_k, with k ranging from 1 to a finite integer N.

The governing equations for the equivalent voltage and current in the kth section are given by:

$$\frac{\partial i_k}{\partial x} = - G_k V_k - C_k \frac{\partial v_k}{\partial t} \qquad (1)$$

$$\frac{\partial v_k}{\partial x} = - R_k i_k - L_k \frac{\partial i_k}{\partial t} \qquad (2)$$

Let us define the following transformation for the dependent variables,

$$V_k = \frac{1}{2} (f_{\alpha k} + f_{\beta k}) \tag{3}$$

$$i_k = \frac{1}{2Z_{ok}} (f_{\alpha k} - f_{\beta k}) \tag{4}$$

where $\quad Z_{ok} = \sqrt{\frac{L_k}{C_k}}$

Substituting equations (3) and (4) into equations (1) and (2), we obtain the following equations for $f_{\alpha k}$ and $f_{\beta k}$,

$$\frac{\partial}{\partial t} f_{\alpha k}(x,t) + U_k \frac{\partial}{\partial k} f_\alpha(x,t) = -a_k f_{\alpha k}(x,t) + b_k f_{\beta k}(x,t) \tag{5}$$

$$\frac{\partial}{\partial t} f_{\beta k}(x,t) - U_k \frac{\partial}{\partial k} f_{\beta k}(x,t) = b_k f_{\alpha k}(x,t) - a_k f_{\beta k}(x,t) \tag{6}$$

where $\quad a_k = \frac{U_k}{2} \left(\frac{R_k}{Z_{ok}} + G_k Z_{ok} \right)$

$$b_k = \frac{U_k}{2} \left(\frac{R_k}{Z_{ok}} - G_k Z_{ok} \right)$$

$$U_k = \frac{1}{\sqrt{L_k C_k}} \tag{7}$$

The coupled partial differential equations given in equations (5) and (6) may be further simplified by transforming the independent variables in the Eulerian approach to that of the Lagrangian approach[4,5]. Thus, using the following relations,

$$\frac{\partial}{\partial t} f_{\alpha k}(x,t) + U_k \frac{\partial}{\partial x} f_{\alpha k}(x,t) = \frac{\partial}{\partial t} f_{\alpha k}(t,t_\alpha) \tag{8}$$

$$\frac{\partial}{\partial t} f_{\beta k}(x,t) - U_k \frac{\partial}{\partial x} f_{\beta k}(x,t) = \frac{\partial}{\partial t} f_{\beta k}(t,t_\beta) \tag{9}$$

Equations (5) and (6) become,

$$\frac{\partial}{\partial t} f_{\alpha k}(t,t_\alpha) = -a_k f_{\alpha k}(t,t_\alpha) + b_k f_{\beta k}(t,t_\alpha) \tag{10}$$

$$\frac{\partial}{\partial t} f_{\beta k}(t, t_\beta) = b_k f_{\alpha k}(t, t_\beta) - a_k f_{\beta k}(t, t_\beta) \tag{11}$$

In these equations, t_α is used to identify a moving observer traveling in the increasing x direction after entering the kth layer at time t_α. $f_{\alpha k}(t, t_\alpha)$ and $f_{\beta k}(t, t_\alpha)$ are the values observed by this observer at time t. Similar interpretations can be given for t_β and the values of $f_{\alpha k}(t, t_\beta)$ and $f_{\beta k}(t, t_\beta)$ observed by the observer traveling in the decreasing x direction after entering the kth layer at time t_β.

Solving equations (10) and (11), we obtain,

$$f_{\alpha k}(t, t_\alpha) = f_{\alpha k}(t_\alpha, t_\alpha) e^{-a_k(t-t_\alpha)}$$

$$+ b_k \int_{t_\alpha}^{t} e^{-a_k(t-t_\rho)} f_{\beta k}(t_\rho, t_\alpha) dt_\rho \tag{12}$$

$$f_{\beta k}(t, t_\beta) = f_{\beta k}(t_\beta, t_\beta) e^{-a_k(t-t_\beta)}$$

$$+ b_k \int_{t_\beta}^{t} e^{-a_k(t-t_\rho)} f_{\alpha k}(t_\rho, t_\beta) dt_\rho \tag{13}$$

Our immediate objective is to solve these equations and to obtain the relations of $f_{\alpha k}$ and $f_{\beta k}$ on the interfaces at different times. This can be achieved with successive substitutions and integrations of equations (12) and (13).

To simplify the analysis, we assume that the propagation velocity U_k is equal in all impedance layers and that the width of each layer is the same. Thus, the trajectory of each moving observer in this case will be a straight line. Figure 1 shows two families of trajectories; trajectories with positive slopes are for observers traveling in the increasing x direction, while trajectories with negative slopes are for observers traveling in the decreasing x direction. It can be seen from this figure that the transit time τ for any moving observer in any impedance layer is equal to $t_j - t_{j-1}$.

After successive substitutions of equations (12) and (13), it can be shown that the relations among $f_{\alpha k}$ and $f_{\beta k}$ are given by the integral equations,

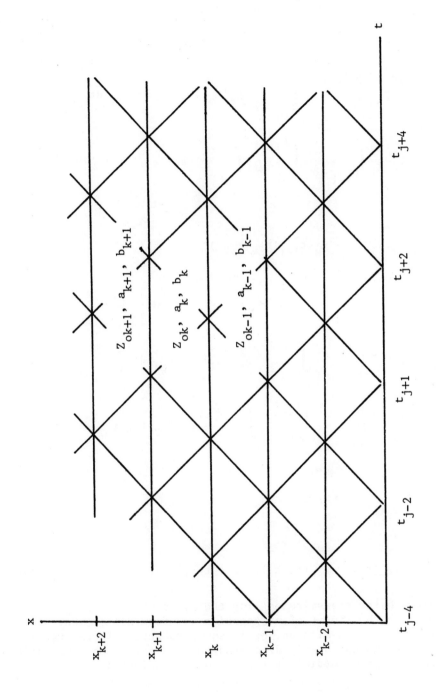

Figure 1. Trajectories of moving observers

$$f_{\alpha k}(t_j, t_{j-1}) = g_k(t_{j-1}) f_{\alpha k}(t_{j-1})$$

$$+ \frac{1}{2} \int_{t_{j-1} - \tau}^{t_j} g_k(t_{\beta 1}) f_{\beta k}(t_{\beta 1}) dt_{\beta 1}$$

$$+ \frac{1}{4} b_k^2 \int_{t_{j-1} - \tau}^{t_j} \int_{t_{\beta 1} - \tau}^{t_{j-1}} g_k(t_{\alpha 1}) f_{\alpha k}(t_{\alpha 1}) dt_{\alpha 1} \, dt_{\beta 1}$$

$$+ \frac{1}{8} b_k^3 \int_{t_{j-1} - \tau}^{t_j} \int_{t_{\beta 1} - \tau}^{t_{j-1}} \int_{t_{\alpha 1} - \tau}^{t_{j-2}} g_k(t_{\beta 2}) f_{\beta k}(t_{\beta 2}) dt_{\beta 2} dt_{\alpha 1} dt_{\beta 1}$$

$$+ \ldots \tag{14}$$

$$f_{\beta k}(t_j, t_{j-1}) = g_k(t_{j-1}) f_{\beta k}(t_{j-1})$$

$$+ \frac{1}{2} b_k \int_{t_{j-1} - \tau}^{t_j} g_k(t_{\alpha 1}) f_{\alpha k}(t_{\alpha 1}) dt_{\alpha 1}$$

$$+ \frac{1}{4} b_k^2 \int_{t_{j-1} - \tau}^{t_j} \int_{t_{\alpha 1} - \tau}^{t_{j-1}} g_k(t_{\beta 1}) f_{\beta k}(t_{\beta 1}) dt_{\beta 1} dt_{\alpha 1}$$

$$+ \frac{1}{8} b_k^3 \int_{t_{j-1} - \tau}^{t_j} \int_{t_{\alpha 1} - \tau}^{t_{j-1}} \int_{t_{\beta 1} - \tau}^{t_{j-2}} g_k(t_{\alpha 2}) f_{\alpha k}(t_{\alpha 2}) dt_{\alpha 2} dt_{\beta 1} dt_{\alpha 1}$$

$$+ \ldots \tag{15}$$

where $g_k(t) = e^{-a_k(t_j - t)}$

$$f_{\alpha k}(t_i) = f_{\alpha k}(t_i, t_i)$$

$$f_{\beta k}(t_i) = f_{\beta k}(t_i, t_i), \quad i = t_{\alpha 1}, \ t_{\beta 1}, \ t_{\alpha 2} \ \cdots \tag{16}$$

According to the transmission line theory, the value of b_k is proportional to the amount of propagation waveform distortion. For distortionless transmission, b_k is equal to zero. Therefore if the waveform is not highly distorted, only a few terms in equations (14) and (15) are sufficient for a good approximation.

If we assume that $f_{\alpha k}$ and $f_{\beta k}$ remain constant over the time interval (t_j, t_{j-1}) which depends on the thickness of the impedance layer in our approximation, the following equations are obtained by carrying out the integrations in equations (14) and (15).

$$f_{\alpha k}(t_j, t_{j-1}) = A_1 f_{\alpha k}(t_{j-1}) + A_2 f_{\beta k}(t_{j-2}) + A_3 f_{\alpha k}(t_{j-3})$$

$$+ A_4 f_{\beta k}(t_{j-4}) + \dots \tag{17}$$

$$f_{\beta k}(t_j, t_{j-1}) = A_1 f_{\beta k}(t_{j-1}) + A_2 f_{\alpha k}(t_{j-2}) + A_3 f_{\beta k}(t_{j-3})$$

$$+ A_4 f_{\alpha k}(t_{j-4}) + \dots \tag{18}$$

where the approximated values of the coefficients are given by,

$$A_1 = e^{-a_k \tau} \tag{19}$$

$$A_2 = \frac{1}{2} \frac{b_k}{a_k} (1 - e^{-2a_k \tau}) \tag{20}$$

$$A_3 = (\frac{1}{2} \frac{b_k}{a_k})^2 e^{-a_k \tau} (e^{-2a_k \tau} + 2a_k \tau - 1) \tag{21}$$

$$A_4 = (\frac{1}{2} \frac{b_k}{a_k})^3 e^{-2a_k \tau} 2(a_k \tau)^2 - 2a_k \tau + 1 \tag{22}$$

. . . .

Results given by equations (19) to (22) have been derived for a lossy non-uniform case. More terms may be included to improve the accuracy. For the lossless case, both a_k and b_k are equal to zero. Thus we have,

$$A_1 = 1$$

$$A_2 = A_3 = A_4 = \dots = 0 \tag{23}$$

Consequently, equations (17) and (18) are simplified to

$$f_{\alpha k}(t_j, t_{j-1}) = f_{\alpha k}(t_{j-1}) \tag{24}$$

$$f_{\beta k}(t_j, t_{j-1}) = f_{\beta k}(t_{j-1}) \tag{25}$$

Once the values of $f_{\alpha k}$ and $f_{\beta k}$ are determined, the equivalent voltage and current can be derived from equations (3) and (4). These are given by the following equations,

$$v_k(t_j) = \frac{1}{2}\{f_{\alpha k}(t_j) + f_{\beta k}(t_j, t_{j-1})\} \tag{26}$$

$$= \frac{1}{2}\{f_{\alpha k-1}(t_j, t_{j-1}) + f_{\beta k-1}(t_j)\} \tag{27}$$

$$i_k(t_j) = \frac{1}{2Z_{ok-1}}\{f_{\alpha k}(t_j) - f_{\beta k}(t_j, t_{j-1})\} \tag{28}$$

$$= \frac{1}{2Z_{ok-1}}\{f_{\alpha k-1}(t_j, t_{j-1}) - f_{\beta k-1}(t_j)\} \tag{29}$$

Equations (27) and (29) were obtained by imposing the conditions that the voltage and the current in the medium must be continuous functions of x. Solving $f_{\alpha k}(t_j)$ and $f_{\beta k-1}(t_j)$ in terms of $f_{\alpha k-1}(t_j, t_{j-1})$ and $f_{\beta k}(t_j, t_{j-1})$, we obtain,

$$f_{\alpha k}(t_j) = T_{k-1,k}f_{\alpha k-1}(t_j, t_{j-1}) + \Gamma_{k,k-1}f_{\beta k}(t_j, t_{j-1}) \tag{30}$$

$$f_{\beta k-1}(t_j) = T_{k,k-1}f_{\beta k}(t_j, t_{j-1}) + \Gamma_{k-1,k}f_{\alpha k-1}(t_j, t_{j-1}) \tag{31}$$

where

$$T_{k-1,k} = \frac{2Z_{ok}}{Z_{ok} + Z_{ok-1}} \tag{32}$$

$$T_{k,k-1} = \frac{2Z_{ok-1}}{Z_{ok} + Z_{ok-1}} \tag{33}$$

$$\Gamma_{k-1,k} = \frac{Z_{ok} - Z_{ok-1}}{Z_{ok} + Z_{ok-1}} \tag{34}$$

$$\Gamma_{k,k-1} = \frac{Z_{ok-1} - Z_{ok}}{Z_{ok} + Z_{ok-1}} \tag{35}$$

These are the transmission and reflection coefficients between two impedance layers Z_{ok} and Z_{ok-1}, which are the same as those of the lossless system. Similar to the lossless case, $f_{\alpha k}(t_j)$ in equation (30) takes the role of the incident wave in the kth impedance region at the interface between the kth and (k-1)th layers. Similarly, $f_{\beta k}(t_j, t_{j-1})$ in equations (30) and (31) is the reflected wave at the interface.

From the analysis given above, it can be seen that the basic differences between the lossless and the lossy cases are given in equations (17) and (18). In the lossless case, the incident and the reflected waves in the kth layer remain unchanged as they propagate. In the lossy case, the incident and the reflected waves are coupled through the lossy medium.

Because the similarity exists between the lossless and the lossy systems, the numerical technique of profile reconstruction used in the lossless medium can be applied to the lossy case. If the values of the incident and the reflected waves at the sending end are given (i.e., $f_{\alpha 1}(t_j)$ and $f_{\beta 1}(t_j, t_{j-1})$), we can simultaneously determine the attenuation and the impedance profiles by using equations (17) to (22) and (30) to (35). This technique of profile reconstruction is independent of applied signal waveforms and the signal processing is entirely performed in the time domain. It should be mentioned that all multiple reflections between layers are automatically included in our formulation.

COMPUTER SIMULATED RESULTS

In this section, we present some computer simulated results from the equations derived in this paper. We consider a simulated medium with the following characteristics,

$$Z_{ok} = 10 \times k$$

$$a_k \tau = .01 \times k$$

$$b_k \tau = .001 \times k, \quad k = 1, 2, 3, \ldots, M$$

Where M is the number of impedance layers used in the approximation. 12 sections were used in our example. Our program is divided in two parts. In the first part of our program, using the characteristic parameters of the medium given above and the interrogating pulse shown in Figure 2, we calcualte the reflected wave $f_{\beta 1}(t_j, t_{j-1})$ on the sending end. The result is given in Figure 3. Figure 4 shows the equivalent voltage wave as it propagates into the medium.

Based on the incident and reflected waves at the sending end, the second part of our program simultaneously reconstructs the profiles of a_k, b_k and Z_{ok}. Maximum error in the simulated results was less than 5 percent.

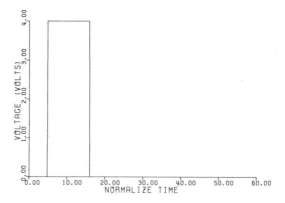

Figure 2.   Plot of Incident Wave

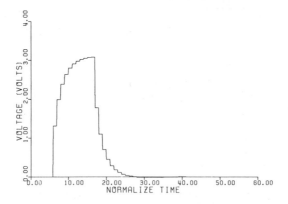

Figure 3.   Plot of Reflected Wave

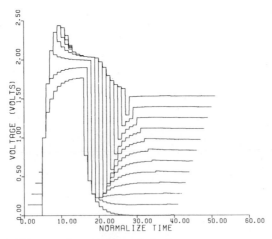

Figure 4.   Plot of Voltage Waves

ACKNOWLEDGMENT

The author would like to thank Mr. B.C. Phan for helping to verify some of the results and Professor Laxpati for his many constructive discussions concerning this paper.

REFERENCES

1. J.P. Jones, Current problems in ultrasonic impediography, in: "Ultrasonic Tissue Characterization," M. Linzer, ed., pp. 253-258. NBS Special Publication 453 (1976).
2. A.C. Kak and F.J. Fry, Acoustic impedance profiling: an analytical and physical model study, in: "Ultrasonic Tissue Characterization," M. Linzer, ed., pp. 231-251. NBS Special Publication 453 (1976).
3. A.K. Jordan and H.N. Kritikos, An application of one-dimensional inverse-scattering theory for inhomogeneous regions, IEEE Trans. Antennas Propagat. 21:909 (1973).
4. C.K. Birdall and W.B. Bridges, "Electron Dynamics of Diode Regions," Academic Press, New York (1966).
5. C.Q. Lee, Analysis of semiconductors with field-dependent mobility, IEEE Trans. Electron Devices, 23:581 (1976).

COMBINED TWO-DIMENSIONAL TISSUE/FLOW IMAGING

E. Aaron Howard,[1] Marco Brandestini,[2] Jeff Powers,[1] Saeed Taheri[1]
Mark E. Eyer,[1] David J. Phillips,[1] and Edward B. Weiler[2]
1. Center for Bioengineering, University of Washington, Seattle, WA. 98195
2. Advanced Technology Laboratories, Inc., Bellevue, WA. 98005

I. ABSTRACT

A functional sectorscan system has been built which provides two-dimensional (2D) images based on a coherent echo/Doppler processor.

The phase and magnitude of the demodulated signal (complex video) are analyzed by means of a fast digital processor. An image field of 256 X 256 pixels is stored in memory. A digital scan convertor and color coding scheme formats the image for standard raster display. The tissue and flow features are distinguished by contrasting color or grey scale assignments. A microprocessor-based operating system provides flexible control of sector format, pulse repetition frequency, sample clock, encoding scheme, etc.

The significant features of the echo/Doppler processor are: The output of the receiver is log compressed and demodulated by a quadrature pair of double-balanced mixers to preserve the phase and magnitude information within a 1 MHz bandwidth. A flash 8 bit A/D convertor is multiplexed between the "real" and "imaginary" components at a 7 MHz rate. A linear approximation (hypotenuse function) is used to calculate the magnitude term. The phase is computed by means of a look-up table. To obtain the Doppler shift for any given point, the derivative $d\theta/dt$ is substituted by the change of phase ($\Delta\theta/\Delta t$) between consecutive transmit/receive cycles.

Initial in vivo trials show that the system is capable of producing a longitudinal view of the carotid artery. Both tissue and flow features are clearly distinguishable.

II. INTRODUCTION AND BACKGROUND

One of our primary objectives is the continued development of combined pulsed Doppler flowmetry and 2D sectorscan imaging. The instrument described herein (Ecomega II) is the latest generation in this development and represents a state-of-the-art device for tissue/flow imaging.

Ecomega II provides two fundamental modalities for imaging. The first is a combination of 2D tissue images with superimposed flow features (B/Q mode), and the second is a combination of tissue and flow features presented in a standard time-motion display (M/Q mode). Both of these modalities have been reported on in earlier publications (Brandestini, et al., 1979; Eyer, et al., 1979). The emphasis in this report is on B/Q mode imaging and a new, coherent echo/Doppler processor.

The use of pulsed Doppler techniques for describing regions of flow (Baker, 1970) has had both peripheral vascular and cardiac applications. Flow maps of vessel lumens (Hokanson, et al., 1972; Curry, 1978) have been made without the aid of 2D tissue images for orientation. While flow features have been extracted, it has not been possible to identify vessel walls or cardiac anatomy.

In contrast to pulsed Doppler techniques, 2D sectorscan images have been used to visualize vascular walls and cardiac anatomy. This technique can provide important tissue visualization, but yields no hemodynamic information. Clearly, the combination of 2D sectorscan and pulsed Doppler is advantageous.

The Duplex echo/Doppler sectorscan systems described by Barber (1974) and Phillips (1978) served to combine the two techniques described above into a single instrument. These systems provided non-storable sectorscans and single sample volume pulsed Doppler flowmetry, thus allowing the user to investigate flow features with the aid of 2D tissue image orientation.

The Duplex scanner went through another evolution by the addition of the digital multigate Doppler (DMD) and the digital field store (DFS) as reported by Brandestini (1978) and Eyer (1978). This new scanner was called Ecomega I and is well reported (Eyer et al., 1979, Brandestini et al., 1979, Howard et al., 1979, and Stevenson et al., 1979). Ecomega I provided the first B/Q and M/Q modalities for clinical trials. Due to limitations in the instrument and exclusive 5 MHz transducer operation, Ecomega I is used most extensively in pediatric cardiology, providing M/Q images.

Ecomega I served to establish the feasibility of combining 2D tissue images and 2D flow features into a superimposed 2D echo/flow image. Many of the problems inherent to such a prototype scanner

have been improved upon in the new Ecomega II scanner described in
this report. The most notable improvement is the new coherent
echo/Doppler detection scheme which, in addition to being a novel
method for processing echo/Doppler information, provides a very
versatile and sophisticated instrument for continuing our tissue/
flow imaging research.

III.   PRINCIPLES OF OPERATION

A.   System

Echomega II consists of two primary subsystems: the amplitude and Doppler detector, and a display media (Fig. 1). A microprocessor provides overall control of the scanner and facilitates
the necessary operator interface. The Digital Echo/Doppler Processor (DED) is the heart of the scanner and will receive the main
emphasis of this report. Much of the technology of the DED is an
evolution of the Digital Multigate Doppler (DMD). The DMD established the feasibility of a multigate Doppler instrument utilizing
a time multiplexed digital processor and thus served as the basis
for developing the DED.

The Image Field Signal Processor (IFSP) is a further development of the Digital Field Store reported by Eyer (1978). The IFSP
provides digital processing of the echo/Doppler data such that
images may be stored, mixed, and eventually displayed by a standard color television monitor. This report will not go into a
great deal of detail regarding the IFSP since it represents primarily a simplification of the DFS rather than new technology.

B.   Digital Controller

The entire scanner is under the control of a central timing
module (digital controller). This module provides all necessary
clock functions. Timing is generated from a master clock (14.318
MHz) with phase coherence maintained throughout the system. Although the reader might expect this central timing to be the only
reasonable approach, it should be noted that Ecomega I was not
entirely synchronous. Thus, a single digital controller for
Ecomega II was a significant improvement.

Relevant timing signals will be discussed as they apply to
the modules described below.

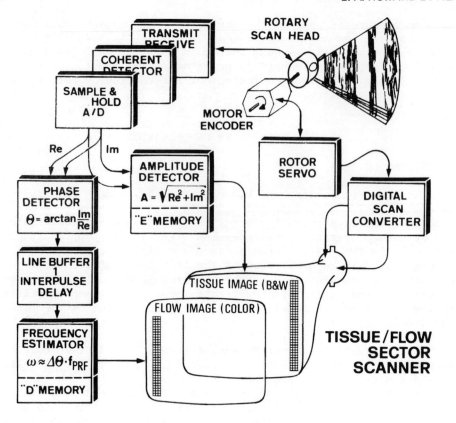

Fig. 1.   Tissue/Flow Scanner Block Diagram.

The diagram illustrates the separation of the processor
into a tissue (echo "E") and a flow (Doppler "D") channel
after detection and digitization. Tissue and flow information is being determined simultaneously and stored into
two separate banks of memory (256 X 256 X 4 each) by
means of the digital scan convertor. The data is then
read from the two memories in a TV line format and the
two fields appear superimposed. The color coding allows
combining of four dimensions (range, angle, intensity,
velocity) into a composite TV picture.

C. RF Section

     The DED processor is based on a phase coherent detection
scheme. Gated bursts of ultrasound are transmitted via a piezoelectric transducer. The same transducer acts as both transmitter and receiver of the backscattered signal. The entire trans-
receiver is designed to operate at 3 or 5 MHz with an overall
bandwidth of 1 MHz. The gated bursts are 3 cycles in duration (at
the respective carrier, 3 or 5 MHz) with a pulse repetition frequency (PRF) dependent upon the depth and zoom selected for the
particular application.

     The received signal is amplified, logarithmically compressed,
and demodulated. The detection is accomplished with a pair of
double balanced mixers. Quadrature reference signals are fed into
the mixers and the result is a pair of quadrature outputs. This
scheme is essential to the direction sensitive frequency estimator
(Brandestini, 1978) and ensures that all phase and magnitude information is preserved (Fig. 2).

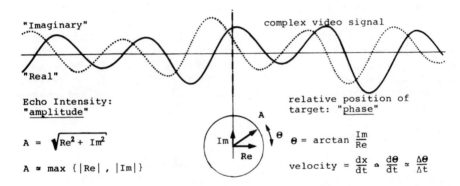

Fig. 2.   Rectangular to Polar Conversion to Derive Magnitude and
          Phase from Orthogonal Projections of Complex Video Signal.

     The complex video signal generated by synchronous quadra-
ture detection consists of two orthogonal components. The
signal contains amplitude and phase information for any
point in time (range). In our system, we convert from
rectangular to polar at fixed sample periods (e.g., 256
over the entire depth range). While the amplitude can be
obtained during a single transmit/receive cycle (mono-
pulse), the Doppler detection requires samples from two
successive cycles (frequency = temporal derivative of
phase).

The output of the demodulator maintains its phase relationship to the master clock for all points along the range of the
sound beam. Thus, this signal is called the range phase signal.
It is the processing of this range phase signal which allows us to
extract both Doppler and echo simultaneously, as the signal consists of "real" and "imaginary" components.

D. Sample and Hold

The DED processor is a discrete-time and discrete amplitude
instrument. The discrete-time format allows a single processing
unit to provide 256 sequential channels of information during each
transmit/receive cycle. The discrete amplitude format allows semiconductor memory storage and digital processing of the complex
video signal.

In order to provide 256 channels of complex video information, the DED must sample the output of the demodulator. This
means that both of the quadrature outputs must be sampled prior
to the analog-to-digital conversion (A/D). A pair of sample/hold
circuits performs this function.

At this point, the complex video is in discrete-time format.
Thus, a single processor can be time multiplexed to perform the
operations necessary to provide 256 channels of echo/Doppler information. The next crucial step is to perform the A/D conversion
necessary to utilize digital techniques for processing and storage.

E. A/D Conversion

Both the "real" and "imaginary" components of the sampled
complex video must be A/D converted. This is accomplished by time
multiplexing a flash 8 bit A/D convertor between the two components. The convertor is operating at 7 MHz.

It is worth reviewing the signal thus far processed. The output of the A/D convertor is a sequential stream of data representing 256 pixels. This sequence is actually 256 pairs of discrete-
time, discrete amplitude data elements. Each pair can be referred
to as a "real" and "imaginary" component. Since the "real" and
"imaginary" components are processed sequentially in time, yet represent data acquired from the same "point" in space, it is important that they be closely spaced in time. In fact, the "real" and
"imaginary" components are never more than 140 nanoseconds apart,
and this represents (nominally) a spatial ambiguity of 216 μm. This
exceeds the resolution capabilities of our current transducers.

F. Echo Detection (Magnitude Construction)

Classic echo processing in ultrasound systems utilizes asyn-
chronous amplitude demodulating schemes such as envelope detectors.
The DED extracts the echo information directly from the sampled
complex video signal. In a linear system, the best representation
of magnitude on a pixel by pixel basis would be:

$$A = \sqrt{RE^2 + IM^2}$$

Since this operation would require excessive dynamic range and the
DED performs log compression in the RF section, it is appropriate
to use an approximation (hypotenuse function) to calculate the
magnitude term. We have found the following function to be
satisfactory:

$$A \simeq \max\ (\ |RE|\ ,\ |IM|\ )$$

G. Doppler Detection

In order to derive velocity (Doppler) information from the
complex video signal, the processor has to determine the phase
angle for every spatial sample. This is achieved by computing the
arctangent of the "imaginary" divided by the "real" portion of the
signal (Brandestini, 1978). Just as velocity can be regarded as
the temporal derivative of position or range, the Doppler-shifted
frequency is represented by the derivative of the phase of the received signal. In our sampled processor, we substitute the derivative by the difference between two transmit/receive cycles. As
shown in an earlier report, this simplification yields results
very close to the true instantaneous frequency.

As the phase shifted signal scattered from the moving blood
cells contains a great deal of random information, temporal or
spatial averaging has to be performed to get an estimate of the
actual actual blood flow velocity. So far, the system averages
only in time (1 ... 20 msec per line); it has therefore been
necessary to collect the 2D flow velocity map over several heart
cycles (Fig. 3). In a real-time flow imaging apparatus, a certain
amount of spatial averaging would have to be incorporated to compensate for the short observation time due to rapid scanning.

H. Image Field Signal Processor (IFSP)

The IFSP accepts image data from the DED and properly formats
for B/Q or M/Q displays. The functions performed include:

- digital scan conversion
- image storage
- color coding
- video interface for RGB or RS170 TV display

The scan conversion is necessary to reproduce the 60° sector currently created by our mechanical scan head. This conversion is accomplished by direct computation of the polar to rectangular equations:

$$X = 3/4 \cdot R \cdot \sin \ + X_o$$

$$Y = R \cdot \cos \ + Y_o$$

where the scaler 3/4 accounts for the 3 x 4 aspect ratio of the display and X_o, Y_o are the offsets representing the "window" of interest.

Images are stored digitally in high speed static RAM fields. The IFSP currently has the capability of storing four separate images with a pixel depth of 4 bits. Plans are being made to expand the pixel depth to 6 bits. Tissue and flow are stored in separate image fields and combined in the readout circuitry. The microprocessor allows the operator to select flow only, tissue only, or composite displays.

Color coding schemes are designed to allow the operator to select predetermined codes to represent tissue and flow. In addition, the operator can create color schemes of his own choosing through alphanumeric keyboard control. The prime objective is to provide contrasting representations for tissue and flow. Typically, tissue is encoded in shades of grey scale while flow is represented by bi-directional color schemes. Thus, four dimensions of information (range, angle, intensity, velocity) are displayed on the TV screen.

The video interface provides all of the necessary timing signals for standard TV synchronization. The primary display mode utilizes separate red, green, and blue (RGB) drives for the TV monitor, thus providing a high quality color image. Also available is an NTSC composite video channel for video tape recording.

IV. PRELIMINARY RESULTS

Fig. 3 is a black and white reproduction of a color image constructed by Ecomega II. It is representative of a two-dimensional B/Q mode image containing both tissue and superimposed flow features. The original color-encoded flow is reproduced here as slightly fainter grey tones within the vessel lumen.

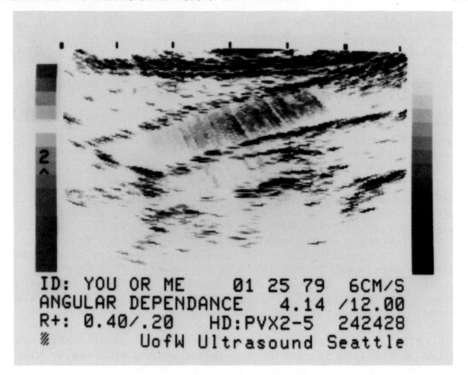

Fig. 3.   2D Image of the Common Carotid Artery Showing Anatomic
          and Superimposed Flow Features

        In addition to the tissue/flow image, supplemental informa-
tion is provided to facilitate ease of interpretation as well as
serve as a reference upon which quantitation of important parameters can be made. Along the top of the image are cm range mark-
ers which, for this case, indicate that the horizontal field of
view is 6 cm at the skin surface. A color bar for the velocity
information is located to the left, while a grey scale bar for detected echo amplitude is located at the right. If color were possible, one would see a "rational red" color scheme for flow velocities away from the pulsed Doppler transducer while those velocities toward the transducer would be displayed in a "rational blue
scheme" ranging from dark blue, through green, and then to yellow.
The phrase "rational color scheme" is intended to denote a color
sequence in which the viewer can readily distinguish relative
velocity magnitudes. Below the image, alphanumerics relate
patient identification and various instrument settings. The 6
cm/s is the velocity magnitude step change from the flow velocity
scale. The 4.14 denotes that the pulsed Doppler PRF is 4.14 KHz
while in the Duplex mode (i.e., alternating pulsed echo with

pulsed Doppler) while the 12.00 denotes the pulsed Doppler PRF in KHz in the Doppler-only mode. This change in pulsed Doppler PRF is essential if accurate velocity information is to be obtained in regions of disease where high velocities are generally present (Phillips et al., 1979). The 0.40 just after "R+:" is the time in seconds after the last R-wave of the ECG and is the time when the two-dimensional tissue image was stored. The 0.20 indicates that the flow velocity image was generated 0.20 seconds after the R-wave. The scan head used was PVX2-5 and the composite video frame stored is number 242428.

The 2D image in Fig. 3 shows a longitudinal sectional of a normal common carotid artery just proximal to the carotid bulb located to the right and out of the field of view. The skin surface is seen anterior with an intervening muscle layer seen at the left between the skin and arterial wall. The two bright targets within the confines of the vessel at the left are reverberant artifacts from sound within the scan head boot. This type of reverberent artifact is readily distinguished from "real" tissue interfaces through slight movements of the scan head while observing the real-time image.

During an examination, the operator surveys local anatomy and listens to the flow character at selected points by manually placing a sample volume (i.e., in Duplex mode) at various sites within the blood vessel. When a location of interest is reached, and while holding the scan head still, a foot switch is pressed. This maneuver puts the system into a Doppler-only mode which "freezes" the last entire two-dimensional tissue image and increases the Doppler PRF to 12 KHz. The pulsed Doppler transducer is then manually swept through the object volume to detect flow velocities, if present, and to display each velocity magnitude according to its proper spatial location and direction. The flow image is constructed at 0.20 seconds past the R-wave, which is near the time at which the peak velocities occur in this portion of the carotid artery.

The two-dimensional flow velocity image is constructed over approximately 2.5 cm of blood vessel. Since a multi-channel pulsed Doppler is employed, only about 40 heart beats were required to generate the flow image. The multi-channel feature makes such two-dimensional flow images practical by significantly reducing the time over that required for the single sample volume approach. The "angular dependence" of the velocity magnitudes as a function of Doppler angle with respect to the vessel axis can be appreciated in this image. The magnitudes are relatively low toward the left, where an "unfavorable" Doppler angle approaches 90°, while, when scanning toward the carotid bulb, the angle becomes more favorable and higher magnitudes are detected. At the present time, the factor of Doppler angle is not taken into account in the

image display, but should be in future studies to provide accurate
information. When combining the anatomic image with the flow velocity image, in the IFSP, differentiation of the two types of information is relatively straightforward. The hypothesis is that the
integration of tissue and velocity information into a single display format can provide more information than could be acquired by
either modality employed separately. The reader is referred to
Brandestini et al., 1979, for color examples obtained from this
system.

While initial clinical trials on presumably normal blood
vessels are encouraging, clinical usefulness for diseased peripheral arteries has yet to be shown. Considerable flexibility is
built into this system in order to modify operating parameters and
output display characteristics, as need be. Problems related to
noise in the Doppler channel and obtainment of higher quality two-
dimensional tissue images are primary areas of future efforts.

ACKNOWLEDGEMENTS

The authors wish to acknowledge the technical assistance of
Jay Borseth, Gordon Kirkendall, George Mahler, John Ofstad, Bob
Olson, and Vern Simmons. We wish to thank Carolyn Phillips for
her work in preparing this manuscript. The technical research was
supported by NIH Grant HL-07293-17.

REFERENCES

Baker, D. W. (1970), Pulsed ultrasonic Doppler blood-flow sensing.
    IEEE Trans. Sonics Ultrasonics 17, 170-185.

Barber, F. E., Baker, D. W., Nation, A. W. C., Strandness, D. E.,
    Jr., and Reid, J. M. (1974), Ultrasonic duplex echo-Doppler
    scanner. IEEE Trans. Biomed. Engr. 21, 109-113.

Brandestini, M. A., Forster, F. K. (1978), Blood flow imaging
    using a discrete-time frequency meter. Ultrasonics Symposium
    Proc. IEEE Cat.# 78 CH1344-1SU.

Brandestini, M. A., Eyer, M. K., and Stevenson, J. G. (1979), M/Q-
    mode echocardiography -- the synthesis of conventional echo
    with digital multigate Doppler. Echocardiology, Third Sympo-
    sium on Echocardiology, Rotterdam, Netherlands, June 1979.

Curry, G. R. and White, D. N. (1978), Color coded ultrasonic dif-
    ferential velocity arterial scanner (Echoflow). Ultrasound
    in Medicine and Biology 4, 27-35.

Eyer, M. K. (1978), A microprocessor based digital scan converter
    and color display system for ultrasonic image presentation.
    Masters thesis, Electrical Engineering, University of Wash-
    ington, Seattle, Washington.

Eyer, M. K., Brandestini, M. A., Phillips, D. J., and Baker, D. W.
    (1979), Color digital echo/Doppler image presentation. Arti-
    cle submitted to Ultrasound in Medicine and Biology.

Hokanson, D. E., Mozersky, D. J., Sumner, D. S., McLeod, F. D.,
    Jr. and Strandness, D. E., Jr. (1972), Ultrasonic arterio-
    graphy: A non-invasive method of arterial visualization.
    Radiology 102, 435-436.

Howard, E. A., Brandestini, M. A., Eyer, M. K., and Weiler, E. B.
    (1979), Color-coded digital echo/Doppler imaging, storage,
    and display. Proc. 4th International Symposium on Ultrasound
    Imaging and Tissue Characterization, pp. 25-28.

Phillips, D. J., Blackshear, W. M., Baker, D. W. and Strandness,
    D. E., Jr. (1978), Ultrasound Duplex scanning in peripheral
    vascular disease. Radiology/Nuclear Medicine 8, 6-10.

Phillips, D. J., Powers, J. E., Eyer, M. K., Blackshear, Jr., W.
    M., Bodily, K. C., Strandness, Jr., D. E., Baker, D. W.
    (1979), Detection of peripheral vascular disease using the
    Duplex scanner III. Article submitted to Ultrasound in
    Medicine and Biology.

Stevenson, J. G., Brandestini, M. A., Weiler, E. B., Howard, E. A.
    and Eyer, M. K. (1979), Digital multigate Doppler with color
    echo and Doppler display - diagnosis of atrial and ventricu-
    lar septal defects. Proc. 52 Meeting of the American Heart
    Association, Part II, Vol. 60, #4.

AN ULTRASOUND IMAGER FOR FETAL RESPIRATORY RESEARCH

A.J. Cousin

University of Toronto
Department of Electrical Engineering
Toronto, Ontario, Canada

INTRODUCTION

As a result of the limited number of useful and measurable indicators available to assess fetal well-being, research efforts are being directed toward measurement of parameters such as fetal breathing movements (FBMs). These rhythmic movements of the fetal chest occur repeatedly during the second and third trimester. Presently, investigations are proceeding at both basic physiology and clinical levels to determine the efficacy of FBM for assessment of fetal condition and as a neonatal predictor.

Most centres now employ real-time B-scan ultrasound techniques to distinguish FBM from artifact but the major difficulties are associated with the reliable quantification of FBM into hardcopy form. Extraction of FBMs from the real-time B-scan data involves the measurement of time differences corresponding to chest movement. As a research tool, it is now being recognized that multiple site measurements are required to understand more fully the nature of FBMs.

Systems existing in various centres range from manual observation of the pre-recorded image from a slow motion video tape recorder to schemes that provide automatic range gating based on a fixed threshold to our present system that provides a hardcopy of FBM that is essentially independent of amplitude variations of the echo complex. In the sections that follow, it is presumed that a good quality video image has been obtained where both proximal and distal chest structures are discernable. Although this requirement has presented some difficulty in the past, it no longer represents a significant problem with the improved array transducer sensitivity and low noise electronics currently available.

SOURCES OF ERROR

There can be four primary sources of error associated with the measurement of FBM from a real-time B-scan instrument. These comprise single-ended measurement error, fixed threshold triggering error, tracking speed/range limit error and fetal rotational error. Each of these potential error contributors will be described briefly in the following subsections.

Single-Ended Measurement Error

Biological artifacts are due to nonbreathing movements of the fetus relative to the transducer located on the maternal abdomen. Systems which measure the movement of a _single_ echo such as that from the distal chest wall with respect to the transducer position, or equivalently to a marker located a fixed time from the transducer, can produce an undesired output for gross fetal movements, maternal respiration, coughing, placental pulsations, etc. In fact, any nonbreathing movement that causes a change in the _absolute_ difference between the transducer and the selected echo will produce an artifactual record.

Fixed Threshold Triggering

In a system employing a fixed threshold trigger, the time at which an echo occurs is registered as the time at which it first exceeds this fixed threshold. Since the echo returns vary in amplitude due to parameters such as time and direction of insonation that are not within the control of the operator, several error mechanisms exist as illustrated in Figure 1.

One effect of a small change in amplitude is shown in Figure 1b. Here the variation in the time at which the echo return first reaches the trigger threshold can be seen to be not only related to the spatial position of the echo complex but also to its amplitude. Therefore a small amount of amplitude modulation on even a stationary echo return will produce "breathing". This type of error has a maximum value of $\lambda/4$ which is reasonably small at 2MHz compared to typical breathing ranges.

An effect due to a larger change in amplitude is shown in Figure 1c. In this case, the amplitude has increased sufficiently to change the wavefront that first meets the threshold level. Under these conditions, the error can be several wavelengths which is easily as large as many breathing signals.

Lastly, an effect specific to range gated techniques is depicted in Figure 1d. In the case shown, only a portion of the echo complex is contained within the range gate. If the complex moves further outside of the range gate window, the recorded movement can contain large errors.

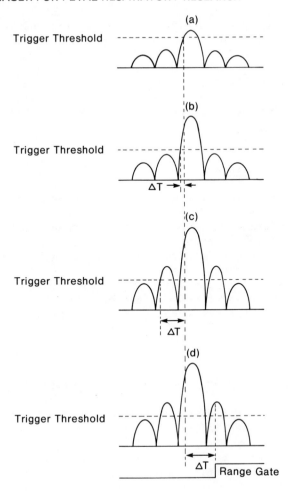

Figure 1.   Illustration of Sources of Error in Fixed Threshold
            Systems. ΔT Represents the Amount of Artifact Produced
            in Each Case.
               (a) Idealized
               (a) Idealized, Rectified Echo Complex
               (b) Effect of a Moderate Increase in Amplitude
               (c) Effect of a Large Increase in Amplitude
               (d) Effect of a Range Gate for an Echo Complex
                   Moving Over the Boundary

Tracking Speed/Range Limit

This error source comes about due to the requirement of tracking
on the chest wall while simultaneously displaying the real-time B-
scan picture. In this situation, one is required to interlace the

ultrasound beam appropriately to obtain sufficiently current information about the moving chest. Although this will be explained
in some detail in a later section, it is important to note that the
update speed and allowable range of echo movement must be compatible
with the speed of the fetal chest. That is, the echo corresponding
to the desired chest structure must not have moved farther than the
range allowed by the electronic timing circuitry between successive
updates.

Fetal Rotational Error

This error source is due to the nonsymmetrical fetal thoracic
anatomy. Since the real-time B-scan picture provides only two
dimensional information, apparent chest movement can be mimicked
by rotational fetal movement even with no concommitant FBM. As the
fetus tilts, its thoracic dimensions can change. Therefore, for a
periodic tilting motion caused perhaps by maternal respirations,
the two dimensional B-scan image will appear to show FBMs. Although
the previous three error sources can be significantly reduced, there
appears to be no straight forward method of eliminating this problem aside from the usual careful scrutiny given to each trace to
ensure proper sense.

METHODS OF ARTIFACT REDUCTION

In the instrument to be described the first three sources of
error have been essentially eliminated. The single ended method
has been replaced by our group and others with a differential method
where the difference between two echo complexes originating on the
fetus is measured and this variation with time used to produce the
FBM recording[1,2]. To maximize the breathing signal, it is customary
to record the difference between the proximal and distal chest walls.
Implementation of this relatively simple concept does much to eliminate most biologic artifact that will appear in single ended
measurements and is best documented in a reference.[3]
The second error source stems fundamentally from the use of a
fixed threshold trigger to determine timing from a signal where
amplitude varies. For the purposes of breathing measurement, the
amplitude of the echo complexes contains no information. Therefore,
this error may be eliminated by triggering on a parameter that is
independent of amplitude. A system that can extract timing information from the zero crossings of a selected echo complex has
been reported previously[4]. It should be noted that by detecting
the time of occurrence of a particular zero crossing, one has
realized a potentially artifact-free method of FBM recording but
at the expense of more complex circuitry that will allow the
selection of that zero crossing as distinct from any other.
A block diagram of a tracking channel based on zero crossing

detection is shown in Figure 2. The incoming echo complex is first
quantized by a bipolar one bit A/D convertor. This block produces
three output states: High if the input is above ε, Low if the
input is below -ε and Zero if the input is between ±ε. ε is a
small number and related to the input noise. The digitally formatted
output is then amplified whenever the gated amplifier receives an
enable pulse from the one shot. This gated output is accumulated
during the one shot time width and appears as the input to the
integrator controlled delay. This block functions as a time delay
that varies in direct proportion to the accummulated integrator
output. In an analog implementation, this delay could be voltage-
controlled by the integrator. The delay begins synchronous with the
main bang pulse and ends with the triggering of the one shot. It
can be shown that for stable operation (i.e. with the integrator
output fixed), the one shot will be centred around a zero crossing
of the input echo. For a moving echo complex, the loop will con-
tinually recenter the one shot pulse around the selected zero
crossing and FBM can be obtained as the variation in the integrator's
output.

The third error source can now be quantified in terms of the
phase-locked tracking scheme described above. Since the tracking
circuit is operating in a sampled data mode, it is important to
consider the maximum amount of movement allowed between successive
updates. To track large excursions of the fetal chest, one would

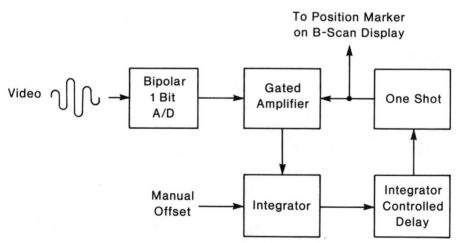

Figure 2. Single Tracking Channel Based on Zero Crossing Detection

like both rapid updates of the desired echo complex and a wide one
shot pulse width. The maximum update rate is determined by the
pulse repetition frequency of the real-time system and is quickly
fixed by the maximum desired observation depth. The one shot pulse
width can be set from a very small value up to λ. However, as this
pulse width increases toward λ, the resulting change in the integrator output for some fixed change in echo complex location varies
as shown in Figure 3. Since it is desirable to track large or
fast movements and to produce a large error correction signal $\lambda/2$
is chosen as the one shot pulse width.

Given these limits, the speed of tracking can be computed.
For a conventional real-time B scan array that is cycling at 30Hz,
a new update on a particular echo complex is available every 33.3ms.
At 3.5MHz, this yields a maximum change of .214mm or a maximum
tracking rate of 6.4mm/sec. Unfortunately, FBMs can correspond to
higher rates than this. Further, in a differential system, two
tracking channels are required and each channel must remain locked
on echo complexes that move due to nonbreathing activity which can
be significantly faster than 6.4mm/sec. Our solution to this speed
limitation is to interlace the B-scan with the desired A-scan line
to achieve an effectively faster update rate while still maintaining
the B-scan image. The details of this interlace method will be
described subsequently.

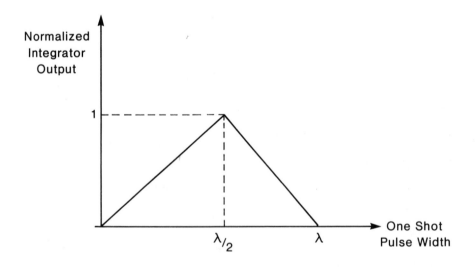

Figure 3.   Integrator Output As as a Function of Pulse Width

SYSTEM DESCRIPTION

A simplified system block diagram is shown in Figure 4. The array transducer is a 64 element linear array that has been customized to excite an arbitrary group of crystals at each main bang pulse. For the present, we are exciting the crystals in adjacent groups of four which yield a total of 61 lines of independent B-scan information. When in the tracking mode, we obtain a faster update rate of a selected line(s) by employing an interlaced pattern so that the crystals are excited sequentially as: 1-2-3-4, 12-13-14-15, 2-3-4-5, 21-22-23-24, 3-4-5-6, etc. This example illustrates two tracking channels (12-13-14-15 and 21-22-23-24) interlaced with the typical real-time B-scan excitation of a sliding group of four crystals. As more tracking channels are required, the update rate for each channel decreases. However, this alternate interlace mode provides a very large increase in tracking speed capability. For the best case of one active channel, the speed of tracking has increased from 6.4mm/sec to 389mm/sec which is more than adequate for fetal breathing research. The tracker outputs are selected according to which dimension is desired from the B-scan image and the resulting output subtracted to obtain the difference over time that represents FBMs.

A block diagram of a tracking channel is shown in Figure 5. This all digital realization eliminates many of the analog error sources associated with a direct analog implementation of Figure 2. The time of occurrence of the zero crossing of one cycle of an echo complex is compared with the output of the one shot. Their time difference and polarity, N, is accumulated and used to adjust the integrator controlled delay. At the start of the next update, this

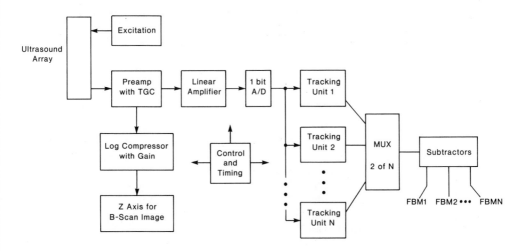

Figure 4.   System Block Diagram

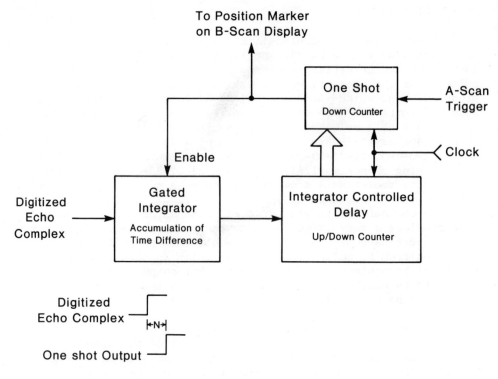

Figure 5. Single Channel Digital Tracking Loop

new delay value is loaded into the one shot which produces a pulse that is shifted by N. If the echo complex has not moved during this time, both integrator inputs will be coincident with the result that the integrator output will not change and the one shot will be triggered at the same point in time.

A further advantage of this digital implementation is that the equivalent gating time can be increased to λ with no decrease in loop gain. This increase doubles the maximum speed of tracking independent of the number of trackers being used.

CONCLUSIONS

An example of the type of clinical quality obtainable with an instrument of this type is shown in Figure 6. This trace represents differential fetal breathing obtained from the thorax. It is to be noted that the tracing is typical of instrument performance and clearly allows detailed analysis of waveshape and other parameters that could not be performed previously.

A further result of the higher tracking speed capability is the quantification of fetal hiccoughs. The tracing in Figure 7 displays a maximum velocity of 11mm/sec.

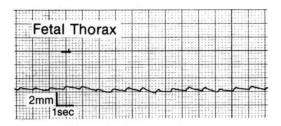

Figure 6. Differential Recording of Fetal Breathing Movement

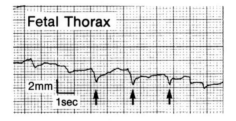

Figure 7. Differential Recording of Fetal HIccoughs

Finally, an interesting intrinsic property of the zero crossing concept to extract imaging information is that it is frequency independent. In the digital implementation presented here, the distance resolution is limited by the finite clock frequency of approximately 25MHz and the finite middle state for the comparator to eliminate noise ($\pm\varepsilon$). Although the ultrasonic wavelength is on the order of .5mm, the resolution in the FBM tracing is on the order of .03mm.

Acknowledgements

The author gratefully acknowledges the financial support of the Hospital For Sick Children Foundation.

REFERENCES

1. K. Lindstrom, K. Marsal, G. Gennser, L. Bengtsson, M. Benthin, and P. Dahl, 1977, Device for Measurement of Fetal Breathing Movements - 1., Ultrasound Med. Biol, 3:143
2. L.W. Korba, A.J. Cousin, R.S. Cobbold, and D. Gare, 1977, An Ultrasonic Fetal Respiratory Movement Monitor, San Diego Biomedical Symposium, February.
3. D.J. Farman, G. Thomas, and R.J. Blackwell, 1975, Errors and Artifacts Encountered in the Monitoring of Fetal Respiratory Movements Using Ultrasound, Ultrasound Med. Biol., 2:31.
4. L.W. Korba, R.S.C. Cobbold and A.J. Cousin, 1979, An Ultrasonic Imaging and Differential Measurement System for the Study of Fetal Respiratory Movements, Ultrasound Med. Biol., 5:139

TRANSESOPHAGEAL ULTRASONIC IMAGING OF THE HEART

+Balasubramanian Rajagopalan, °Eugene P.
DiMagno, +James F. Greenleaf, °Patrick T.
Regan, *James Buxton, *Philip S. Green,
and +J. William Whitaker

+Department of Physiology and Biophysics and
°Gastroenterology Unit, Mayo Clinic,
Rochester, Minnesota, and *Bioengineering
Research Center, SRI International, Menlo
Park, California

INTRODUCTION

In echocardiology ultrasound backscattering is
used to visualize heart structures. The various real-
time cardiac scanners give detailed dynamic pictures
of the moving boundaries of cardiac anatomy [1,2,3,4].
The intensity of ultrasonic backscattering from within
the myocardial tissue may vary with the normal and
ischemic status of the heart tissue [5,6]. Balasubramanian Rajagopalan, et al., [7] first reported that
regional myocardial perfusion could be imaged using
ultrasonic backscattering when suitable contrast agents
were arterially introduced. Using a 10 MHz high resolution B-scanner that was placed directly on the surface of the heart of anesthetized dog after a left
thoracotomy, the authors showed that the backscattering
from the myocardial mass was enhanced significantly if
an ultrasound contrast agent such as indocyanine green
was injected into the left ventricle or at the aortic
root. This enhancement of backscattering was observed
to be related to regional blood flow since after depri-
vation of blood supply to a region of the myocardium,
the intensity of backscattering from the deprived
region did not exhibit enhancement with the injection
of the contrast agent. This effect may be explained
by postulating that microbubbles of air in the contrast
carried by the blood reach the coronary arterioles and

get trapped at different sites depending upon their
sizes. These microbubbles slowly dissolve into the
blood with the concomitant disappearance of enhanced
ultrasound backscattering.

The enhanced backscattering seemed to have a frequency dependence since attempts to observe similar
phenomenon in intact persons with a commercial scanner
using low frequency transducers (2.5 MHz) were not
successful. One possible method to investigate enhanced backscattering from myocardium in intact animals, using high frequency ultrasound with a small
penetration depth is to study the heart from the esophagus. A unique opportunity to pursue these ideas
arose when an ultrasonic endoscope became available to
the investigators. This ultrasonic imaging instrument
is a modified fiber optic endoscope developed by SRI
International [8] and was specifically designed for
abdominal examinations. It was developed to overcome
the problems of poor resolution and bowel gas that
may plague abdominal ultrasonic scans. It has been
used safely and successfully to view portions of the
cardiovascular, gastrointestinal and genitourinary
tract from within the esophagus and stomach of dogs
[9,10].

The aims of this study were to assess the potential for transesophageal ultrasonic imaging of cardiac
structures, and to determine whether contrast enhanced
backscattering from myocardium could be observed in
intact animals.

Materials and Methods

Five dogs weighing between 6-12 kg were anesthetized with intravenous Nembutal (sodium pentobarbital,
25 mg/kg). Anesthesia was maintained by the intermit-
tent intravenous injection of pentobarbital. In the
case of one dog, in which it was necessary to record
stop action images of the heart without the respiratory motion, the respiration was maintained via a
cuffed endotracheal tube with a mechanical ventillator.
In the remainder of the experiments the respiration was
unassisted. Two catheters (#6 French, internal diameter 1.19 mm) for injecting the contrast agent were
introduced into the left and right ventricles via the
femoral artery and vein respectively. The placements
of the catheters were checked fluoroscopically by injecting x-ray contrast agent (Renovist). Pressure

gauges (Statham-P23DD) connected to the catheters continuously monitored the pressures except during the
injection of the contrast material or flushing of the
catheters with ringer solution.

The SRI ultrasonic imaging system (Figure 1) used
in this study has a high frequency (10 MHz) ultrasonic
transducer array mounted at the end of an ACMI FX-5
side viewing gastroscope. The transducer can be
placed in the esophagus, stomach or duodenum, and
therefore the proximity of the internal organs to the
probe facilitates obtaining high resolution images.
Air can be removed through the standard aspiration
port of the endoscope and a good acoustic contact can
be maintained between the probe and the esophageal
mucosa. The rigid tip of the endoscope is 80 mm long
and 13 mm in diameter and houses the ultrasound probe.

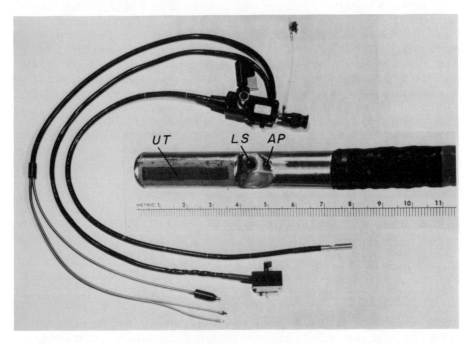

Figure 1 - The ultrasonic endoscope. Modified ACMI
-FX 5 side viewing endoscope with the insert showing
the rigid tip containing the ultrasonic transducer
(UT), the aspiration port (AP) and the light source
(LS). The ultrasonic transducer consists of 10 MHz,
64 element linear array capable of generating a high
resolution real-time image 3 cm wide and 4 cm
deep.

The ultrasound system generates a rectilinear B scan
using a 10 MHz, 64 element linear array. Submillimeter resolution in the axial and lateral dimensions
over the entire image is obtained by dynamic focusing
on both transmission and reception. A cylindrical
acoustic lens provides focusing of the beam in the
direction orthogonal to the image plane. The resulting real-time images correspond to a 3 cm wide by 4
cm deep region of tissue at a repetition rate of 30
frames per second. A digital scan converter provides
television signals for video taping as well as operator
adjustable grey scales. Because of the orientation
of the linear array in the esophagus, the probe generates longitudinal images of the heart. The real-time
images were recorded using a 2" videotape recorder for
later analysis.

     Several experiments were performed with the in-
jection of contrast material, usually indocyanine
green (Hynson, Westcott & Dunning Inc., Baltimore,
Maryland) into the left and right ventricles to identify the cardiac structures. The small field of view
relative to the cardiac structures imposed some difficulties in readily identifying cardiac anatomy. But
after an initial learning period we were able to identify left and right chambers of the heart, mitral
valve, aorta, aortic bulb and pulmonary artery with
relative ease.

     The enhanced backscattering from myocardial tissue
was studied with the contrast injection through the catheter tip placed at the aortic root. The B scan images
of the left atrial myocardium and part of the left ventricular myocardium were continuously recorded while
indocyanine green was injected into the aortic root.

     To comprehend the three-dimensional structure of
the heart it is necessary to obtain several longitudinal cross sections simultaneously. Rotating the
probe and recording different longitudinal cross-sections is not an adequate procedure because of cardiac
motion. To circumvent this problem, we performed one
experiment in which temporary cardiac arrest was
achieved by electrically stimulating ($\simeq$ 10 volt square
pulses, 2 msec. long, 20 pulses per sec.) the exposed
vagi. This produced cardiac arrest for approximately
10 sec. and during this period several longitudinal
cross sections were obtained by rotating the endoscope
about its long axis.

RESULTS AND DISCUSSIONS

Anatomical Identification

        For anatomical orientation we constructed a
schematic diagram of a longitudinal cross section of
the dog heart as it is depicted in most of our ultrasonic B scan images (Figure 2). In this orientation
the head of the dog is to the right and the top and
bottom of the picture correspond to the dorsal and
ventral aspects of the dog respectively. The ultrasonic scanner is introduced from the right side and
lying inside the esophagus produces a longitudinal
cross-sectional images that approximately correspond
to the areas enclosed by the dotted lines in the
Figure 2. Frames A and B correspond to two different
ultrasonic images that can be obtained by positioning
the endoscope at two different regions of the esophagus. To cover the full length of the heart it is
necessary to traverse the endoscope along the esophagus. By rotating the endoscope about its long axis
longitudinal cross sections that are not in the plane
of the diagram can be imaged.

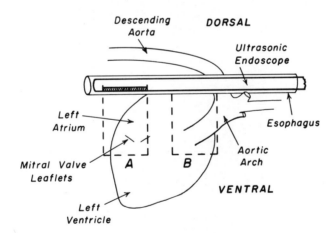

Figure 2 - Schematic diagram of a longitudinal view
from esophagus of canine heart using the ultrasonic
endoscope. Frames A and B correspond to B scans
from two different regions of the esophagus.

In the ultrasonic studies we first identified
the positions of the catheters in the cardiovascular
structures by x-ray fluoroscopy with radiopaque contrast injection and by monitoring pressures. Then the
catheter was ultrasonically visualized and 5 cc of
indocyanine green dye injected.

Identification of the aorta. The catheter could
be identified within the aorta as parallel echo dense
lines within the pulsating large vessel (Figure 3).
In this particular study the catheter was introduced
from the carotid artery. A clot formed at the end of
the catheter could be seen moving with arterial pulsations in the video recordings.

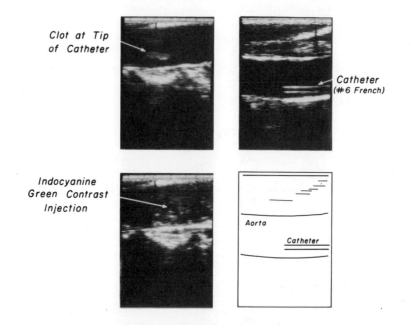

Figure 3 - Transesophageal B scan of canine aorta
in vivo. The top right frame shows the catheter
in the aorta. The image on the top left shows a
clot that has formed at the end of the catheter and
the left bottom image shows the intense backscattering from the contrast injected into the aorta.

Cardiac chamber identification. In longitudinal
view of the heart the right ventricle could be identified before and after the injection of indocyanine
green dye through the right heart catheter (Figure 4).
The intense scattering from the dye at the lower right
hand corner of the second frame identifies this region
of the image as the right ventricle. Since the catheter tip was close to the pulmonary valve in the right
ventricle the contrast echoes should be originating
in the right ventricular outflow tract. This ultrasonic image corresponds to the frame B illustrated in
Figure 2.

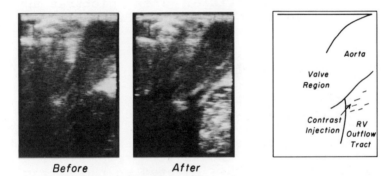

Figure 4 – Transesophageal B scan of canine heart
in vivo before and after injection of indocyanine
green into the right ventricle. These pictures
approximately correspond to frame B in Figure 2.

In views of the left ventricle corresponging to
frame A in Figure 2 a limited field of view extending
only about a cm below the mitral valve was obtained.
However, the catheter, part of the left ventricle and
the closing and opening of the mitral valve with the
heart beat were clearly seen in the real-time images.
Unfortunately the still frame reproductions do not do
justice to the informative dynamic images one obtains
with this ultrasonic endoscope. With the injection of
green dye, the contrast could be observed coming out
of the catheter tip and moving towards the right of
the field of view. By moving the endoscope to position 'B' in Figure 2, the flow of the contrast material
past the aortic bulb could be observed (Figure 5).

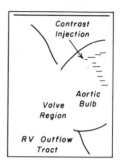

Before   After

Figure 5 - Transesophageal B scan of canine heart
in vivo before and after injection of indocyanine
green into left ventricle. These pictures approximately correspond to frame B in Figure 2.

Contrast Enhancement of Myocardium

     To evaluate contrast enhancement within the myo-
cardium the endoscope was positioned to image the
area that approximately corresponded to frame A in
Figure 2 and was rotated about its axis to view the
left atrial wall. The images were recorded continuously while green dye was injected at the aortic
root. Similar to the observations reported earlier
[7], the myocardium got much brighter after the contrast injection (Figure 6). The endo and epicardial
boundaries which were only faintly visible before
injection of the contrast could be distinguished
clearly with the enhanced brightness of the myocardium.

     To attempt to image different longitudinal cross
sections at the frame 'A' position of Figure 2 the
heart beat was stopped by vagal blocking, and the different longitudinal cross sections of the stopped
heart were obtained by rotating the endoscope about its
axis. Five such cross sections obtained by successive
rotations ($\simeq$ 6°) of the endoscope are shown in Figure 7.
These images correspond to instantaneous cross-sectional
images of the left atrium at a specific moment in the
cardiac cycle. With a more precise orienting mechanism
for the endoscope and more contrast enhanced images
the three-dimensional reconstruction of this portion
of the heart should be possible using techniques reported in the literature [11].

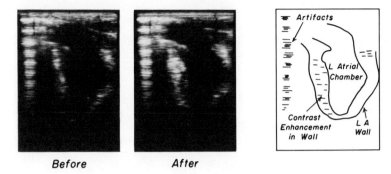

Before   After

<u>Figure 6</u> - Transesophageal B scan of canine left atrium in vivo before and after the injection of indocyanine green at the aortic root. These pictures approximately correspond to Frame A in Figure 2. Enhancement of the backscattering from regions of myocardium is seen in the B scan image on the right.

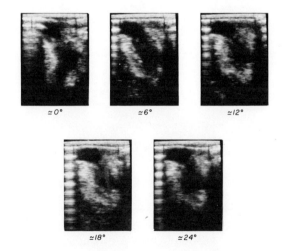

Figure 7 - Transesophageal B scans of vagally blocked canine left atrium in vivo obtained with incremental rotations about the axis of ultrasonic endoscope. The approximate angles of clockwise ratations are given under each frame. The 0° picture approximately corresponds to Frame A in Figure 2.

CONCLUSION

        Conventional transcutaneous cardiac scanners have
some distinct limitations. It is often difficult to
obtain heart images of diagnostic quality in patients
with chronic obstructive pulmonary disease, barrel
chest, obesity or in elderly patients. Even under
ideal circumstances because of the ribs it is impossible to observe all parts of the heart. From Figures
8 and 9 it appears that in the human an esophageal
ultrasonic scanner will have unobstructed view of the

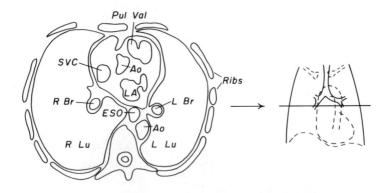

Traced from: *Atlas of Cross-Sectional Anatomy*,
Ed by S A Kieffer and E R Heitzman, Harper & Row 1979

Figure 8 - Schematic diagram of human transaxial
cross section at the level of the sixth thoracic
vertebra looking cephelad. This particular cross
section which is close to the base of the heart
is important for the studies of congenital anomalies
of outflow tracts. This picture shows that the
left and right bronchi (L Br and R Br) are not
obstructing the view of the heart from the esophagus
(ESO). AO:Aorta; SVC:Superior Vena Cava; LA: Left
Atrium; Pul Val: Pulmonary Valve;L Lu: Left Lung;
R Lu: Right Lung.

whole heart. The proximity of the heart to the esophagus allows the possibility of using high frequency

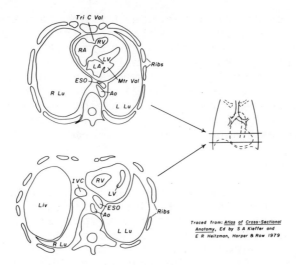

Figure 9 - Schematic diagrams of human transaxial cross sections at the level of seventh (top) and eighth (bottom) thoracic vertebrae, looking ceph- alad. At the level of the seventh vertebrae the left and right atrium (LA, RA), the ventricles (LV, RV), mitral valve (Mtr Val) and the tricuspid valve (Tri C Val) can be ultrasonically viewed from the esophagus (ESO) with out obstruction by the lungs (R Lu and L Lu). The bottom picture is closer to the apex of the heart and here again the esophagus has unobstructed view of the heart. Ao: Aorta; IVC: inferior vena cava; Liv: Liver.

transducers which in turn produce high resolution images. The instrument used in these experiments was not specifically designed for imaging the heart and hence it has the limitation of small field of view even though the images are of submillimeter resolution. It is clear from the results of our experiments and work by others [12] that a transesophageal ultrasonic scanner specifically designed for cardiac imaging will yield useful pictures of the heart. One can envision collecting several transverse cross sections simultan- eously using a two-dimensional array and synthesizing dynamically varying three-dimensional cardiac struc- tures. In addition because of the high resolution capability of such a scanner there is a potential for imaging portions of the coronary arterial tree.

The backscattering enhancement demonstrated with the contrast injection suggests the possibility of imaging myocardial perfusion. Several other ultrasonic contrast agents which can be introduced intravenously are undergoing experimental evaluation [13]. Transesophageal ultrasonic cardiac imaging combined with a suitable contrast agent that can be administered intravenously should prove to be a significant tool in the diagnosis and management of ischemic heart diseases.

ACKNOWLEDGEMENTS

The authors thank Mr. Eric A. Hoffman, Dr. Thomas Behrenbeck and Mr. Julijs K. Zarins for their experimental assistance. We express our appreciation to Delories C. Darling, Elaine C. Quarve, Marge C. Fynbo, Pat J. Sargent, Marge A. Engesser and Leo Johnson for their assistance in the preparation of this manuscript and the figures. This work was supported by the following grants: CB 74136 and HV-7-2928.

REFERENCES

1. Kloster, F. E., J. Roelandt, F. J. Ten Cate, N. Bom, and P. G. Hugenhotz: Multiscan echocardiography. II Techniques and initial clinical results. Circulation 48:1075-1084, 1973.
2. Griffith, J. M., and W. L. Henry: A sector scanner for real-time two-dimensional echocardiography. Circulation 49:1147-1152, 1974
3. Kisslo, J., O. T. Van Ramm, and F. L. Thurstone: Cardiac imaging using a phased array ultrasound system. II. Clinical technique and application. Circulation 53:262-267, 1976.
4. Tajik, A. J., J. B. Seward, D. J. Hagler, D. D. Mair, and J. T. Lie: Two-dimensional real-time ultrasonic imaging of the heart and great vessles: Technique, image orientation, structure identification, and validation. Mayo Clinic Proceedings 53:271-303, 1978.
5. Balasubramanian Rajagopalan and J. F. Greenleaf: 1977 Unpublished results.
6. O'Donnell, M., D. Bauwens, J. W. Mimbs, and J. G. Miller: In vivo detection of acute myocardial ischemia in the dog by quantitative ultrasonic backscatter. Fourth International Symposium on ultrasonic imaging and tissue characterization, June 18-20, Gaithersburg, Maryland.

7. Balasubramanian Rajagopalan, J. F. Greenleaf, P. A.
    Chevalier, and R. C. Bahn: Myocardial blood flow:
    visualization with ultrasonic contrast agents.
    Acoustical Imaging 8:719-729, 1980, edited by
    A. F. Metherell. Plenum Publishing Corporation.
8. Buxton, J., E. DiMagno, D. Wilson, J. Suarez, P.
    Regan, R. Hattery, and P. Green: Initial results
    in the development of an ultrasonic endoscope.
    Proceedings of the 24th Annual Meeting of AIUM,
    1:121, 1979.
9. DiMagno, E. P., J. L. Buxton, P. T. Reagan, R. R.
    Hattery, D. A. Wilson, J. R. Suarez, and P. S.
    Green: Canine intragastroesophageal ultrasonog-
    raphy. Clinical Research 27:628A 1979.
10. DiMagno, E. P., J. L. Buxton, P. T. Reagan, R. R.
    Hattery, D. A. Wilson, J. R. Suarez, and P. S.
    Green: The Ultrasonic Endoscope (Lancet - to be
    published).
11. Harris, L. D., R. A. Robb, T. S. Yuen, and E. L.
    Ritman: Display and visualization of three-dimen-
    sional reconstructed anatomic morphology: Exper-
    ience with the thorax, heart, and coronary vascu-
    lature of dogs. Journal of Computer Assisted
    Tomography 3:439-446, 1979.
12. Hisanaga, K., and A. Hisanaga: A new transesoph-
    ageal radial scanner using a rotating flexible
    shaft and initial clnical results. Proceedings of
    the 24th Annual Meeting of the AIUM, 1:122, 1979.
13. Tyler, T. D., A. H. Gobuty, J. Shewmaker, and N.
    F. Maklad: In vivo quantitation of induced ultra-
    sonic contrast effects in the mouse. Ninth Inter-
    national Symposium on Acoustical Imaging, December
    3-6, 1979, Houston, Texas. Acoustical Imaging,
    Vol. 9. Plenum Press.

ULTRASONIC IMAGING: ITS PLACE IN MEDICAL DIAGNOSIS WITH RESPECT TO

RADIONUCLIDE SCANS AND X-RAY COMPUTED TOMOGRAPHIC STUDIES

Mahfuz Ahmed, Ph.D., M.D., Michael Grossman, M.D.

Department of Radiological Sciences, University of
California Irvine Medical Center, Orange, CA 92668

ABSTRACT

In this paper we shall present the clinicians' view point
regarding the usefulness of ultrasonic images in making a diagnosis.
The diagnostic information in an ultrasound image is compared to
that present in images obtained through other modalities such as
conventional radiography, x-ray planar tomography, x-ray computed
tomography and radionuclide scans. Examples of situations which
dictate the clinical use of one or other of these modalities, and
the role played by ultrasound in making the diagnosis are presented.
Areas of research which may improve the diagnostic information content and make ultrasound more competitive with x-ray computed tomography and radionuclide scans are discussed.

INTRODUCTION:

When confronted with a medical problem, the clinician wishes
to learn the correct diagnosis as quickly as possible and with the
least possible risk, discomfort, and cost to the patient. He begins his investigation with a detailed history and physical examination. Often this is all that is needed to reach a diagnosis
and begin treatment. If further information is required, the
physician may proceed to simple laboratory tests and non-invasive
imaging procedures, reserving more invasive diagnostic procedures
for problems which cannot be otherwise resolved.

"Non-invasive" imaging modalities include conventional radiography and planar tomography, ultrasound, radionuclide imaging,
and x-ray computed tomography. Computerized reconstruction techniques using ultrasound or radioisotopes are in the developmental

569

stages and may well see wide clinical use in the future.

Temporal and economic factors prohibit the use of all available procedures. Usually the clinical history and physical examination will indicate the most reasonable next step, with further investigations being suggested as additional information becomes available. Let us illustrate this by an example.

> Example: A 65-year old man complains of left flank pain, and history and physical examination suggests a problem in the left kidney. This suspicion is strengthened by the discovery of blood in the urine on microscopic examination.

The cheapest, most convenient, and usually most rewarding imaging study in this case is an x-ray examination of the kidneys using a radiographic contrast agent given intravenously. This study is called an Intravenous Urogram (IVU) and a representative film is shown in Figure 1A. This was considered normal. Planar tomography (Figure 1B) done at the same time was also felt to be normal.

Because the symptoms and laboratory abnormality remained unexplained, computed tomography (CT) was obtained to study the kidneys further, and also to search for a cause for the problem elsewhere in the abdomen. The slice shown in Figure 1C was highly suggestive of a simple cyst in the left kidney, but the diagnosis was not completely certain. Since the most likely alternative diagnosis would be a malignant tumor, further investigation was indicated. Ultrasound (Figure 1D) confirmed the cystic nature of the lesion and permitted conservative treatment.

This example illustrates an orderly progression of diagnostic tests. Ultrasound might have preceded CT in this case, but would not have been as useful in excluding intra-abdominal disease because of the interference of interposed bowel gas. Thus, both CT and ultrasound played an important role in the management of this patient. We shall now assess the diagnostic role of ultrasound by comparing it with two other important imaging modalities viz. x-ray computed tomography and radionuclide scans.

COMPARISON OF THE MODALITIES

There is little doubt that ultrasound is well suited to the examination of a wide variety of soft tissue lesions involving such diverse areas as the eye, the heart, the pancreas, and the extremities. In the initial enthusiasm that attended the early clinical recognition of the value of ultrasound, it was even suggested that the technique would eventually replace conventional radiography. This, however, is very unlikely because ultrasound

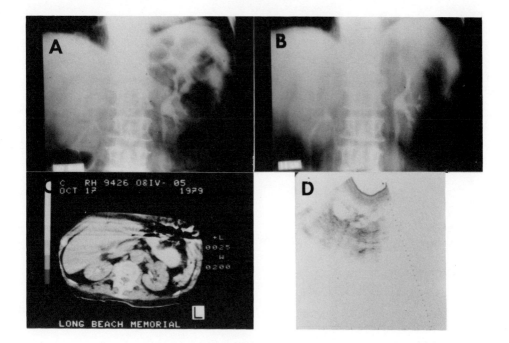

Fig. 1 (A) Intravenous urogram (IVU) of the patient discussed in
the example, which was considered normal. (B) Planar
tomography done at the same time which also was felt to
be normal. (C) CT scan shows a cystic lesion in the left
kidney but the diagnosis is not completely certain. (D)
Ultrasound confirms the cystic nature of the lesion.

is limited by a number of fundamental problems.

The first one is that ultrasound is attenuated so greatly by
gas or bone that these materials effectively prevent examination
of any structure they shield. Thus, lesions within the skull or
lungs cannot be seen, and gas in the bowel often hampers examination of the abdomen. Another problem is the relatively poor
resolution, particularly in the azimuthal plane, as shown in Table 1.
Axial resolution, as is well known, is limited by wave length,
while lateral resolution depends on the effective beam width and
is an order of magnitude worse.[1] It may be possible to improve the
azimuthal resolution to a level comparable with the axial resolution,
but further improvement is unlikely because of the inverse relationship between resolution and penetration. Higher resolution requires
higher frequency (narrower pulse), but the energy at higher frequencies suffers greater attentuation per unit length travelled
resulting in lesser depth penetration.

TABLE I

IMAGING MODALITY	RESOLUTION
1. Conventional Radiography	0.2 mm
(limited by film-screen combination)	
2. X-ray CT	
Spatial resolution	2 mm
Contrast resolution	1-2%
3. Ultrasound	
Axial (at 3 MHz)	1-3 mm
Azimuthal (at 3 MHz)	6-8 mm
4. Nuclear Medicine	10 mm at
(depends on depth)	10 cm depth

In many ways, CT is the most versatile radiologic tool at the present time. It is not hampered by gas or bone, so all areas of the body are accessible. Although initially effectively limited to transverse sectional imaging, CT is theoretically capable of producing an image of any desired plane as long as sufficient data has been collected. Considerable progress has been made in this area over the past few years.

CT has the best contrast resolution of any available imaging modality, and the precise attenuation value determined may at times be diagnostic. In addition to excellent inherent contrast discrimination CT also permits the adjunctive use of exogenous contrast materials. No comparable contrast agents are available for use with ultrasonic imaging.

As regards spatial resolution, CT at relatively low dose levels is second only to conventional film radiography. Although the axial resolution of the ultrasonic images (see Table 1) is comparable to that of x-ray CT at the present time this is likely to change in favor of x-ray CT with the advent of larger, faster computers, and more sophisticated algorithms. This is because the current resolution of CT is not limited by the wavelength of x-rays (which is several orders of magnitude smaller than the wave length of ultrasound used for imaging) but rather by dose considerations and the number of image points that can be scanned and computed within a reasonable time.

One relative drawback of CT is that it involves the use of ionizing radiation. The dose for an "average" examination is highly variable, but might be considered in the range of 2-5 rads. This is similar to the dose delivered in a standard fluoroscopic study such as an upper gastrointestinal series, and should be of little concern except in pregnancy. Furthermore, it is likely that new developments will reduce the dose considerably.

The cost of a CT study is another limitation, at least at present. An abdominal CT ranges from $300-$400, while ultrasound usually costs $100-$150 for a similar study. Cost must be viewed in perspective, however. CT is the only modality that can scan the entire trunk with high resolution when evaluating a patient with lymphoma, for example. The $350 study may obviate more expensive, time-consuming, and hazardous procedures such as lymphography and/or exploratory surgery.

Let us now consider another imaging modality used extensively in diagnosis, viz. radionuclide scans. This is the oldest and best established of the modalities under consideration. It involves administering a tracer element tagged with a radioactive compound and mapping its distribution within the body with an external detector. Anatomic information can be obtained in those areas where the isotope distribution is normally relatively uniform within an organ. Resolution is poor, however, and a lesion of 3-4 cm may not be detected as, for example, deep within the liver.

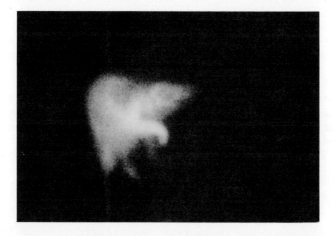

Fig. 2   Excretion of Tc^{99m} labelled PIPIDA into the biliary system demonstrating a partially occluded distal common bile duct. The duct may also be dilated but this is uncertain due to the poor resolution of this modality. The poor filling of the gallbladder in the early excretion phase is normal.

The major advantage of nuclear medicine is its ability to provide physiologic information. Because the chemical compound to be
detected can be varied almost at will, the distribution of activity
can be made to reflect physiologic functions such as renal or biliary excretion, myocardial ischemia, or disruption of the blood-
brain barrier. Figure 2 is an example of the use of radionuclides
to evaluate the function of the biliary system.

THE ROLE OF ULTRASOUND IN DIAGNOSIS

As mentioned, ultrasound is an extremely valuable tool for
study of soft tissue structures. It is highly sensitive in differentiating among soft tissues and in particular excels in demonstrating fluid-filled structures such as the gallbladder, pregnant
uterus, cardiac chambers, and abnormal fluid collections such as
abscesses, hematomas, or fluid collections in the chest or abdomen.

Often the history and physical will suggest ultrasound as the
primary diagnostic modality. This would be the case, for example,
when a fluid-filled lesion such as an abscess is suspected in an
area not obscured by bone or gas.

Ultrasound is especially useful in pregnancy where ionizing
radiation would be a major concern, not only because of its unique
ability to demonstrate intrauterine structures, but also because
of its apparently innocuous nature.[2] Figure 3 is an ultrasound
scan of the abdomen of a pregnant woman. There is a defect in the
baby's abdominal wall and the fetal bowel has passed through the
opening and lies outside his abdomen. This quite unexpected finding led to delivery by Caesarian section and repair of the problem
within a few hours of birth. If this abnormality had not been
detected, an attempt at vaginal delivery might have been fatal.

Figure 4 is a scan through the heart of a patient with congenital heart disease. A defect in the intraventricular septum is
clearly seen. Ultrasound is presently the only non-invasive means
of demonstrating such cardiac malformations, though recent work in
the area of computed tomography is very promising.

AREAS FOR FURTHER RESEARCH IN ULTRASOUND

It has been shown above that despite significant inherent
limitations, ultrasound has applications in a wide variety of
situations, and remains the method of choice in some. Research
and development in some areas could expand its applications and
usefulness in clinical diagnosis. These are:

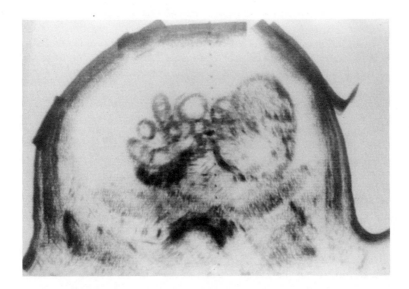

Fig. 3   Ultrasonic scan of the abdomen of a pregnant woman showing
fetal bowel outside the fetal abdominal wall.

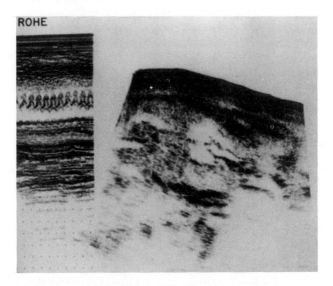

Fig. 4   Ultrasonic scan showing a defect in the intraventricular
septum of a patient with congenital heart disease.

1. Quantitative Information

Quantitative information regarding tissue absorption or
scattering of sound, similar to the CT number, may be
most helpful. Some workers believe a characteristic
"signature" of different tissue types can be determined,
but no definite "signature" has been found as yet.

2. Development of an Ultrasonic Contrast Agent

Intravenous and oral contrast materials have proved invaluable in other areas in imaging, but no useful ultrasonic contrast has yet been developed. The use of microbubbles as a contrast agent has been tried,[3] but is of
limited applicability.

3. Physiologic information

Other than blood flow and limited cardiac dynamics, ultrasound provides very little physiologic information. Agents
analogous to radionuclides used in cardiac, renal and
gallbladder scans, which would significantly change the
ultrasonic properties of organ systems need to be developed.

4. Volume Image

Although the ultrasonic scans demonstrate the structures
in a plane in great detail, it is at times desirable to
have a volume image of the entire area to better visualize
the extent of an abnormality. Volume images, possible
with the use of computer graphics, would be very helpful.
It is doubtful that holography, in its currently developed
stage, would be of any significant benefit.[4]

REFERENCES

1. J. F. Havlice and J.C. Taenzer, "Medical Ultrasonic Imaging:
   An Overview of Principles and Instrumentation", Proc. IEEE, 67
   620 (1979).
2. F. J. Fry, "Biological Effects of Ultrasound-A Review", Proc.
   IEEE, 67: 604 (1979).
3. F. W. Kremkau, R. Gramiak, E.L. Carstensen, P.M. Shah and D. H.
   Kramer, "Ultrasonic Detection of Cavitation at Catheter Tips",
   Radiology, 110: 177 (1970).
4. M. Ahmed, K. Y. Wang and A. F. Metherell, "Holography and its
   Application to Acoustic Imaging", Proc. IEEE, 67: 466 (1979).

ULTRASOUND AND PATHOLOGY

H. E. Melton, Jr. and Elsa B. Cohen

Department of Pathology
The Medical College of Wisconsin
Milwaukee, WI 53226

INTRODUCTION

It seems to be a current assumption, when one considers how
best to develop ultrasound to examine biological tissues, that the
higher the resolution, that is, the nearer one can approach the
level of resolution offered by light microscopy, the likelier one
is to make significant observations about such tissues. While the
ultrasonic examination of tissues at a cellular level will no
doubt provide important data and observations,[1,2] we feel that it
is a mistake to overlook the importance of examining tissues ultrasonically at a level of resolution generally within the range
afforded by the unaided eye. The practice of anatomic pathology
entails the examination of tissues both with the naked eye
(grossly) and with the light microscope in order to determine
presence, extent and nature of disease. In general, the presence,
extent, and nature of the disease is recognized by the presence,
extent, and pattern of disruption of normal structures.

An experienced pathologist expects that he should be able to
make a diagnosis - that is, a predictive statement about the likely
natural course of the disease - most of the time on the basis of
the gross appearance of the tissue. In these cases he uses the
microscopic appearance of the tissue simply as a confirmation of
his gross impressions. In short, the gross examination of tissues
is already known to be immensely clinically important in a well
established discipline, anatomic pathology; therefore, the body of
information from this discipline can be utilized as a standard to
ensure a clinically useful level of performance in ultrasonic
imaging.

Requirements and Considerations

The fundamental requirement for ultrasonic imaging of tissues at a level of resolution generally within the range afforded by the unaided eye is a limit of resolution equal to the limit of visual acuity which is taken here to be 0.1 mm. Of course it is known that many structures of particular interest are larger than 0.1 mm. Thus the limit of resolution or visibility becomes important in relation to the size of structures of importance. However, the limit of resolution alone gives little assurance that linear, extended or planner structures will be imaged because of the important effects of ultrasonic speckle.[3-5] The false interruptions caused by speckle in an otherwise continuous structure is unacceptable in general; these interruptions produce the impression of an alteration of structure, a hallmark for the presence of disease. Even the appearance of structures of small extent, isolated or aggregated, may be uncorrelated by the actual structure.[6,7]

The means for realizing any desirable limit of resolution are a suitable operating frequency and a focused aperture. If the aperture is circular, the limit of lateral resolution, d, is taken to be [8]

$$d = 0.61 \ \lambda \ (F^2+r^2)1/2/r. \qquad\qquad 1$$

Similar relations can be found for other aperture shapes but the proportionality between resolution and wavelength, λ , is unchanged whereas the accounting for aperture size, r (radius for a circular aperture), for focal distance, F, and for any constants (0.61 for the circular aperture) is changed. Were it not for the attenuation of ultrasound by tissues most any desirable wavelength could be selected. Thus, the sizes of tissue specimens to be scanned places very important constraints on the selection of operating frequency. Hereafter consideration is limited to pulse-echo ultrasonic imaging since the orientations of the planes of visualization are closely analogous to the orientations of the planes exposed on cut surfaces during examinations for pathologic anatomy. At this point then a number of important aspects of diagnostic ultrasonic imaging merge and thereby establish the principal criteria for the performance of ultrasonic imaging systems for pathologic anatomy.

One aspect of importance is the amount of attenuation and its relation to the spectral characteristics of ultrasonic pulses. For most purposes the total attenuation, A, of the ultrasonic pulse returned from the furthermost tissue structure must not be below the equivalent input noise, n, of the receiver (no signal, no picture). However if the ultrasonogram is to contain any

dynamic range of signal at the furthermost range, then the return
pulse should be above the noise. Since most usual types of displays
are capable of displaying about 20db of dynamic range, the furthermost echo should be about 10 times larger than the equivalent
input noise of the receiver. Now if the echo-level of a glass
block or other suitable reflector (approximately 100% reflection)
in water is taken as T, the echo expected from a fat-muscle
interface is about T/10 in the absence of attenuation and

$$A = \ln [T/(100n)]. \qquad\qquad 2$$

By including the spectral shift of ultrasonic pulses caused by the
frequency dependent attenuation of soft tissues[4] A can be
substituted:

$$A = 2a [f_o - a B^2/K^2], \qquad\qquad 3$$

where

$$a = \int_o^R \alpha(z) \, dz,$$

R is the range of depth of penetration, $\alpha(z)$ is the one-way
attenuation coefficient, f_o is the center frequency, operating
frequency, associated with the transmitted pulse in the absence
of attenuation, B is the one-way bandwidth associated with the
pulse and K^2 is a constant equal to 2.77. By letting

$$a = \alpha R \qquad\qquad 4$$

where α is an average attenuation coefficient, Equation 3 may be
substituted and written as follows:

$$2 \alpha R [f_o - \alpha R B^2/K^2] = \ln [T/(100n)]. \qquad\qquad 5$$

From this Equation an expression for f_o can be obtained for
$R = R_m$, the maximum range:

$$f_o = \frac{\ln [T/(100n)]}{2 \alpha R_m} + \alpha R_m B^2/K^2 \qquad\qquad 6$$

The presence of B in this expression is dependent on f_o in
practice for given transducer material, mounting, backing and
matching plate, i.e. fractional bandwidth or Q can be considered
a constant dependent on material and construction. Thus,

$$f_o = \frac{K^2 Q^2}{2 \alpha R_m} \left[1 - \sqrt{1 - 2 \ln [T/(100n)]/(K^2 Q^2)} \right] \qquad 7$$

where $B = f_o/Q$. It should be clear that Equation 7 is based solely
on the tissue's attenuation, the transmit strength, transmit-
transducer-Q, and receiver sensitivity (its equivalent input noise).
The limit of lateral resolution can now be set by the aperture
but it too is affected by the spectral shift:

$$d = 0.61 (c/f_c) (F^2 + r^2)^{1/2}/r, \qquad 8$$

where

$$f_c = f_o - 2 \alpha RB^2/K^2 \qquad 9$$

and $R = F$ for attenuation over the total range. If the same
transducer is used for receiving then the spectral shift is half
as great, i.e.

$$f_c = f_o - \alpha RB^2/K^2$$

At this point at least multiple frequency receiving arrays
are worthy of consideration from the standpoint of spectrum
matching,[4] sensitivity to deeply located structures. This has
been investigated as one means whereby acoustic output and patient
exposure can be reduced in clinical diagnostic imaging. It has
also been shown that multifrequency arrays can be used for real-
time speckle-reduction[5] in addition to increasing sensitivity to
layered structures and to scatter.[4] This latter feature has been
explored previously using a very broad band transducer for detection of ischemia and infarction in myocardium:[9] needless to say
such aspects of alteration of tissues are of importance in
pathologic anatomy.

One additional feature of pulse-echo imaging which has been
used routinely in the clinical scanning is the generally superior
range resolution (superior to lateral resolution). When extended
linear structures are oriented nearly normal to the line-of-sight
of the scanning transducer in the scan plane the structures are
more clearly visualized than when some other orientation is used.
In fact, as an aside, one good way of assessing lateral performance of a clinical scanner in vivo is to scan vessels oriented
toward or along the line-of-sight of the scanning transducer.
This generally superior range resolution can be used to advantage
in pathologic anatomy since most tissue planes of interest tend
to be normal to the line-of-sight. This advantage, however, is
moderated by the effects of speckle.

The visibility of a thin tissue layer in range depends on the thickness of the layer and on its acoustic velocity and density. If the layer is embedded in a different but otherwise homogeneous medium then the largest echo from the layer occurs when the thickness of the layer is one quarter wavelength (for fractional bandwidths of between 30 to 40%); the layer can be resolved in range when its thickness is one wavelength or more.[4,10] For layers of 0.1 mm of thickness with a velocity range of from 1500-1900 m/s, the frequency range for maximum echo ($\lambda/4$) is from 3.75-4.75 MHz. In order to resolve this layer a frequency range of from 15 to 19 MHz is required. In addition, the frequency requirement for specimens of 2 cm thickness having an average attenuation coefficient of 1 db/cm/MHz (0.115 Neper/cm/MHz) is approximately 20 MHz for transmit-transducer Q of 2.5 and for T/(100n) = 1000. The limit of lateral resolution then depends on aperture size; the depth-of-field depends on the means for focusing.[8] The most important point is that frequencies as low as 3 to 5 MHz may be useful for ultrasonic depiction of pathologic anatomy, i.e. 0.1 mm layers may be visible but not resolved. Resolution of these structures necessitates a large jump in frequency; the maximum usable frequency is higher still; aperture then determines lateral resolution; depth-of-field by the means for focusing; to say nothing of the additional considerations necessary for speckle reduction. Because of these several approaches for ultrasonic depiction of pathologic anatomy we felt that our studies should be staged along the lines of these approaches (1) by using range-visibility criterium, (2) by using range resolution criterium, (3) by using lateral resolution criterium and (4) by using the findings for speckle reduction.

The work described here is concerned with stage one; namely, using range visibility for ultrasonic depiction of pathologic anatomy. Therefore, we have chosen to examine a tissue which is layered and a lesion which is characterized by the disruption of the normal layers. We have chosen a single tissue and a single lesion. The tissue is colon and the lesion is carcinoma of the colon which arises in the mucosa and destroys the underlying layers of the colon as it invades them.

METHODS

Samples of colon taken at autopsy were opened to be relatively flat. They were then fixed in Formalin by using the generally accepted ratio of 20 volumes of Formalin to 1 volume of tissue. These fixed specimens were examined both visibly and ultrasonically. Ultrasonic examination of the specimens was done by using several different ultrasonic scanners. The operating frequencies ranged from 2 MHz to 5 MHz with limit of lateral resolution of from 3 mm to 1 mm, approximately; range resolution

also varied from approximately 1.0 mm to 0.3 mm.   A linear sequential scanner was used at operating frequencies of 2, 3 and 5 MHz;
an annular-array scanner was used at operating frequencies of 2,
3, 4 and 5 MHz; and a single transducer scanner was used at
operating frequencies of 2.25, 3.5 and 5 MHz.

    Specimens were scanned in a waterbath at room temperature
(20°-23°C).   All scans were made with the mucosa toward the
transducer (s).   Where possible shallow compounding was used to
reduce the effects of speckle so that layers would not be interrupted
by the speckle pattern.

RESULTS AND DISCUSSION

    An example of an ultrasonic scan of a specimen of normal
colon is shown in Figure 1.   This scan was made using a fixed-
focus transducer (4 cm) operating at 5 MHz and having a one-way
bandwidth of 2 MHz approximately.   The lateral resolution near the
focus was 1.5 mm approximately; the range resolution was about
0.3 mm.   The mucosa, submucosa and muscularis propria are clearly
shown.   The muscularis mucosae is visible as a second bright line
in the mucosa; however, it is not resolved since it is only about
0.1 mm thick.   The second bright line is much thicker than the
muscularis mucosae by at least a factor of 3.   Moreover, this
layer is best seen when the lines-of-sight of the scan are nearly
normal to this layer.   A few small lobules of pericolic fat are
seen frequently though none are shown in this particular scan.

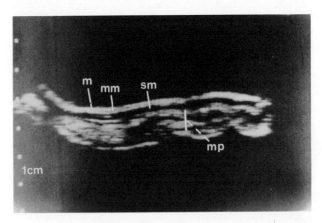

Figure 1.   Scan of specimen of normal colon; m = mucosa, mm =
muscularis mucosae, sm = submucosa and mp = muscularis propria.

The same layers are identifiable in Figure 2 (a) and (b).
As the layers are followed from the normal portions of the
specimen to the portion containing the lesion they become dis-
rupted. This disruption signals the presence of cancer which has
spread from the mucosa, across the line of the muscularis mucosae,
across the submucosa and to the muscularis propria. The depth
of invasion of the cancer into the bowel wall is readily
appreciated in both the macrosection and in the ultrasonic scan.
The muscularis mucosae is visible in a part of the normal region
as a second bright line in the mucosa just as' in the normal scan
of Figure 1. The scanner used for Figure 1 was also used for
Figure 2 (b). The plane of the macrosection was taken near to
the scan plane of Figure 2 (b). Part of the normal region was
removed in order to make this macrosection. Thus a small normal
region is shown in (a) whereas a much larger normal region is
shown in the scan in (b).

As mentioned in the introduction, we planned to conduct our
study in several stages: 1) by meeting the requirements for
range visibility; 2) by meeting the requirements for range
resolution; 3) by meeting the requirements for lateral resolution;
and 4) by using speckle reduction techniques. The observations
presented above constitute the first stage. Namely, they seem to
confirm the prediction that although every structure in a layered
tissue cannot be resolved we can nevertheless visualize many of
them. In particular, we are reasonably certain that we are able
to distinguish the muscularis mucosae from the rest of the mucosa
and from the submucosa. We are, in addition, quite certain that
we are able to distinguish the mucosa from the submucosa, and the
submucosa from the muscularis propria; furthermore, we feel, on
the basis of the observations presented above, that we can
readily recognize disruptions in these layers, with the possible
exception of the muscularis mucosae. In short, these results
show that at least in this system, layers as thin as 0.1 mm can
be visualized ultrasonically. Our own observations, taken in
conjunction with those in which skin layers are visualized
ultrasonically and their disruption reliably predicted[11], seems
to justify our now proceeding to the next stages of our
investigations.

If we are able to achieve both a range and a lateral resolution of 0.1 mm, we shall then be able to examine tissues ultrasonically for the presence and disruption not only of layered
structures, but also for the presence and disruptions of structures which are small in all their dimensions; for example,
lymph nodes. The achievement of such resolution will of course
imply that speckle has been successfully reduced.

(a)

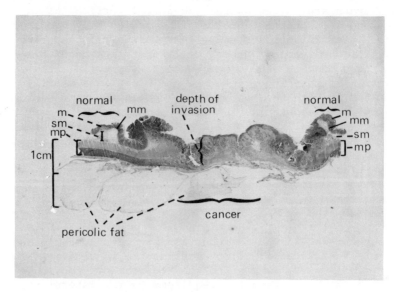

(b)

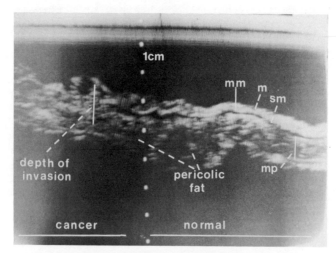

Figure 2.   (a) picture of macrosection taken near the scan plane
of (b).   Some of the normal region was removed in order to make
the section.   A Movat stain was used on this section.   (b) scan
taken through lesion and including large normal region.   Labeling
is the same as for Figure 1.

In all of these stages, we are attempting to capitalize upon a vast and well-established body of data. That is, we are testing whether anatomic features already known to be biologically significant can be detected with sufficient detail and reliability by using ultrasonic techniques. If indeed they can be, the way is then open for detecting these same features ultrasonically in individual situations where the currently-available methods are either inefficient or otherwise unsatisfactory. As an example of the first kind of situation, it is often inefficient to dissect the adipose tissue in a specimen in order to find lymph nodes; scanning might be faster. If both methods were equally reliable, scanning would therefore be preferable. As an example of the second kind of situation, it may be highly desirable to know before surgery how deeply a carcinoma has involved the rectal wall, since one may be able to avoid removing the rectum completely if the cancer is only superficially invasive.[12] However, it is not currently possible to assess the depth of invasion before surgery very accurately. If scanning offered a reliable, noninvasive method of assessing the depth of invasion, it ought to prove quite useful.

REFERENCES

1. L. W. Kessler, The Sonomicroscope, 1974 Ultrasonics Symposium Proceedings: 735-737 (1974)

2. R. A. Lemons and C. F. Quate, A Scanning Acoustic Microscope, 1973 Ultrasonics Symposium Proceedings: 18-21 (1973)

3. C. B. Burckhardt, Speckle in Ultrasound B-Mode Scans, IEEE Trans. Sonics and Ultrasonics, SU-25:1-6 (1978)

4. H. E. Melton, Jr., On Improvement of Sensitivity or Reduction of Dose in Medical Ultrasonic Imaging, 1979 Ultrasonics Symposium Proceedings: (In press) (1979)

5. R. Entrekin and H. E. Melton, Jr., Real-time Speckle Reduction in B-Mode Images, 1979 Ultrasonics Symposium Proceedings: (In press) (1979)

6. J. C. Bamber and R. J. Dickinson, Computer Simulation of the Process of Formation of an Ultrasonic Image, Fourth International Symposium on Ultrasonic Imaging and Tissue Characterization: 105 (1979)

7. J. G. Abbot, "Speckle Reduction Techniques in Ultrasound B-mode Images," Ph.D. Dissertation, Duke University (1979)

8. H. E. Melton, Jr. and F. L. Thurstone, Annular Array Design and Logarithmic Processing for Ultrasonic Imaging, Ultrasound Med. Biol. 4:1-12 (1978)

9. M. O'Donnell, D. Bauwens, J. W. Nimbs and J. G. Miller, Integrated Ultrasonic Backscatter as an Approach to Tissue Characterization in vivo, 1979 Ultrasonics Symposium Proceedings: (In press) (1979)

10. J. M. Reid, Challenges and Opportunities in Ultrasound, in "Ultrasonic Tissue Characterization", M. Linzer, ed., NBS Special Publication 453, Washington, D.C. (1976)

11. R. E. Goans and J. H. Cantrell, Jr., Ultrasonic Characterization of Thermal Injury in Deep Burns, Third International Symposium on Ultrasonic Imaging and Tissue Characterization: 41-46 (1978)

12. B. C. Morson, H. J. R. Bussey and S. Samoorian, Policy of Local Excision for Early Cancer of the Colorectum, Gut 18:1045-1050 (1977)

IN VIVO QUANTITATION OF INDUCED ULTRASONIC

CONTRAST EFFECTS IN THE MOUSE *

Thomas D. Tyler, Allan H. Gobuty, Jeff Shewmaker, and
Nabil F. Maklad

Department of Diagnostic Radiology
University of Kansas Medical Center
Kansas City, Kansas 66103

ABSTRACT

The ability of certain aqueous solutions to enhance ultrasonic
echogenicity from the intact mouse *in vivo* was examined. Substances
which in solution increase the velocity of sound as a function of
increasing concentration were found to intensify the "contrast" of
whole mice visualized sonographically. Increase in contrast was
determined by photometric measurement of the average integrated
screen luminance of the area of the screen representing the mouse.
The ability to enhance screen luminance correlated with the ratio
of the slope of the change in speed of sound *vs.* concentration to
the maximum non-lethal drug dose attainable, as determined by acute
LD_{50} studies. Neither body temperature nor osmolarity changes
appeared to be significant contributing factors in the enhancement
of "contrast" produced by the agents tested. These results indicate
that compounds with a high slope of the change of speed of sound
vs. concentration in solution and with low toxicity can act *in vivo*
as ultrasonic "contrast" agents when given systemically.

INTRODUCTION

Differentiation of normal from abnormal tissue by ultrasonic
examination can be difficult. Since, although normal and abnormal
tissues can differ in metabolic, morphologic or physiologic characteristics, they may have similar ultrasonic properties. Agents with
ultrasonic properties different from normal tissue may act to en-

* Supported by National Cancer Institute Contract #NO1-CB-84236

hance the contrast between normal and abnormal tissue when administered systemically. This would be due to unequal distribution of
the agents between normal and abnormal tissues, much as in nuclear
medicine. Subsequently, differences between abnormal and normal
tissues would become easier to visualize.

There have been few reports regarding use of ultrasonic contrast
agents. Earlier studies tested the use of scattering agents introduced into the blood stream which resulted in changes in contrast.
These agents include indocyanine green (Gramiak and Shah, 1968), gas
bubbles (Ziskin et al., 1972) and most currently, collagen spheres
(Maklad et al., 1979B). An alternate method used to enhance ultrasonic contrast required external heating of tissue. This resulted
in detectable interfaces that would not have otherwise been acoustically observable (Sachs et al., 1976).

Recently, a new approach was described by Ophir et al., 1979
and Maklad et al., 1979A. They described the use of solutions which
had been shown to increase the velocity of sound in water, and which
when introduced into dog kidneys, enhanced their ultrasonic image or
"contrast". It was proposed that the contrast effect observed was
due to an enhancement of the acoustic impedance mismatch between the
vascular and extravascular beds. Augmentation of the impedance mismatch subsequently increased the scattering of ultrasonic waves.
(See Ophir et al., 1979.)

The purpose of this study was to attempt to quantify the contrast-enhancing ability of certain compounds which increase the velocity of sound in water as a function of concentration. The efficacy
of this contrast ability would be limited by the maximum non-lethal
doses attainable during their acute administration. Therefore, the
acute median lethal dose for each of these compounds was also determined.

MATERIALS AND METHODS

Animals

Male, ICR mice (Spartan Labs, Inc., Haslett, MI) weighing approximately 20-28 g were used. Mice were maintained on a 12-12 hour
light-dark cycle and allowed access to food and water ad libitum.

Compounds

Solutions of all agents tested were prepared from commercially
available reagent grade chemicals in a physiological (0.9%) saline
solution.

Acute Intravenous Toxicity

Bolus injections of each potential contrast agent were made via the tail vein at a constant volume/weight ratio of .01 ml/gm body weight. Injections were made between 8:00 a.m. and 12:00 p.m. Determination of lethality was made 24 hours post-injection. At this time the median lethal dose (LD_{50}) was calculated by the Thompson (1947) method.

Contrast-Quantitation: Apparatus

An ADR real-time scanner (model #2130, ADR Ultrasound, Tempe, AZ) with a 3.5 MHz transducer was used to scan the mice. Screen luminosity was measured with a Tektronix J16 digital photometer (Tektronix, Inc., Beaverton, OR) using an uncorrected probe (Tektronix #J6504) having a 120° angle of acceptance. The probe was located within an opaque Plexiglas^{R} tube so that the tip of the probe was 75 mm from the surface of the screen. The tube was located in a light-protective housing which covered the entire scanner screen and held the tube in contact with the center of the screen. The diameter of the Plexiglas^{R} tube (54 mm) corresponded to the length of the mouse body being visualized. To further reduce background light, the screen was covered with cardboard except for a central rectangular opening of 60 x 30 mm. This region was slightly larger than that occupied by the image of the mouse. The photometer probe would therefore measure the average integrated intensity of light from the region of the screen corresponding to the entire visualized mouse (contrast).

Contrast-Quantitation: Procedure

Mice were anesthetized with Innovar-Vet C II (Pitman-Moore, Inc., Washington Crossing, NJ). Abdominal and thoracic hair was removed with a depilatory. The skin was then coated with mineral oil. The mouse was placed longitudinally on the transducer so as to allow scanning of the median plane of the body. Rectal temperature was monitored using a YSI temperature controller (#71A, Yellow Springs Inst. Co., Inc., Yellow Springs, O) with a temperature probe (YSI #423). Drugs were infused via the tail vein with a Harvard infusion pump (#2681, Harvard Apparatus, Millis, MA). With the mouse in position, the best possible image was obtained. Screen luminosity as measured by photometer was then adjusted to a relative reading of 40. A reading of 40 corresponded to a light output of .058 foot lamberts. Each additional unit increment represented an increase in light output of .00145 foot lamberts. Infusion of the solution was then begun and continued for 15 minutes. Screen luminosity and rectal temperature were monitored throughout the infusion period.

Data Analysis

Data were generally analyzed by one-way analysis of variance and the Student–Newman–Keuls test for significance among means.   If necessary, the difference between two means at the same infusion time interval were analyzed using Student's $\pm$ test.

Acute Toxicity

These results are seen in Table I under the column headed "Median Lethal Dose" (LD_{50}).   1-Lysine monohydrochloride was the least toxic agent (LD_{50} = 17.49 mmole/kg) while sodium citrate was the most toxic agent tested (LD_{50} = 1.08 mmole/kg).   The toxicity of sodium citrate is thought primarily to be due to a severe hypo-calcemia from citrate ion chelation of plasma calcium (Cooper et al., 1973).   In an attempt to decrease this toxicity, another potential contrast agent, calcium gluconate, was added to a sodium citrate solution.   The results shown in Table I indicate that the addition of calcium gluconate (2.32 mmole/kg) to a sodium citrate solution produced an approximately six-fold increase in the LD_{50} value for sodium citrate (6.05 mmole/kg).

TABLE I

SLOPES FOR SPEEDS OF SOUND[1,2], MEDIAN LETHAL DOSE (LD_{50}) VALUES, AND ESTIMATED

RANKINGS OF POTENTIAL *IN VIVO* CONTRAST CAPABILITIES OF SEVERAL COMPOUNDS

Compound	Slope $(m/s/mole)^{[a]}$	LD_{50}[3] $(mmole/kg)^{[b]}$	Rank[4]
1-Lysine Monohydrochloride	125.20	17.49	1.000
Calcium Gluconate	129.76	2.32	.635
Sodium Citrate	180.20	6.05	
Calcium Disodium Edetate	155.00	5.957	.422
Calcium Gluconate	129.76	> 2.32[5]	.137
Sodium Citrate	180.20	1.08	.089

[1] McWhirt, 1979.

[2] Slope of speed of sound in distilled water *vs.* concentration at 37 °C in meters/sec/mole.

[3] Agent administered *via* tail vein as bolus injection.

[4] Ranking estimated from slope (*a*) x LD_{50} (*b*) normalized to value for 1-lysine monohydrochloride (2.190 m/sec/kg).

[5] Calcium gluconate insoluble at higher concentrations.

In Vivo Contrast Enhancement

The acute LD_{50} value for each agent tested (Table I) established relative equivalent lethal doses as administered. For example, injection of the LD_{50} (17.49 mmole/kg) of 1-lysine monohydrochloride would be expected to produce the same lethal results as
the injection of the LD_{50} (1.08 mmole/kg) of sodium citrate. The
maximum increase in the velocity of sound that these agents could
produce acutely *in vivo* would then reflect their respective LD_{50}
values. The actual molar drug amounts introduced into the mice
would vary considerably, however. The relative potential abilities
of these agents to enhance ultrasonic contrast would then be indicated by the ratio of the slope for speed of sound *vs.* the LD_{50} for
each agent (Table I).

Greater accumulation of the agents would be possible (with
subsequent greater contrast enhancement) if they were infused into
the mice, since infusion of the test agents over several minutes
was less toxic than when the agents were given as a bolus. To compare the contrast capabilities of these compounds, it was necessary
to determine a maximum, non-lethal infusion rate for an equivalent
LD_{50} "value" for all of the drugs tested. A preliminary study was
made to determine this infusion rate. The combination of calcium
gluconate and sodium citrate was used since they had each been
shown to produce contrast enhancement *in vitro* (Ophir *et al.*, 1979)
and together as a combination *in vivo* (Maklad *et al.*, 1979B). Using
the previously described procedure, it was found that an infusion
rate of 1.6 ml/hr produced significant enhancement of screen luminance while observable toxic effects did not occur until after 25
minutes of infusion (data not shown). At a 1.6 ml/hr rate of infusion, mice would receive approximately twice the "LD_{50} amount"
over the 15 minute infusion period. At a slower infusion rate
(1.0 ml/hr) no significant increase in contrast was found, while at
higher infusion rates (2.2 and 2.8 ml/hr) toxic reactions and/or
death were commonly observed during the first 15 minutes of infusion. Later infusion of the other test agents for 15 minutes at
1.6 ml/hr also showed few, if any, toxic signs.

The results of the contrast enhancement study are shown
(Fig. 1). Screen luminance increased with time in all cases. The
increase in luminance associated with the infusion of saline was
not significant when compared to the zero-time infusion value. When
compared to their corresponding saline values, screen luminance
values for the infusion of sodium citrate (28 mg/ml) were also not
significantly increased.

Maximum screen luminance or contrast occured at the 15 minute
infusion time for all of the drugs tested. At this time the relative ability of these agents to enhance screen luminance roughly
correlated with the ranking predicted in Table I. However, there

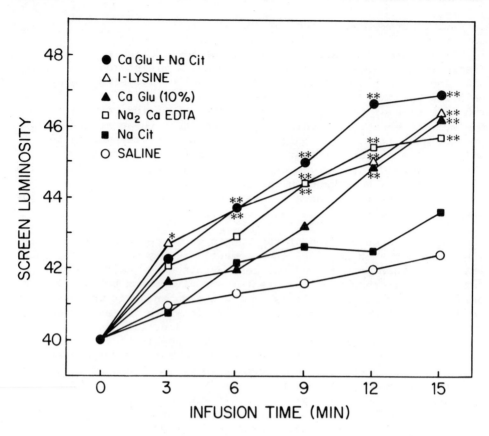

Figure 1. Effects of contrast agents on screen luminance when infused into the tail vein of mice visualized by a real-time ultrasound scanner. Mice were infused with either l-lysine hydrochloride
(320 mg/ml; closed square), sodium citrate (28 mg/ml; open square),
calcium gluconate (100 mg/ml; open circle), sodium citrate plus calcium gluconate (156 mg/ml and 100 mg/ml, respectively; closed circle), calcium disodium edetate (223 mg/ml; closed triangle) and
physiological saline (9 mg/ml; open triangle). Mice were infused
at a rate of 1.6 ml/hr for 15 minutes. Procedure for measurement
of screen luminance described in text. Luminance values represent
mean of 8-38 mice. Asterisks indicate difference between indicated
value and its appropriate saline control ($^*P<.05$; $^{**}P<.01$).

was found an exact correlation of the <u>onset</u> of significant enhancement of screen luminance induced by each agent (Fig. 1) with the
predicted rank of contrast capability (Table I).

Role of Body Temperature on Screen Luminosity

        The velocity of sound in water has been positively correlated

with temperature (Barthel, 1952 and 1954). It was therefore necessary to examine the role, if any, of alterations in mouse body temperature upon the screen luminance changes previously observed. The
results (Fig. 2) indicate that mouse rectal temperature after saline
infusion and after infusion of calcium disodium edetate (223 mg/ml)
did not increase a statistically significant amount over their own
zero-time temperatures. Rectal temperatures of both saline and calcium disodium edetate infused mice were also not statistically different at any corresponding infusion times. In contrast, infusion
of calcium disodium edetate (223 mg/ml) in the same experiment produced a significant (P<.01) enhancement of screen luminance over
the corresponding saline-induced luminosity values at the 9, 12 and
15 minute measurement times.

Effect of Drug Osmolarity on Screen Luminosity

The "contrast" agents used to increase screen luminance of ul-
trasonically-scanned mice might have acted via an osmotic effect.
For example, infusion of a highly osmotic solution into the vascu-
lature would produce a net influx of fluid into the vascular system
from the interstitial or intracellular compartments. This would
result in an enlargement of the diameters of the microvasculature
making it become significantly echogenic via enhanced scatter of
ultrasonic waves. Enhancement of screen luminance would then occur.
To test this hypothesis, mice were ultrasonically scanned while being
infused with either normal physiological saline (9 mg/ml) or with a
sodium chloride solution (88 mg/ml) of osmolarity equal to that of
the combined sodium citrate (156 mg/ml) plus calcium gluconate (100
mg/ml) solution. If infusion of the combination of calcium gluconate
plus sodium citrate had produced a significant increase in contrast
due to increased plasma osmolarity, then infusion of the sodium
chloride solution (88 mg/ml) would also be expected to produce a
substantial increase in screen luminance. The results shown (Fig. 3)
indicate that infusion of the hypertonic solution of sodium chloride
had no significant enhancement effect upon the screen luminance.

DISCUSSION

The results of this study indicate that intravenous administration of several compounds in aqueous solution can produce significant
contrast (echo) enhancement *in vivo*. The increased contrast was
quantitatively measured as the increase in sonographic screen image
luminance. The ability of the substances tested to enhance contrast
was shown to be dependent upon: 1) ability to increase the velocity
of sound in water and 2) the maximum non-lethal dose which could be
administered.

A correlation was found between predicted "contrast" ability
rank (Table I) and the observed onset of significant screen luminance enhancement (Fig. 1). This correlation might be expected

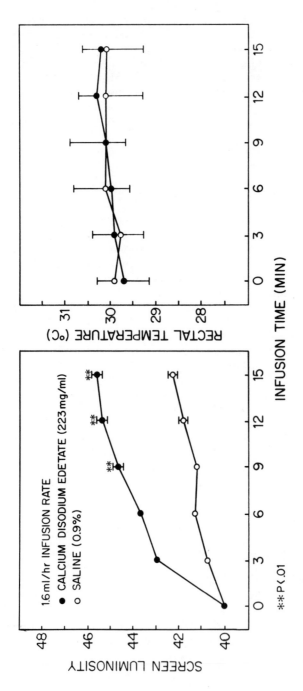

Figure 2. Role of body temperature in the contrast effects induced by infusion of calcium disodium edetate (223 mg/ml; closed circle) when compared to its saline infused control (9 mg/ml; open circle). Simultaneous screen luminance measurements of mice visualized by a real-time scanner and rectal temperature measurements were made. Mice were infused at a rate of 1.6 ml/hr for 15 minutes. Procedure for measurement of screen luminance described in text. Both luminance and rectal temperature values represent mean ± S.E.M. of 9-10 mice. Asterisks indicate difference between indicated value and its appropriate saline (** P<.01).

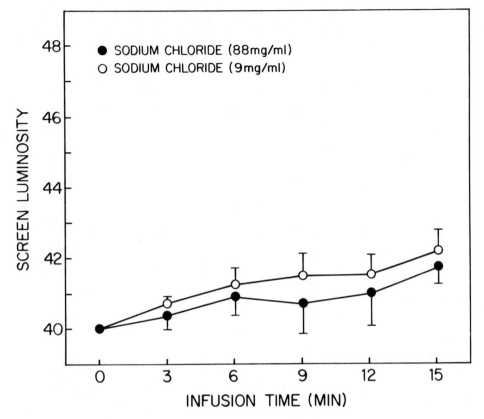

Figure 3.   Effect of infusion of two different osmolarities of sodium chloride upon screen luminance of mice visualized by a real-
time scanner.   Sodium chloride (9 mg/ml - open circle and 88 mg/ml -
closed circle) was infused at 1.6 ml/hr for 15 min.   Procedure for
measurement of screen luminosity described in text.   Luminance
values represent mean ± S.E.M. of 7-10 mice.

rather than a correlation between rank of predicted contrast ability
with observed maximal contrast effect.   Differential metabolism of
each of the agents tested would be expected to alter the amount of
agent per unit time in the mouse during infusion.   Since no assays
were made of the actual concentrations of these agents in the mouse,
it is unlikely that their actual concentrations reflect the total
amount of agent infused by the end of the infusion period.   Thus,
agents which were metabolized faster might not have enhanced screen
luminance as would have otherwise been expected.   Regional distribution of the infused agents may also affect the maximal screen
luminance, either by concentrating the agent in a small region that
is (for example, the kidney) or is not (for example, the brain) capable of being scanned.   In the case of a region such as the kidney,

the limited resolution of a 3.5 MHz transducer may not be able to
detect or the sonograph screen may not accurately represent, the increased echogenicity of a small mouse kidney as contrast agent levels
accumulate, with time, during the infusion. In the case of the brain
region, a contrast agent may be differentially accumulated here decreasing its concentration in the "scannable" region of the mouse.

The onset of a significant increase in screen luminance might
then be a more accurate reflection of the contrast ability of the
agents tested using the procedure described. Onset of augmentation
of contrast would probably be less influenced by metabolic degradation of the contrast agent, by sensitivity to contrast enhancement
as detected by the 3.5 MHz transducer and by image representation
on the ultrasonic scatter screen.

Other mechanisms by which the agents tested may have caused
increased contrast were examined. Infusion of a highly hypertonic
solution of sodium chloride failed to show a role for osmolarity,
per se, in the contrast-enhancing effects of the agents tested. It
appeared unlikely then, that the contrast effects seen were due primarily to an osmotic effect. Also of interest was the effect of the
agents tested upon body temperature, since contrast effects are dependent upon temperature (Sachs et al., 1976). Infusion of calcium
disodium edetate did not significantly increase body temperature nor
change it significantly from that of saline-infused control mice.
However, it significantly increased screen luminance over saline
control levels. These results suggest that the increased screen
luminance observed was not due to body temperature change, either.

These results indicate that a quantifiable ultrasonic contrast
enhancement can be produced in vivo. The ability to predict contrast capability depended, at least in part, upon the lethality of
the compounds as used and their ability to increase the speed of
sound in solution. In addition, the method described offers a simple procedure for the preliminary in vivo testing of other potential
ultrasonic contrast agents.

ACKNOWLEDGEMENTS

We thank R. E. McWhirt for his assistance in establishing the
slopes of the velocity of sound in the solutions used and Gladys K.
Cage for skillful technical assistance. Dr. J. Ophir provided many
helpful suggestions in preparing this manuscript.

REFERENCES

Barthel, R., 1952, A precise recording ultrasonic interferometer and
        its application to dispersion tests in liquids, J. Amer.
        Acoust. Soc., 24:8.
Barthel, R., 1954, Sound velocity in some aqueous solutions as a

function of concentration and temperature, J. Amer. Acoust.
Soc., 26:227.
Cooper, N., Brazier, J. R., Hottenrott, C., Mulder, D. G., 1973,
Myocardial depression following citrated blood transfusion,
Arch. Surg., 107:756.
Gramiak, R. and Shah, P. M., 1968, Echocardiography of the aortic
root, Invest. Radiol., 3:356.
Maklad, N. F., Ophir, J., Gobuty, A., Tyler, T., McWhirt, R. E. and
Rosenthal, S. J., 1979A, Aqueous solutions as ultrasonic
"contrast" agents, 65th Sci. Assembly Ann. Meeting Radiol.
Soc. N. Amer. #390, Atlanta.
Maklad, N. F., Ophir, J., Gobuty, A., Tyler, T., Rosenthal, S. J.,
Spicer, J. and Betzen, M., 1979B, Collagen microspheres as
ultrasonic "contrast" agents, 65th Sci. Assembly Ann. Meeting
Radiol. Soc. N. Amer., #394, Atlanta.
McWhirt, R. E., 1979, "Speed of Sound Measurements in Potential Con-
trast Agents for Use in Diagnostic Ultrasound," M. S. Dissertation, University of Kansas, Dept. Physics and Astronomy.
Ophir, J., McWhirt, R. E. and Maklad, N. F., 1979, Aqueous solutions
as potential ultrasonic contrast agents, Ultrasonic Imaging,
1:265.
Sachs, T. D., Anderson, P. T., Grimes, R. S., Wright, Jr., S. J.,
and Danaghy, R. M. P., 1976, TAST: A non-invasive tissue
analytic system, in "Ultrasonic Tissue Characterization,"
M. Linzer, ed., National Bureau of Standards Spec. Publ. 453:
153, U. S. Government Printing Office, Washington, D. C.
Thompson, W. R., 1947, Use of moving averages and interpolation to
estimate median-effective dose, Bact. Rev., 11:115.
Ziskin, M. C., Bonakdapour, A., Weinstein, D. P. and Lynch, P. R.,
1972, Contrast agents for diagnostic ultrasound, Invest.
Radiol., 6:500.

A TUTORIAL ON UNDERWATER ACOUSTIC IMAGING

Jerry L. Sutton

Naval Ocean System Center

San Diego, California   92152

INTRODUCTION TO UNDERWATER ACOUSTIC IMAGING

The purpose of underwater acoustic imaging is to produce two dimensional images of underwater objects that are somehow recognizable, or at least useful. Underwater acoustic imaging systems are generally useful for either classifying objects or observing the details of objects, usually from some form of underwater vehicle. For example, acoustic imaging systems are useful in differentiating mines from rocks, coral heads, and garbage on the ocean bottom, and in general, differentiating between objects that warrant further investigation and the many uninteresting objects that are in the ocean. Acoustic imaging systems are also useful for inspecting or examining objects when water turbidity precludes the use of closed circuit television or other optical means of viewing.

THE NEED

Why bother with acoustic imaging systems when underwater television cameras already exist? Although optical visibility ranges can sometimes reach 30-60 meters in very clear waters such as those of the Caribbean, most ocean waters are much more turbid. Deep ocean water (undisturbed) typically has 6-15 m visibility, while near-shore waters typically have 1-6m. Within harbors, estuaries, and in general wherever man disturbs the environment, visibilities are generally in the 0-1 m range. This includes even the deep ocean and the Carribean when man is working on the bottom, stirring up clouds of sediment. Thus, optical visibility is often most limited just when it is most needed. Because acoustic energy more easily penetrates the mud and silt that causes optical turbidity, the range of acoustical systems

is generally larger than that of optical systems. However, the best
resolution capability of acoustic imaging systems is usually
significantly lower than that of optical systems. This is because the
useful wavelengths of sound for underwater imaging are much longer
than optical wavelengths. However, using higher acoustic
frequencies, large aperatures and clever system designs, underwater
acoustic imaging systems having performances comparable to optical
systems are possible. None of the underwater acoustic imaging
systems built to date, however, have achieved this performance.

On the other hand, acoustic imaging is generally considered to
have higher resolution and shorter range than most sonar systems.
The distinction made here between sonar and acoustic imaging systems
is a difficult one to make. Both share many techniques, hardware
implementations, and physical properties, and they provide similar
kinds of information. Yet, they are not really the same. One
difference comes from their intended purpose: the goal of a
conventional sonar system is to indicate WHERE something is located,
while the goal of underwater acoustic imaging is to indicate WHAT it
looks like. As will be seen later, the distinction dissappears
almost entirely in the case of high resolution sonars. In general,
acoustic imaging systems tend to fill the performance gap between
sonar and underwater optics.

PROPERTIES OF UNDERWATER ACOUSTIC IMAGING SYSTEMS

Imaging Dimensions

Underwater acoustic imaging, in common with other forms of
acoustic imaging, such as those used in medical and non-destructive
testing, seeks to form two dimensional images of objects using sound
waves. High reslution sonar also produces horizontal two dimensional
images, either as a polar plot of range versus bearring on a plan
position indicator (ppi) display, or in a similar rectangular format
as created by side look sonars. Acoustic imaging systems generally
(although not strictly) produce orthoscopic images, that is in a
vertical plane similar to that used in optical imaging systems; that
is, reflected energy versus vertical and lateral directions at some
range. This is the familiar imaging format from optics. In fact,
the goal is usually to present an image that is similar to an optical
image in appearance. In at least one case, positions of all three
dimensions have been displayed as a 3-dimensional display.

Acoustic Performance

Typical characteristics of underwater acoustic imaging systems
are:

Frequencies	100kHz--2MHz
Wavelengths	.075cm--1.5cm
Apertures	10cm--1m
Resolution	0.1--2degrees
Ranges	1m--100m
Depths of Field	1cm--50m

The values of these characteristics are somewhat different from those in the other fields of acoustic imaging, such as medical and non-destructive testing. There are major differences in range spread of interest, geometries that are practical, and techniques that can be used. Because of these differences, the techniques and results from one field are not directly transferrable to another. Notice that the units of resolution are different from those normally encountered in medical/non-destructive testing systems, namely minimum resolvable distances. Underwater imaging generally uses the angular resolution subtended by a minimally resolvable dimension at some range. This is due to the great spread of ranges that are of interest in these applications. Underwater systems typically operate in a power-limited mode, limited either by available electrical power, power conversion capacity or non-linear acoustic limit. In this last case, power is lost by conversion to harmonic frequencies. Acoustic power density is not the concern it is in medical systems.

The parameters listed above are not independent of each other. The wavelength λ is, of course, dependent on the frequency f, λ = c/f where c is the speed of sound, (approximately 4860 ft/sec or 1500 m/sec in sea water). The angular resolution in radians, α, is related to λ and the receiving aperture diameter, D, by $\alpha = \lambda/D$. The maximum ranges are limited at the upper frequencies by the attenuation coefficient of seawater (0.3 dB/meter at 1MHz), and at higher power levels by the non-linear response of water. Finally, the depth of field of the imaging system depends on the duration of the acoustic signal, the properties of the acoustic receiver, and the focusing capability of the imaging system. It can be quite short when limited by the acoustic pulse length to, say, 10 cycles of the sound signal. It has an upper bound determined by the focusing properties and aperture size of the system.

Capability

Focusing is the process of correcting for curvature of the sound field reflected from objects at ranges relatively short with respect to the aperature size. The far field, the region in which spherical waves radiating from a point source can be treated as plane waves and no focusing is necessary, begins at a range of D^2/λ. Underwater acoustic imaging systems generally work at ranges inside this range, called the Fresnel region, which extends down to a range of D or 2D from the aperture. In this region, some account must be taken of the sound wave curvatures in order to realize the desired resolutions.

Hence, most underwater acoustic imaging systems require some ability
to focus.

UNDERWATER ACOUSTIC IMAGING LIMITATIONS

Several aspects of the underwater environment and the expected
usage of underwater acoustic imaging systems impose certain
restrictions on system design. They include the nature of underwater
objects, underwater noise and clutter phenomena, geometries
available, timeliness required, and system noise.

Targets

A characteristic of underwater objects is that they are
partially transparent. Sound nearly always penetrates objects,
reflecting from inner and back surfaces to produce multiple echoes
and interfering internal reflections. This effect, called "ringing",
provides additional information about the target while complicating
the interpretation of the acoustic image.

A second property of underwater objects is specularity. Natural
objects tend to be rough with respect to an acoustic wavelength.
Rough objects tend to produce diffuse images that are easy to
recognize. Many man-made objects, however, which are usually the
ones of interest, tend to be fairly smooth with respect to a
wavelength of sound. They produce specular or mirror-like
reflections which have very high contrast. The specular nature of
most objects of interest makes recognition and inspection difficult.
The problem of specularity is generally worse when the acoustic
source is located near the receiver in a reflective geometry, the
geometry generally imposed on underwater systems by practical
considerations. The problem of specularity can be appreciated if one
imagines trying to examine a very shiny object in a dark room using
only a head-mounted flashlight. Generally what is seen are small
images of the flashlight reflecting from small parts of the surfaces
of the object. Some relief can be obtained by shining the flashlight
from several widely spaced directions. Similarly, one method of
alleviating specularity effects underwater uses several acoustic
sources spaced far apart to provide more diffuse illumination, in
effect adding up many specular returns from different parts of the
object to form a more identifiable image. This scheme may not be
practical if underwater acoustic imaging is to be performed from some
platform, such as a manned or remote control underwater vehicle. In
this case, the lateral distances necessary for widely spaced sources
are generally unavailable because the cross sectional area of a
vehicle is limited by the need for drag reduction. Appendages that
might extend out to the side to hold acoustic sources could become
snagged in cables or other debris, and so are heavily discouraged by
vehicle operators. Hence, only a few sources placed near the

receiver is the general rule and specularity remains a problem.

Motion

In most cases, an underwater acoustic imaging system will be
attached to a moving platform or carried by a diver. In either case,
because the system cannot rely on a stationary geometry, the imaging
technique must be fairly insensitive to platform motion.

Packaging

In order to operate in the often hostile environment of the
underwater environment, especially in the ocean, an acoustic imaging
system must be rugged, corrosion resistant and able to withstand
pressure. Except for the ambient pressure of water, the most severe
stresses are not applied when the system is actually in use.
Vibration stresses are applied during transit to the operating site
aboard a surface vessel. Shock stresses are worst during launching of
the entire platform/sensor system through the air-sea interface due
to wave slap and ship roll. And of course, in an ocean environment,
corrosion must be taken into account. Again, fixed installations or
fresh-water applications may escape some or all of these impediments
to acoustic imaging system design. In general, the techniques which
compensate for these environmental hazards affect system size, weight
cost, and the availability of system design options.

UNDERWATER ACOUSTIC IMAGING SYSTEM ARCHITECTURES

In this section, several techniques for realizing underwater
acoustic imaging systems will be described.

Sonar Acoustic Imaging

For some time, underwater sonars have provided a form of
acoustic imaging in certain geometries. Both sector scan and
side-looking sonars produce basically the same form of acoustic image
by transmitting a beam of sound at a grazing angle to the bottom, as
shown in figure 1. When no object is present, the bottom backscatters
the energy as a more-or-less uniform background. When an object is
present and rests above the bottom, it intercepts and reflects some
of the sound earlier than would the bottom. Hence, more energy is
reflected to form a bright spot at the range of the object.
Meanwhile, no energy is reflected by the bottom within the shadow of
the object. As the beam is scanned across the field of view, the
beam intersects different parts of the object. Hence; the shadow of
the object appears on the display screen just beyond the bright spot.
The bright spot indicates the presence of something, the shadow
indicates the shape of it. An example of a SLS image of a sunken
vessel is shown in figure 2.

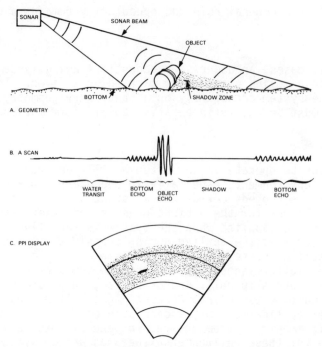

Fig.1 Sonar imaging: (a) geometry showing acoustic return from bottom and object, but none from shadow behind object; (b) A-scan or oscilloscope trace of received sonar signal versus time; (c) resulting sonar shadow image from many narrow sonar beams as shown in plan position indicator (ppi) display.

Fig.2 Side look sonar acoustic image of a sunken vessel. (Courtesy Klein Associates, Inc.)

   The synthetic aperture side look sonar produces essentially the
same results while using a slightly different technique. This scheme
uses a wide beam and, by means of accurate navigation, coherently
combines the returns from several transmit cycles to form each
display beam. Thus, the effect is similar to that of the side look
sonar. All of these sonar systems use time to obtain the second
dimension of the image. As with any other form of acoustic imaging,
the displays from these kinds of systems require time and training to
interpret well.

Array Geometry

   The purpose of an imaging array is to sample the acoustic field
over some spatial extent. A straightforward example is a two
dimensional acoustically sensitive plate which produces a
continuously varying measurement of the sound field. An equivalent
array, from an information theory point of view, is a planar two
dimensional regularly sampled, filled array such as that shown in the
upper left of figure 3. In order to reduce the number of hydrophones
requred, a sampled array may be spatially undersampled by spacing the
hydrophones at distances greater than a half wavelength. To prevent
aliasing however, the illuminating sound must be confined to a field
of view (FOV) with half angle θ given by $\sin \theta = \lambda/2d$, where d is the
interhydrophone distance. This is not an easy task, as projector

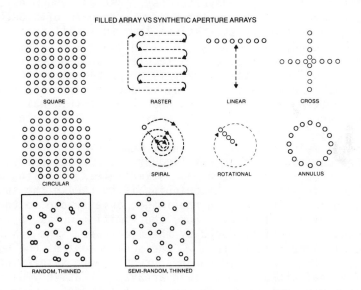

Fig.3 A few simple synthetic aperture patterns to synthesize the
filled apertures shown, a random array and a partially random array.

sidelobes falling outside the FOV can reflect off objects, thus producing cleared images on top of the desired images within the FOV.

Synthetic aperture techniques are also used in acoustic imaging. Note that the term "synthetic aperture" means far more here than the narrow usage it has attained with respect to synthetic aperture side scan sonar. The side scan synthetic aperture technique is in fact one special case of a broader concept that refers to the simulation of a filled array by the clever use of some smaller array. Examples include a line array scanned perpendicular to its length, a point transducer scanned in a raster, and many other schemes, a few of which are shown in figure 3. In some systems, the source is scanned in one direction while the receiver scans in another direction. Whatever scheme is used, the idea remains the same: replace the large array of receiving transducers with a smaller number of sensors in some scanning arrangement. The price to be paid for the savings in transducers may include increased scan time (which can lead to motion blurring and slower frame rate), or lower signal-to-noise ratio. Sophisticated signal processing techniques can compensate for some of these performance costs, as long as the added complexity and cost are managable.

The regularly spaced arrays may also be replaced by thinned, randomly spaced, arrays of hydrophones. In contrast with the regularly spaced, well behaved side lobe responses of regularly spaced arrays, random arrays have irregularly spaced sidelobes of unpredictable amplitudes. Nevertheless, just as random noise will exhibit a reproducable RMS amplitude, an aggregate of sidelobes of a random array obey similar statistical laws in a well behaved manner. The resulting array can be shown to exhibit almost the same performance characteristics as its regularly spaced counterpart. For example, resolution is still inversely proportional to the size of the aperture. The semi-random array is derived from the random array merely by removing any hydrophones that lie within a half wavelength of another such that they do not receive independent information.

"Thinned" means that significantly fewer hydrophones are used to sample the aperature than in a filled array. Consequently, the aperture is greatly under sampled. Nevertheless, aliasing does not occur. The main effect of the thinning is to reduce the spatial signal-to-noise ratio (S/N) in the image, chiefly due to the smaller number of samples collected. Thus, the limiting noise in such arrays is the spatial noise introduced by the sidelobes. The S/N is approximately proportional to $\sqrt{M}$, the square root of the number of hydrophones. Thus, more hydrophones produce more samples and lower noise. The real attraction of such arrays is that very high resolution can be obtained with the use of a modest number of hydrophones. Research has only begun in this area,[1,2] but it is conceivable that some of the synthetic aperture techniques can be

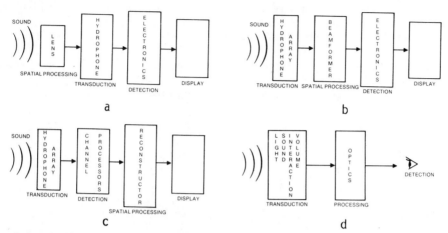

Fig.4 Basic forms of acoustic imaging. Except for Bragg, each performs same basic processes, but in different orders. a. Lensed; b. Beamformed; c. Holographic; d. Bragg.

combined with the random array concept to effect significant savings in system hardware.

SPATIAL CONFIGURATIONS

Acoustic imaging systems suitable for use underwater can have several possible configurations. Four fundamental approaches to acoustic imaging (figure 4) have been used in underwater imaging applications. In all three approaches, the same operations are performed: spatial processing to derive an image from the acoustic field; transduction to change from acoustic energy to electrical energy; detection to convert high frequency signals to an observable, near-DC image signal; and some form of display. The approaches differ mainly in the order in which these steps are performed. Lensed systems use an acoustic lens or reflector to focus the high frequency sound image onto an acoustic retina, much like a TV camera focuses light images onto an optical retina. That is, these kinds of systems perform spatial processing first, followed by transduction and detection. Beamformed systems perform the spatial processing electronically following transduction, eliminating the need for an acoustic lens. Then comes detection. In the holographic approach, transduction and detection are performed first to produce an acoustic hologram. Then the spatial processing (sometimes called reconstructing the hologram) is performed last, allowing the processing to be performed at baseband rather than at the acoustic frequency. The technologies available for each of the functions and certain physical constraints determine which particular type of system is best in a given application. A fourth type of system is

Bragg imaging, wherein laser light interacts directly with the water
volume to produce observable optical images. This technology has not
been used because it is not as sensitive as piezoelectric transducer
systems. It has been claimed to be more sensitive[3] but this has
not been demonstrated in underwater applications. Liquid surface
holography, wherein interfering acoustic beams form the ripples of an
acoustic hologram at an air-liquid interface, is not used for similar
reasons. Thus, the primary kinds of systems suitable for use
underwater are:

    lenses
    beamformers
    holography

Each of these will be discussed in more detail.

Lensed Acoustic Imaging

    Lensed acoustic imaging systems operate in a manner very
analogous to optical imaging systems. In addition to acoustical
analogs of the familiar single and multiple element thin lenses found
in optics, spherical acoustic lenses have also been developed and
tested with good results.

    Properties of lensed acoustic imaging systems that are important
for underwater applications are their simplicity, inherently
broadband nature, and their size and bulk. The system design is
fairly simple and readily understood by analogy to geometrical
optics. The acoustic array and processing electronics can be fairly
simple since they only need to detect the amplitude (or intensity) of
the incident acoustic energy. On the other hand, a lens requires
some distance between itself and the focal plane, which generally
implies size and bulkiness. This is especially true of thin lenses,
less so of spherical lenses.

Beamformed Acoustic Imaging

    Beamforming systems are slightly more difficult to understand
than systems, but are nonetheless very appealing. The essence of
beamforming is the combination of the acoustic signals from several
transducers into beams, followed by detection of the energy in each
beam. The entire process usually operates at the acoustic frequency,
unless the signals are mixed to a lower intermediate frequency (IF).
These systems vary from a single pencil beam scanned in a raster
fashion across the field of view to multiple pre-formed beams that
observe the entire field of view at one time. They encompass actual
physical beams of sound transmitted into the water as well as the
separation of received signals into electronic paths that represent

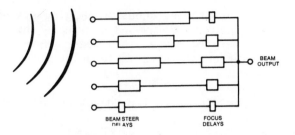

Fig.5 Delay and sum beamformer block diagram. The delays on the left form a beam at one angle; the delays on the right correct for wavefront curvature.

virtual beams. By combining the acoustic signals from many acoustic transducer elements, acoustic beams of various shapes (usually pencil or fan beams) can be generated in the water and scanned over the field of view. As the beams scan across the field of view, the received intensity is measured and displayed in a corresponding geometry. Three basic classes of beamformers have been used: (1) delay and sum, (2) phased array, and (3) correlation.

    Delay and Sum Beamformer: The delay and sum beamformer is best typified by time delays and summing amplifiers, the delays for each image element being chosen to compensate for the acoustic path delay. For example, consider the geometry of a regularly spaced line array as shown in figure 5. For $\theta=0$ (array steered broadside), all of the acoustic wavefronts arrive simultaneously. In this case, no delays are applied. When the sound arrives from some other angle, such as that shown in the figure, the wavefront arrives at the first transducer then at the next, and so forth on down the array. In order for the signals to arrive at the outputs of each of the receiver channels at the same time, succeedingly shorter time delays must be introduced at each subsequent channel. A different set of time delays is required for each beam. To form one beam in one direction requires N time delay modules. To form P such beams simultaneously looking in different directions requires PxN time delays, or alternatively N delays having P taps each. It has the desirable property of being inherently a broadband system. The details of design of delay-and-sum beamformers have been explored in the design of linear beamformers for underwater sonars, and have been presented widely in the literature. The primary problem with this type of system is the unavailabilty of high speed, short term storage devices--such as CCD's or bulk constant delay lines--which have the appropriate speed, linearity, dynamic range, and uniformity from device to device. With the increasing availability of good charge coupled devices (CCD's), this technique may come into greater usage.

    Focusing at other than one or a very few intermediate ranges is difficult for most beamformed acoustic imaging systems, since some form of variable quadratic delay and sum phase shift (or time delay) is necessary with varying range. This means variable taps in delay line beamformers, or possibly variable delay rates in CCD implementations.

Phased Array Beamformer: The phased beamformer is a narrow-band approximation to the delay and sum type. It generally uses some form of properly phased reference signal to sample or mix with the incoming acoustic signal to achieve the proper signal addition for beam formation. Implementations often use a delay line of some sort to obtain reference signals with the desired phase at each hydrophone (figure 6). Each hydrophone signal is multiplied by the corresponding reference signal and added to the rest of the products. Only those signals arriving at the angle being examined will have phases that add coherently after multiplication by the reference waves. Inserting a sine wave into a uniformly tapped delay line produces the desired reference waves whose phases vary linearly with distance for one scan angle. Inserting a slowly varying frequency (slow in that essentially only one frequency is present at all taps of the delay line at any one time) causes the reference signal phases to vary linearly with time, thus scanning out a variety of angles.

Focusing can be achieved with the same beamformer by using a fast chirp. Here, fast means that the frequency chirp changes frequency faster than the signal can propagate from one end of the

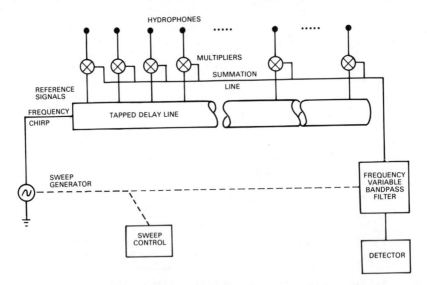

Fig.6 Delay line phased array beamformer. Only sound arriving at one angle to the array will have the signal phases that match the reference phases present at the delay line taps at any one reference frequency. As the frequency sweeps, so does the receive beam.

phase forming delay line to the other. Thus, there is a linear frequency shift and hence a quadratic phase shift present along the delay line. Focal range is controlled by just how fast the chirp is swept, and hence, by how much phase variation is present in the delay line at any one time.[4,5]

Chirps have been sent down delay lines made with many technologies, including surface acoustic waves (SAW),[4] Charged Coupled Devices (CCD), shift registers,[5] and conventional delay lines, as well as generated in Read Only Memories (ROM's). Several of these concepts have been explored for imaging purposes in laboratory systems, but none have been built for actual underwater use.

Correlation Beamformers:

The third type of beamformer, the correlation type, relies on the fact that a planar sine wave arriving at some angle to an array of sensors produces a sinusoidal pattern of amplitudes across the array at any one instant. By looking for those spatial sinusoids at a variety of spatial frequencies using matched spatial correlators, a multi-beam beamformer results. A simple form of such a correlator is just a series of resistors, each attached to one hydrophone and each having a value corresponding to one point of one sinusoid extending across the array (figure 7). The outputs of all the resistors are

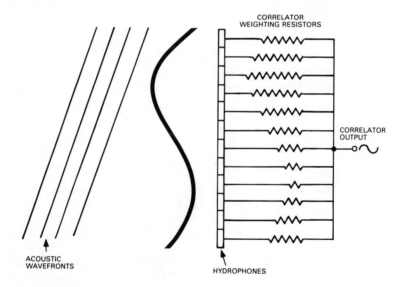

Fig.7 Correlation beamformer. At any one instant, a plane wave has a spatial sinusoidal amplitude across the array which can be detected by the sum of appropriately weighted resistors.

summed together so that when the right plane wave is incident on the array, all of the outputs add with the proper amplitudes to form a signal When plane waves arrive from other angles, the amplitudes are at random values and the resulting sum tends towards zero. This of course, implies that the correlation beamformer is necessarily narrow band.

Focusing might be difficult in this implementation, however, as some form of variable phase shift or amplitude multiplication dependent on the selected focal range must be applied to the incoming signal.

Difficulty with all beamforming techniques is that virtually all of the circuitry must operate at the acoustic frequency (unless converted to an intermediate frequency, a process which has its own problems). At lower acoustic frequencies and in synthetic aperture systems, this is not a serious drawback. As the frequency goes up, or as the number of parallel circuits increases, howewever, so generally does the power required to run the high frequency electronic circuits, especially integrated circuits. Total system power can be quite high (100's to 1000's of watts) with completely filled arrays at high frequency.

Holographic Acoustic Imaging

Holographic acoustic imaging is generally a narrow bandwidth, phase sensitive technique that usually processes all of the "beams" simultaneously. This doesn't mean that al parts of the hologram must be collected simultaneously, as long as certain stability requirements are met. Since holography is a phase sensitive technique, stability generally requires that a holographic acoustic imaging system either employ many parallel acquisition channels, maintain extreme stability during the acquisition period (on the order of a tenth wavelength), or use some fairly sophisticated means to achieve the same result, such as inertial navigation to detect the motion combined with motion compensating signal processing to remove its effects. However, it can be as physically compact as a beamformed system since no lens is used, and the signal processing to perform the image reconstruction can be conveniently separated from the data collection. One advantage of this separation is that existing high speed, general purpose signal processors such as digital minicomputers and optical processors can be adapted for use, relieving some of the need for specialized hardware. Another is that the processing electronics can be located somewhere other than in the hydrophone array, thus relieving some of its bulk. This is because the acoustic hologram is a minimal amount of information that fully describes the narrow bandwidth signals at each hydrophone. The array, in effect, freezes a representation of the acoustic field (the hologram) present at the face of the array. This representation is then sent to a signal processor which can evaluate all of the data contained in the hologram simultaneously in the most efficient manner possible.

In holographic acoustic imaging, the sound field of a narrow band signal is spatially sampled by a hydrophone array and is immediately converted to a stable set of DC values called a hologram. That is, the signal present at each point in the receiving array is mixed with a reference signal of the same frequency to produce a DC component. This value represents the term A cos Ø, where A is the acoustic amplitude and Ø is the phase relative to the reference signal. In some systems (actually preferrable in most) there may actually be two numbers, A cos Ø and A sin Ø, that can be combined and interpreted to be the complex field value at the aperture position.

In acoustic holography, image reconstruction can be accomplished in one of at least three ways: (1) optical diffraction, (2) backward propagation, and (3) Fourier/Fresnel transformation.

Optical Diffraction: In optical diffraction, a coherent light beam from a laser is bounced off the acoustic hologram as it e8ists on the surface of the water or is refracted through the hologram that has been copied onto some form of transparent surface, much as in the reconstruction of optical holograms.[6] The results are nearly identical to those in optical holography as well. There is a zero order bright spot, a real image and an out-of-focus conjugate image. If the spatial sampling rate in the receiving aperture is undersampled, the conjugate image will lie on top of the real image and obscure it. The departure from optical holography is that the difference in optical and acoustical wavelengths causes the z (range) dimension to be greatly magnified with respect to the lateral dimensions in the reconstructed image. This means that only a short depth of field can be in focus at any one time.

Backward Propagation: This consists of solving the inverse integral equation to the acoustic propagation equation at a specific range.[7] The result is exact and works even in the very near field of the acoustic array. The acoustic image can be found on a digital computer by
(a) Taking the Fourier transform of the array data,
(b) Multiplying the result by a factor

$$\exp[\frac{j2\pi z}{\lambda}\sqrt{1-(\lambda f_x^2) -(\lambda f_y^2)} \,]$$ (1)

(c) Taking the inverse Fourier transform of the result.

Fourier/Fresnel Transform: In the Fourier/Fresnel transform technique of reconstruction, use is made of the Fresnel approximation to the acoustic propagation equations. These equations imply that a spatial Fourier transform of the spatial hologram data results in an image. Since most acoustic imaging is of objects in the moderately near field (the so-called Fresnel region), a quadratic phase factor preceeds the Fourier transform to compensate for wavefront curvature. This is sometimes called the Fresnel focus factor, since it focuses the image at the desired range. The primary advantage of using this approximation is that a fast Fourier transform (FFT) algorithm can be implemented on a digital minicomputer to perform the image reconstruction rapidly. The result is that the acoustic image can be found by

(a) Multiplication of the array data by a term (the focus factor)

$$\exp[\frac{j\pi}{\lambda z}(x^2+y^2)]$$ (2)

(b) Taking the Fourier transform of the result.

In each case, holography replaces the large number of parallel signals that correspond to the acoustic field by a relatively small set of values--the hologram--and then processes them in a single processor, thus ensuring uniformity and possibly economy.

SIGNAL PROCESSING TECHNIQUES

Signals

Three classes of signals have been used in underwater acoustic imaging systems:

     narrowband
     broadband
     coded coherent

Most sonar and most of the present underwater acoustic imaging systems use narrow bandwidth acoustic signals. Narrow bandwidth signals are used partly because single frequencies are easier to generate with existing acoustic transducer designs, but primarily because they allow significant simplifications in signal processing and system design. This is because these systems approximate time delay by the phase shift of a sine wave signal. Narrow bandwidths have the undesirable property of speckle, a problem shared by optical holography with lasers. Basically, speckle is the mutual interference of the signal from one part of the target with that from another part. This gives rise to bright and dark areas in the image that are on the same order of size as the details of the object being imaged, but are not related to any details on the object. Hence, they tend to be very distracting and confusing.

When using broad bandwidth signals, the speckle patterns due to the different frequency components tend to cancel each other, leaving more uniform and representative image information. In some systems the signal processing techniques are independent of frequency over a wide bandwidth. The use of broadband signals is limited to delay and sum type beamformers and lenses.

Coded coherent signals fall somewhere in between the other two. These are very carefully generated signals, such as FM sweeps or binary coded biphase, whose coherence is maintained through the imaging process. The broader frequency content tends to remove speckle as in broadband imaging. The coherent nature of the signals tends to retain the signal processing simplification properties of the narrowband signals.

Processing Techniques

There are a variety of minor subjects relating to signal processing in underwater acoustic imaging, some of which are worth discussing briefly.

Complex Acoustic Field: In optics, the only sensors available respond to the intensity. As a result, the equtions of optical holography rely on a term $A \cos \emptyset$ which gives rise to two images (real and conjugate) during reconstruction. In acoustics, however, the signal can be sensed, amplified and processed to produce both $A \cos \emptyset$ and $A \sin \emptyset$ terms which form a complex acoustic hologram that reconstructs to only one image. These terms are drived by nultiplying the signal by $\sin wt$ in one channel of electronics and by $\cos wt$ in another. This technique, also called "quadrature" or "IF" processing, is also useful in beamforming techniques as well as in acoustic holography.

Separability of Dimensions: This refers to a property of the equations of beamformed and holographic acoustic imaging that the terms relating to the X dimension are separable from those dealing with the Y dimension. This results in the ability to stack two sets of linear beamformers, one set following the other, to achieve a two dimensional beamformer. For an array of N by M hydrophones, one set of N-point beamformers (M of them) forms beams in one direction, while another set of M-point beamformers (N of them) uses the outputs of the first set of beamformers to form beams in the other direction.

This technique works with all of the beamforming schemes, and with Fourier/Fresnel holography as well.

Image Signal-to-noise Ratio: The sources of noise in underwater acoustic imaging systems can be classified as either spatial or temporal.[8,9] Temporal noise includes ambient acoustic noise and thermal noise in system electronics. Of the two, thermal noise appears to be predominant, especially in the first stage of hydrophone amplification. Ambient noise is at a minimum at about 500kHz, the center of the region for underwater acoustic imaging. The temporal S/N can be improved by averaging signals for a longer period of time, T, and the improvement expected is approximately the square root of T. Time-bandwidth compression techniques, such as coded coherent signals, are another technique of achieving this without suffering a loss in time resolution due the longer signal.

Spatial noise includes clutter (bubbles or objects that are in the scene but not wanted) and spatial variations introduced by the imaging system. Spatial noise due to clutter does not appear to be a significant noise source in most cases. The limiting noise in most systems realized so far has been self-induced spatial noise. In lensed systems, this arises from non-uniformity between different channels of acoustic sensors and signal processing. A uniform illumination of the sensing array results in a more or less uniformly lit image, punctuated by bright and dark spots or streaks. The response amplitudes will exhibit a distribution about some average value. The ratio of the spread to the average is the S/N.

A similar result is found in holographic and beamformed systems. A distribution of amplitude variations can be expected between hydrophone/processors similar to that in the lensed case. In addition, there would also be a distribution of phase responses and, in the case of holography which produces DC values, in DC offset values. In either holography or beamforming, the processing is such that the output is a linear combination of the inputs. Hence, by conservation of energy laws, it can be shown that the resulting S/N in the image is directly proportional to the S/N in the input. Moreover, if the input spatial noise represented by these variations is Gaussian distributed, the output noise is probably also Gaussian distributed. The output noise is not necessarily equal to the input noise, however. Each image point is some linear combination of all of the input signals. Hence, to the extent that the input spatial noise is random, the output noise is reduced by the square root of the number of input signals averaged in the linear combination, just as the temporal noise is reduced by the number of independent input samples averaged. In sonar terms, this gain in S/N is called the directivity index or array gain. Thus, output S/N is inversely proportional to the number of input samples. A similar result can only be true in lensed systems when the spatial sampling is much finer than the resolution. Thus holographic and beamformed systems have the potential for greater S/N performance than lensed systems.

In large arrays, this array gain effect will probably swamp out temporal noise effects. In very large, high resolution, regularly spaced arrays, there will probably be more output S/N than can be used. This is where the thinned random array pays off. The number of sensors determines the S/N performance. Independently, the aperture size determines the resolution. And there is no link between the two specifications as there is in regular arrays.

Non-linear Processing: There is a disparity between the wide dynamic range (high contrast) of specular acoustic images found underwater and the usable dynamic range of most display devices. This makes a simple mapping of input dynamic range to output dynamic range unpractical. Moreover, the human eyeball cannot use wide dynamic ranges either. Hence, it makes sense to develop some form of transfer characteristic between input and output that improves the observability of acoustic images. Simple clipping is one (crude) form of such a function. Just what transfer characteristics make sense for underwater acoustic images has not been determined.

Interlacing: This is the process of interleaving several sub-images to form a larger image. Often, the time averaging or integrating aspect of the human eye is used to add up the sub-images, just as on a standard television set. This is a processing technique that goes hand-in-hand with certain of the synthetic aperture techniques described previously. The chief advantage is that the signal processing task can be split into smaller pieces, just as the data acquisition task is. This can mean fewer processing channels, smaller amounts of data to process at one time, and more even • workloads for the processing components. The chief limit to such schemes is the smearing effect of relative motion between object and sensor.

Displays

By far the most popular display device in underwater imaging systems is, of course, the cathode ray tube (CRT). Other displays have been used, such as direct viewing of optical trains (Bragg and surface holography) and light emitting diodes. The first is fragile and inappropriate for most underwater imaging techniques and applications. The latter generally suffers from cost and uniformity problems. Since the usual goal of acoustic imaging is to produce images comparable to normal visual images, and since the imaging systems functionally resemble a conventional TV system, it should not be surprising that the CRT is generally chosen. If certain three dimension acoustic imaging schemes pay off, there will probably be some new display devices introduced to better take advantage of the technique. For the time, even this function has been handled by pseudo 3D display techniques on a flat CRT.

EXAMPLES OF UNDERWATER ACOUSTIC IMAGING SYSTEMS

The following paragraphs describe several examples of underwater acoustic imaging systems that have been developed to date. The examples are certainly not exhaustive, but it is believed that they are fairly representative of what has been done in this technology area.

EMI Ultrasonic Camera

EMI, Ltd. of Middlesex, England has developed[10] a small, portable acoustic imaging system using Sokoloff tube technology anc a single element thin acoustic lens, as shown in figure 8. The system uses a single element thin lens made of plastic to focus the acoustic image on the plate of the sensor. Their Ultrsonic Image Converter Tube consists of a thin plate of piezoelectric quartz that is scanned by an electron beam, just as a television camera is. The relatively fragile quartz plate is isolated from the ambient pressure of the water by a plastic plate. Acoustic energy is coupled from the plate to the quartz by a small liquid volume. The display is a CRT. Its specifications are as follows:

frequency	2 MHz +/- 0.25 MHz
resolution	1.0 deg.
field of view	40 deg.
range	50 m
detection threshold	0.001 watt/sq. m
size	360 x 120 x 80 mm
weight	2 kg

Figure 9 is an acoustic image obtained with the EMI camera of a small anchor.

NCSC Multiple Element Lens

An example of lensed acoustic imaging with multi-element thin acoustic lenses has been developed at the Naval Coastal Systems Center in Panama City, Florida by Henry Warner and Don Folds. The system, shown in figure 10, consists of a 1 meter diameter temperature-corrected, six element, acoustic lens on a movable focusing framework, a rotating line array of hydrophones and their corresponding signal amplification/detection circuitry, and a display screen. Operation of the system is a straightforward analogy to an underwater TV camera. The lens focuses the acoustic energy onto the receiving plane, while the rotating line receiver scans out the image to the display. Four separate acoustic projectors transmit acoustic

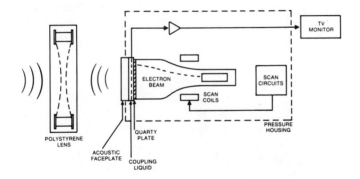

Fig.8 Block diagram of EMI lensed ultrasonic camera.

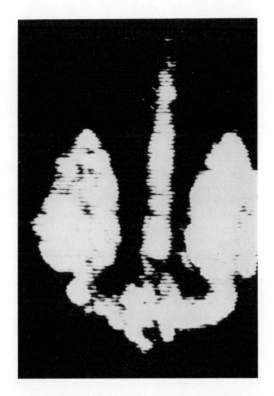

Fig.9 Acoustic image of anchor obtained with EMI camera.

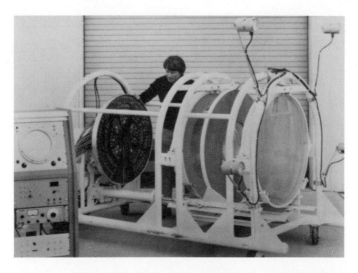

Fig.10 A focused underwater acoustic imaging system using a
6-element temperature corrected lens and a rotating line array of
receivers. The four projections at the front are acoustic sources.

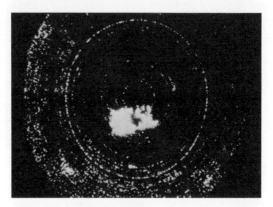

Fig.11 Acoustic image of the 18-inch tall letters "NCSL" made of
pipes taken at 30 meters with a focused system.

energy into the water to obtain diffuse illumination. System
specifications include:

frequency	8690 kHz
resolution	0.12 deg
range	150 m
frame rate	1 per sec
array size	100 in line

Figure 11 is an image of the letters NCSL on the display tube.
The letters are 45 cm tall and at a range of 30 m.

Harmuth FFT Correlation Beamformer

An example of a correlation type of beamforming acoustic imaging system has been developed and tested in a test tank by Dr. Henning Harmuth of The Catholic University of America in Washington D.C.[11] The system was designed to show feasibility of the techniques involved. Figure 12 indicates the basic beamformer circuit in this system. Each letter on the left represents a hydrophone input. The terms on the right are various linear combinations of the hydrophone signals. Rather than combining the outputs of many single (spatial) frequency correlators of the type shown in figure 7, this

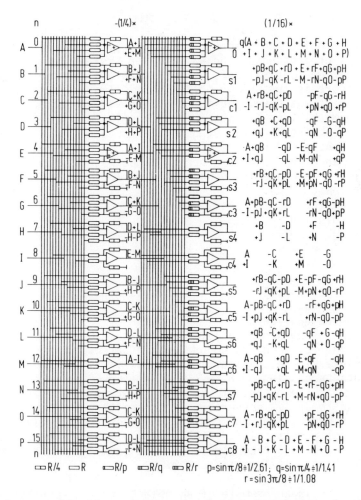

Fig.12 A generalization of the correlation beamformer is this analogy to a Fast Fourier Transform (FFT) where resistors provide multiplication of signals, operational amplifiers provide addition.

circuitry performs a high frequency analog (i.e. operational amplifiers and resistors), analogy of a fast Fourier transform (FFT). The FFT is a computationally efficient but equivalent process of the Fourier transform, which in turn is an efficient means of determining how much of each of many (spatial) frequencies is present in a signal. The FFT algorithm calls for multiplying each data sample by various constants, followed by various additions. The differential operational amplifier configuration shown in the schematic performs just this function admirably. With some care, it can be verified that each pair of terms on the right of figure 12 is one of the terms in an FFT.

NOSC Holographic Acoustic Imaging System

Work in holographic acoustic imaging systems for underwater viewing has progressed at the Naval Ocean Systems Center in San Diego by Jerry Sutton, Ben Saltzer, Newell Booth and Jim Thorn. One product of that effort is the holographic Acoustic Imaging System (AIS).[12,13] It was developed for use as a viewing aid aboard deep diving submersibles. The system is a regularly sampled, filled array with completely parallel quadrature channel processors. It was designed to have moderate resolution and moderate range to operate in the performance gap between sonar and underwater television cameras. The resulting hardware is shown in figure 13. System specifications are as follows:

frequency	642 kHz
resolution	0.3 deg
range	30 m
field of view	11 deg
array size	48 x 48
aperture	60 cm
projector power	250 watts acoustic
frame rate	1 per 2 sec
operational depth	3700 m

Inside the housing are parallel electronic channel processor circuits, one for each hydrophone, which detect the complex acoustic hologram in a single snapshot. Figure 14 is a block diagram of the signal/data processing path in the AIS. The DC value is held in analog form at the output of the integrator until it can be multiplexed to an A/D convertor and transmitted to the control panel for further processing. The control rack for the AIS (figure 13 a.) holds control panels, display CRT's, and a general purpose digital minicomputer. Image reconstruction uses Fourier/Fresnel techniques in a dedicated digital minicomputer.

Figure 15 is an example of an object and its acoustic image obtained with AIS, where the object was 4 m away, in ocean water which had an optical visibility of about 0.3 m.

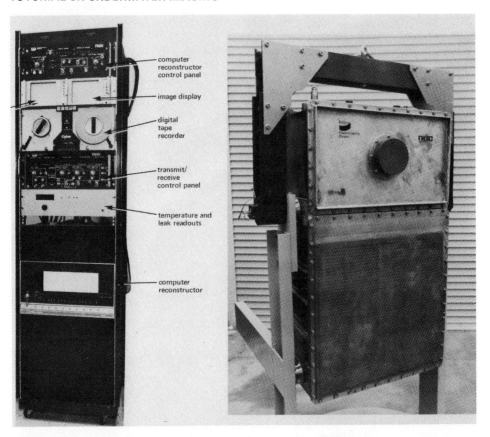

Fig.13 Holographic acoustic imaging system (a) control electronics with minicomputer image reconstructor, and (b) underwater array (2'Wx3'Hx1.5'D) with 48x48 filled array of hydrophones, separate acoustic projector and electronics housing.

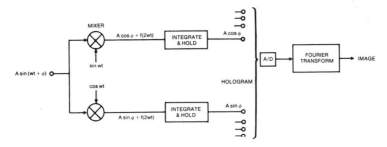

Fig.14 Complex holographic signal processing channel.

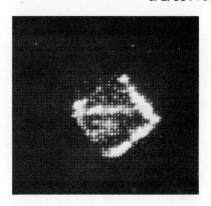

Fig.15 (a) Steel plate, 30 x 30 cm. (b) Acoustic image at 4 m.

OKI Synthetic Aperture Holographic System

A newer system, related to the holographic acoustic imaging system just discussed, but using a synthetic aperture technique, has been developed at OKI Electric Industry Company in Japan.[14] Figure 16 is a photograph of the imaging system, showing both the underwater array and the two racks of processing equipment. An image is obtained by pulsing the transmit array 64 times, introducing just enough phase shift with each pulse to shift the transmit pattern by one pixel, or resolution cell. The receiving and processing electronics then produce a portion of the image using a digital computer and FFT processor. Each sub-image is an array of 32 x 32 image pixels spaced eight pixels apart in each dimension. The 64 sub-images thus produced are interleaved to display the full image on a TV monitor. System specifications include:

frequency	200 kHz
resolution	0.4 deg
field of view	40 deg
receive array size	32 x 32
receive aperture	32 cm
transmit array size	4 x 4
transmit aperture	1 m
projector power	200 watts total
frame rate	1 per 2 sec

Figures 17 and 18 are acoustic images of a bicycle and diver, respectively, taken in the ocean alongside a pier.

Bendix Research Labs Synthetic Aperture Holography

Pat Keating and Roger Koppleman of the Bendix Research Labs, Southfield Michigan, have used a small 20 x 20 element hydrophone array for a number of experiments in underwater acoustic imaging.

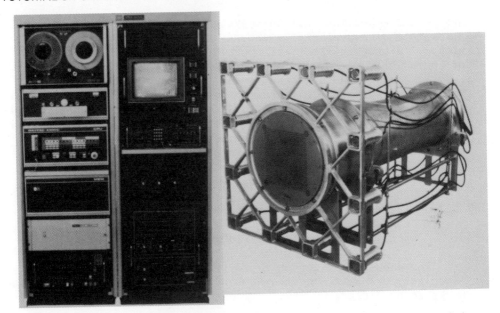

Fig.16   Synthetic aperture holographic acoustic imaging system (a) electronics and display rack, and (b) 32x32vreceiving array imbedded in a 4x4 transmitting array (1 meter on a side).

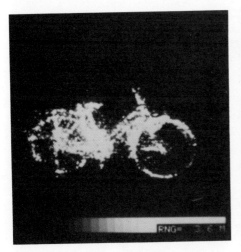

Fig.17 Synthetic aperture acoustic image at 3.6 meters.

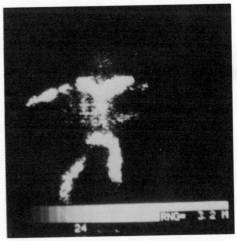

Fig.18 Synthetic aperture acoustic image of diver at 3.2 m.

One of the embodiments of this system is shown in figure 19. It is an example of a synthetic aperture holographic acoustic imaging system.[15] The details of its operation are very similar to those of the NOSC AIS described above, since that system was based on this 20 x 20 array. In this case, an array of four transmitting transducers was used to increase the system resolution by a factor of two in each direction. Each projector was pulsed in turn, followed by the collection of a hologram at the 20 x 20 array. A factor was then applied to each sub-hologram before processing the spatial combination of the four of them together. Specifications for the system are as follows:

frequency	250 kHz
resolution	1.0 deg.
field of view	40 deg.
range	10 m
array size	180 x180 mm

The acoustic images obtained from this small array are not especially dramatic, but they do indeed indicate that the technique works as predicted.

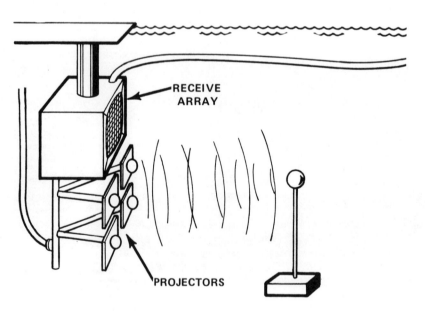

Fig.19 Synthetic holographic system using 20 x 20 array and 4 x 4 transmitters.

Bendix 3D Holographic System

This system is very similar to the one described above.[16]
The four projectors may be used or not. The primary difference is in
the acoustic channel processor that appears in figure 20. Also, a
biphase encoded acoustic signal is used in order to perform time
bandwidth compression. The biphase encoding chosen was one of the
Barker codes, all of which exhibit a single very sharp
autocorrelation peak. In this system, several holograms are
effectively gathered in quick succession and stored in the channel
processor memory following one transmit pulse. A computer, after
sorting out the time compressed signals then processes each of the
holograms independently. It displays the resulting collection of
range slice images in a pseudo 3D format on a CRT as shown in figure
21. Operator-controlled rotations of the image enable him to
appreciate the 3D nature of the acoustic image.

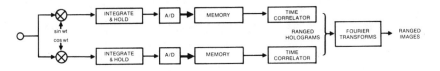

Fig.20 Holographic 3D system signal processing channel with integral
memory. The time correlator is software in the minicomputer that
performs the FFT processing.

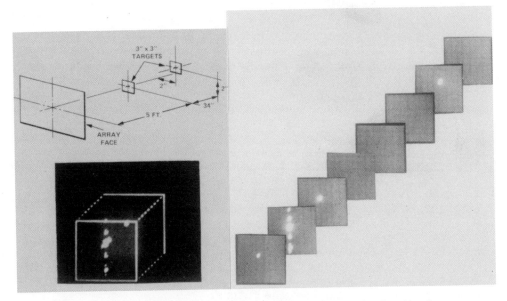

Fig.21 a. Target arrangement and one form of pseudo 3D display.
b. Another pseudo 3D display that separates the image range slices.

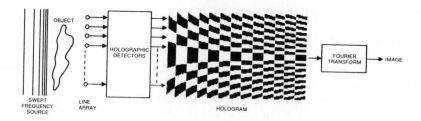

Fig.22 Block diagram of frequency diversity technique of acoustic
holography showing a binary hologram of a rectangular object.

Farhat Frequency Diversity Imaging

A very interesting variation on the holographic process has been
described by Farhat and others.[17,18,19] In this technique
(figure 22), only a line array of hydrophones is used. The other
dimension is obtained from a swept frequency sound source. As the
frequency changes, the line array is sampled to produce a hologram
that can be reconstructed by conventional means. A binary hologram
of a rectangular object is shown in the figure. This technique is a
good example of how to replace costly sensors with time and
frequency, both of which are relatively cheap. Only computer
simulations of this technique have been reported so far.

SUMMARY

This report has sought to present a brief overview of the
complex field of underwater acoustic imaging. There are many possible
techniques, technologies and strategies for designing and building
underwater acoustic imaging systems, and a few of them have been
tried by various groups at various times. In general, the resolution
performance of underwater acoustic imaging systems is lower than for
optical systems--when the water is clear. But since their range,
especially in turbid waters, is significantly greater than for
optical systems, there is probably a valuable role for them to play
in the underwater world. At the present time, however, there does not
appear to be the strong desire for such systems, and so there is not
as much driving interest and fiscal support to develop such systems
as there is in the medical and non-destructive testing areas.
Additionally, the severe environmental requirements that underwater
systems must meet makes development of any system more expensive.
Nevertheless, research in this area does continue at a slow pace.
Improvements in techniques, in technologies, and in matching system
performance with human needs should lead to useful underwater
acoustic imaging systems in the future.

REFERENCES

1. B. D. Steinberg, "Principles of Aperture and Array System Design, Including Random And Adaptive Arrays," John Wiley & Sons, NY, (1976).

2. J. V. Thorn, N. O. Booth, J. C. Lockwood, Random and Partially Random Acoustic Arrays, submitted to Journal of Acoustic Society of America, November 1979.

3. R. A. Smith, Raman-Nath Imaging, in "Acoustical Holography," Vol. 6, ed. by N. O. Booth, Plenum Press, New York, (1975), pp. 363-383.

4. J. F. Havlice, G. S. Kino, J. S. Kofol, and C. F. Quate, An Electonically Focused Acoustic Imaging Device, "Acoustical Holography," Vol. 5, ed. by P. S. Green, Plenum Press, New York, (1974), pp. 317-334.

5. J. W. Young, Electronically Scanned and Focused Receiving Array, in "Acoustical Holography," Vol. 7, ed. by L. W. Kessler, Plenum Press, New York, (1977), pp.387-403.

6. A. F. Metherell, H. M. A. El-Sum, L. Larmore, "Acoustical Holography," Vol. 1, Plenum Press, New York, (1969).

7. J. P. Powers, LCDR D. E. Mueller, A Computerized Acoustic Imaging Technique Incorporating Automatic Object Recognition, in "Acoustical Holography," Vol. 5, ed. by P. S. Green, Plenum Press, New York, (1974), pp. 527-539.

8. J. V. Thorn, Gain and Phase Variations in Holographic Acoustic Imaging Systems, in "Acoustical Holography," Vol. 4, ed. by G. Wade, Plenum Press, New York, (1972), pp. 569-581.

9. J. L. Sutton, J. V. Thorn, J. N. Price, The Effect of Circuit Parameters on Image Quality in a Holographic Acoustic Imaging System, in "Acoustical Holography," Vol. 5, ed. by P. S. Green, Plenum Press, New York, (1974), pp. 573-590.

10. R. T. Clayden, P. H. Brown, A Simple, High Definition Ultrasonic Imaging System for the Location and Inspection of Submerged Objects in Turbid Conditions, in "Proceedings of the Oceanology International '78, Brighton" Conference, March 1978.

11. H. Harmuth, Generation of Images by Means of Two Dimensional, Spatial Electric Filters, Advances In Electronics and Electron Physics, Vol.41, Academic Press, New York, (1976), pp.167-248.

12. J. V. Thorn, N. O. Booth, J. L. Sutton, B. A. Saltzer, Test and Evalutation of an Experimental Holographic Acoustic Imaging System, Naval Undersea Center Technical Publication 398, Nov. 1974.

13. J. L. Sutton, J. V. Thorn, N. O. Booth, B. A. Saltzer, Description of a Navy Holographic Underwater Acoustic Imaging System, in Acoustical Holography, Vol. 8., ed. by A. F. Metherell, Plenum Press, New York, yet to be published.

14. K. Nitadori, An Experimental Underwater Acoustic Imaging System Using Multi-beam Scanning, in Acoustical Holography, Vol. 8, op. cit.

15. P. N. Keating, R. F. Koppelman, Holographic Aperture Synthesis via a Transmitting Array, in "Acoustical Holography," Vol. 6, op. cit. pp. 485-506.

16. R. F. Koppelmann, P. N. Keating, Three-Dimensional Acoustic Imaging, in "Acoustical Holography," Vol. 8, op. cit.

17. N. H. Farhat, T. Dzekov, E. Ledet, Computer Simulation of Frequency Swept Imaging, in "Proceedings of the IEEE," September 1976, pp. 1453-1454.

18. N. H. Farhat, M. S. Chang, J. D. Blackwell, D. C. K. Chan, Frequency Swept Imaging of a Strip, in "1976 Ultrasonics Symposium Proceedings," IEEE cat   76 CH1120-5SU.

19. D. C. K. Chan, N. H. Farhat, M. S. Chang, J. D. Blackwell, New Results in Computer Simulated Frequency Swept Imaging, in "Proceedings of the IEEE," August 1977, pp. 1214-1215.

DIGITAL RECONSTRUCTION OF ACOUSTIC HOLOGRAMS IN

THE SPACE DOMAIN WITH A VECTOR SPACE APPROXIMATION

Hua Lee, Carl Schueler, Glen Wade, and Jorge Fontana

Department of Electrical and Computer Engineering
University of California
Santa Barbara, California 93106

ABSTRACT

We have developed a computer-assisted ultrasonic underwater
imaging system. Until recently, we used back-propagation to reconstruct images. This method, accurate both in the near and far
fields of the object, is an inverse filtering technique that operates in the spatial frequency domain and requires the taking of two
Discrete Fourier transforms (DFT's).

Here, we present a method based on back-projection, an alternative reconstruction technique which operates in the space domain
and reconstructs images without any DFT's. Back-projection is computationally as fast as a single DFT, and it is accurate in both
the near and far fields. This new algorithm works because the spatial propagation of a point-source is known, and any object is a
linear combination of point-sources. We, therefore, can construct
a spatial propagation matrix which maps the object-wave field into
a field at the receiver plane. To use the received data to obtain
the object wave field requires the inverse matrix. The back-
projection algorithm applies an approximation technique from vector-
space algebra to estimate the inverse matrix and reconstruct an
acceptable image of an unknown object.

We not only present computer simulations which demonstrate the
feasibility of back-projection, but we also use a modified version
of the above algorithm to correct simulated data that has been degraded by motion of the object during data acquisition. Results
from simulated data are compared with those from experimental data.

I. INTRODUCTION

This paper describes back-projection, a space-domain method
of reconstructing scanned acoustic holograms which yields the same
image resolution as is obtained using frequency-domain processing.
Back-projection offers the advantage that reconstruction can begin
as soon as the first data element is received. This enhances the
possibility of dynamically programming a scanned hologram reconstruction for real-time imaging.

II. THEORY

A. Back-propagation

We review back-propagation briefly for comparison with the
new imaging technique. Our theoretical development is generally
applicable, but we use a specific experimental situation to compare
the performance of back-projection and back-propagation. Our experimental acoustic imaging system consists of a 3300 liter water
tank and a 40 cm. linear array of 100 acoustic sensors (Fig. 1).
We insonify an object under water with 0.4 msec tone bursts of 250
kHz sound at the rate of 50 bursts each second. The reflected a-
coustic field just below the water surface is measured by sweeping
the sensor array perpendicular to its length along the tank to
synthesize a rectangular aperture. For simplicity, we limit our
computations to the case where a one-dimensional object is to be
imaged. The receiving electronics sequentially scans the 100 sensor outputs at a rate of 50 per second and range gates the received
signal at each sensor to eliminate echoes from the tank walls.
From the data, which is stored on disk in a PDP-11/45 computer, we
can apply reconstruction algorithms to obtain an image [1].

The back-propagation algorithm is an inverse filtering scheme
that can be used when the receivers are in the near or the far
field of the object to be imaged [2]. A digital Discrete Fourier
transform (DFT) is applied to the received data, to get into the
spatial frequency domain. Then an inverse filter is applied to
backward propagate each plane wave by the correct phase factor to
obtain the phase distribution of spatial frequencies at the object
plane. The image is finally obtained by an inverse DFT. The
drawback is that all the data must be obtained before processing
can begin.

B. Back-projection

This new technique for reconstructing acoustic holograms is
based on the assumption that the holographic data at the receivers
results from a linear spatial operation on the object function to
be imaged. With this assumption, we start with the following
expression.

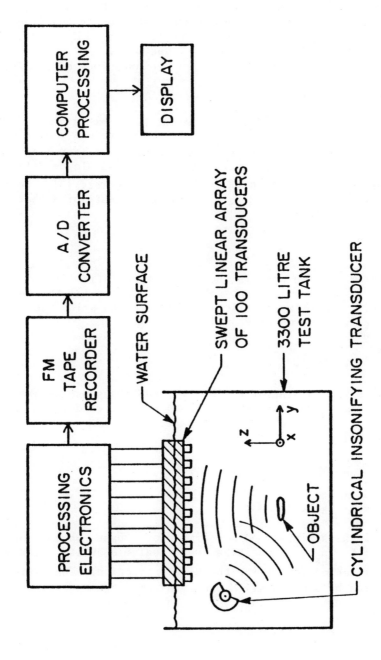

Figure 1. Ultrasonic underwater imaging system with computer image reconstruction

$$Y = PX \tag{1}$$

(1) represents a linear input-output system. P is a spatial linear operator that transforms an unknown spatial vector function X into a measured spatial data vector Y. We let P be an $M \times N$ matrix relating the $N \times 1$ vector X to the $M \times 1$ vector Y.

Eq. (1) can be broken down into M expressions of the form:

$$y_k = P_k X \tag{2}$$

where $k = 1, \ldots, M$. X continues to be the $N \times 1$ vector in (1), and y_k is the k-th projection of (1), and is therefore one scalar element of Y. Since P_k is a row vector, it has rank one (assuming that P_k is not the zero vector). Therefore, it is possible to define a pseudo-inverse of P_k, even if P_k is not invertible [3]. We define the pseudo-inverse by finding a suitable estimate of X that we call $\hat{x}_k$ which is given by:

$$\hat{x}_k = P_k^\dagger y_k . \tag{3}$$

The vector $\hat{x}_k$ that we want is the one that satisfies (2), and has minimum norm among all the possible solutions to (2) that we might select [4]. These two restrictions allow us to write the pseudo-inverse, $P_k^\dagger$, in terms of P_k [5].

$$P_k^\dagger = P_k^* [P_k P_k^*]^{-1} \tag{4}$$

where P^* is the adjoint of P, that is, the transpose of the complex conjugate of P [6]. To simplify the notation, we define

$$n_k = 1/[P_k P_k^*] \tag{5}$$

so that we have

$$\hat{x}_k = P_k^* n_k y_k . \tag{6}$$

Eq. (6) yields a minimum norm estimate of X based on only one data element of Y. n_k is a positive, real scalar quantity which we call the normalization factor for the k-th projection. The estimate of X is completed by summing all M back-projected estimates:

$$\hat{X} = \sum_{k=1}^{M} \hat{x}_k$$

$$= P^* N Y \tag{7}$$

where N is a diagonal $M \times M$ matrix with n_k's along the diagonal.

$P^* N$ is the back-projection operator that carries the measured data Y back to the image $\hat{X}$.

III. RECONSTRUCTION OF DIGITAL HOLOGRAMS

A. Stationary object

To reconstruct an acoustic hologram with back-projection, we identify (1) for the acoustic propagation case, and then form the pseudo-inverse operator. The back-projection algorithm operates as fast as a single DFT to obtain the image.

The approximate Rayleigh-Sommerfeld integral [7], in the simple case of a small aperture and distant acoustic source, is

$$y(u_2) = \frac{1}{j\lambda} \int X(u_1) \frac{e^{j\frac{2\pi}{\lambda}r_{12}}}{r_{12}} \, du_1 \tag{8}$$

where λ is the wavelength of sound , $j = \sqrt{-1}$,

u_1 represents the source distribution coordinates.

u_2 represents the received distribution coordinates.

r_{12} is the distance from a point in u_1 to a point in u_2.

In discrete form, (8) becomes:

$$
\begin{aligned}
y_k &= P_k X \\
&= \sum_{\ell=1}^{N} \frac{e^{j\frac{2\pi}{\lambda}r_{k\ell}}}{j \, r_{k\ell}} \, x_\ell
\end{aligned}
\tag{9}
$$

The k-th estimate of X for the case of reconstruction of data from our linear acoustic array is given by (6), with y_k given by (9). The ℓ-th element of $P_k^{\dagger}$ in (6) is:

$$P_{k\ell}^{\dagger} = j \frac{n_k}{\lambda r_{k\ell}} e^{-j\frac{2\pi}{\lambda}r_{k\ell}}$$

where $n_k = \lambda^2 / \sum_{i=1}^{N} (1/r_{k_i}^2)$.

The k-th estimate, $\hat{x}_k$, may not yield a good estimate of X. Upon combining all M $\hat{x}_k$'s, we get an estimate of X which is almost identical to the back-propagated image. Figure 2 illustrates the estimate $\hat{x}_{50}$, from the 50-th data element, y_{50}, of a 100 element data vector Y. The data was obtained by computing the response of the 100 element acoustic receiving array to a point source 54 cm. below the center of the array. Figure 2 does not represent a good estimate of a point image, and each data element of Y will yield a similar result. However, if we combine all 100 such estimates according to (7), phase cancellation produces Figure 3. For comparison, Figure 4 is the result of applying the back-propagation algorithm to the same data.

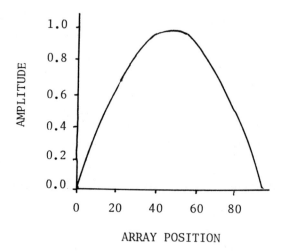

ARRAY POSITION

Fig. 2. Back-projection estimate for simulated point source data. Result of estimate from 50th element of received data vector of length 100.

Although the results of applying the back-projection (Fig. 3) and back-propagation (Fig. 4) algorithms to simulated point source data appear to yield almost identical results, the results are not exactly alike. This can be seen more easily from Figures 5 and 6. Fig. 5 resulted from reconstruction of experimental data using the back-propagation algorithm, and Fig. 6 from reconstruction of the same data, but with back-projection. The width of the point image peak is approximately the same in both figures, but the details of the side-lobes betray the fact that different reconstruction methods were used.

Figures 7 (back-projection) and 8 (back-propagation) each show the image of two point sources in close proximity, and both show approximately the same resolution. Although the back-propagation algorithm yields resolution that depends on the wavelength and aperture size, the back-projection algorithm can yield better

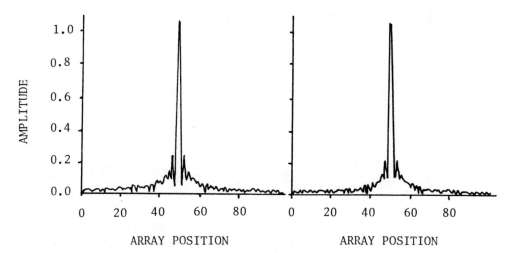

Fig. 3.   Back-projection esti-
          mate for simulated
          point source data, us-
          ing all 100 element of
          the data.

Fig. 4.   Back-propagation image
          of point source using
          the same data as that
          for Fig. 3.

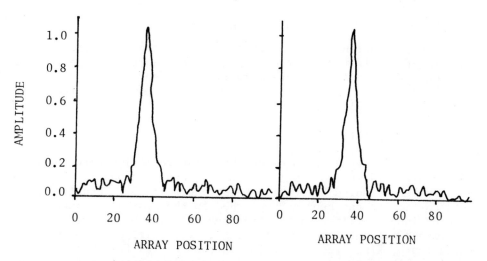

Fig. 5.   Back-propagation of ex-
          perimental data from a
          point-source located 54
          cm. below the center of
          the 100 element acous-
          tic-imaging system in
          the water tank (Fig. 1).

Fig. 6.   Back-projection esti-
          mate of point-source
          image using the same
          data as for Fig. 5.

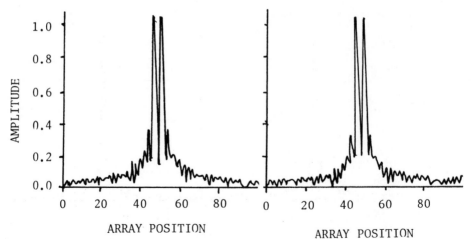

Fig. 7.   Back-projection esti-
          mate of image of two
          point-sources using
          simulated data.   The
          sources are under sen-
          sors 47 and 51 of the
          acoustic array, and are
          1.6 cm. apart.   The
          sound wavelength is 0.6cm.

Fig. 8.   Back-propagation of the
          same data used to ob-
          tain the result shown
          in Fig. 7.

or worse resolution depending on the correlation among the row
vectors of P. If the row vectors of P are uncorrelated (orthogonal)
then we will obtain an exact image in the absence of noise. The
more correlation that exists among the row vectors of P, the worse
the resolution will become.

B. Motion Correction

     The image degradation caused by relative motion between source
and receiver is an important problem in scanned acoustic holography,
and has been studied by several researchers [8-10]. Usually, motion correction has been accomplished in the spatial frequency domain by inverse filtering the transfer function of the motion. For
non-linear motion, inverse filtering is very difficult to apply.
Furthermore, since reconstruction in the spatial frequency domain
requires all of the data to be obtained prior to the start of processing, digital reconstruction when motion is present will not
attain real-time speed unless a more efficient technique is found.

     Back-projection offers the advantage of being able to start
reconstruction as soon as the first piece of data is obtained.
Furthermore, no matter how complicated the relative motion between
the object and the receivers may be, only the motion that occurs
during the transit time of the acoustic signal from the object to

each receiver element has to be accounted for to provide motion correction for that signal. Therefore, a piece-wise uniform motion approximation may be used even in the non-linear case, which should simplify the computations.

To implement motion correction with the back-projection algorithm, note that motion transforms the distance r from the source to the receiver, into a function of time. If $r = [(y - y_0)^2 + (z - z_0)^2]^{1/2}$, (Fig. 1 coordinates) and the source moves in the y direction with uniform speed v, then we can write:

$$r(t) = [(y-y_0 + v[t-t_0])^2 + (z-z_0)^2]^{1/2}$$

where t_0 = time of start of array scanning, and t = time at which the specific receiver of interest is interrogated. We then replace r by $r(t)$ in $P_k^\dagger$, and apply the back-projection algorithm without further modification.

Figure 9 shows the image of a simulated moving point source estimated by back-projection without motion correction. Figure 10 shows the result of the application of the algorithm after modification by the procedure outlined in the previous paragraph. Figures 11 and 12 are similar, except that the motion degradation is more severe. The simulated point source moved half the aperture length in the time the receivers were scanned. In our water tank,

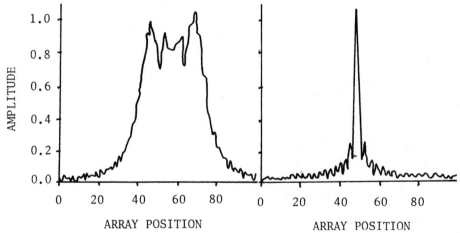

Fig. 9
Back-projection of simulated data
for a point-source moving uniformly
in the 7-direction of Fig. 1. The
speed is such that the source moved
a fifth of the length of the array
in the time it takes to scan the
array once.

Fig. 10
Back-projection of the
same data used to pro-
duce Fig. 9, but with
motion correction.

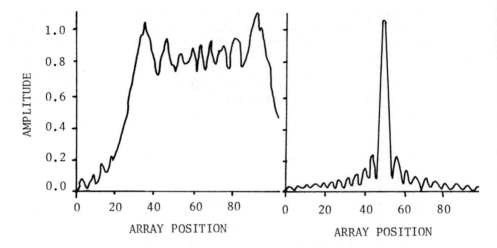

Fig. 11. Back-projection of simulated data for a point-
source moving in the y-
direction of Fig. 1.
Here, the speed is such
that the source moved
half the length of the
array in the time it
takes to scan the array
once.

Fig. 12. Back-projection of
the same data used to
obtain Fig. 11, but
with motion correc-
tion.

this would correspond to a speed of 10 cm./sec. This is a motion
of approximately 16 sound wavelengths per scan, and more than an
order of magnitude greater than it was recently thought would ir-
revocably ruin the image obtained by an acoustic holographic system
[8].

The procedure for correcting non-linear motion would be conceptually indistinguishable from that used to create Fig. 10 and
12. Since the correction is done simultaneously with the imaging,
and can be done as the data is taken, real-time correction and
imaging appears to be a possibility.

IV. SUMMARY

A. The back-projection method yields images with comparable
resolution to back-propagated data. However, back-projection only
approximates the object to be imaged using the pseudo-inverse
method. Where the resolution of the back-propagation method depends on the spatial frequency content of the received data, the
resolution of the back-projection method depends on the correlation among the rows of the propagation matrix P. P* N P is a

square matrix with unity diagonal elements, but is not necessarily the identity matrix. Only when the row vectors in P are orthogonal will an exact image be obtained.

B. The main advantage of the back-projection method lies in the independence of the data set. Each data element can produce an estimate independently of the others. It is very convenient not to have to wait to obtain the entire data set to start processing. Therefore, the back-projection method may allow scanned acoustic holographic imaging in real-time.

C. In many cases of interest, we may find that the round-trip operator, $P^* N P$, is a positive semidefinite matrix. Therefore, back-projection may be modified to produce a recursive filtering algorithm which will yield improved resolution over back-propagation [11].

ACKNOWLEDGEMENTS

The authors wish to express their appreciation to the Computer Systems Laboratory at the University of California at Santa Barbara for providing the computer facilities to perform the work reported. We also thank Teddi Potter of CSL for preparing the manuscript.

V. REFERENCES

[1]   C. F. Schueler, J. Fontana, and G. Wade, "Ultra-sonic Underwater Imaging System with Computer Image Processing and Reconstruction", Proceedings of Ultrasonics International '79, Graz, Austria, 15-17 May, 1979.

[2]   J. L. Sutton, "Underwater Acoustic Imaging", Proc. of the IEEE, Vol. 67, No. 4 (1979), pp. 554-566.

[3]   D. G. Luenberger, Optimization by Vector Space Methods, John Wiley & Sons (1969), p. 163.

[4] Ibid., pp. 161.

[5] Ibid., pp. 165.

[6] Ibid., pp. 150.

[7] J. W. Goodman, Introduction to Fourier Optics, McGraw-Hill, (1968), pp. 44.

[8] H. D. Collins and B. P. Hildebrand, "The Effects of Scanning Position and Motion Errors on Hologram Recording", Acoustical Holography, Vol. 4, Plenum Press, 1972, pp. 467-501.

[9] L. Schlussler, A. E. Coello-Vera, G. Wade, and J. Fontana, "Motion Limitations of an Acoustic Holographic System Utilizing a Scanned Linear Array", Optical Engineering, Vol. 16, No. 5, Sept.-Oct. 1977, pp. 426-431.

[10] N. B. Tse, L. Schlussler, J. Fontana, and G. Wade, "Computer-Corrected Reconstruction of Acoustic Holograms of Non-uniformly Moving Objects", 1977 Ultrasonics Symposium Proceedings.

[11] D. G. Luenberger, Introduction to Linear and Non-linear Programming, Addison-Wesley, 1973, pp. 148.

DATA ACQUISITION SYSTEM FOR COMPUTER

AIDED COHERENT ACOUSTIC IMAGING

John P. Powers, Major Reid Carlock,
USMC and LT Rod Colton, USN

Department of Electrical Engineering
Naval Postgraduate School
Monterey, California 93940

ABSTRACT

A coherent acoustical imaging is under design and test. This
system utilizes precisely spaced samples of the amplitude and
phase of an ultrasonic field as data input for the computer processing and display sub-system. Two designs of the data acquisition system are described: one a raster scan and the other a
single sweep of a linear array. Design goals are to achieve a
256 x 256 array of amplitude and phase data recorded on a digital
cassette recorder. Both techniques use microprocessor control of
the data flow and sampling. The relatively slow motion of the
raster scan presents few problems in achieving results and control
was easily implemented with an 8748 microprocessor. The speed
requirement in sampling, measuring and converting the data in the
linear array case however present a more difficult challenge requiring a more complex microprocessor system described in detail
in the paper.

INTRODUCTION

The imaging system[1] under study consists of a computer that
can perform backward wave propagation and other image processing
operations on coherent ultrasonic data. The computer generated
image is displayed on graphical output devices in an interactive
fashion. The data acquisition portion of the system, the subject
of this paper, needs to provide a square array of data samples
that measure the amplitude and phase of the ultrasonic field in
question. Data field sizes are operator controlled with repre-

643

sentative sizes being 64 x 64, 128 x 128, and 256 x 256. Sample spacing is controlled by a precision scanner with a usual sample spacing of .75 mm (one-half wavelength at the 1 MHz operating frequency). Scan geometries, speed and automatic operation are controlled by limit switches and a hardware controller. One option for the future is to bring this portion of the system also under microprocessor control.

Currently two modes of scan geometry are under active investigation. The first is a comparatively slow raster scan over the data field by a single point receiver. This scan is useful with static objects for the production of data to be used in the study of processing options and interactive techniques on the computer side of the system. The second scan is a single sweep of a large linear array with high speed sampling and multiplexing of the data columns producing the square data array. Other scan geometries are also easily implemented in this system; in fact, one of the primary advantages of a computer based system is the flexiblility of the technique to investigating novel geometries and/or processing options.

Microprocessors help to achieve additional flexibility in the data acquisition system. In the present system the devices control the data flow through the digitizers, synchronize the electronic hardware and control the recording devices. The addition of moderate memory capacity allows for full buffering of the data before recording and/or transmission to the computer. While current microcomputer systems do not the have speed or data handling capability required for a true stand-alone imaging system, rapid rate of improvement and predicted performance levels holds the promise of small inexpensive imaging systems based on microcomputers vs the present-day minicomputer systems.

SYSTEM FOR RASTER SCAN

The goals of the system[2] designed to work with the slow speed raster scan include control of data flow (Fig. 1) from the A/D converters to the recording devices, data conversion, between binary, BCD and ASCII formats, and interfacing to the various peripheral devices. To illustrate the requirements the following features of various peripheral devices are listed:

Line printer--This device has an 'RS-232-like' interface (hence allowing the interface to also be used for CRT or video displays) that requires ASCII characters to be transmitted at a 2400 baud rate with a proper parity bit and two stop bits inserted after the seven data bits.

Teletype terminal--This device requires a 25 ma current loop
drive with ASCII data at a 110 baud rate. Again proper parity
bits must be inserted into the data stream. (The Teletype
paper tape capability has also provided a backup recording
capability to the cassette recorder.)

Digital cassette recorder--This device, providing 2.2 megabits
of storage requires straight binary data at CMOS voltage
levels at a rate of 23 bits/sec. Start/stop signals and
read/write controls must also be provided since the recorder
operates only when data is present.

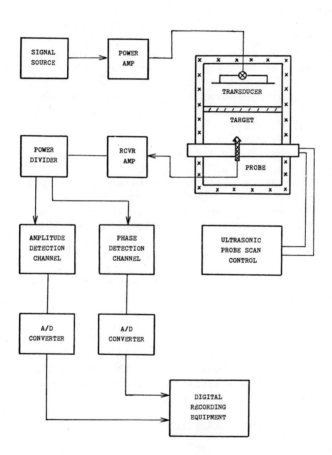

Figure 1. System block diagram

The system implementation was based on an Intel 8748 micro-computer with various output drivers, an I/O address decoder and status displays. The system was programmed and debugged using an Intel PROMPT 48 development system in an emulate mode to simulate the microprocessor with the actual I/O hardware connected to the development system extender card. The software structure, chosen for enhanced flexibility, uses an executive program under operator control through a touch pad keyboard to call the appropriate sub-routines. The subroutines are performed as triggered by the sample signal generated by the raster scan equipment.

As an example, consider the case where one wishes to write the data on the Teletype unit (and punch a paper tape). The operator presses the keys for the Teletype program and begins the scan process. The analog values of the amplitude and phase are sensed continuously and fed to the A/D converters. At the sample signal the A/D converters sample and hold the values and perform the A/D conversion. The microprocessor is also triggered by the sample pulse and waits for the end-of-conversion signal from the A/D units (a nominal $4\,\mu$ sec.) When the pulse is received, the micro-processor reads the data from the amplitude A/D unit on its I/O ports, converts the binary data to BCD for internal manipulation, converts the BCD data to ASCII, adds a start baud, two stop bauds and a parity baud as required by the Teletype, and presents the data stream to the proper I/O port by use for the current loop driven to the Teletype. The process is then repeated for the phase data before the next sample. Sampling time is limited to 1 sample/sec by the response time of the Teletype (especially the carriage return and line feed mechanics).

Software reprogramming and additions are easily accomodated. In fact the Teletype unit was added to the system at a late date due to a failure of the cassette recorder mechanism. The micro-processor was easily reprogrammed, debugged and the unit was quickly added to the system once the TTL to 25 ma current loop adapter was designed and implemented. Additional testing and calibration options have been included throughout the software to test the various hardware elements and data manipulations in the system.

SINGLE SCAN LINEAR ARRAY SAMPLING[3]

A single lateral scan of a linear array can produce the data matrix in time of a single row scan. The requirement for a rapid-fire sampling of the column element requires comparatively fast processing electronics. If a 256 element is sampled every 500 milliseconds, for example, the sample interval is approximately 2

milliseconds. Faster sweep times reduce the sampling time accordingly. While these rates are not fast electronically much of the
phase and amplitude sensing as well as the microprocessor control
had to be redesigned to accommodate this speed. The sampling rate
is also much higher than the recording rate or transmission rate
of the peripherals necessitating the requirement for an electronic
memory to buffer the data until ready for recording. Since 256 x
256 x 8 bit memories are needed for both the amplitude and phase,
two 64K byte memories are required. Not available on the microprocessor chip with only a 1K RAM, a memory expansion capability
was one system requirement. It should also be noted that only
dummy data was tested since the large linear array will not be
built until the raster scan technique has been thoroughly tested.

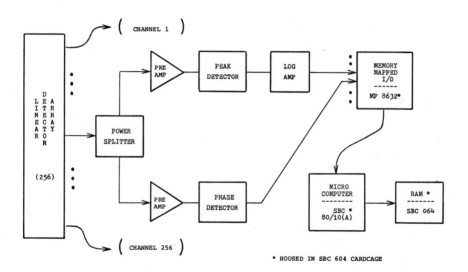

Figure 2.   Block diagram of multichannel detector system

Two alternative systems were considered. The first (Fig. 2)
required multiple amplitude and phase detectors whose output was
multiplexed and read into memory by the microprocessor system.
The second (Fig. 3) multiplexed the samples of the 1MHz signal
onto a single channel which measured the amplitude and phase of
the successive samples and read these into the microprocessor I/0
ports for digitizing a memory assignment. A trade off study of
the electronic costs show a 12:1 advantage for the latter system
with a cost of only $7,000.

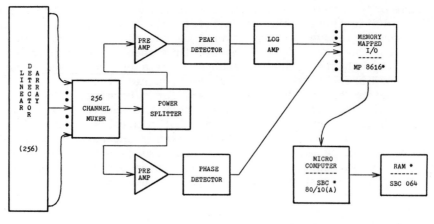

Figure 3.   Block diagram of multiplexed single detector system

        As shown in Fig. 2 the primary elements of this system are the
multiplexers, phase detector, amplitude detector and
microprocessor system.   Each of these will be described in some
detail.

        The primary component of the multiplexer is a sixteen channel
CMOS switch.   As shown in Fig. 4, 16 devices are cascaded to provide full coverage.   An 8 bit counter sucessively enables each
device and switches in the channels through an on-chip decoder.   A
seventeenth device multiplexes the output of the sixteen others.
Primary considerations in checking the switching performance were
switching transition time and crosstalk effects on the amplitude
and phase transmission performance.   Switching transition times
were typically measured as 1 μsec presenting no problem.   Crosstalk effects are more complicated since both amplutude and phase
can be affected.   Changing adjacent channel amplitudes over a 60
dB range caused a worst case channel transmission change of
2.5 dB.   Phase distortion was minimized by the choice of channel
load levels.   A 100 Ω load produced the least phase distortion
(less than .3°) but requires buffer amplifiers to interface with
the A/D convertor.   The ON channel attenuation was also maximized
to -10 dB with this load but could be made up with the buffer amplifiers.   This ON channel attenuation was also found to be signal
dependent but a signal independent range (± 2 dB) was found to
exist for signal values between -40 dBV and 0 dBV.   Larger dynamic ranges are possible by including a nonlinear correction in
the computer processing but, lacking this refinement, the multi-

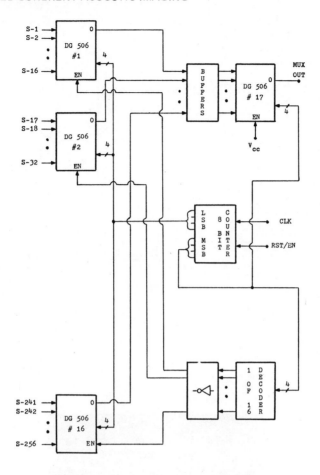

Figure 4.   Multiplexer subsystem

plexer restricts the system's dynamic range to 40 dB. (Other components have proven to have at least 50 dB dynamic range.)

The high speed phase detector circuit[4] shown in Fig. 5 uses
voltage comparators for limiting and digital logic for determining
phase. Figure 6 shows the response (lower trace) of the phase
detector circuit to the burst shown in the upper trace. A .2 msec
delay is encountered which must be accounted for in triggering the
A/D converters.

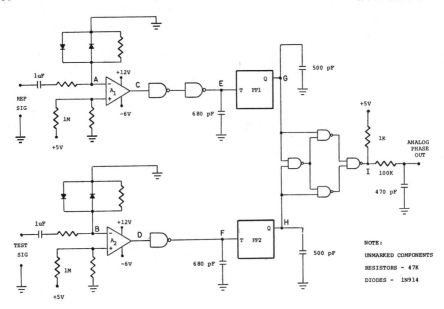

Figure 5.   Phase detector circuit

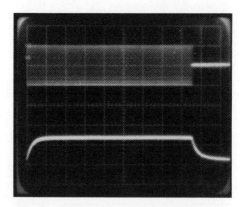

Horizontal Scale: 0.2 ms/div   Vertical Scale: 2 V/div

Figure 6.   Phase detector response. Upper trace: Signal burst.
            Lower trace: Phase detector transient response

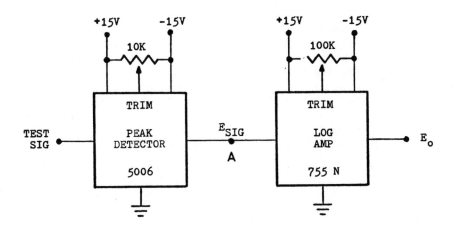

Figure 7.   Amplitude detector circuit

The amplitude detector circuit (Fig.7) is a commercial peak
detector (Optical Electronics 5006) driving a log amp providing
dynamic range compression. The peak detector can measure peak-
peak variations from − 33 dBV up to +17 dB within a 2% error.
Only several microseconds are needed for transition times--well
within requirements for the linear array system.

The microprocessor system is built around an Intel SBC
80/10(A) single board computer with a Burr-Brown 8616-AO analog
input-output board to provide the A/D conversion (within 44μ secs)
and memory loading as directed by the 80/10. The memory of the
80/10 is augmented by the SBC-064 RAM extension board with 64K
bytes of memory. The SBC 80/10 is capable of addressing two of
these memory boards over the Intel Multibus interface. A
Tektronix development system was used to program a power-up
sequence, a write operation where simulated data for a 128 x 128
array was sucessfully written into memory through the multiplexer
and I/O board, and a read routine where data can be transferred
from the electronic memory onto a digital cassette recorder.
Other options for reading the data are also easily programmable
(e.g., telephone modem, CRT display, line printer, etc.)

SUMMARY

The primary advantage of using microprocessor control of the data acquisition system is flexibility. Since the system is a testbed for research investigations in acoustic imaging, this flexibility is extremely important in being able to reconfigure the system to test new ideas. Similarly the capability of driving various peripheral devices with separate drive requrements is important. Both systems described in this paper have been designed and tested. The slower raster scan system is currently providing data for the study of various computer processing options in acoustic imaging. While our present research studies do not require the high speed data acquisition system its applicaiton is known and its feasibility has been demonstrated.

ACKNOWLEDGEMENTS

This research has been supported by the National Science Foundation under Grant No. ENG-77-21600.

REFERENCES

1. J.P. Powers, J.R.Y. De Blois, R.T. O'Bryon and J.W. Patton, "A Computer Aided Ultrasonic Imaging System" Acoustical Holography Vol. 8, A.F. Metherell, Ed, pp. 233-248, Plenum Press, New York, 1979.

2. R.A. Colton, "A Microprocessor-based Digital Data Acquisition Controller for a Computer Aided Acoustic Imaging System", Master's Thesis, Naval Postgraduate School, Monterey, California, 1979 (Available from Defense Documentation Center, Alexandria, VA 22314).

3. R.O. Carlock, "Analog and Digital Hardware Development for a Microcomputer Controlled Data Acquisition System for Acoustical Imaging" Master's Thesis, Naval Postgraduate School, Monterey, California, 1979. (Available from the Defence Documentation Center, Alexandria, VA 22314).

4. Electronic Circuit Designer's Casebook, p 38, Electronics, McGraw Hill.

SEISMIC IMAGING

Fred J. Hilterman

Seismic Acoustics Laboratory
University of Houston
Houston, Texas 77004

ABSTRACT

Seismic time sections ideally represent a vertical cross-section
of the earth. By using reflection amplitude and character on these
time sections, lateral events are picked from trace-to-trace and
these events are then interpreted as geologic boundaries between
formations which have different acoustic impedances.

In order to increase the signal-to-noise ratio on the time
sections and, at the same time, provide a velocity distribution profile of the earth, multi-fold data acquisition techniques are employed.
In these field techniques, all reflecting interfaces below a common
surface point are specularly illuminated, at least six times, by
varying the offset distance between the source and receiver. Multi-
fold data are acquired for those source-receiver pairs which are
centered around a common surface point.

Processing techniques, some standard to image processing, such
as deconvolution, stacking and gain correction, are then implemented.
However, because the source and receiver are omni-directional, energy
reflecting or diffracting from points not vertically beneath the
common surface point can obscure the geophysicist's interpretation.
These non-vertical features, as fortune will have it, are normally
the zones of economic value. The processing algorithms, which place
this non-vertically traveling energy in its true spatial position,
are called Migration Techniques. They are imaging processes, which
are anologues to a Kirchhoff or backward wave propagation reconstruction, that is a form of synthetic aperture imaging.

Because of the massive amounts of data involved, very efficient

algorithms have been developed for migration. Only minute differences
are evident between the time-domain (finite difference or backward
propagation summing) and frequency-domain methods.

Currently, seismic research in total of 3-D imaging is very
active in areas of data acquisition, processing, display and interpretation.

INTRODUCTION

The concepts of imaging or reconstruction are constantly evoked
by exploration geophysicists because without them the interpretation
of the geological structure from the seismic time sections would be
erroneous. If the reconstruction is not done on the computer, it is,
at best, mentally approximated by the geophysicist. In fact, his
view of an ideal or reconstructed seismic section guides the processing flow of the field data.

Whenever there is only gentle geological structure under the
seismic profile line, a geophysicist might finish his interpretation
without performing a computer migration (reconstruction) on the data.
In other words, the seismic data are judged to be close enough to
the image (geologic structure) so that computer migration would not
significantly enhance his interpretation. However, hydrocarbons
have a tendency to accumulate in geological structures which do not
resemble their counterpart on the raw seismic data. Therefore, let's
investigate how migration enhances the geophysicist's interpretation
of the seismic data by, first, reviewing what constitutes the ideal
reflection model.

THE IDEAL MODEL

In the simplest model of the reflection process, the seismic
wavelet, generated from a near surface energy source, propagates
downward in the earth and reflects from all formations which are
vertically beneath the common source-receiver position at the surface.
This model assumes that the source is a plane wave, there is no
transmission or attenuation loss and the geological formations are
horizontal. With these stipulations, the earth parameters that affect
the reflection of the seismic wavelet are the density and compressional wave velocity of each medium (Figure 1).

However, the seismic wavelet does not reflect due to an acoustic
impedance, but rather reflects when there is a difference in acoustic
impedances between two formations as is illustrated in Figure 2.
In this reflection model, a plane wave, with an amplitude of 1,
approaches the first acoustic impedance interface, a portion of the
downward wave reflects upward with an amplitude R_1, and this wave
returns to the surface with a two-way traveltime of T_1. The original
wave continues to travel downward, still with an amplitude of 1,

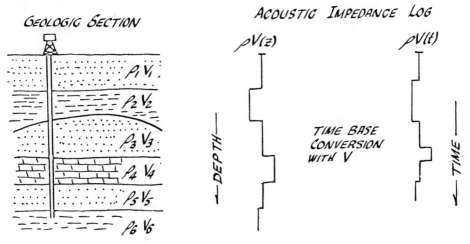

Fig. 1.   Earth parameters affecting seismic waves.

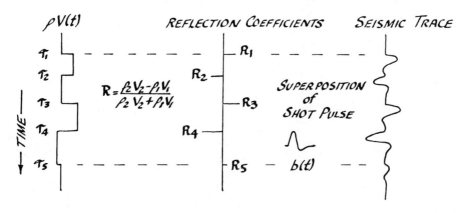

$$s(t) = R_1 b(t-T_1) + R_2 b(t-T_2) + R_3 b(t-T_3) + R_4 b(t-T_4) + R_5 b(t-T_5)$$

Fig. 2.   Acoustic impedance log to seismic trace.

until it reaches the second inferface and a reflection is again generated.   Besides not accounting for energy loss, an unrealistic
assumption in this model is the infinite bandwidth of the propagating
seismic wavelet.   It is physically impossible for an earth particle
to be displaced from its rest position and then return without any
oscillation around the equilibrium state.   Thus, the concept of the
seismic wavelet enters the reflection process.   Instead of recording
the reflection coefficient spikes as the seismic trace, a seismic
wavelet is reflected from each of the boundaries and the superposition
of the reflected wavelets is recorded as the seismic trace.

The simple mathematical description of the seismic trace in Figure 2 is for a discrete geological model and this needs to be converted to a continuous model. The equivalent of the reflection coefficient time series in the continuous case is called the reflectivity function f(t) and its relationship to the earth parameters is r(t) = ½dln(pv)/dt. Note that r(t) = (reflection ciefficient/time-sample interval). With the above definition of the reflectivity function, the seismic trace is described by the convolutional model r(t) = s(t) * b(t). This is the ideal seismic trace that governs the processing flow of the field data. If non-horizontal geological structures are evident on the seismic data, then a form of migration is necessary in the processing stage to reach the ideal seismic trace so that a correct interpretation of an area can be done. The field methods by which s(t) is obtained are described in the next section.

Data Acquisition

A schematic of the main categories in the data acquisition phase of seismic reflection exploration is illustrated in Figure 3.

The selection of field parameters in each of the five categories

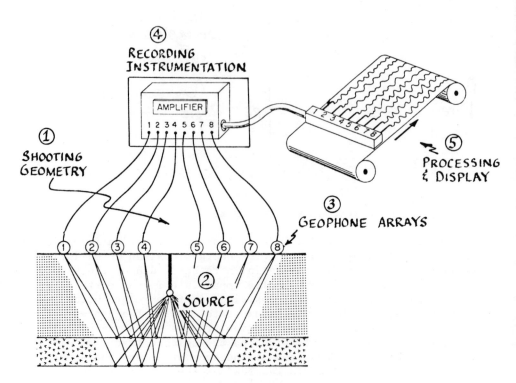

Fig. 3. Main categories of the seismic data acquisition system.

is heavily dependent upon, not only the geologic objective, but also the acoustic impedance constrast and structure of the geological section above the objective formation.

For land acquisition, both surface sources and dynamite are potential energy systems. A surface source, which can even be used in residential areas, is a truck-mounted vibrator, VIBROSEIS*, which sends out a 15-sec FM pilot signal. In order to compress the time duration of the propagating wavelet, the recorded signal is correlated against the original pilot to yield a zero-phase seismic wavelet. VIBROSEIS sources have a controllable frequency band which yields a fairly predictable seismic wavelet for further resolution enhancement.

The geophone arrays are designed to cancel coherent source noise which is trapped in the near surface layer. Normal dimension of the array are 12 geophones (or receivers) spaced 15-ft apart. If the near surface layers contain many scatters, areal arrays, instead of in-line linear arrays, will be employed to cancel the omni-directional traveling surface waves.

Most recording systems have the capability of digitizing at .001, .002 and .004sec sample intervals for 48 different channels. By using a floating-point amplifier and recording a mantissa and exponent, the true amplitude of the geophone signal is maintained. To appreciate the amount of data recorded, a conventional VIBROSEIS crew will record approximately .25 billion seismic amplitudes per day.

Depending upon the area being explored, the field data will be processed according to a specific flow which, hopefully, will lead to the simple convolutional model described above. The typical data-processing routines and their functions are outlined in Table 1. A description of several of these processes will be given later.

If the conditions exist in a particular area, such that the subsurface is relatively flat, the reflection coefficients on an average are small (<0.04), and the near-surface is not highly disturbed (such as it would be if there were glacial till), then single-fold continuous profiling could be an adequate shooting geometry. This is shown in the upper portion of Figure 4. Typical shotpoint spacing is 1/4 mile with 110-ft spacing between geophone arrays. The cable between shotpoint 1 and 2 connects the geophone arrays to the recording truck and is used by both shotpoints; thus, a 2-to-1 overlap of surface coverage to subsurface coverage results in a continuous subsurface profiling system.

In the bottom of Figure 4, the processing sequence called Normal Moveout Correction (NMO) is shown. Normal moveout corrections

_____

*Trademark of Conoco Oil Company

STEPS IN DATA PROCESSING
(modified from H.G. Beggs)
TABLE 1.

DATA REDUCTION

PROCESS	FUNCTION
Demultiplex	Put data in trace sequential formats.
Gain Recovery	Multiple data by floating-point exponent.
Editing	Remove bad data traces.
Summing	Attenuate source and random noise.
Crosscorrelation	VIBROSEIS source energy compression
Gain Function	Amplify deep events relative to shallow

GEOMETRIC CORRECTIONS

CDP Sort	Arrange traces according to common depth point.
Datum Static	Vertical time correction for elevation differences.
NMO Correction	Horizontal time correction

DATA ENHANCEMENT

Mute	Deaden high amplitude noise spikes.
Stack	Attenuate random and other types of coherent noise.
Band-Pass Filter	Attenuate noise having frequencies outside of signal band.
Notch Filter	Attenuate narrow frequency band noise within signal band.
Deconvolution	Compress input seismic wavelet, attenuate near surface ringing.
Trace Equalization	Amplify weak reflection events relative to strong events.

ADDITIONAL PROCESSING

Migration	Move events to true spatial position to shot and receiver stations.

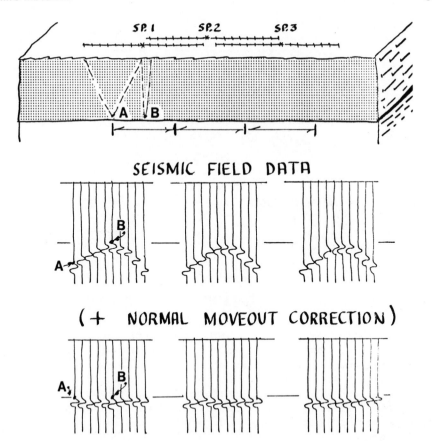

SEISMIC FIELD DATA

(+ NORMAL MOVEOUT CORRECTION)

Fig.4.   Single-fold split continuous shooting and subsequent NMO.

remove the extra traveltime to a particular reflection that the far
traces have over the near traces due to their oblique travelpath.
The NMO-connected trace represents the output from pseudo source-
and-receiver pairs that are at the same location, midway between the
actual source and receiver locations. These NMO-corrected traces,
therefore, simulate zero-offset source-receiver shooting, ZSR.

        In order to increase the S/N ratio and also to attenuate the
effects of certain internal bounces of energy within a section of
the earth, the multi-fold profiling system was developed.

        This system, shown in Figure 5, is essentially a method of
illuminating one reflection point many times by recording with various offsets between source and receiver. Of course, this point may
be at various depths, so in reality, we have a common depth line
being illuminated. For example, if four different offsets are used

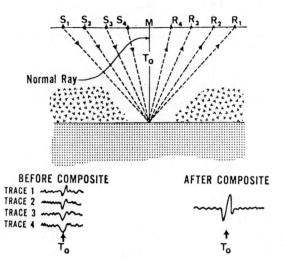

Fig. 5. Multi-fold seismic reflection profiling.

in the field for each common depth point, the coverage is called
"4-fold" CDP (Common Depth Point). The amount of traces involved
in the stack may be as high as 48 in some of the conventional shooting programs. Notice that a NMO correction was performed on the
four input traces in Figure 5 before they were composited. In complex geological areas, this NMO correction and stacking develops
problems because the geology and raypaths are 3-D, but the stacking
and NMO corrections are done for a 2-D environment.

The shooting geometry in Figure 5 is simulated in the data-
processing phase by sorting the field traces into a CDP format.
Normally, one shot is recorded into 96 geophone arrays rather than
the 1-shot 1-receiver arrangement as is illustrated.

Results of multi-fold vs. single-fold shooting are shown in
Figure 6. The improvement of the 12-told over the single-fold is
evident. These record sections are from South Mississippi. The
geology in the area places the geological tops of the Midway formation at 1.09 sec., the Selma Chalk at 1.25 sec. and the Smackover
at 3.20 sec.

It is interesting to note that the 12-fold stacked section required NMO corrections which in return required an estimate of the
velocity profiles as a function of depth for each output trace.
This velocity inversion problem is solved by assuming the raypaths
follow essentially straight line paths and the traveltimes, t_x,
are governed by the hyperbolic relationship of $t_x^2 = t_o^2 + x^2/v^2$.

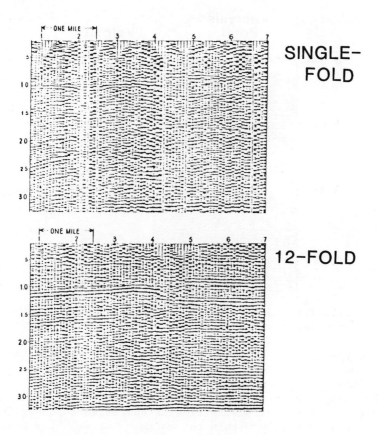

Fig. 6. Multi-fold vs. Single-fold (Modified from Courtier and
         Mendenhall.)

The traveltime, t_0, is the ZSR two-way traveltime to the same reflector that the trace, with a source-receiver offset of x, is
reflecting from at a two-way traveltime of t_x. The velocity, V,
is expressed as a function of t_0.

    Now, with the hyperbolic assumption, the CDP traces (as would
have been gathered in the shooting arrangement shown in Figure
5) can be searched for maximum reflection coherence between traces,
while the velocity is varied at a particular t_0. We might think
of this as a statistical velocity inversion.

    Figure 7 contains a velocity profile for the CDP location
shown on the surface of the time section. The velocity plot is
actually 3-dimensional where the coherence amplitude, represented
by the Graussian-shaped curves, is the 3rd out-of-plane dimension. By
picking the maximum coherence measurements and connecting the picks,
a velocity profile is created. The maximum coherence measurements

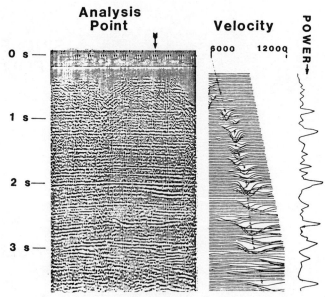

Fig. 7.   Velocity Analysis by Coherence Measurements.   (Modified
          from Taner and Koehler.)

should correlate timewise to reflection events beneath the analysis
point.

     Normally, the propagational velocity increases with depth, so
any decrease of velocity with time is suspected to be derived from
multiple bounce energy within the earth.   The velocity profile is a
good quality control check on the processing and field acquisition
techniques.   For example, abnormally long duration seismic wavelets
appear as duplicate picks on the velocity profile with the same
velocity as the primary portion of the seismic wavelet.

ZSR-CDP Images

     As indicated before, the stack section sometimes is a good
image of the geological object and no further processing is required.
A few geological examples and their corresponding seismic time sections will illustrate this concept.   In all of the models, it is
assumed that a ZSR data-acquisition system was used or the equivalent
data-processed section was derived.

     In Figure 8, the time section is in focus and a true interpre-
tation of the geological object will be obtained if the average
velocity is used to convert the traveltimes t_1, t_2 and t_3 to depth.

     Whenever the geology has dip, then a slightly out-of-focus
image is observed, as is shown in Figure 9.   Note that the specular
reflection from the $30°$ dipping plane is occurring on traces 30-34,

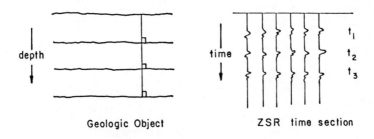

Fig. 8.   Horizontal Reflector.

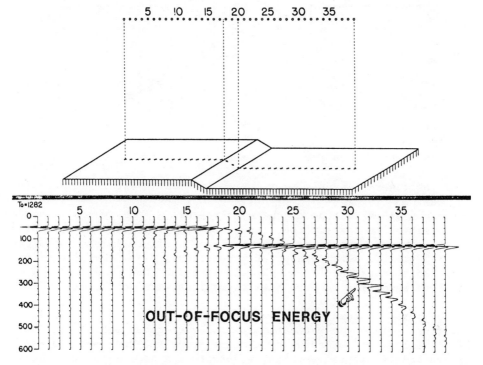

Fig. 9.   Reflection Profile from a Dipping Interface. (From
          Hilterman).

while the dipping plane is actually under the CDP locations 17-20.

While the time section in Figure 9 still relates,somewhat,
to the structure of the geology, the time sections taken over the
structures, shown in Figures 10 and 11, are slightly more complex
to interpret and could be misleading.

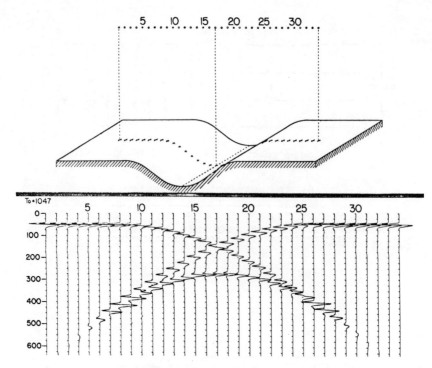

Fig. 10.   Reflection Profile from a syncline (From Hilterman.)

The coherent events on the time section in Figure 10 are often called the "buried focus" effect, which eminates from the fact that curvature at the bottom of the syncline has a focus which is buried with respect to the surface. Many oil drilling programs were based on the apparent anticline or hill shown in Figure 10 and, needless to say, yielded very disappointing results. Also, the size of the syncline appears smaller on the time section than it actually is. Negative anomalies tend to heal themselves with depth.

Figure 11 shows the seismic response due to an anticlinal structure. The over estimation of the geological anomaly is caused by the spherical spreading of the wavefront and the fact that the reflection is plotted vertically beneath the source-receiver location, rather than in its true spatial location.

What we would like to do is process Figures 9-11 so that they are in focus and the resulting image is a good representation of the object. Geophysicists call this process Migration. In essence, examining Figure 12, the energy segment on Trace 31 needs to be moved (migrated) to the position shown by the superimposed dotted estimate of the true object shape. There are several graphical and computer algorithms to perform this operation. The basis of

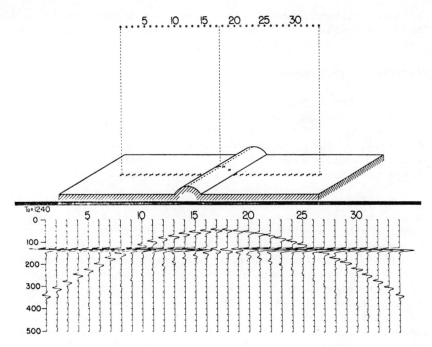

Fig. 11.   Reflection Profile from an Anticline. (From Hilterman.)

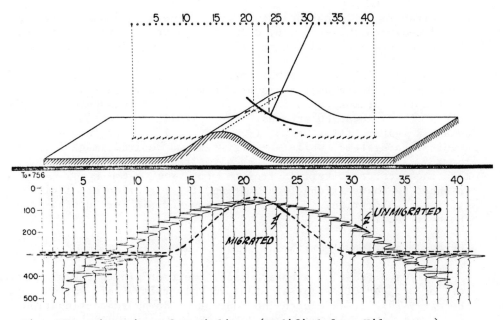

Fig. 12.   Migration of Anticline. (Modified from Hilterman.)

the three current wave equation migration techniques is called
"downward continuation", which is discussed in the following sec-
section.

Migration - By Downward Continuation

    In Figure 13, the most obvious observation is that the closer
the source-receiver pair is to the boundary, the more in-focus the
time section is. If the source-receiver location were just above
the boundary, then the seismic time section would be a duplicate
of the geological depth section. This presents a hypothetical,
data-acquisition process that is illustrated in Figure 14.

    Imagine that the earth could be cut vertically and gridded
on an even basis according to two-way traveltime. At each grid
location place a source-receiver pair, where the receiver records
only upcoming waves (a decomposition of the wave equation). Now,
fire all sources simultaneously and record only the reflection amplitude that arrives immediately after the shot. If this is done
and the source-receiver pairs are more densely spaced, the resulting amplitudes, plotted in the cells where they were gathered,
would be a sampled-approximation of the hatched-marked geology
shown in Figure 14. That is, it would be a time section consisting
of ideal seismic traces. If there is not a reflector immediately
beneath a given source-receiver pair, then a zero amplitude is
recorded.

    Although we do not have this type of data-acquisition system,
we can generate these data by doing an analytic continuation of a
wavefield or a downward continuation of the recorded surface data.

    The cube in Figure 15 has the axes of t, z and x. The z-axis
refers to the depth that the source-receiver pairs are buried, while
the t-axis contains the seismic amplitudes that the source-receiver
pair at (x,z) location records. The seismic time section recorded,
by a ZSR system, falls on the t-x plane at z = 0 (surface data-
acquisition system), while the migrated and desired image falls on
the z-x plane at t = 0. Thus, an algorithm is needed that will
move the recorded wavefield from the top face to the front face;
in short, we have Kirchhoff summation, f-k migration and finite-
difference migration. These will be covered, in more depth, in
the next section.

Current Migration Algorithms

    The wave equation algorithms owe their origin to the pioneering
work of Dr. Jon Claerbout at Stanford University (Ref. 1), who, in
the early 1970's, developed the concept of downward continuation
of seismic data. Although his work was initially done for 2-D

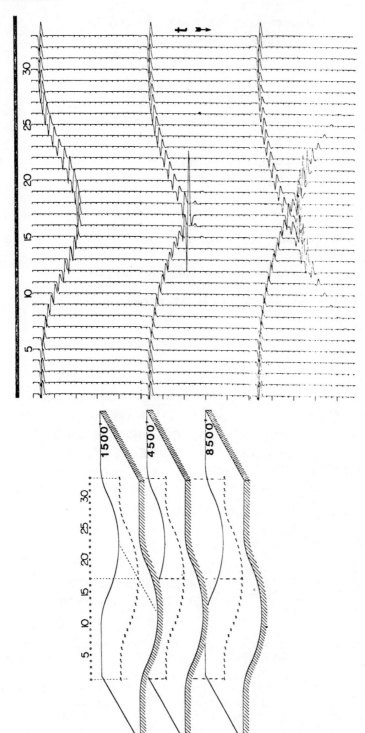

Fig. 13. Seismic sections over three synclines buried at a depth of 1500, 4500 and 8500 feet. (Modified from Hilterman.)

1.) PLACE SOURCE-RECEIVER PAIRS AT GRID POINTS IN VERTICAL SECTION OF EARTH

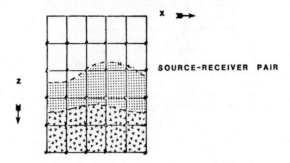

2.) FIRE ALL SOURCES SIMULTANEOUSLY

3.) RECORD UPGOING WAVES ONLY AND RETAIN FIRST TIME SAMPLE, t - O, AT EACH GRID

4.) RESULT IS DEPTH MIGRATED SECTION - FOCUSED IMAGE

Fig. 14.   Hypothetical data acquisition for focused images.

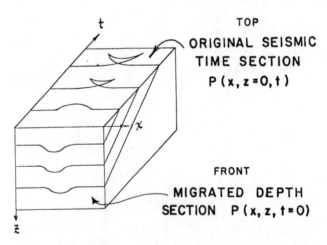

Fig. 15.   Relationship of original field $P(x, z = 0, t)$ to migrated (imaged) field $P(x, z, t = 0)$.

reconstruction, and this is all that we will show in this paper, the same philosophy has been extended to 3-D reconstruction for each of the wave equation approaches.

    The f-k migration algorithm, in Figure 16, starts with a 2-D
Fourier Transform with respect to x and t of the surface-recorded
data.   The solution of the subsequent differential equation reveals
the classical technique of the propagation of the angular spectrum
which is well known to the image processors.

1.) WAVE EQUATION

$$\frac{\partial^2 P}{\partial x^2} + \frac{\partial^2 P}{\partial z^2} = \frac{1}{V^2}\frac{\partial^2 P}{\partial t^2}$$

2.) 2D-FOURIER TRANSFORM WRT x & t

$$\frac{\partial^2}{\partial z^2}P(k_x, z, k_t) = -\left\{\left(\frac{k_t}{V}\right)^2 - k_x^2\right\}P(k_x, z, k_t)$$

3.) SOLUTION OF (2)

$$P(k_x, z, k_t) = P(k_x, 0, k_t)\, e^{\,iz\left\{(k_t/v)^2 - k_x^2\right\}^{1/2}}$$

PROPAGATION OF ANGULAR SPECTRUM

4.) INVERSE FOURIER TRANSFORM (3) , SET t=0

$$P(x, z, t=0) = \frac{1}{(2\pi)^2}\iint dk_x\, dk_t\; P(k_x, 0, k_t)\, e^{\,iz\sqrt{(k_t/v)^2 - k_x^2}}\, e^{\,ik_x x}$$

CHANGE IN VARIABLE EXPRESSES THIS IN EXACT 2D FOURIER TRANSFORM

Fig. 16.   Basic mathematical relationship of f-k migration.

By taking the 2-D inverse Fourier Transform of Eq 3 and setting

the time variable equal to zero before evaluation of the double integral, the desired wavefield on the left front of the cube in Figure
15 is obtained, as shown by the left-hand side of Eq 4 in Figure 16.
This is the migrated depth section. Stolt, 1978, has shown that Eq
4 can be expressed as an exact 2-D Fourier Transform by performing a
simple variable transformation, thus, allowing application of the
2-D FFT for economical processing.

The inverse application of the Kirchhoff integral equation was
developed by French, 1975, and the mathematical subleties, as shown
in Figure 17, were developed by Schneider, 1978. Basically, the
field recorded data is time-advanced rather than retarded to a subsurface location. In fact, the inner integral in Figure 17 does not
start until the advanced two-way traveltime, from the surface aperature point to the subsurface point being reconstructed, has been
reached. Then all amplitudes on the surface-recorded trace after
this time can contribute to the subsurface point of reconstruction.
However, the weighting factor in the integral contains a singular
point and the weighting factor must be bandlimited before application.

The final scheme is finite-difference migration, which is
shown in Figure 18. Rather than applying a finite-difference algorithm to the exact two-dimensional scalar wave equation, a more stable
differential equation is derived and subsequently finite-differenced.

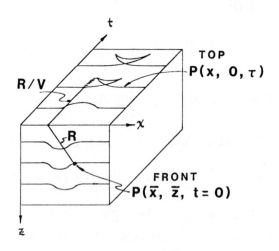

$$P(\bar{x},\bar{z},t=0) = -\frac{1}{\pi}\frac{\partial}{\partial z}\int dx \int_{R/v} d\tau\, \frac{P(x,0,\tau)}{\sqrt{\tau^2 - R^2/v^2}}$$

ig. 17.   Application of Kirchhoff integral for migration.

FINITE DIFFERENCE MIGRATION

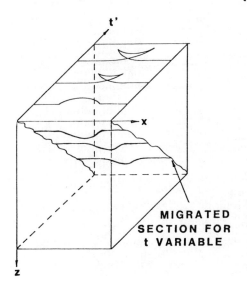

WAVE EQUATION :

$$\frac{\partial^2 P}{\partial x^2} + \frac{\partial^2 P}{\partial z^2} = \frac{1}{V(x,z)} \frac{\partial^2 P}{\partial t^2}$$

GENERAL SOLUTION :

$$e^{\,i\omega(t - x\sin\theta/V - z\cos\theta/V)}$$

APPROXIMATE SOLUTION (15°) :

$$e^{\,i\omega(t - x\theta/V - z/V + z\theta^2/2V)}$$

MOVING COORDINATE SYSTEM

$$t' = t - z/V$$

NEW WAVE EQUATION

$$\frac{\partial^2 P}{\partial z \partial t'} = \frac{V}{2} \frac{\partial^2 P}{\partial x^2}$$

MIGRATED
SECTION FOR
t VARIABLE

Fig. 18.   Finite-differencing the 15° approximation of the 2-D wave
           equation for migration.

The new wave equation is the basis for the 15° wave equation and
the more advanced methods, such as 45° wave equation migration allow
higher order terms in the approximation of cosθ and sinθ. One of
the huestic benefits of the moving coordinate system and the tri-
gometric approximations is that the wavefield from the top face does
not have to be moved diagonally across the cube and to the front face,
but has a more direct path to the diagonal plane. This reduces the
numerical dispersion inherent in differencing algorithms.

Migration Examples

    An example of a commercially available migration package is
illustrated in Figure 19. The geological interpretation of the time
sections is definitely easier on the migrated version. However, it

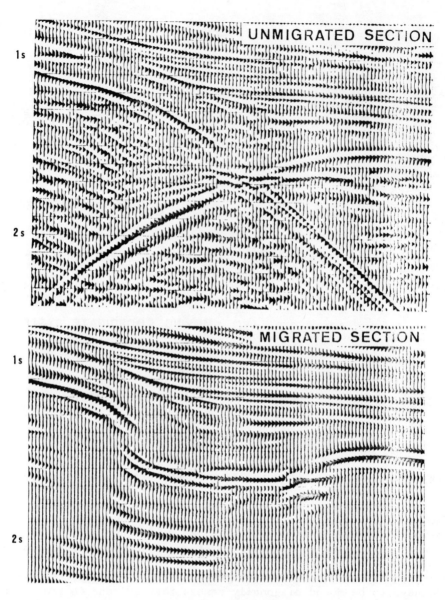

Fig. 19.   Commercial migration and raw seismic data.   Courtesy of
          CGG.

is difficult, sometimes, to evaluate a migration scheme when the
input is not exactly known.   Also, 3-D effects, such as reflection-
diffraction events which orginate from out-of-plane of the seismic
profile, might obscure the evaluation.   Because of this, several
physical models and their corresponding time sections were generated

at the University of Houston's Seismic Acoustics Laboratory (SAL) and were used for migration evaluation. These time sections and the designs will be discussed.

Figure 20 depicts a low velocity (3100 ft/s) RTV silicone model resting on a high velocity (9000 ft/sec) plexiglass base. The model was submerged in a water tank (water velocity is 5000 ft/s).

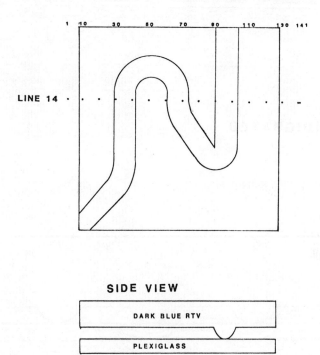

SIDE VIEW

Fig. 20.   Physical model used to generate synthetic seismic sections. (From Duffy.)

A constant-offset profile (simulates ZSR) was gathered and is shown in the upper section of Figure 21. The time-converted object is superimposed on the Figure for comparison. In the lower portion of Figure 21, is shown the migrated time section along with the super-imposed object.

At first, it appears that a perfect migration was not accom-plished because the three synclines on the lower section appear to go through the plexiglass surface. This is because no low velocity compensation was entered into the migration program to account for the lateral change in the structure. The problem that one continues to encounter is that you need to know the exact answer in order to obtain a perfect image. If the exact velocity profile were known

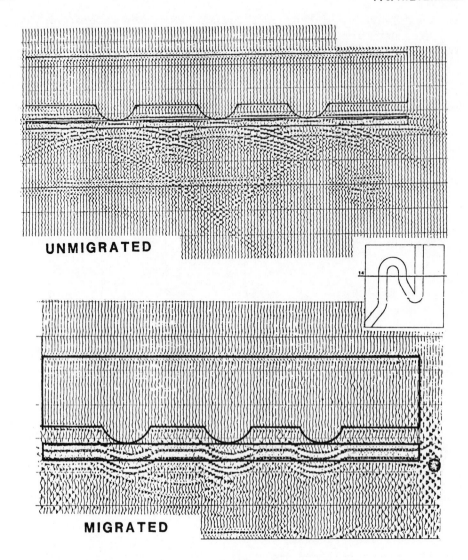

Fig. 21.   Seismic sections, raw (top) and migrated (bottom), over a
           low velocity structure.   (From Duffy.)

before hand, then a migration would not be necessary because the
answer is already known.   This is a practical problem often encountered.

In order to appreciate the problems encountered when a 3-D
object is seismically profiled and processed as a 2-D feature,
several seismic profile lines were run over an oblong basin.

The basin, which has two different radii curvature, both buried, is shown in Figure 22. Two seismic sections are presented in Figures 23 and 24, along with their corresponding 2-D migrated sections. Note that the data in Figure 23 is for a profile line that was gathered in a principal plane, while that in Figure 24 was a profile line gathered, both oblique and offset to a principal plane.

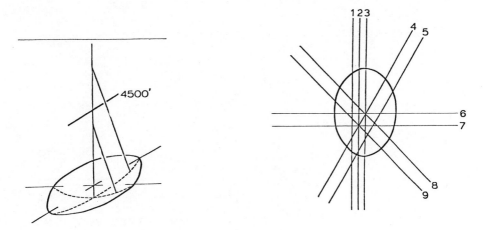

Fig. 22.   Geometry of oblong basin.

The best 2-D migration, across this basin, was obtained for the principal plane line which is shown in Figure 23. Even this was poor, as the ringing portion in the middle of the basin indicates. For lines which traversed the basin obliquely, an absolutely erroneous interpretation results as is shown in the lower portion of Figure 24. There appears to be a fault on the migrated section and an additional sediment infill in the basin. All this was caused by doing a 2-D migration on a 3-D prospect. A partial 3-D migration, in Figure 25, shows the improvement of 3-D over 2-D migration. Here the basin is interpretable.

Conclusions

The seismic reflection process by itself does an accurate job of focusing the geologic section as long as the horizons are flat. In fact, by using multi-fold coverage, a statistical velocity profile is generated.

However, for dipping or curved surfaces, the time sections are not a good estimate of the geologic structure and the focusing process of migration is necessary to place the reflection-diffraction energy back into its true spatial position. Without migration, anticlines appear too large on the time sections and synclines appear too small.

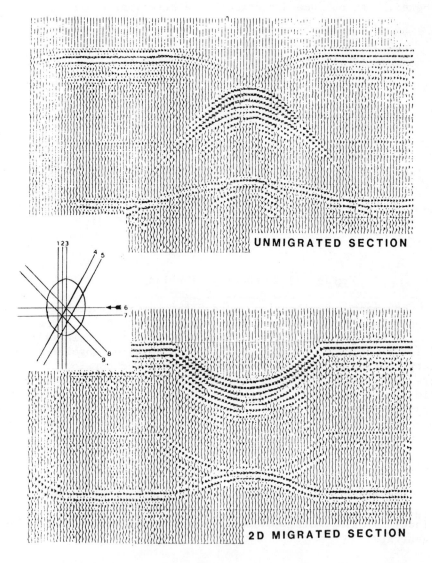

Fig. 23.   Principal plane seismic section across oblong basin. Top
           is raw seismic section; bottom is migrated seismic section.

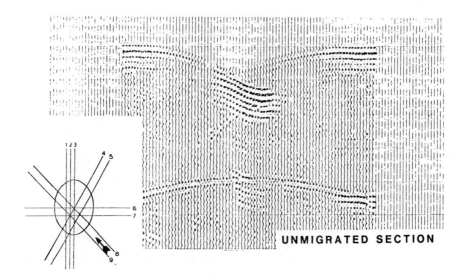

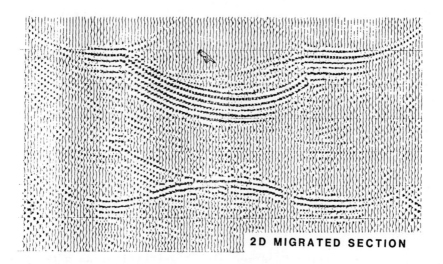

Fig. 24.   Seismic section gathered obliquely across oblong basin.
Top is raw seismic section; bottom is migrated seismic
section.

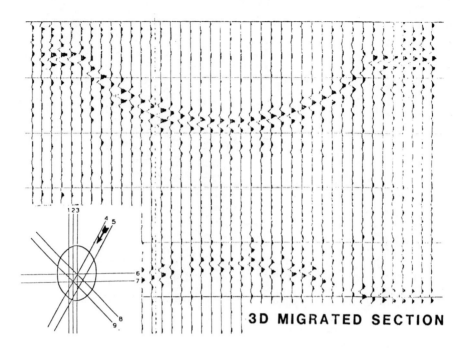

Fig.   25.   3-D migrated seismic section for oblique profile line
             across oblong basin.

     The migrated process is an imaging procedure which utilizes
the wave equation by downward continuing the wavefield measurement
at the surface of the earth.   There are three efficient algorithms
for downward continuing the wavefield which are f-k, Kirchhoff and
finite difference.   Basically, only minute differences are evident
between these three methods.

     Finally, care must be taken in interpreting 2-D migrated data
that comes from 3-D structures, it is not a simple problem of the
data being slightly misplaced in the 2-D migrated section, but it
is a total misinterpretation, as was illustrated by the basin data.

REFERENCES

Claerbout, J.F., 1976, Fundamentals of geophysical data processing;
     McGraw-Hill, 274 p.
Courtier, W.H. and Mendenhall, H.L., 1967, Experiences with multiple
     coverage seismic methods, Geophysics, V. 32, no. 2, p. 230-258.
Duffy, Jr., R.E., 1980, Seismic coal modeling constrained by deposi-
     tional environment, Univ. of Houston MS. publication.
French, W.S., 1975, Computer migration of oblique seismic reflection
     profiles, Geophysics, v. 40, no. 6, p. 961-980.

Hilterman, F.J., 1970, Three-dimensional seismic modeling, Geophysics,
     v. 35, no. 6, p. 1020-1037.
Schneider, W.A., 1978, Integral formulation for migration in two and
     three dimensions, Geophysics, v. 43, no. 1, p. 49-76.
Stolt, R.H., 1978, Migration by Fourier Transform, Geophysics, v. 43,
     no. 1, p. 23-48.
Taner, M.T. and Koehler, F., 1969, Velocity Spectra-Digital computer
     derivation and applications of velocity functions, Geophysics,
     v. 34, no. 6, p. 859-881.

A FREQUENCY-DIP FORMULATION OF WAVE-THEORETIC MIGRATION

IN STRATIFIED MEDIA

Stewart Levin

Western Geophysical Company of America
Houston, Texas

Introduction

Techniques employing the explicit recognition of angular dip have been used since the inception of seismic exploration (Rieber, 1936). Traditionally, plane wave or collimated beam steering algorithms for the migration (i.e. imaging and focusing) of reflected seismic energy have been based upon ray-theoretic arguments; we assume that coherent events on a time section are smooth wavefront arrivals which can be associated with subsurface reflectors by simple ray tracing. Focused and kinked arrivals, lacking a well-defined normal ray direction, do not satisfy this ray-theoretical model and their treatment becomes undesirably dependent upon the particular "local plane wave" decomposition employed by the algorithm.

The failure of ray-theoretic migration to accommodate such wave-theoretic effects as buried foci is well known. Indeed one of the major reasons for the ready acceptance of wave-theoretic migration techniques, such as finite-difference migration or frequency domain algorithms, was the marked improvement in their handling of ray-theoretic failures. For this reason some researchers, such as Ottolini et al (1978) in their work on common-offset migration, have in recent years sought beam steering migration techniques that rest upon wave-theoretic bases. In this paper we derive and implement an explicit-dip, zero-offset migration algorithm for stratified media by reformulating frequency domain migration. We also indicate how to generalize our approach to higher dimensional problems.

681

We apply the algorithm to some synthetic and real data and obtain, albeit at significantly greater computational cost, results of comparable quality to the conventional migration approach from which ours was derived. Both the computational cost and the inherent problems in extension to more realistic velocity models lead us to conclude that our approach, while interesting, is not a practicable addition to conventional seismic data processing.

The Frequency-Dip Formulation

Our starting point is the phase-shift migration approach introduced by Gazdag (1978) and Bolondi et al (1978). Briefly, theirs is an implementation of the single square-root, one-way wave propagator of Claerbout (1976) in the temporal frequency and spatial wavenumber domain. For a stratified medium, wavefield extrapolation is accomplished with the equation:

$$P(x,t,z) = \iint e^{i\omega t + ikx} \, e^{i\omega\Phi} \, \tilde{P}(k,\omega) \, dk \, d\omega \tag{1}$$

where $P(k,\omega)$ is the double Fourier transform of the surface wave field $P(x,t,z=0)$, and

$$\Phi(k,\omega,z) = \int_0^z \left(\frac{4}{v(s)^2} - \frac{k^2}{\omega^2} \right)^{\frac{1}{2}} ds \tag{2}$$

The migrated subsurface image is realized by setting $t = 0$.

Computing the shift factor Φ requires some effort, but for each fixed extrapolation depth z, it depends only upon the ratio of spatial to temporal wavenumber, i.e. only upon the time dip of the corresponding monochromatic plane wave (Fig. 1). We will call this ratio the Snell coordinate,

$$p = k/\omega \quad . \tag{3}$$

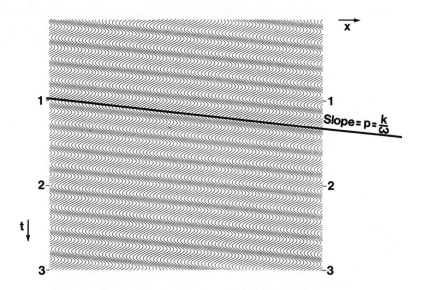

Figure 1. Real part of the harmonic plane wave $\exp(i\omega t - ikx)$. The curves of constant phase are straight lines of slope k/ω.

When we introduce p into equation (1) we obtain

$$P(x,t,z) = \iint e^{\,i\omega\left[t + px + \Psi\right]} |\omega| \; \widetilde{P}(\omega p, \omega) \; dp \, d\omega \;\;, \qquad (4)$$

where

$$\Psi(p,z) = \int_0^z \left(\frac{4}{v(s)^2} - p^2 \right)^{\frac{1}{2}} ds \;\;. \qquad (5)$$

Regrouping in equation (4) produces

$$P(x,t,z) = \iint e^{\,i\omega\left[t + px + \Psi\right]} \; \widetilde{Q}(p,\omega) \; d\omega \, dp \qquad (6)$$

with

$$\widetilde{Q}(p,\omega) = |\omega| \; \widetilde{P}(\omega p, \omega) \qquad (7)$$

One may now recognize the inner integration in equation (6) as a one-dimensional Fourier transform and so introduce the function

$$Q(p,t) = \int e^{i\omega t}\ \tilde{Q}(p,\omega)\ d\omega \tag{8}$$

to obtain the strikingly simple wavefield extrapolation equation

$$P(x,t,z) = \int Q(p, t + px + \Psi)\ dp \quad . \tag{9}$$

In physical terms this is a three part extrapolation process:

(1)   Plane wave decomposition of the surface wavefield
(2)   Migration of each constant-dip component by simple time shifting
(3)   Superposition of the independently migrated components

Such a procedure is the basis of the wavefront charts of Hagedoorn (1954) and more recently the dip-domain migration of Robinson and Robbins (1978) and the slant-midpoint migration of Ottolini et al (1978). Indeed our explicit dip migration approach fits in the gap between the former, ray-theoretic, post-CDP-stack algorithms and the latter, wave-theoretic, pre-stack technique.

Among the three stages of the extrapolation process, wave theoretical aspects are most evident in the initial global plane wave decomposition of the recorded wave field. Formulae similar to equations (8) and (9) appear in fields such as tomography and radio astronomy (Mueller et al, 1979). One of the first applications was in the work of Bracewell (1956) on strip scans. There the problem was one of constructing an intensity image map of solar microwave emission from average intensities recorded on a radio telescope with a long, thin aperature. In tomography a three-dimensional picture is reconstructed from flat X-ray or ultrasonic recordings taken at various orientations about the object of interest.

The connection between such recordings and the Fourier domain is revealed in the Projection-Slice Theorem due to Radon (1917). Simply put, this theorem says that transmitting a plane wave through an object will give a projection whose one-dimensional Fourier transform is a slice of the two-dimensional Fourier transform of the object (or more precisely its absorptivity). In two dimensions it may be formulated as follows:

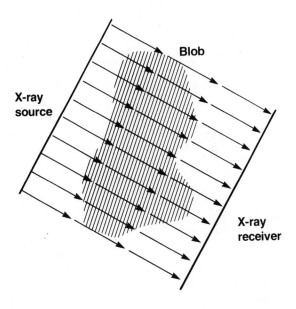

Figure 2. In tomography, projections taken at various angles through an object are used to image its interior.

Definition Let $P(x,t)$ be an integrable function. For a given slope p we define the p-stack of P, denoted P_p , by

$$P_p(t) = \int P(x, t - px)\, dx \quad .$$ (10)

Theorem (Projection Slice) Let $\widetilde{P}(\omega)$ and $\widetilde{P}(k,\omega)$ be the Fourier transforms of $P_p(t)$ and $P(x,t)$ respectively. Then

$$\widetilde{P}_p(\omega) = \widetilde{P}(\omega p, \omega) \quad .$$ (11)

Proof:

$$\iint e^{-i\omega t} \, P(x, t - px) \, dt \, dx \qquad\qquad (12)$$

$$= \iint e^{-i\omega[t + px]} \, P(x,t) \, dt \, dx \qquad\qquad (13)$$

$$= \iint e^{-i\omega t} \, e^{-i\omega px} P(x,t) \, dx \, dt \quad . \qquad\qquad (14)$$

This theorem allows one to reconstruct the transform, and hence the original function, from its projections. One particular method used by DeRosier and Klug (1968), Bracewell (1956), and others is to discretize equation (4) in the special case z=0 (i.e. for the recorded data.) Mersereau and Oppenheim (1974) give an excellent guide to this type of image reconstruction with particular regard to problems of discretization and noise. We shall return to their results shortly.

The Projection-Slice Theorem provides an alternate method of computing the function Q of equation (8) without recourse to the Fourier domain. Recalling that a multiplication in the frequency domain corresponds to a convolution in the time domain, we may, using equations (7) and (11), interpret Q as the convolution of an operator having Fourier transform $|\omega|$ with the p-stack of the surface data. Factoring

$$|\omega| = i \, \text{sgn}(\omega) \cdot (-i\omega) \qquad , \qquad\qquad (15)$$

this operator becomes the cascade of a first derivative and a Hilbert transform. Ramachandran and Lakshminarayanan (1971) have obtained an efficient convolutional approximation of this frequency scaling operator in the time domain.

In geophysical literature, p-stacks are often termed slant stacks. Typically they are used to simulate plane wave sources and are a special case of the more general wave stack proposed by Schultz and Claerbout (1978). The slant stack (involving linear moveout) is closely related to the Radon transform (Ludwig 1966) which is obtained by rotation and projection.

The second part of the extrapolation procedure, time shifting of each constant-dip component, is in fact the implementation of Snell's Law for a horizontally layered medium. Snell's Law can be written

$$p\,v(z) \;=\; 2\,\sin\theta(z) \qquad , \qquad (16)$$

where θ is the local inclination of the planar wave front to the horizontal. (The factor of two is included to account for the two-way traveltime between the surface and the reflector.) Substituting (16) into equation (6) gives the time shift as

$$\Psi \;=\; \int_{0}^{z} \frac{2\cos\theta(s)}{v(s)}\,ds \qquad , \qquad (17)$$

the integral of the vertical component of the slowness (reciprocal velocity) of a plane wave of ray parameter p.

The last part of the extrapolation process composites the independently migrated plane wave components back as a final image. Equation (9) does this by expanding each migrated dip component into a full plane wave section and adding the results in accordance with the principle of superposition.

Implementation

Because observable seismic energy occupies only a portion of frequency space, we restrict our attention to a fan-shaped region in the ω-k domain indicated schematically in Figure 3. This provides a two-fold advantage. First dip-filtering becomes a simple matter of adjusting the fan width. Second, we avoid the non-sequential storage accessing required by algorithms that must rotate data in order to form an image.

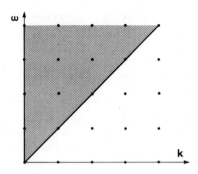

Figure 3. Unaliased non-evanescent seismic energy appears in a fan in the transform domain. We sketch only the first quadrant here; the rest is obtained by reflection across the axes.

Within the chosen fan we select a series of radial slices. This selection, which controls the accuracy of our plane wave imaging scheme, involves three choices:

(1)   The number of dip slices.

(2)   The distribution of the slices in the $\omega-k$ plane.

(3)   The interpolation algorithm used to evaluate the transform along these slices.

Our own experience and the results of Mersereau and Oppenheim mentioned above, indicate that the number of transform slices needed for a good reconstruction is approximately the number of input traces. This "conservation of information" criterion sets a lower bound on the number of dip slices we need to process and so limits the computational efficiency of frequency-dip migration. Indeed the accuracy with which the surface time section can be reconstructed from our Q function sets an upper limit on the quality of the migrated section. In Figure 4 we illustrate how dip undersampling can degrade imaging. The major artifacts produced by this are long, faint ghost tails appearing at some distance from the actual events. As the number of dips increases the amplitudes of these tails decrease; the more frequent interference between them causes the spurious amplitudes to appear as grainy background noise. Compared to typical seismic information, this imaging noise has fairly high frequency and so can be reduced by high-cut filtering.

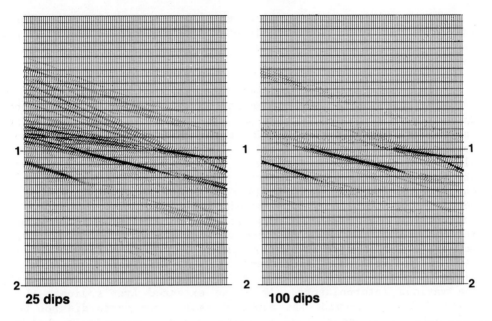

25 dips   **100 dips**

Figure 4. A 96-trace input processed using a) 25 dips and b) 100 dips.
The dip undersampling in the former shows up in the ghost tails.

While the results above suggest the number of slices to choose,
they give no direct indication of where to choose them. Actually,
almost any reasonable selection of dips can be used in the
reconstruction. For migration we have opted to sample hyperbolic
diffraction patterns at uniform spatial increments. Letting the
hyperbola be characterized by

$$t^2 = \tau^2 + (2\Delta x/v)^2 \qquad , \qquad (18)$$

where t is two way traveltime, v is velocity, and Δx is offset, we
can differentiate implicitly to obtain

$$\frac{dt}{dx} = \frac{4\Delta x}{v^2 t} \qquad . \qquad (19)$$

Rearranging and simplifying we have the well-known relation
involving the true dip angle θ :

$$2\Delta x = v\tau \tan\theta(z) \qquad . \qquad (20)$$

This result says that uniform spatial sampling of the hyperbola
is equivalent to uniform sampling of the tangent of the true dip
angle. In particular, the greater the dip the greater the angular
sampling density.

Having chosen where to place the slices, we lastly need to
evaluate the Fourier transform along them. In general these slices
do not pass through many of our original ω-k grid points and
values must be interpolated. Under the standard assumption of data
bandlimited within the temporal and spatial Nyquist limits,
elementary Fourier analysis says the transform may be evaluated
analytically with high order trigonometric polynomials. By
choosing to interpolate laterally at each temporal frequency we
reduce our problem to one-dimensional interpolation. For
computational purposes we restricted our studies to two-sample
interpolators. Taking a cue from the F-K migration of Stolt (1978)
we have tried a geometric interpolation algorithm reported by Lynn
(1977), as well as linear and nearest neighbor interpolation.
Figure 5 shows reconstructions of a synthetic time section using
the various interpolators. While we expected Lynn's linear phase
interpolator to provide the superior image, our tests did not bear
this out. We found that it produces stronger ghost tails than our
other two choices. Our explanation is that this geometric
interpolator better preserves higher frequency plane wave
components and so produces less smoothing and cancellation in the
composite image.

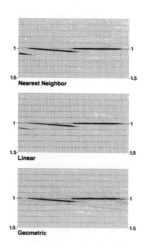

Figure 5. Image reconstruction using different transform interpolators. The features are taken from 125 dip reconstructions of a
120-trace input synthetic.

Once the constant dip slices are obtained we scale them (by the Jacobian of the angular mapping) and apply an inverse Fourier transform. This forms the function Q of equation (8). Finally we perform the integration using quadratic interpolation to evaluate the integrand. The luxury of quadratic interpolation was chosen because its availability on our array processors made it a computationally attractive choice; virtually identical results are obtained with simple nearest neighbor quadrature.

Examples

In Figure 6 we show some stages in the migration of a dipping-segments model. Panel (a) shows the time section generated by a Kirchhoff diffraction response integration. Panel (b) shows a dip-filtered surface reconstruction using the same dip decomposition parameters used in its subsequent migration. The dip filtering chosen rejected information with spatial dips greater than 35 degrees. Panel (c) shows the ω-k amplitude spectrum of the input data, and (d) displays the function $Q(p,t)$. The wrap-around along the time axis arises in the discrete Fourier transform method used to obtain Q. If p-stacks had been used to compute Q directly from the time section such information would appear at negative time. Finally, in panels (e) and (f) we display a frequency-dip migration and, for comparison, a conventional phase-shift migration of the line. As expected, the faint ghost tails from the surface reconstruction have migrated into similar artifacts on the frequency-dip image.

Our second example is actual seismic data from the Gulf of Mexico shown in the top of Figure 7. Migration velocities ranged from a water velocity to 9100 ft/sec at depth. These data contain fairly gentle dips; they are included primarily to illustrate how the algorithm images in the presence of recorded seismic noise, a major problem in seismic work. The bottom pair of panels show results of migrating this line. On the left is the frequency-dip migration while the right displays the corresponding phase-shift result. The two sections appear much the same both in their treatment of seismic reflections and noise. In the upper right hand corners we do find some difference, however. There the frequency-dip algorithm has added an underlying grainy background to the image. As noted previously, this is a typical artifact of the frequency-dip process and may be attenuated by judicious high-cut frequency filtering.

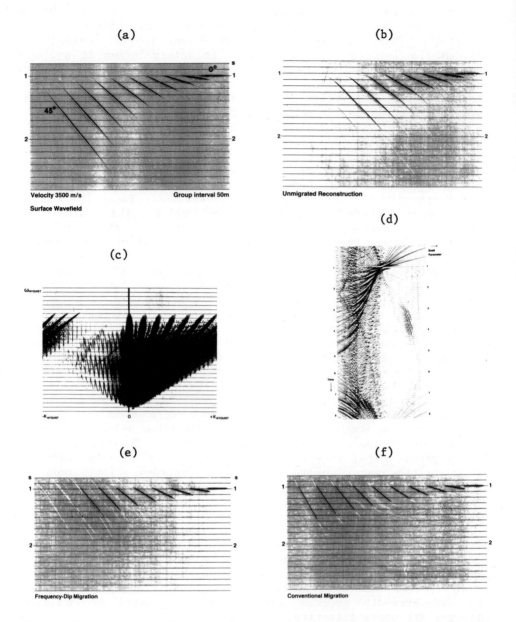

Figure 6. Synthetic frequency-dip migration example. The 500-trace input section is in a). Panel b) shows a dip filtered frequency-dip reconstruction of a). The transform of a) is shown in c) and the function Q is in d). The final panels contain a frequency-dip migration (e) and a conventional migration (f).

(a)

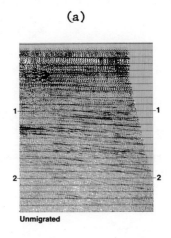

Unmigrated

(b)   (c)

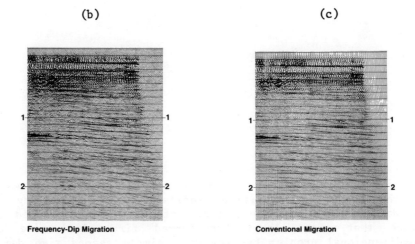

Frequency-Dip Migration   Conventional Migration

Figure 7.   Panel b) is a frequency-dip migration of the 350-trace
seismic data of a).   Panel c) shows the corresponding conventional
migration.

Extensions

No new practical problems are introduced when frequency-dip migration is formulated in higher dimensions. For migrating data from a rectangular x,y surface grid, the substitutions

$$\kappa^2 = k_x^2 + k_y^2 \quad , \quad \text{(21a)}$$

$$\rho^2 = p_x^2 + p_y^2 \quad , \quad \text{(21b)}$$

and

$$dk_x dk_y d\omega = \omega^2 dp_x dp_y d\omega \quad \text{(21c)}$$

will transform the basic three-dimensional phase-shift extrapolation equation

$$\widetilde{P}(z = \zeta) = e^{i\omega \Phi(\kappa, \omega, \zeta)} \, \widetilde{P}(z = 0) \quad \text{(22)}$$

into the frequency-dip formulation

$$P(x,y,t,z) = -\partial_{tt} \iint P_{p_x p_y}(p_x, p_y, t + x p_x + y p_y + \Psi(\rho \, z)) \, dp_x dp_y \quad , \quad \text{(23)}$$

where Φ and Ψ are given by equations (2) and (4), respectively, and the double p subscripting indicates p-stacking along both of the designated coordinates, i.e. integration over tilted planes. Notice that the Hilbert transformed first derivative of the two-dimensional problem has become a simple second time derivative in the three-dimensional equations.

If, in a different context, one reinterprets x and y in equations (21) through (23) as independent source and receiver coordinates, the wavefield extrapolation equations for the impulse response function P(x,y,t,z) are obtained by substituting

$$\frac{1}{2}\left[\Phi(2k_x, \omega, z) + \Phi(2k_y, \omega, z) \right] \qquad \text{for} \qquad \Phi(\kappa, z)$$

and $\frac{1}{2}\left[\Psi(2p_x, z) + \Psi(2p_y, z) \right]$ for $\Psi(\rho, z)$

in equations (22) and (23). Setting t=0 and x=y specializes the new equations to produce a subsurface image P(x,x,0,z). These choices provide a frequency-dip formulation of what is known in geophysical literature as "migration before stack". In comparison, the migration of Ottolini et al (1978) mentioned before employs p-stacking in only the offset direction where the midpoint $\frac{1}{2}(x + y)$ is constant, and reverts to a more conventional algorithm to process the resulting suite of slant-midpoint gathers.

We have not implemented these formulae numerically but expect that the same computational criteria will apply to these algorithms also.

Conclusions

Frequency-dip migration is a wave-theoretical technique that employs the explicit recognition of dip. Our algorithm combines a plane wave imaging scheme familiar in tomography with a downward extrapolation that honors Snell's law for stratified media. It is easily generalized to three-dimensional migration as well as migration before stack. The before-stack extension is similar but not identical to slant-stack migration procedures that have appeared in the literature.

Unfortunately, we pay a significant additional computational cost in order to achieve an accuracy comparable to that of conventional wave-theoretic algorithms, such as frequency domain or finite difference migrations. This cost, inherent in our explicit dip decomposition, can be significantly reduced only by sacrificing part of the wave-theoretic accuracy of the formulation.

An even more serious shortcoming is the difficulty of extending our algorithm to handle lateral velocity variation. This problem, shared with other frequency domain approaches, arises because a plane wave is no longer planar after extrapolation across a zone of lateral velocity variation. This suggests that any plane wave imaging migration is most useful in a before-stack formulation where recording geometry limits the beam dispersion caused by departure from uniform velocity gradient. We remark that most of the similar work on slant-stack migration has dealt with before-stack formulations.

In view of its difficulty with lateral velocity variation as well as its computational cost we do not consider frequency-dip migration to be a useful addition to conventional seismic data processing. Both of these limitations may be thought of as arising from the same source - representing the data in terms of basis functions that are non-zero over most or all of the computational domain. Such a representation leads to very non-sparse matrix operations for image reconstruction and, in the case of velocity inhomogeneity, for image extrapolation. Non-sparse matrix operations usually involve hefty computations; only in a few fortuitous cases, such as the discrete Fourier transform, can recognizable symmetries be exploited to reduce this load. For our imaging scheme we have not found such computational simplifications. Moreover, even if such were found, each serious departure from a plane-layered earth model would likely require the development of a corresponding downward continuation algorithm. As a result, we judge it highly unlikely that we could efficiently downward extrapolate more than a handful of seismic models with the frequency-dip approach.

References

Bolondi, G., Rocca, F., and Savelli, S., 1979, A frequency domain approach to dimensional migration: Geophysical Prospecting, v. 26, (1978); p. 750-772.

Bracewell, R.N., 1956, Strip integration in radioastronomy: Australian Journal of Physics, v. 9, p. 198-217.

Claerbout, J., 1976, Fundamentals of Geophysical Data Processing: San Francisco, McGraw-Hill international series in the earth and planetary sciences, p. 199-208.

DeRosier, D.J., and Klug, A., 1968, Reconstruction of three-dimensional structures from electron micrographs: Nature, v. 217 p. 130-134.

Gazdag, J., 1978, Wave equation migration with the phase shift method: Geophysics, v. 43, N. 7 p. 1342-1351.

Hagedoorn, J.G., 1954, A process of seismic reflection interpretation: Geophysical Prospecting, v. 11, p. 7-127.

Ludwig, D., 1966, The Radon transform on Euclidean space: Communications on Pure and Applied Mathematics, v. 19 p. 49-81.

Lynn, W., 1977, Implementing F-K migration and diffraction: Stanford Exploration Project Reports, v. 11 , p. 9-28.

Mersereau, R.M., and Oppenheim, A.V., Digital reconstruction of
    multidimensional signals from their projections: Proceedings
    of the IEEE, v. 62 n. 10, p. 1319-1338.

Mueller, R.K., Kaveh, M., and Wade, G., 1979, Reconstructive
    tomography and applications to ultrasonics: Proceedings of the
    IEEE, v. 47, n. 4, p. 547-587.

Ottolini, R., Clayton, R., and Claerbout, J., 1978, Migration with
    slant-midpoint   stacks: Presented   at   the   48th   Annual
    International   Meeting   of   the   Society   of   Exploration
    Geophysicists; San Francisco.

Radon, J., 1917, Uber die bestimmung von funktionen durch ihre
    integralwerte   langs   gewisser   mannigfaltigkeiten   (On   the
    determination of functions from their integrals along certain
    manifolds), Berichte Saechsische Akademie der Wissenschaften,
    v. 69 p. 262-277.

Ramachandran, G.N. and Lakshminarayanan, A.V., 1971, Three
    dimensional   reconstruction   from   radiographs   and   electron
    micrographs: II. Application of convolutions instead of
    Fourier transforms: Proc. Nat. Acad. Sci., v. 68, n. 9, p.
    2236-2240.

Rieber, F., 1936, Visual presentation of elastic wave patterns
    under various structural conditions:, Geophysics, v. 1, p.
    196-218.

Robinson, J.C., and Robbins R. R., 1978, Dip-domain migration of
    two-dimensional seismic profiles: Geophysics, v. 43, n. 1, p.
    77-93.

Schultz, P.S., and Claerbout, J.F., 1978, Velocity estimation and
    downward continuation by wavefront synthesis: Geophysics, v.
    43 n. 4, p. 691-714.

Stolt, R.H., 1978, Migration with Fourier transforms: Geophysics,
    v. 43, n. 1, p. 23-48.

LIMITATIONS IN THE RECONSTRUCTION OF THREE-DIMENSIONAL SUBSURFACE IMAGES

T. Smith, K. Owusu, and F. Hilterman

Seismic Acoustics Laboratory
University of Houston
Houston, Texas   77004

Seismic reflection data (CDP-Common Depth Point shooting) are normally collected on the earth's surface in such a way that as many as 24 different source and receiver pairs share the same midpoint location. These ASR (Arbitrary Source and Receiver position) data are processed in order to simulate CSR (Coincident Source and Receiver position) data. Image reconstruction for geologic structures are commonly applied by extrapolating these CSR data into the earth.

Several assumptions about the seismic imaging process are examined both theoretically and experimentally.

For the ASR configuration, the scattered field has been computed for several simple models using the boundary diffraction wave theory, as formulated by Maggi and Rubinowicz. Using this theory the ASR data are remarkably similar to data computed using a CSR configuration at the bisector angle image point.

CSR data simulated by processing ASR data, either theoretical or experimental, are more complicated than theoretical CSR data. There are situations where there is considerable waveform distortion within a CDP gather because of the interactions of reflections and diffractions. The problem of interference of conflicting dip data is here illustrated.

INTRODUCTION

Seismic reflection data (CDP-Common Depth Point shooting) are collected by recording signals where as many as 24 different source and receiver combinations are symmetrically disposed about

a point (called the midpoint) on the earth's surface. The pairs
of sources and receivers are usually layed out along a line and
organized so that there is an ordered range of offsets between
them. The reflections from subsurface surfaces of acoustic impedance discontinuities give valuable information about the shape
of these surfaces and about the acoustic properties of the material between the discontinuities.

Migration is a portion of the analysis of these data that
attempts to construct the geometric arrangement of these reflecting surfaces. This process is usually made after velocities in
the ground have been estimated. The velocity distribution in the
earth is used to convert the recorded ASR (Arbitrary Source and
Receiver position) data to CSR (Coincident Source and Receiver
position) data. The common solution involves the assumption of
hyperboloid time moveout surfaces which are dependent on velocity
and independent of source-receiver separation. When the reflecting surface dips, a dip-dependent velocity function may be used
if the angle between the recording line and the strike of the
plane is known everywhere (Levin, 1971).

Since this much a priori information is seldom available,
imaging the offset data directly may be preferable. If it may be
assumed that the raypaths to-and-from a point diffractor are
straight, the Kirchhoff summation migration may be used to move-
out-correct the downgoing and upcoming raypaths separately.
Imaging in this manner is dip-independent and has been presented
elsewhere (Gardner and Kotcher, 1978).

This paper will discuss the diffraction effects which give
rise to waveform distortion within the CDP gather. Experimental
results will show amplitude changes within the gather, and synthetic models will show phase changes as well. Gathers of data
will be stacked using velocities appropriate for flat and dipping
planes. Where recorded reflection data from these planes interfere, we demonstrate that the estimate of a coincident source-
receiver wavefield may be questionable.

Half-Plane Model

Diffraction wave theory using boundary integration, was
numerically compared with the solid angle approach (Trorey, 1970;
Hilterman, 1970 and 1975). Theoretically and numerically, they
are consistent. In Figure 1, a reproduction of the diffraction
response of a half-plane using the boundary integration method
is shown. The half-plane is parallel to the observation plane
and 2500 feet below it. The edge of the plane is 5000 feet from
the origin. The large arrow points toward the time trace where
a coincident source-receiver is immediately above the edge. The
amplitude of this event is half as large as the reflection from

the plane. Positive and negative diffraction arrivals are seen to
the right and left of the arrow.

To test boundary wave diffraction computations (SAL Report,
1979) against earlier results, impulse-response traces were computed for the model. These data were convolved with a 48 msec.
symmetrical wavelet (bandpass 20–40 Hz) and are shown in Figure 2.
A spherical divergence gain correction has been made in both Figures 1 and 2.

In Figure 3, the source and receiver have been separated by
1250 feet; advancing the receiver location 625 feet and retarding
the source location 625 feet in relation to the midpoint coordinate. The specular reflections are delayed slightly because of
the extra travelpath length when the source and receiver are
separated. The positive and negative diffraction events fall on
a curve that is slightly flatter than before and is no longer
hyperbolic.

Monocline Model

In order to understand the diffractions recorded when source
and receiver are separated, a simple two-dimensional monocline
model has been tested. The model consists of a flat reflecting
plane at a depth of 4000 feet on the left and another flat reflecting plane at a depth of 3650 feet on the right (see Figure
4). The plane connecting the two half planes has a dip of slightly more than 30 degrees. The velocity of the medium between the
observation plane and the model is 10,000 ft/sec. The dotted
line on Figure 4 encompasses the surface location of the specular
reflections from the dipping plane.

The coincident source-receiver diffraction responses are
displayed in Figure 5. Once again, the impulse response traces
have been convolved with the 48 msec symmetric wavelet. The
trace spacing is 150 feet, and the synthetic data for this model
are plotted at a fixed gain. The section is strongly controlled
by the diffraction responses of the left and right half planes.
There is a noticeable distortion of waveform due to the interaction of these diffractions with the diffractions from the dipping
strip. Because of the narrow bandwidth of the wavelet, resolving
these superpositional effects is difficult.

These interfering diffraction effects are more easily understood by studying the model impulse response data rather than the
filtered data. While the impulse responses could never be recorded by a physical experiment, they do serve to reveal the
mechanism by which real waveforms are distorted. They might also
serve the purpose of showing what data would look like if the
wavefield could be perfectly recorded.

The model impulse responses in Figure 6 demonstrate that the specular reflections from the dipping plane fall on the three traces centered about the - 2000 feet coordinate. For traces with coordinates less than - 2500 feet, the positive diffraction arrivals from the dipping strip and the negative diffraction arrivals from the lower flat half-plane may be seen. Traces with coordinates between - 1500 and 0 feet have two diffractions (lower pointer) and one reflection. The positive diffraction associated with the upper flat half-plane (upper pointer) is stronger than its negative diffraction counterpart because of the extra positive diffraction contribution from the dipping strip.

Common depth point gathers were computed for midpoint coordinates of - 1800, 0, 1800 feet. The source-receiver separation ranged from 0 to 4000 feet. The filtered and impulse response gathers are presented in Figures 7 and 8. For the most part, the waveforms in Figure 7 are uniform in shape and do not appear to vary with offset.

The impulse responses in Figure 8 reveal a different story. While the filtered gather at coordinate −1800 feet looks remarkably uniform, the impulse response diffractions for the lower event in Figure 8 (pointer) reveal a significant variation with offset. What appeared to be a specular reflection from the dipping plane is actually a closely spaced interference of the specular reflection with both a positive diffraction from the upper flat plane edge and a negative diffraction from the lower flat plane edge.

Rectangular Plate Model

A rectangular plate model has been used to investigate the diffraction and specular reflection patterns from a simple three-dimensional body. A horizontal plate with side lengths of 2000 and 4000 feet was positioned 1000 feet below the observation plane. A velocity of 5000 feet/sec was used.

In the first set of calculations, four lines of fixed source-receiver separation were chosen. The separation distance was 1250 feet, and the sources and receivers were aligned with the direction of recording. The midpoint spacing was 250 feet and the line spacing was 1000 feet. The line locations are plotted in Figure 9. In Figure 10, the impulse responses have been convolved with the symmetric wavelet described previously. A spherical divergence correction has been applied.

On line 1, the only detected diffractions are those due to the left edge, parallel to the line direction. Line 2 passes over the left edge of the model, and so the reflections from the edge are half amplitude. The flat events in Lines 3 and 4 are

reflections from the plate. A hint of a side diffraction from
the left edge is visible on Line 3 (pointer).

The model impulse responses are presented in Figure 11. Now
on Line 2, the left edge diffraction is easily recognized (upper
pointer), and the right side diffraction (lower pointer) is also
identifiable. On Line 4, these two side diffractions add constructively (right pointer).

A synthetic shot gather has also been computed for the model.
The source was placed above the center of the left edge, and a
line of 41 receivers was placed along the long axis of the model.
The receiver locations are identified in Figure 12. Figure 13
shows the filtered responses for these receivers. The data are
plotted with a spherical divergence correction. The strongest
events are due to the diffraction from the left edge and from the
specular reflections over the plate. It is obvious that there is
a complicated interaction of diffraction effects, but with the
narrow bandwidth of the wavelet, their identification is difficult.

In Figure 14, the impulse responses for the shot gather reveal a complicated superposition of diffraction events. The diffractions from the right edge fall on a symmetric hyperbola whose
apex falls on Trace 21 (center pointer). It's polarity is negative until Trace 37 (right pointer). Beyond this, it's polarity
is positive. Between Traces 5 and 37, the first arrival is a
positive specular reflection. The left edge diffraction has
negative polarity and arrives at the same time as the right edge
diffraction on Trace 21. Beyond Trace 21, the left edge diffraction has the latest arrival time. The side diffractions become
visible on Trace 6 (left pointer) and maintain negative polarity
on all traces. At Trace 21, the side diffractions arrive before
the left and right edge diffraction. Near Trace 37, they arrive
after the right edge diffraction.

Wrench Fault Model

The boundary diffraction wave theory is compared to experimental data collected in our laboratory. The map diagram in Figure 15 shows the orientation of an RTV model as it rested in our
water tank. The model represents a stylized version of a wrench
fault, in which a monocline is terminated against a fault edge.
In the figure, the lower quarter-plane is in the upper right
corner of the diagram, and on the left and bottom is the upper
three quarter-plane. The difference in elevation between the
two planes is 1000 feet. The observation plane is 5300 feet
above the upper surface.

A set of 8 parallel CDP lines were run over the model. Each

line comprised 121 midpoints spaced 100 feet apart, and at each
midpoint a set of 6 source-receiver positions were used, ranging
in offset from 2000 feet to 7000 feet. The line spacing was 1000
feet. Lines 3 and 5 were selected and their location are indicated in the figure (see pointers). The results in this section
are all plotted with fixed gain.

Near- and far-trace common offset gathers for Line 3 are
shown in Figure 16. The upper and lower flat planes are represented by the flat reflections on the left and right, respectively. Diffraction from the right edge of the lower plane is just
detectable. Note the amplitude thinning at the break between the
upper flat plane and the dipping plane.

The same type of gathers on Line 5 are shown in Figure 17.
There is now a new event above the reflection from the lower
edge. These are side diffractions from the fault edge along the
upper plane. There is also a noticeable weakening of the reflection from the lower plane, since the line is near the fault.

The boundary diffraction wave theory was used to simulate
the experimental data. The near- and far-trace offset gathers
are presented in Figure 18. The filter operator chosen for these
data was taken from the first trace of Line 3 of the experimental
data.

The amplitude dimming associated with the joint between the
upper flat plane and the dipping plane is similar to that run on
the experimental data. Note that the interference of edge diffractions with the specular reflection causes a leggy appearance
in one portion of the data (upper pointer). The dead zone (lower
pointer) is due to destructive interference. These phenomena are
observed in both the synthetic and experimental data.

In Figure 19, the numerical modeling of Line 5 is presented.
The event above the lower plane reflection (pointer) is due to
the side diffraction from the upper plane edge. There is a remarkable agreement with the experimental data (Figure 17). Note
that in both cases, the side diffraction causes an arrival time
waviness and an amplitude fluctuation.

In Figure 20, Lines 3 and 5 were retun assuming that the
source and receiver were coincident. Except for the small time
delay, the features of these data are quite similar to the experimental data (Figures 16 and 17, Top). These wavefields would
be appropriate to use with an imaging method that assumed the
coincident source-receiver configuration.

The Midpoint and Bisector Image Points

Numerical computations using this diffraction theory can be quite time consuming, and so, some effort should be given to finding acceptable approximations for the line integral. Recently (Berryhill, 1977; Trorey, 1977) it has been suggested that a coincident source-receiver image point could be used to approximate the diffraction response for separated source-receivers. This method places the image point on a line that connects the minimum time point on the edge to the midpoint of the line joining source and receiver. The image point on this line would also be chosen so that the traveltime would be the same as the traveltime for separated source and receiver. (See Figure 21).

In another choice, a line is chosen which bisects the angle between the lines drawn from source and receiver to the edge (Gardner and Kotcher, 1978; Ilukewitsch, 1977). As in the midpoint approximation, the image point is chosen to fall on the line so that the traveltime is equivalent to the traveltime with source and receiver separated.

Using the bisector approximation, the near- and far-trace gathers for Line 3 were computed and shown in Figure 22. They are quite similar to the results from before (Figure 18).

The experimental CDP gathers from midpoint 31 to 40 on Line 3 are shown in Figure 23. This portion of the line represents the transition zone from a positive diffraction to a specular reflection from the left portion of the dipping plane. Within a gather, the amplitude falls off with offset because of the extra travel distance. From gathers 31 to 40, the amplitude increases as the specular reflection point is reached.

In Figure 24, the numerical simulations exhibit an amplitude balance and waveform shape that is quite similar to the experimental data.

In Figure 25, the 10 near-offset and far-offset traces from each set of data have been replotted. The left column are experimental data and the right column are synthetic data. Notice that there is an amplitude buildup on both the experimental and synthetic near-trace gathers. The amplitude buildup is not present in the two far-offset gathers in the second row.

It is clear that, within this transition zone, some mechanism is causing certain traces in a CDP gather to change and not others. For this waveform bandwidth, the observed effect is mainly an amplitude variation.

To understand the change in waveform within a CDP gather

due to a diffraction effect, consider a gather above the diffracting edge of a half plane, as in the upper portion on Figure 26.
The raypath for each of the source-receiver pair is a diffraction
raypath involving the edge. If the half plane is rotated and
shifted up dip, the inside traces will record two arrivals, one a
diffraction from the edge and the other a specular reflection from
the half plane (pointer). For a far-offset trace, the specular
reflection path does not exist.

A situation could also arise when the far-offset traces have
a specular raypath and the near-offset traces do not. This case
would arise when the half plane extends up dip and the position
of the edge will determine how many far-offset traces will have a
specular reflection. In Figure 27, the unfiltered synthetic model
impulse responses show that, for sufficient source excitation
bandwidth, there is not only an amplitude change within the
gather, but also a change in phase as well. On the near-offset
traces, the impulse responses grade from a positive diffraction
near CDP 31 to a combination of specular reflection and negative
diffraction near CDP 40 (upper pointer). On the far-offset traces
the impulse responses are all positive diffractions (lower pointer).

Stacked Wavefields

The tank CDP gathers in Line 3 were stacked assuming a constant scaled water velocity (Figure 28, top). The specular reflections from the upper and lower flat portions of the model
stack in-phase and are quite similar to those seen in the synthetic estimate of the coincident source-receiver wavefield (Figure 20, top). These data were also stacked with a velocity that
accounts for the known dip (15.3 degrees) of the model (Figure 20,
bottom).

The specular reflections from the dipping plane may be seen
to stack (Figure 20, bottom) in-phase. The reflections from the
flat portions of the model are not as strong as before because of
destructive interference. In addition to the amplitude dimming,
there is also an apparent polarity reversal between the two sections. The stack section with the higher stacking velocity has
the appearance of delaying the flat-lying reflections by approximately half a wavelength, and so gives the appearance of deepening
the event by approximately 400 feet. This mispositioning of the
reflector is difficult to detect on the section because the reflection grades smoothly into a diffraction (pointer).

The sometimes difficult selection of proper stacking velocity functions is well known. It is made less ambiguous by increasing the number of traces in the gather. In addition to the
signal-to-noise improvement, there is the benefit that finer

spatial offset sampling lessens the risk of mis-identification of the primary arrival in the presence of multiples and aliasing effects. The stack multiplicity cannot resolve the problem of interference where the velocity function is truly multi-valued however. In Figure 29, the two stacked sections between CDP 50 and 90 have been replotted on a larger scale. Within this CDP range the data from the dipping plane and the flat plane should actually be decomposed and stacked independently. As seen in the figure, choosing one velocity function properly stacks one wave-field at the expense of the other.

CONCLUSIONS

The boundary wave theory calculations have been shown to accurately predict experimental data collected with our tank model. The temporal bandwidth of the source waveform appears to play a critical role in the resolution of interfering diffraction phenomena. It has been shown that there can be significant wave-form variations within a CDP gather, so that even if the arrivals within the gather could be time-aligned, the estimation of a co-incident source-receiver trace could be in error. The more well known problem of interference of conflicting dip data is also demonstrated.

REFERENCES

Berryhill, J. R., Diffraction Response from Nonzero Separation of Source and Receiver, Geophysics, Vol. 42, No. 6, pp. 1158-1176, 1977.

Gardner, G. H. F. and Kotcher, J. S., Improvements in Dip-Analysis, Velocity Analysis by the Use of Kirchhoff Summation, published in Modeling and Migration, A Symposium, Dallas Geophysical Society, 1977.

Hilterman, Fred J., Three-Dimensional Seismic Modeling, Geophysics, Vol. 35, No. 6, pp. 1020-1037, 1970.

Hilterman, Fred J., Amplitude of Seismic Waves - A Quick Look, Geophysics, Vol. 40, No. 5, pp. 745-762, 1975.

Ilukewitsch, A. G., Three-Dimensional Seismic Modeling with Source-Receiver Offset, M.S. Thesis, University of Houston, 1977.

Levin, F. K., Apparent Velocity from Dipping Interface Reflections, Geophysics, Vol. 36, No. 3, pp. 510-516, 1971.

Seismic Acoustics Laboratory Progress Review, "Boundary Diffraction Wave", pp. 373-377, May 7, 1979.

Trorey, A. W., A Simple Theory for Seismic Diffractions, Geophysics, Vol. 35, No. 5, pp. 762-784, 1970.

Trorey, A. W., Diffractions for Arbitrary Source-Receiver Locations, Geophysics, Vol. 42, No. 6, pp. 1177-1182, 1977.

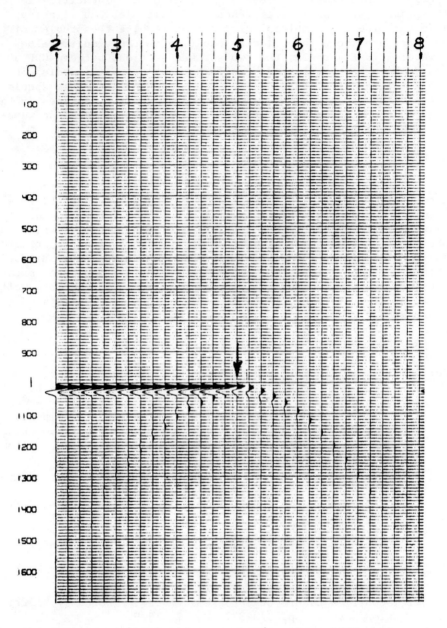

Fig. 1 (after Trorey) Coincident source/receiver diffraction
response over a half-plane using a "wide band" wavelet.

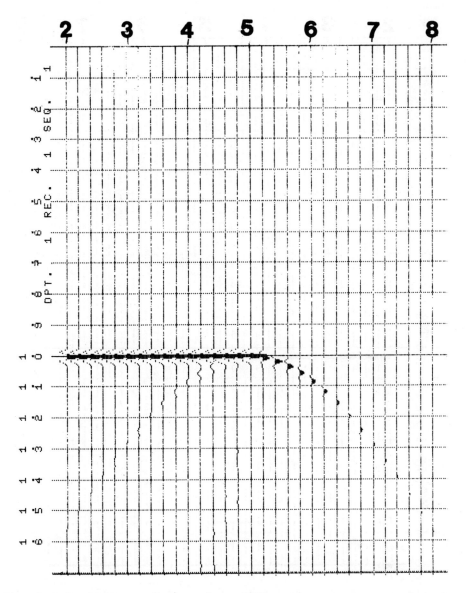

Fig. 2 Coincident source/receiver diffraction responses using the
Maggi-Rubinowicz line integral and a 10-40 Hz. wavelet.

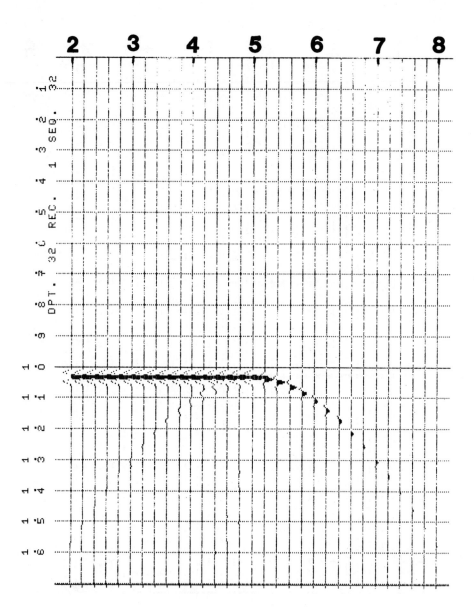

Fig. 3 Diffraction responses where the source to receiver separation
       is 1250 ft.

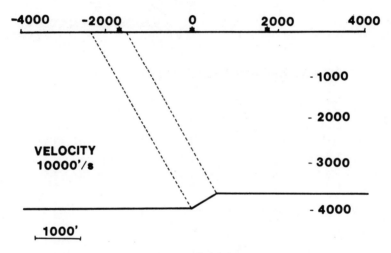

Fig. 4 Two-dimensional depth model of a step with a displacement of
350 ft. The dotted lines project the limits of specular
reflections to the surface.

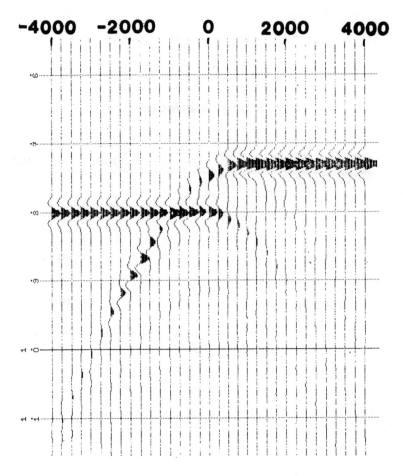

Fig..5 Coincident source/receiver diffraction responses over the
monocline model using a 10-40 Hz. wavelet.

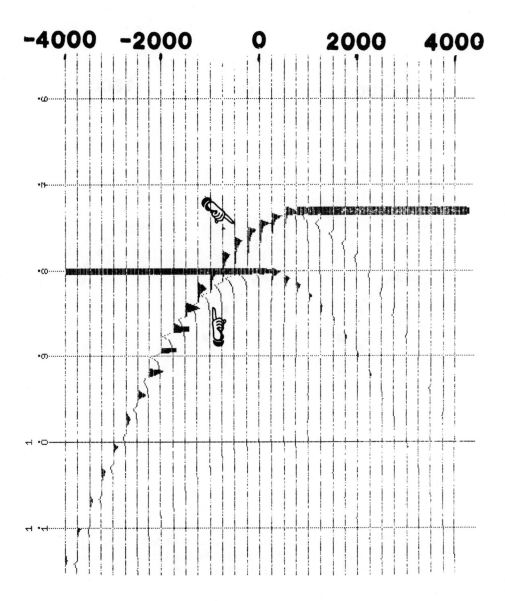

Fig. 6 Diffraction responses when source and receiver are coincident
and not bandlimited by a source wavelet.

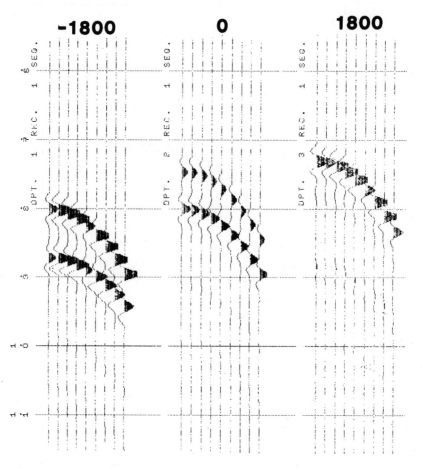

Fig. 7 Three CDP gathers over the monocline model using a 10-40 Hz.
wavelet. Midpoint coordinates are indicated. Source/rec-
eiver separation ranges from 0 to 4000 ft.

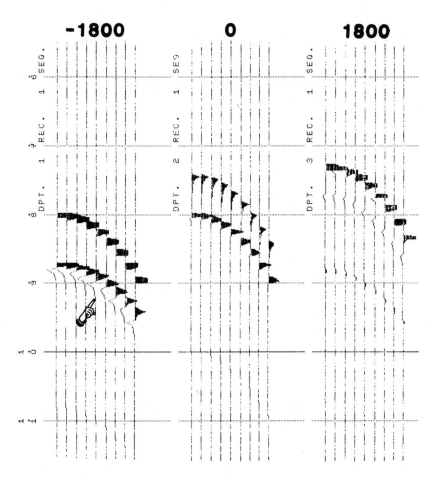

Fig. 8 Three CDP gathers without the bandlimiting wavelet. Note
the change of the negative (left) diffraction event with
increasing source to receiver separation.

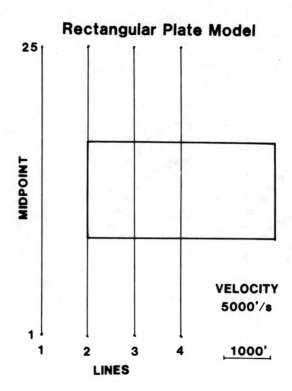

Fig. 9 Map view of a rectangular plate model at a depth of 1000 ft., indicating the location of 4 lines of data.

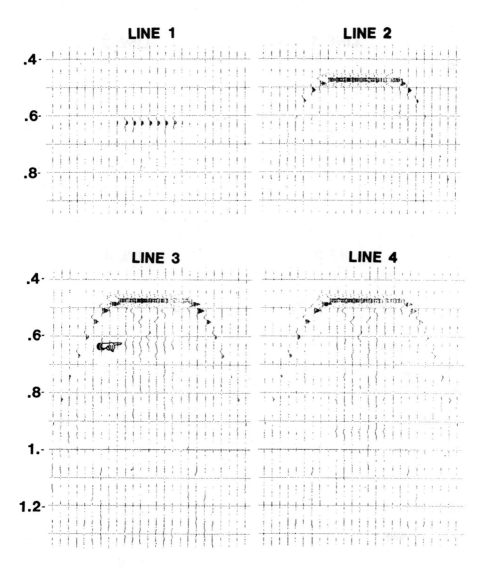

Fig. 10 Bandlimited diffraction responses for 4 lines over the
rectangular plate. The source to receiver separation
is 1250 ft.

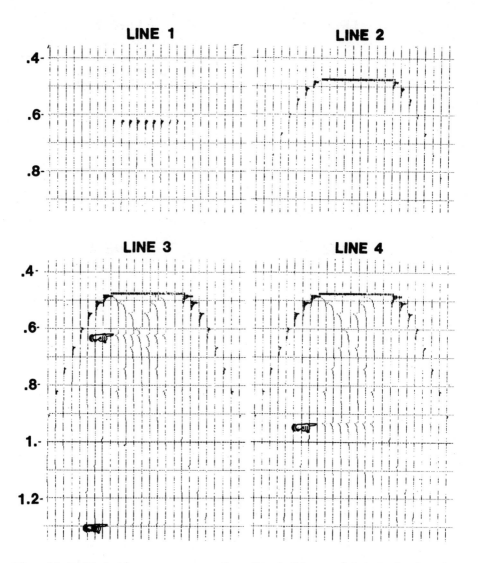

Fig. 11 Diffraction responses for the 4 lines without the band-
        limited wavelet. The sideswipe diffractions on line 3
        add constructively on line 4.

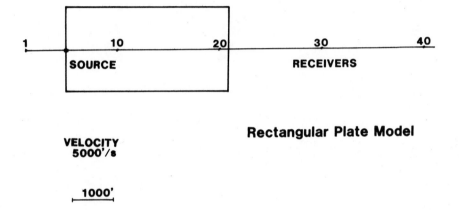

Fig. 12 Location map of 41 receiver positions over the rectangular
plate using a fixed source position over the left edge.

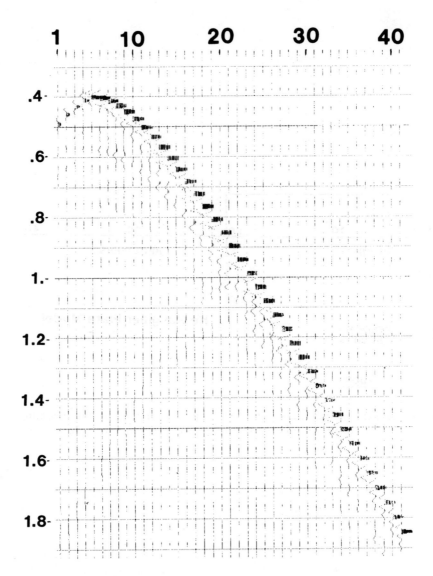

Fig. 13 Diffraction responses over the rectangular plate using a
10-40 Hz wavelet. The source and receiver are coincident
on the fifth trace from the left.

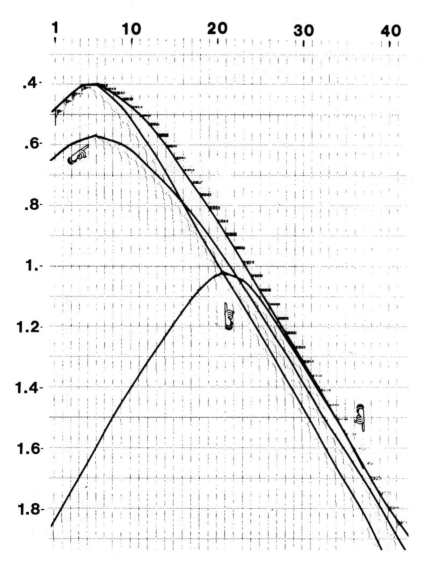

Fig. 14 Diffraction responses without filtering. The diffraction
events are discussed in the text.

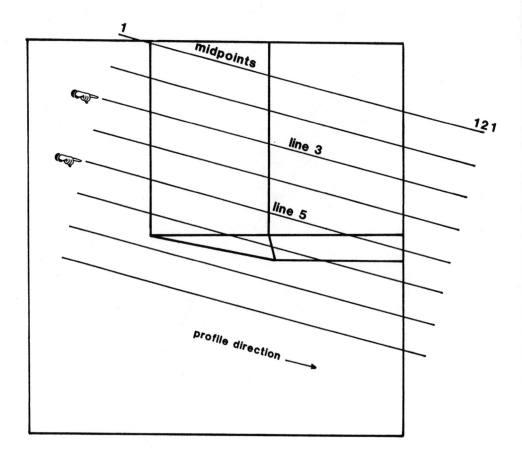

Fig. 15 Map view of a wrench fault model and the location of lines
        3 and 5. The lower plane of the monocline below the right
        portion of line 3 abuts a wrench fault below modpoints 100
        on line 5.

Line 3 Tank Data

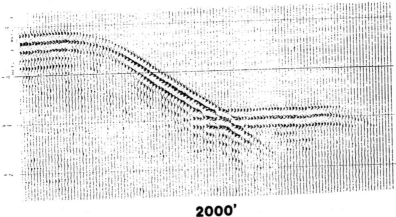

2000'

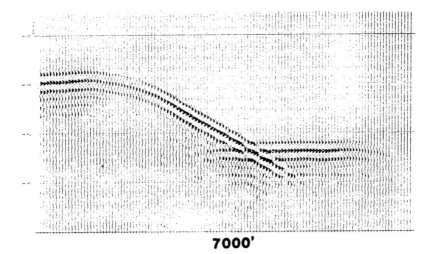

7000'

Fig. 16 Gathers of traces with 2000 ft. and 7000 ft. source/rec-
eiver separation along line 3, from the tank experiment.

Line 5 Tank Data

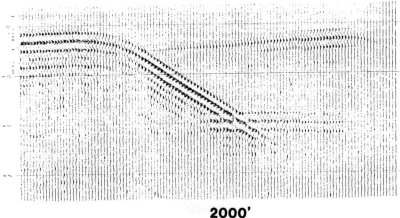

2000'

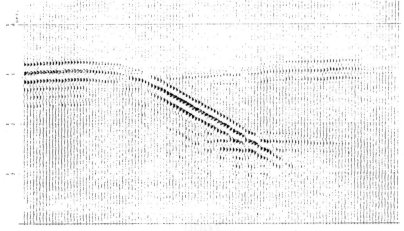

7000'

Fig. 17 The same gathers along line 5.

Line 3 Synthetic Data

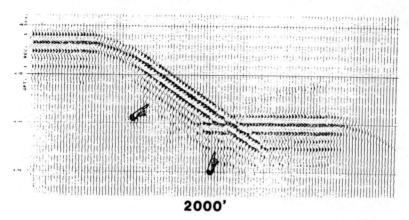

2000'

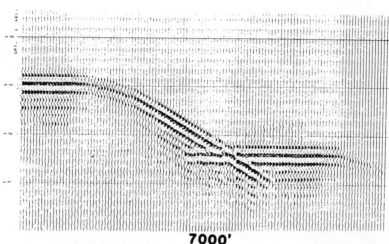

7000'

Fig. 18 Numerical simulation of the experimental data on line 3. The diffraction responses have been filtered with the reflection event on the upper left trace in Fig. 16.

Line 5 Synthetic Data

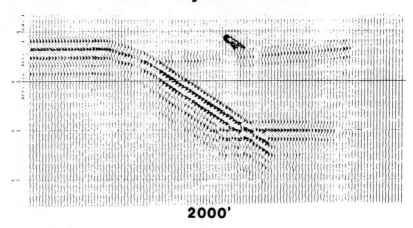

2000'

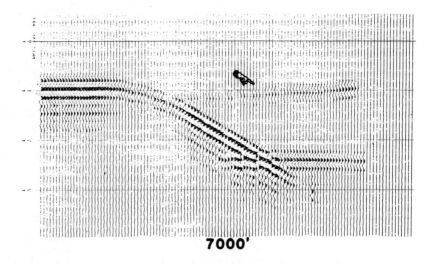

7000'

Fig. 19 Numerical simulation of line 5.

Line 3 Synthetic Data

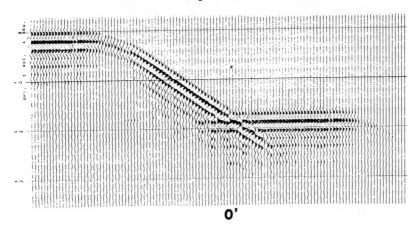

O'

Line 5 Synthetic Data

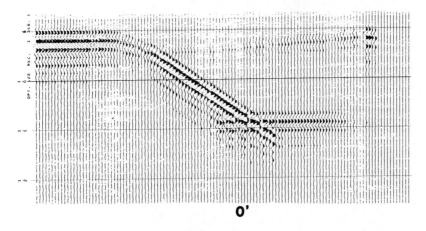

O'

Fig. 20 Numerical simulation of coincident source/receiver data
along lines 3 and 5.

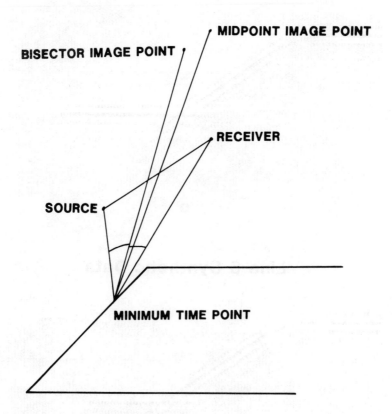

Fig. 21 Bisector and midpoint positions for approximating the diffraction response of an edge. The responses, with source
and receiver separated, may be approximated by placing
both at one of these positions (and simplifying computations).

Line 3 Synthetic Data
Bisector Approximation

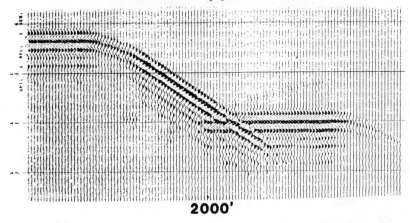

2000'

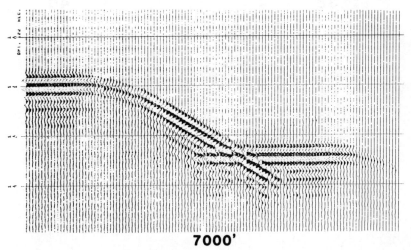

7000'

Fig. 22 Numerical simulation using the bisector approximation along
line 3.   Compare with Fig. 18.

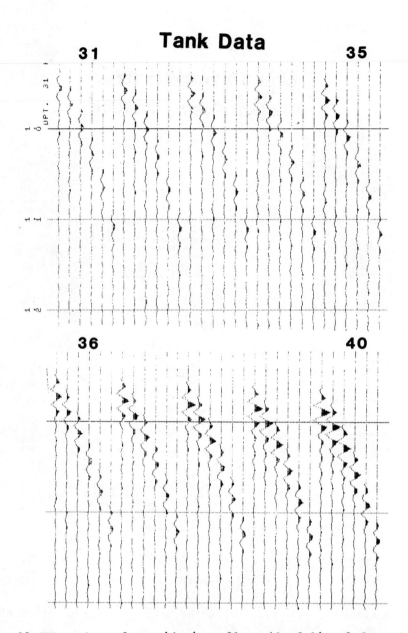

Fig. 23 CDP gathers from midpoints 31 to 40 of line 3 from the
tank experiment. Source/receiver offsets ranged from
2000 ft. to 7000 ft. (scaled).

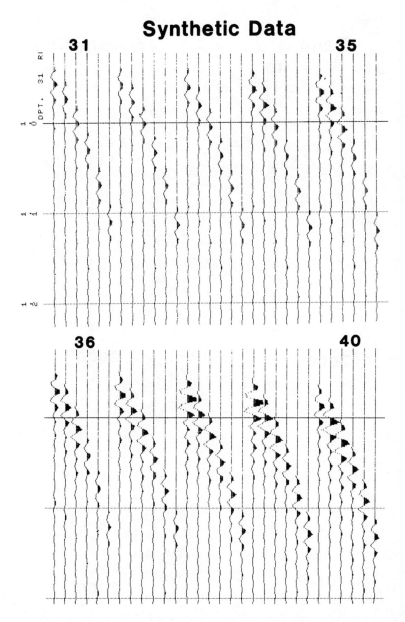

Fig. 24 CDP gathers for the same midpoints using the line integral.

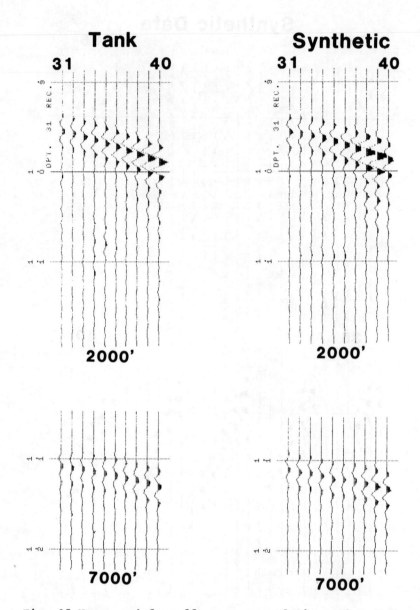

Fig. 25 Near- and far-offset traces of Figs. 23 and 24.

CDP Waveform Distortion

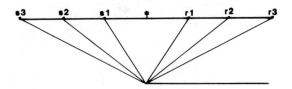

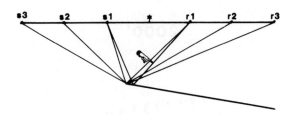

Fig. 26 The upper diagram depicts how the edge diffraction is
recorded on all traces in the CDP gather over the edge.
The lower diagram shows that by rotating and shifting the
half-plane, the near-offset receivers record a diffraction
and specular reflection, while the far-offset receivers
record only a diffraction.

Synthetic

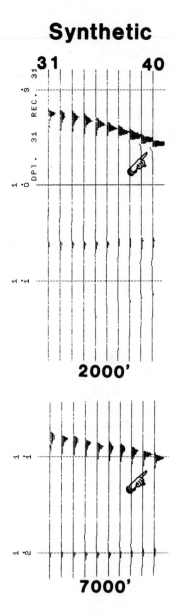

Fig. 27 The numerical diffraction responses from the right column
of Fig. 25 without a bandlimiting wavelet.

Line 3 6-Fold Stack

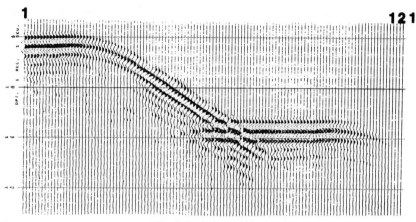

VSTACK 11928'/s

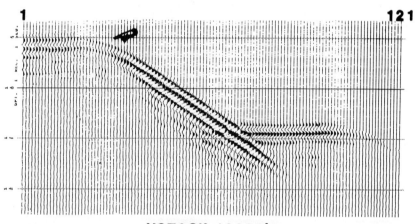

VSTACK 12366'/s

Fig. 28 Estimation of a coincident source/receiver wavefield by
stacking the 6 traces/midpoint using a scaled waver velocity (upper) and a velocity adjusted for the dipping
plane (lower). Compare to the coincident wavefield in
Fig. 20.

Line 3 6-Fold Stack

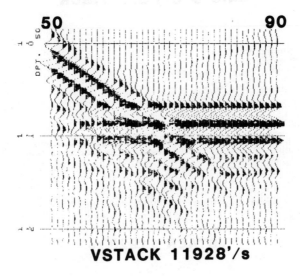

VSTACK 11928'/s

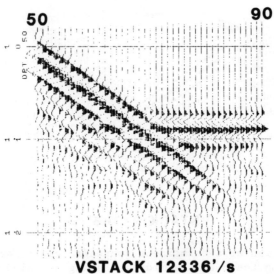

VSTACK 12336'/s

Fig. 29 An enlargement of a portion of Fig. 28 revealing the differences in the estimate of the coincident source/receiver
wavefield.

COMPARISON OF SOME SEISMIC IMAGING TECHNIQUES

Thomas T. Hu, Keith Wang and Fred Hilterman

Seismic Acoustics Laboratory
University of Houston
Houston, Texas 77004

ABSTRACT

Scalar wave theory is applied to solve three-dimensional
seismic imaging problems. Different ways of implementing the
solutions lead to various migration techniques where the areal
distribution of subsurface reflection coefficients is obtained.
Current widely used migration techniques are the Kirchhoff summation method, the frequency domain method and the finite difference method. Other algorithms include the use of Fresnel imaging,
Fraunhofer imaging, the lensless Fourier transform holography and
the wave vector diversity concepts to generate subsurface images.

Fundamental differences and similarities among these algorithms are compared on a theoretical basis. The mathematical
basis of each of the above-mentioned techniques is cast into a
similar integral form for comparison and implementation. Medium
homogeneity is assumed in the analysis for simplicity. A set of
synthetic time sections is generated; the migrated sections using
these different migration algorithms are compared. Major differences in the images are observed with some explanations.

INTRODUCTION

Migration is a procedure which transforms surface recorded
seismic sections into subsurface images, with the scatterers
located in their true spatial positions. The principle of seismic
migrations, which are basically synthetic aperture imaging techniques, is applicable to other areas of application of acoustical
imaging. For example, see the paper by Corl and Kino[1] in this
volume. Some of the current migration techniques are identified

as: Kirchhoff summation, Fresnel imaging, Fraunhofer imaging,
frequency domain and finite difference migrations. Other techniques such as the wave vector diversity method (WVD) and the
lensless Fourier transform holographic imaging (LFTH) are also
capable of producing subsurface images.

This paper compares the above-mentioned migration techniques
by examining their theoretical concepts as well as judging their
algorithms towards being practically implemented. The purpose is
to illustrate their similarities and differences. Limitations
and aberrations associated with each technique are discussed and
should be considered in order to choose appropriate techniques
for various applications. The basis of all these techniques is
the solution to the three-dimensional scalar wave equation, with
subsurface medium homogeneity being assumed.

The underlying concept of most of the migration procedures
is referred to as the downward extrapolation concept,[2] in which
case all field data are converted into the CDP format in order to
simulate zero source-receiver offset data. The reflection boundary
is considered as if it were covered by sources which fire simultaneously at time t=0. The field data can simulate this hypothetical case if the equation for two-way travel time is cast into an equation involving one-way travel time of the upgoing wave
field p(x,y,z,t). This can be done by halving the field velocity,
or by scaling down the stacked section scale. Different migration
approaches are then used to project these surface data back to
any depth z. The reflectivity, or the source strength, is found
to be the wave field at the depth z at the instant t=0.[2]

In the following section, these migration techniques are
briefly described.[2,4,5,6,7,8,9] After being analyzed mathematically, the seven techniques can be grouped into four; each represented in an equivalent form by a spatial integration over a
limited-size surface aperture so that the accuracy for each method
can be compared. Differences in these techniques are illustrated by
the flow chart diagrams, the equivalent migration formulae and
the results of implementation on synthetic data.

THEORY

Kirchhoff Migration

The Kirchhoff migration as described in Schneider's article,[2]
is a convolutional operation, operating strictly in the time-space
domain. The propagation of a seismic wave within subsurface
half-space is described by the homogeneous scalar wave equation,

$$p_{xx} + p_{yy} + p_{zz} - \frac{1}{c^2} p_{tt} = 0 \qquad (1)$$

with the boundary condition being known,

$$p(x,y,z_0 = 0,t) = p_\Sigma(\bar{x},\bar{y},0,t) \qquad (2)$$

where $p(x,y,z_0,t)$ is the wave field at $Q(x,y,z_0)$, the migrated depth point, and $p_\Sigma(\bar{x},\bar{y},0,t)$ is the data collected at $\bar{Q}(\bar{x},\bar{y},0)$, a point on the surface aperture. The geometry is shown in Figure 1.

The solution to this boundary-value problem is expressed in the form of time and space integrals of Green's functions.[2] The choice of the free surface Rayleigh-Sommerfeld Green's function leads to the Kirchhoff integral formula, with time being set to zero,

$$p(x,y,z_0,0) = -\frac{1}{2\pi}\frac{\partial}{\partial z_0}\iint d\bar{x}d\bar{y}\ \frac{p_\Sigma(\bar{x},\bar{y},0,R_k/c)}{R_k} \qquad (3)$$

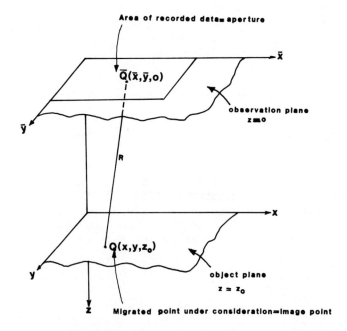

Fig. 1   Geometry of notations used in defining
the distance from $\bar{Q}$ to Q

where

$$R_K = \sqrt{(\bar{x}-x)^2 + (\bar{y}-y)^2 + z_0^2}$$

The Kirchhoff formula enables one to express $p(x,y,z,t = 0)$ at depth $z = z_0$ in terms of the surface wavefield p_Σ. The formula is thus regarded as the analytical formulation of Huygen's principle for acoustic waves underground. The field $p(x,y,z_0,t)$ is derived from the sum along a delay hyperbloid which is taken from the surface time traces $p_\Sigma(\bar{x},\bar{y},0,t + R_k/c)$. The sum of these delayed traces, adjusted by a spreading weighting factor $1/R_k$, is placed at the apex of the hyperbloid. The partial derivative with respect to depth variation provides phase correction factors.

Fresnel Migration

For objects of sufficient depth comparing to the wavelength λ, i.e., when

$$z_0^3 \gg \frac{\pi}{4\lambda} [(\bar{x}-x)^2 + (\bar{y}-y)^2]_{max}$$

the distance between Q and $\bar{Q}$ is approximated by the first two terms in its binomial expansion

$$R \approx R_{FN} = z_0 + \frac{1}{2z_0} [(\bar{x}-x)^2 + (\bar{y}-y)^2] \tag{4}$$

Thus, the migration formula (3) is approximated in this case by,

$$p(x,y,z_0,0) = -\frac{1}{2\pi} \frac{\partial}{\partial z_0} \iint dxdy \frac{p_\Sigma(\bar{x},\bar{y},0,R_{FN}/c)}{z_0} \tag{5}$$

where the phase delay is approximated by R_{FN}/c and the amplitude spreading factor is approximated by $1/z_0$. It is interesting to note that the difference between the time delay, R_{FN}/c, and the delay at apex is $(1/t_0)(x/c)^2$ which is identical to the normal moveout (NMO) correction term except for a factor of $1/2$.

Fraunhofer Migration

If the condition

$$z_0 \gg \frac{\pi}{\lambda} (x^2 + y^2)_{max}$$

is met, a further approximation can be made from Eq. (3). In this case, the phase delay is R_{FH}/c for p_{Σ}, so that the migration formula reads

$$p(x,y,z_0,0) = -\frac{1}{2\pi}\frac{\partial}{\partial z_0} \iint d\bar{x}d\bar{y} \frac{p_{\Sigma}(\bar{x},\bar{y},0,R_{FH}/c)}{z_0} \qquad (6)$$

where

$$R_{FH} = z_0 + \frac{1}{2z_0} [\bar{x}^2+\bar{y}^2-2x\bar{x}-2y\bar{y}] \qquad (7)$$

Both the Fresnel and the Fraunhofer migration techniques entail approximations when observation points fall into certain distances away from the object, i.e., the Fresnel and the Fraunhofer zones. The concepts are derived from Goodman,[3] but his treatment was on forward propagation, while here one deals with the backward propagation case.

FK Migration

The FK migration method solves the wave equations in the frequency domain.[4,5] The spectrum of the wave field for a given frequency, ω, is

$$P(k_x,k_y,z_0,\omega) = \int\limits_{-\infty}^{\infty}\!\!\!\int\!\!\!\int p(x,y,z_0,t)e^{-j\omega t}e^{j(k_x x+k_y y)}\, dx\, dy\, dt \quad (8)$$

Thus, the Helmholtz equation is expressed as

$$P_{zz}(k_x,k_y,z_0,\omega) = -(k^2-k_x^2-k_y^2)P(k_x,k_y,z_0,\omega) \qquad (9)$$

The solution describing down-going waves (backward propagation) is

$$P(k_x,k_y,z_0,\omega) = P(k_x,k_y,0,\omega)e^{j\sqrt{k^2-k_x^2-k_y^2}\,z_0}$$

$$= P(k_x,k_y,0,\omega)\cdot H \qquad (10)$$

where H is the operator characterizing the propagation process. The source strength at the point $Q(x,y,z_0)$ is expressed as,

$$p(x,y,z_0,t)\Big|_{t=0}$$

$$= \frac{1}{(2\pi)^3} \int\limits_{-\infty}^{\infty}\!\!\!\int\!\!\!\int P(k_x,k_y,0,\omega) e^{j\sqrt{k^2-k_x^2-k_y^2}z_0} e^{-j(k_x x+k_y y)}$$

$$\cdot \; dk_x dk_y d\omega \tag{11}$$

By the change of variables,

$$q = \sqrt{k^2-k_x^2-k_y^2} \tag{12}$$

and letting,

$$P'(k_x,k_y,0,q) = P(k_x,k_y,0,c\sqrt{q^2+k_x^2+k_y^2})\,\frac{cq}{\sqrt{q^2+k_x^2+k_y^2}} \tag{13}$$

the final result is given as

$$p(x,y,z_0,0) = \frac{1}{(2\pi)^3} \int\!\!\!\int\!\!\!\int P'(k_x,k_y,0,q) e^{+j(qz_0-k_x x-k_y y)}\, dq dk_x dk_y \tag{14}$$

The process is summarized in Figure 2 in its flow chart form.

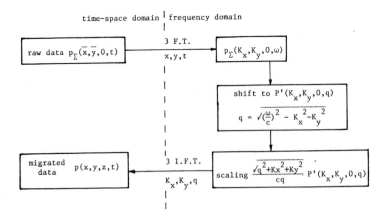

Fig. 2   F K Migration

The migration steps taken in the FK method are equivalent to the time-space operator as used in Kirchhoff migration. This is true because,

$$p(x,y,z_0,t)\Big|_{t=0}$$

$$= F^{-1}\{F[p_\Sigma(\overline{x},\overline{y},0,t)]\cdot e^{j\sqrt{k^2-k_x^2-k_y^2}z_0}\}\Big|_{t=0}$$

$$= p_\Sigma(\overline{x},\overline{y},0,t)*F^{-1}\{e^{j\sqrt{k^2-k_x^2-k_y^2}z_0}\}\Big|_{t=0}$$

$$= -\frac{1}{2\pi}\frac{\partial}{\partial z_0}[\frac{p_\Sigma(\overline{x},\overline{y},0,t+\frac{R_k}{c})}{R_k}]_{t=0} \tag{15}$$

The proof of the last step is in Appendix (A).

Finite Difference Migration

This technique treats the wave migration as an initial-value problem. Starting with the wave field measured on the surface, the unknown wave field is extrapolated step by step, using the downward extrapolation concept.

Different degrees of approximations are taken based on the exact dispersion relation,

$$k_z = (k^2-k_x^2-k_y^2)^{1/2} \tag{16}$$

or, equivalently the second order wave equation,

$$P_{zz} = \frac{1}{c^2}P_{tt}-P_{xx}-P_{yy} \tag{1}$$

For the case of 15-degree operator, only the first two terms in the binomial expansion of Eq. (16) are taken. The 3-D operator is then split into two 2-D operators, one in the x-direction and another in the y-direction. Each 2-D operator corresponds to a differential equation.

In the x-direction,

$$k_z = k - \frac{k_x^2}{2k} \tag{17}$$

the equivalent differential equation being

$$P_{xx} - \frac{2}{c} P_{tz} - \frac{2}{c^2} P_{tt} = 0 \tag{18}$$

In the y-direction,

$$k_z = - \frac{k_y^2}{2k} \tag{19}$$

the equivalent differential equation being

$$P_{yy} - \frac{2}{c} P_{tz} = 0 \tag{20}$$

Notice that the two differential equations contain derivative terms with respect to z only to the first order, making themselves easier to be solved in a numerical way within a moving reference frame.[6]

The 45-degree operator is based on the dispersion relation where a third term has been included. The split of the 3-D operator into two 2-D operators can be done by a least-mean-square optimization process.[7] Again, corresponding to each operator, a differential equation is found and solved numerically.

The flow chart of 3-D finite difference migration is summarized in Figure 3. In this paper, only the 3-D 15-degree migration is considered to show the link between finite difference method and other techniques. In addition to the error due to numerical discretization, the accuracy of the 15-degree operator is limited to the paraxial approximation it takes in the frequency domain. It is proved in Appendix B that the equivalent time-space operator for single frequency is now

$$h_\omega = \frac{e^{jkz_0}}{j\lambda z_0} e^{j \frac{k}{2z_0} (x^2 + y^2)} \tag{21}$$

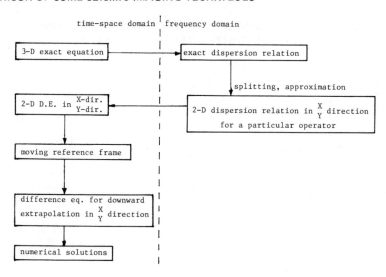

time-space domain | frequency domain

Fig. 3   3-D Finite Difference Migration

which can be recognized as the far-field approximated Rayleigh-Sommerfeld diffraction operator. Further, the broad-band operator, h, convolving with the surface data results in an integral-form migration formula that is used in Fresnel imaging (Eq. (5)). Therefore, the 15-degree finite difference migration can be said to be mathematically equivalent to the Fresnel imaging method.

The comparisons among the Kirchhoff migration, the FK migration and the finite difference techniques are listed in Table I.

Wave Vector Diversity

The wave vector diversity (W.V.D.) method uses dynamic temporal or spatial frequency variations in obtaining 3-D object information.[8,9] The basic concept is that the diffraction patterns of an object under illuminations of different temporal frequencies or different projection angles are different. Various illumination angles correspond to various spatial frequency components of the wave. If weak scattering among diffractors is assumed, the variation of diffraction patterns due to diversified wave vectors with respect to a stationary receiver can be applied to collect 2-D data as if the receiver scans. This data acquisition scheme resembles itself to the synthetic aperture technique used in radar and sonar systems. It resolves the problem in long wave holography where a large sized imaging aperture is needed. In fact, a simple receiver array could be employed to get 3-D

Table I.   Fundamental differences among Kirchhoff,
FK and Finite Difference methods

migration technique features	Kirchhoff	FK	finite difference
accuracy	exact	exact	approx. (Fresnel)
dip handling limit	theoretical all dip angles	theoretical all dip angles	migrate dips up to 15° or 45°
applicability for velocity variations	applicable to vertical and lateral variations	modified version applicable for vertical but only gentle lateral variations	applicable to vertical and lateral variations
migration noise	˙random noise tends to be coherent, might cause wrong interpre- tation ˙produces higher migration noises		˙random noise is left random ˙produces low migra- tion noise
others	˙allows weighting and muting procedure according to dip or coherency ˙generally show more event continuity	preserves the waveform characters after migration	

images that exhibit resolutions that are better than the classical
Rayleigh criterion.

Analyzing W.V.D. is equivalent to performing the Kirchhoff
migration in the opposite direction. As seen in Figure 1, the
upgoing wave scattered by the object instead of downgoing wave is
considered. Assume both the source and the receiver are located
in the far field from the object, that is, $|z_0| \gg 2\pi/\lambda |\sqrt{x^2+y^2}|_{max}$
and $|z_0| \gg 2\pi/\lambda |\sqrt{\bar{x}^2+\bar{y}^2}|_{max}$, the field spectrum at $\bar{x}$-$\bar{y}$ plane is

$$P_\Sigma(\bar{x},\bar{y},0,\omega) = \frac{-1}{j\lambda z_0} e^{-jkz_0 - j\frac{k}{2z_0}(\bar{x}^2+\bar{y}^2)}$$
$$\cdot \iint dxdy\, P(x,y,z_0,\omega) e^{j\frac{k}{z_0}(x\bar{x}+y\bar{y})} \tag{22}$$

The relationship between $P_\Sigma(\bar{x},\bar{y},0,\omega)$ and $P(x,y,z_0,\omega)$ are
thus through a simple Fourier transform since the Fraunhofer con-
dition is met. By expressing $P(x,y,z_0,\omega)$ in terms of $P_\Sigma(\bar{x},\bar{y},0,\omega)$
and performing a temporal Fourier transform, the object informa-
tion is obtained (See Appendix C),

$$p(x,y,z_0,t) = \frac{-1}{2\pi} \frac{\partial}{\partial z_0} \int\!\!\int \frac{p_\Sigma(\bar{x},\bar{y},0,t+\frac{R_W}{c})}{z_0} \, d\bar{x}d\bar{y} \qquad (23)$$

where

$$R_W = z_0 + \frac{\bar{x}^2+\bar{y}^2-2x\bar{x}-2y\bar{y}}{2z_0}$$

the migrated section is

$$p(x,y,z_0,0) = \frac{-1}{2\pi} \frac{\partial}{\partial z_0} \int\!\!\int \frac{p_\Sigma(\bar{x},\bar{y},0,R_W/c)}{z_0} \, d\bar{x}d\bar{y} \qquad (24)$$

which is the same as in Fraunhofer imaging (Eq. 7).

Lensless Fourier Transform Holography

The lensless Fourier transform holography (LFTH) is a technique that utilizes holographic concepts in reconstructing seismic images.[10] The process involves in a temporal Fourier transform, two spatial Fourier transforms and a phase compensation.

The flow chart of LFTH is shown in Figure 4. Surface data $p_\Sigma(\bar{x},\bar{y},0,t)$ are temporally Fourier transformed to show their spectral components at a specific temporal frequency. The phase portion of the gather is then mixed with a reference phase factor which varies with the receiver location. This step acts as if a hologram is made at the recording aperture using a reference wave generated from a synthetic point source. This broadband reference source point is located at (x_c,y_c,z_0) (See Figure 5).

The component generated for the interference pattern contains the product of the object amplitude, the object phase factor and the conjugate of the reference phase factor. This is the term of interest and is reconstructed to obtain the images by going through two spatial Fourier transforms and a temporal transform. The aperture size on the x-y plane is prearranged so that ordinary FFT algorithms can be applied.[10]

The LFTH approach provides us seismic images with some inherent aberrations. The reconstructed image is a good representation of the original object only in regions where the spherical reference wave has a similar curvature with the object wave at the recording plane. For those migrated points that are not near enough to the reference point, the reconstructed images are distorted. In order to limit the aberrations, more than one

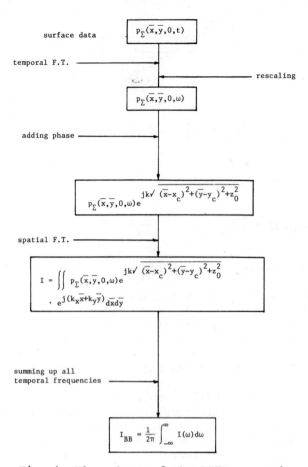

Fig. 4   Flow chart of the LFTH operation

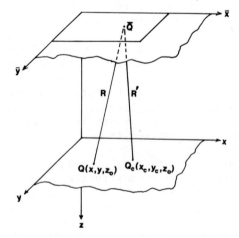

Fig. 5   Geometry of points used in LFTH

synthetic reference points must be used.

If the broadband LFTH image is defined to be,

$$I_{BB} = p(x,y,z_0,0)$$

$$= \frac{1}{2\pi} \iiint P_{\Sigma}(\bar{x},\bar{y},0,\omega) e^{jk\sqrt{(\bar{x}-x_c)^2+(\bar{y}-y_c)^2+z_0^2}-jk_x x-jk_y y}$$

$$\cdot \, d\bar{x}d\bar{y}d\omega \tag{25}$$

then

$$I_{BB} = \frac{1}{2\pi} \iiint P_{\Sigma}(\bar{x},\bar{y},0,\omega) e^{jk[\sqrt{(\bar{x}-x_c)^2+(\bar{y}-y_c)^2+z_0^2} - \frac{\bar{x}x}{z_0} - \frac{\bar{y}y}{z_0}]}$$

$$\cdot \, d\bar{x}d\bar{y}d\omega$$

$$= \frac{1}{2\pi} \iiint P_{\Sigma}(\bar{x},\bar{y},0,t) e^{jkR_{LF}} e^{-j\omega t} \, dt \, d\bar{x}d\bar{y}d\omega$$

$$= \iiint P_{\Sigma}(\bar{x},\bar{y},0,t)\delta(t- \frac{R_{LF}}{c}) \, dt \, d\bar{x}d\bar{y}$$

$$= \iiint P_{\Sigma}(\bar{x},\bar{y},0, \frac{R_{LF}}{c}) \, d\bar{x}d\bar{y} \tag{26}$$

where

$$R_{LF} = \sqrt{(\bar{x}-x_c)^2+(\bar{y}-y_c)^2+z_0^2} - \frac{\bar{x}x+\bar{y}y}{z_0}$$

Compared with the operator of Kirchhoff migration, several differences are observed. Firstly, the time delay for each trace is different. Secondly, the amplitude scaling factor is not included. Finally, the result of integration is not differentiated with respect to z_0.*

COMPARISON OF TECHNIQUES

Table II gives a summary of the equivalent time-space operators in integral forms for various techniques. If we are allowed to compare these equivalent operators all in the time-

---

*However, such operations could be readily accomodated in the LFTH technique.

Table II.   Summary of time-space operators

Migration Technique	Transform Domain	Solution Exactness	Integral Form of p(x,y,z_0,0)	Factor R	Group
Kirchhoff	Space-time	exact	$\frac{-1}{2\pi}\frac{\partial}{\partial z_0}\iint d\bar{x}d\bar{y}\,\frac{p_\Sigma(\bar{x},\bar{y},0,R/c)}{R}$	$\sqrt{(\bar{x}-x)^2+(\bar{y}-y)^2+z_0^2}$	I
Fresnel	Space-time	approx.	$\frac{-1}{2\pi}\frac{\partial}{\partial z_0}\iint d\bar{x}d\bar{y}\,\frac{p_\Sigma(\bar{x},\bar{y},0,R/c)}{R}$	$z_0+\frac{1}{2z_0}((\bar{x}-x)^2+(\bar{y}-y)^2)$	II
Fraunhofer	Space-time	far-field approx.	$\frac{-1}{2\pi}\frac{\partial}{\partial z_0}\iint d\bar{x}d\bar{y}\,\frac{p_\Sigma(\bar{x},\bar{y},0,R/c)}{z_0}$	$z_0+\frac{1}{2z_0}(\bar{x}^2+\bar{y}^2-2\bar{x}x-2\bar{y}y)$	III
FK	frequency	exact	$\frac{-1}{2\pi}\frac{\partial}{\partial z_0}\iint d\bar{x}d\bar{y}\,\frac{p_\Sigma(\bar{x},\bar{y},0,R/c)}{R}$	$\sqrt{(\bar{x}-x)^2+(\bar{y}-y)^2+z_0^2}$	I
Finite Difference (15°)	Space-time	approx.	$\frac{-1}{2\pi}\frac{\partial}{\partial z_0}\iint d\bar{x}d\bar{y}\,\frac{p_\Sigma(\bar{x},\bar{y},0,R/c)}{z_0}$	$z_0+\frac{1}{2z_0}((\bar{x}-x)^2+(\bar{y}-y)^2)$	II
Lensless F.T.H.	temporal frequency	approx.	$\iint d\bar{x}d\bar{y}\,p_\Sigma(\bar{x},\bar{y},0,R/c)$	$\sqrt{(\bar{x}-x_c)^2+(\bar{y}-y_c)^2+z_0^2}-\frac{xx+yy}{z_0}$	IV
W.V.D.	frequency	far-field approx.	$\frac{-1}{2\pi}\frac{\partial}{\partial z_0}\iint d\bar{x}d\bar{y}\,\frac{p_\Sigma(\bar{x},\bar{y},0,R/c)}{z_0}$	$z_0+\frac{1}{2z_0}(\bar{x}^2+\bar{y}^2-2\bar{x}x-2\bar{y}y)$	III

space domain, Kirchhoff migration is equivalent to the FK method, 15 degree finite difference technique is equivalent to Fresnel imaging and the W.V.D. is equivalent to Fraunhofer imaging. Thus, all seven techniques can be considered as classifiable into four groups. Kirchhoff and the FK methods belong to Group I, conceptually the most accurate migration techniques. Group II, consisting of the 15-degree finite difference technique and the Fresnel imaging technique, is an approximate approach. Group III involves further mathematical approximations and includes the wave vector diversity and the Fraunhofer imaging techniques. Lensless Fourier Transform Holography belongs to a fourth group, with an error factor depending on the location of the object point relative to the reference point.

Comparison of the basis of the "delay and sum" curves which the migration formulae reveal may very well indicate the fundamental differences among all seven techniques. Effects due to other factors such as the spreading loss and the vertical distance are not considered because they are comparatively minor to the delay "R" factor and could be accomodated.

As an example, Figure 6 shows a 2-D case where a LFTH reference point is located on the z-axis and at the depth of 5000 feet. The object points at the same depth has a lateral offset x, or an offset angle θ. Again with the assumption that the surbsurface medium is homogeneous, the diffraction curve due to point object Q as recorded along x axis is a hyperbola whose apex coincides with Q. The delay curves used in the migration techniques are plotted in both Fig. 7 and Fig. 8. The abscissa, x_1, represents

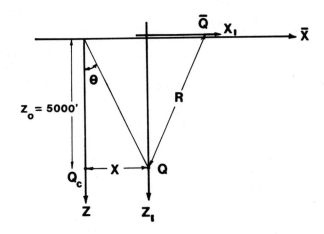

Fig. 6   An example for analyzing delay curves

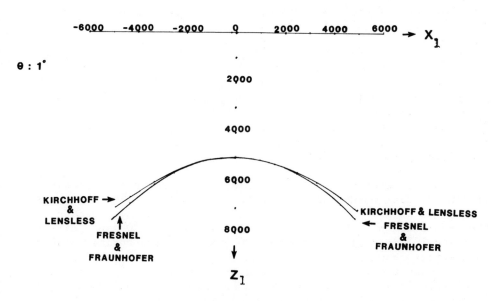

Fig. 7   Delay curves for θ = 1°

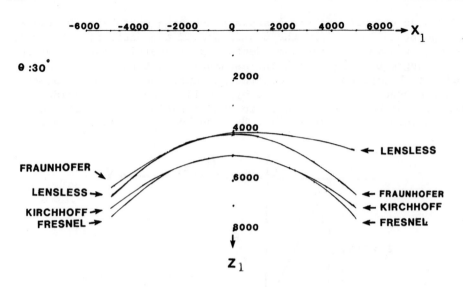

Fig. 8   Delay curves for θ = 30°

the coordinate difference between $\bar{x}$ and x. The ordinate, z_1, passes
through object location. The summation curve for Kirchhoff migration coincides with the diffraction curve, therefore, is considered as an aberration-free standard. Any deviation from this curve
would cause aberrations in the final migrated images.

The case of small offset angle (θ=1°) is examined in Figure
7. The curve for LFTH is very close to the standard hyperbolic
Kirchhoff curve, while Fresnel and Fraunhofer curves are a bit
off because they are parabolas. Figure 8 shows the curves for a
larger offset angle (θ=30°). In addition to shape differences,
these curves have dissimilarities in their z-axis intersections.
Fresnel imaging still migrates to the correct depth of 5000' while
Fraunhofer imaging and LFTH place the migrated energy to erroneous
positions, which result in a large portion of the image aberrations.
The LFTH curve is a hyperbola tilted to the left side because of
the linear term in the "R" factor.

A set of synthetic data is collected for demonstrating migration results using formulae listed in Table II. The object is
assumed to be a dipping triangle with one corner positioned on
the z-axis at the depth of 5000'. Dip angles are 30° in y direction and 10° in x direction. The CDP data are synthesized within
an aperture shown in Figure 9, with 100' receiver separation both
in the x and the y directions. One line section at the x-coordi-
nate of 100' is presented in Figure 10 to show the diffracted
energy location. The actual cross-section of the diffractor
location is shown as a tilted line segment at the upper left

portion.   The subsurface velocity is assumed to be 10,000 ft/sec.

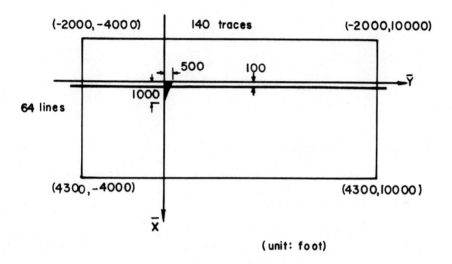

(unit: foot)

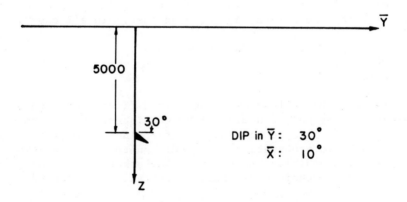

Fig. 9   Geometry for collecting synthetic data

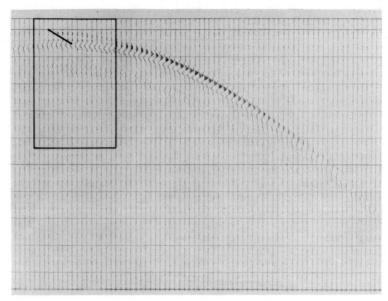

Fig. 10   A time section collected along a line with
          coordinate X = 100.   The tilted solid line
          segment at the upper left is the diffractor.
          The framed block is to be migrated and blown
          up in Fig. 11.

The same portions of the migrated sections are shown in Figure 11.   Part (A) is the result of applying Kirchhoff migration
formula.   The tilted solid line represents the actual diffractor
location.   In this case, the diffracted energies collapse very
well, although the spreading effect has not been taken care of
in our actual migration algorithm.   In Part (B) the Fresnel
method migrated section has worse resolution simply because a 15°
operator is applied for the 30° dipping case.   In all cases, a
migration velocity of 10,000 ft/sec is used.   Results for other
migration techniques are to be investigated.

CONCLUSIONS

This is a theoretical exercise toward finding out fundamental
differences among selected migration techniques.   All techniques
are cast into comparable mathematical formulae for migration, or
known as backward propagation.   In addition to the theoretical
dissimilarities exhibited in the formula, other limitations may
appear for each technique as it is practically implemented.

In this analysis, media homogeneity is assumed.   Based on
this assumption, the summation curves for various migration techniques are compared.   An example of Kirchhoff and Fresnel mi-

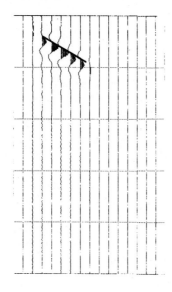
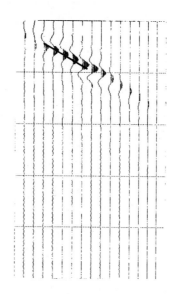

(A) Kirchhoff   (B) Fresnel

Fig. 11 Migrated time sections

grated sections are given to show that the computer program is
working. The differences of these techniques in terms of calculation are merely a few statements within the computer program,
although selecting appropriate aperture function is much more
complex a step in cases other than Kirchhoff and Fresnel migrations.

ACKNOWLEDGEMENT

The authors wish to acknowledge Mr. Luh-Cheng Liang for his
help in running computer programs, and Ms. Michele Maes for typing the manuscript.

REFERENCES

1.   P. D. Corl, and G. S. Kino, "A Real-Time Synthetic-Aperture
     Imaging System", Acoustical Holography, V. 9, Wang Ed.,
     Plenum Press, New York, pp. 000-000.

2.   W. A. Schneider, "Integral Formation for Migration in Two
     and Three-Dimensions", Geophysics, V. 43, pp. 49-76,
     (1978).

3.   J. W. Goodman, <u>Introduction of Fourier Optics</u>, McGraw-Hill,
      New York, (1968).

4.   R. H. Stolt, "Migration by Fourier Transform", <u>Geophysics</u>,
      V. 43, pp. 23-48, (1978).

5.   G. Bolondi, F. Rocca, and S. Savelli, "A Frequency Domain
      Approach to Two-Dimensional Migration", <u>Geophysical Pros-
      pecting</u>, V. 26, pp. 750-772, (1978).

6.   J. F. Claerboud, <u>Fundamentals of Geophysical Data Process-
      ing</u>, McGraw-Hill, New York, (1976).

7.   D. Risto and W. Houba, "3-D Finite Difference Migration",
      Presented at the EAEG 41st Meeting, Hamburg, (1979).

8.   N. H. Farhat, and C. K. Chan, "Three-Dimensional Imaging by
      Wave-Vector Diversity", Presented at the 8th Symposium
      on Acoustical Imaging, Key Biscayne, Fla., May 1978.

9.   P. Reghunathan, "Zero-Source-Receiver-Offset Imaging with
      Wave-Vector Diversity", Master's Thesis, University of
      Houston, (1980).

10.   T. Morgan, G. Fitzpatrick, F. Hilterman, K. Wang, "Lensless
      Fourier Transform Holography as Applied to Imaging of Re-
      flection Seismic Data", <u>Acoustical Holography</u>, V. 9,
      Wang Ed., Plenum Press, New York, pp. 000-000.

APPENDIX A

Derivation of Space-Time Operator for FK Migration

The frequency domain operator of wave propagation is

$$H = \exp(j\sqrt{k^2 - k_x^2 - k_y^2}\; z_0) \tag{A-1}$$

where k is the wave number, k_x and k_y are the x-and y- components of wave vector $\vec{k}$, z_0 is the propagation distance in z-direction. The equivalent operator in the time-space domain, h, is obtained by performing two spatial Fourier transforms and one temporal Fourier transform on H. In the following, F_s, F_T and F_{sT} are used to denote the double spatial Fourier transform, the temporal Fourier transform and the triple spatial-temporal Fourier transform, respectively. Let h_ω be the result of the double spatial Fourier transform

$$h_\omega = F_s(H) = \frac{1}{(2\pi)^2} \iint\limits_{-\infty}^{\infty} dk_x dk_y\, e^{j\sqrt{k^2 - k_x^2 - k_y^2}\; z_0 - j(k_x x + k_y y)} \tag{A-2}$$

If new variables r, ϕ, ρ and θ are introduced such that

$$x = r \cos \phi \qquad\qquad y = r \sin \phi$$

$$k_x = 2\pi\rho \cos \theta \qquad\qquad k_y = 2\pi\rho \sin \theta$$

then

$$h_\omega = \frac{1}{2\pi} \int_0^{\infty} \rho d\rho\; e^{j\sqrt{k^2 - 4\pi^2\rho^2}\; z_0}\; J_0(2\pi\rho r) \tag{A-3}$$

where $J_0(u)$ is the Bessel function of zero-th order.

By using the table of transformations,* it is easy to show that

---

*See, for example, "Tables of Integral Transforms", V. 2, A. Erdelyi, ed., McGraw-Hill, New York, 1954, and Flesher and Elliott, "Acoustical Aperture Diffraction in a Transversely Moving Medium", Acoustical Holography, V. 7, L. Kessler ed., Plenum Press, New York, pp. 611-630.

$$h_\omega = [-j \frac{z_0}{\lambda} + \frac{z_0}{2\pi\sqrt{x^2+y^2+z_0^2}}] \frac{e^{jk\sqrt{x^2+y^2+z_0^2}}}{x^2+y^2+z_0^2} \tag{A-4}$$

$$= \frac{-1}{2\pi} \frac{\partial}{\partial z_0} [\frac{e^{jkR}}{R}]$$

where

$$R = \sqrt{x^2+y^2+z_0^2} \tag{A-5}$$

Finally,

$$h = F_{ST}\{H\}$$

$$= F_T\{h_\omega\}$$

$$= \frac{-1}{2\pi} \frac{\partial}{\partial z_0} [\frac{\delta(t+\frac{R}{c})}{R}] \tag{A-6}$$

APPENDIX B

Derivation of the Space–Time Operator for 15–Degree Approximated

Finite Difference Method

Derivation for the 15–degree approximation operator is in parallel to the one for FK migration operator (see Appendix A). The approximated frequency domain operator is

$$H = \exp[j(k - \frac{k_x^2 + k_y^2}{2k})z_0] \qquad (B-1)$$

After applying similar changes of variables and Hankel transform pairs, the spatial Fourier transformed operator is

$$h_\omega = \frac{ke^{jkz_0}}{j2\pi z_0} e^{j\frac{k}{z_0}(x^2 + y^2)} \qquad (B-2)$$

Thus

$$h = F_T\{h_\omega\}$$

$$= -\frac{1}{2\pi} \frac{\partial}{\partial z_0} \{\frac{\delta(t + \frac{R}{c})}{z_0}\} \qquad (B-3)$$

where

$$R = z_0 + \frac{x^2 + y^2}{2z_0}$$

APPENDIX C

Derivation of the Space-Time Operator for Wave Vector Diversity
Method

Considering the upgoing wave, or the forward propagated
wave, the relationship between the surface data spectrum
$P_\Sigma(x,y,0,\omega)$ and the migrated data spectrum $P(x,y,z_0,\omega)$ is depicted in Equation (22). Expressing $P(x,y,z_0,\omega)$ in terms of
$P_\Sigma(x,y,0,\omega)$ gives

$$P(x,y,z_0,\omega) = \frac{1}{j\lambda z_0} \iint d\bar{x}d\bar{y}\ P_\Sigma(\bar{x},\bar{y},0,\omega)$$

$$\cdot\ e^{jkz_0 + j\frac{k}{2z_0}(\bar{x}^2 + \bar{y}^2) - j\frac{k}{z_0}(\bar{x}x + \bar{y}y)} \qquad\qquad (C-1)$$

or

$$p(x,y,z_0,t) = \frac{1}{2\pi}\int d\omega\ P(x,y,z_0,\omega)e^{j\omega t}$$

$$= \frac{-1}{2\pi}\frac{\partial}{\partial z_0}\iint d\bar{x}d\bar{y}\ \frac{P_\Sigma(\bar{x},\bar{y},0,t+\frac{R_W}{c})}{z_0}$$

where

$$R_W = z_0 + \frac{\bar{x}^2 + \bar{y}^2 - 2\bar{x}x - 2\bar{y}y}{2z_0}$$

LENSLESS FOURIER TRANSFORM HOLOGRAPHY AS APPLIED TO IMAGING

OF REFLECTION SEISMIC DATA

T. Morgan, G. Fitzpatrick, F. Hilterman, and K. Wang

Seismic Acoustics Laboratory
University of Houston
Houston, Texas 77004

INTRODUCTION

Application of lensless Fourier transform holography (LFTH) to the reconstruction of seismic data has led to a better understanding of the processes involved, and to the solution of a number of problems related to the reconstruction of multifrequency images.

Theoretical activities have been aimed at the comparison of LFTH to other imaging techniques through an analysis stemming from the Huygens-Fresnel diffraction formula, which can be linked to the more classical approach of optical holography (see Gabor, 1948; Leith and Upatnieks, 1962; Stroke, 1964 and 1965). A Fresnel analysis technique, used as an approximate representation of LFTH, facilitated a better understanding of the spatial frequencies encoded on the hologram. Also, the nature of image size scaling, due to a change of the temporal frequency (or depth or velocity) of reconstruction, and the characteristics of spatial and temporal frequency effects resulting in image aberrations are better known.

Practical application of these insights has resulted in a successful interpolation scheme that allows fast Fourier transform (FFT) reconstruction of scaled images. This has permitted image summation over a band of frequencies, which has produced improved reconstructions by the elimination of out-of-plane image distortion.

THEORETICAL DEVELOPMENT

Wavefield From An Impulsive Point Source

We begin by describing the waveform set-up by an impulsive

761

point source;

$$A(\bar{x},\bar{y},\bar{z})\ \delta(t-\frac{R_O}{C_O})$$

where,

$$A(\bar{x},\bar{y},\bar{z})=\frac{1}{R_O}= [z_O^2 + (x_O-\bar{x})^2 + (y_O-\bar{y})^2]^{-1/2}.$$

This represents an expanding spherical wave in a medium of constant velocity C_O. The notation for a recording array in the $\bar{x},\bar{y},\bar{z}$ - system is shown in Figure 1.

The Fourier Transform of the wavefield on the aperture Σ (plane $\bar{x}$, $\bar{y}$, and $\bar{z} = 0$) to the temporal frequency domain is

$$\int_{-\infty}^{+\infty} A(\bar{x},\bar{y},o)\delta(t-\frac{R_O}{C_O})e^{-j\omega t}dt = A(\bar{x},\bar{y},o)e^{-j\frac{\omega}{C_O}R_O} \quad , \qquad (1)$$

which will be designated as,

$$B(\bar{x},\bar{y},o;\omega) = A(\bar{x},\bar{y},o)e^{-j\frac{\omega}{C_O}R_O} \quad . \qquad (2)$$

The expression (2) describes a plane wave reference hologram.

The above described impulsive point source can be used to represent a reflection point in common-depth-point reflection seis-mology or a point object in acoustical imaging. A series of geolo-gic interfaces, or an extended object, may then be represented by a supperposition of the contributions of many such reflection points or point objects.

For simplicity we will deal with only one impulsive point source located somewhere in the object space, that is in the volume beneath the aperture plane $z = 0$. We will not, however, assume that its location is known. This means that the entire object space (or at least a large portion of it) must be reconstructed in order to assure that our impulsive point source has been imaged.

The data recorded on the aperture Σ is in the form of zero source-receiver offset time traces, which is analogous to the pulse echo mode in acoustical imaging with a single receiver recording for a single pulse. (It should also be noted that the actual signal generated in the field is band limited.)

Monochromatic Reconstruction

The Huygens-Fresnel diffraction formula can be used to recon-struct a total three dimensional depth section or image from the data recorded on the x,y plane, yielding the impulsive point source in its original position on the x_O, y_O plane. For an infinite aperature all reconstruction points, other than the single object point of our model, will be zero. The reconstruction of the plane containing the point source occurs when we progagate the surface wavefield down to a depth $z = z_0$, we have (from Goodman, 1968, equation 3-26):

$$B(x_0,y_0,z_0;\omega) = \frac{\omega}{j2\pi C_0} \iint_\Sigma B(\overline{x},\overline{y},o;\omega)\frac{e^{j\omega\frac{R_0}{C_0}}}{R_0} \cos(\theta)\,d\overline{x}\,d\overline{y}. \qquad (3)$$

The above integration over Σ must be performed for every output point on plane z_0. Since, for our example, this is the only plane that contains any non-zero image information, it will be the only one considered. For more complicated objects, which have vertical as well as lateral extend, the process must be repeated for a number of different depths.

In order to make the process more manageable, some simplifying assumptions will be made. First, it is assumed that the angle θ (see Figure 1) is less than $15°$, which will occur for sufficiently large z with respect to the maximum aperture dimension. Therefore, $\cos\theta \simeq 1$ and $R_0 \simeq z_0$. Incorporating the small angle approximation yields for equation (3)

$$B(x_0,y_0,z_0;\omega) = \frac{\omega}{j2\pi C_0 z_0} \iint_\Sigma B(\overline{x},\overline{y},o;\omega)\; e^{j\frac{\omega R_0}{C_0}}\,d\overline{x}\,d\overline{y}. \qquad (4)$$

We will now study the effects of propagating with an erroneous phase factor. That is, we will compare the phase term of (4) for (x_0,y_0,z_0) with a phase term appropriate for a neighboring location (x,y,z) a distance $(\Delta x,\Delta y,\Delta z)$ away. A relationship between the two will be formed, allowing the formulation of a reconstruction process that approximately compensates for the differences between the erroneous and proper phase terms. The distance R (see Figure 2) can be defined as,

$$R = [z^2 + (x-\overline{x})^2 + (y-\overline{y})^2]^{1/2}. \qquad (5)$$

Shifting the coordinate frame laterally to the reference position of the erroneous propagator (see Figure 3) via the following transform,

$$\tilde{x} = \overline{x}-x \qquad (6)$$
$$\tilde{y} = \overline{y}-y$$

yields for R_0 and R,

$$R_0 = [(z+\Delta z)^2 + (\Delta x-\tilde{x})^2 + (\Delta y-\tilde{y})^2]^{1/2} \qquad (7a)$$
$$R = [z^2 + \tilde{x}^2 + \tilde{y}^2]^{1/2} \qquad (7b)$$

where $\Delta x = x_o - x$; $\Delta y = y_o - y$ and $\Delta z = z_o - z$.

Expanding relations (7a and b) gives,

$$R_o = z + 1/2\,\frac{\tilde{x}^2 + \tilde{y}^2}{z} + \left[1/2\frac{(\Delta x^2 - 2\Delta x \tilde{x})}{z} + 1/2\,\frac{(\Delta y^2 - 2\Delta y \tilde{y})}{z} + \right.$$

$$\left. \frac{\Delta z^2}{2z} + \Delta z \right] + \varepsilon_o \qquad\qquad (8a)$$

$$R = z + 1/2\,\frac{\tilde{x}^2 + \tilde{y}^2}{z} + \varepsilon \qquad\qquad (8b)$$

where ε_o and ε represent higher order terms. Term by term comparison of expressions (8a) and (8b) yields the result:

$$R_o = R + \frac{\Delta x^2 + \Delta y^2}{2z} - \frac{\Delta x \tilde{x} + \Delta y \tilde{y}}{z} + \Delta z \left(1 + \frac{\Delta z}{2z}\right) + (\varepsilon_o - \varepsilon) \qquad (9)$$

where we have now expressed R_o in terms of R.

Substitution of equation (9) into the exponential phase factor of equation (4) gives:

$$B(x_o - x, y_o - y, z_o; \omega) = \frac{\omega}{j 2\pi C_o z_o} \iint\limits_\Sigma B(\tilde{x}, \tilde{y}, o; \omega) e^{j\frac{\omega}{C_o} R}\, e^{j\frac{\omega}{2C_o z}(\Delta x^2 + \Delta y^2)}$$

$$e^{j\frac{\omega}{C_o}\Delta z\left(1 + \frac{\Delta z}{2z}\right)} e^{j\frac{\omega}{C_o}(\varepsilon_o - \varepsilon)} e^{-j\frac{\omega}{C_o z_o}(\Delta x \tilde{x} + \Delta y \tilde{y})}\, d\tilde{x}\, d\tilde{y}$$

$$(10)$$

Noting that $d\tilde{x} = d\overline{x}$ and $d\tilde{y} = d\overline{y}$.

If the contribution of the higher order terms in equations (8a & b) is small, which will be true when Δx and Δy are small with respect to z, the forth exponential on the right of equation (10) can be neglected. It should be noted that the resulting approximation is more accurate than the paraxial approximation which leads to Fresnel imaging. This is due to the fact that some of the higher order terms, namely those in $\tilde{x}$ and $\tilde{y}$ alone will cancel in the difference $\varepsilon_o - \varepsilon$. All terms in x_o, y_o; x, y; and all cross terms will not be accounted for and thus lead to errors. (For a discussion of specific kinds of errors see Meier, 1965.)

Terms inside the integrals of equation (10) which do not depend on the variables of integration can be transferred to the image side, allowing us to define a "perturbed image,"

$$\hat{B}(x_o - x, y_o - y, z_o; \omega) = \frac{\omega}{j 2\pi C_o z_o} \iint\limits_\Sigma \left[B(\tilde{x}, \tilde{y}, o; \omega) e^{j\frac{\omega}{C_o} R} \right]$$

$$e^{-j 2\pi (f_{\tilde{x}} \tilde{x} + f_{\tilde{y}} \tilde{y})}\, d\tilde{x}\, d\tilde{y} \qquad\qquad (11)$$

where,

$$\hat{B}(x_o-x, y_o-y, z_o; \omega) = B(x_o-x, y_o-y, z_o; \omega) \ e^{-j\frac{\omega}{2C_o z_o}(\Delta x^2 + \Delta y^2)}$$

$$e^{-j\frac{\omega}{C_o}\Delta z(1+\frac{\Delta z}{2z})} \tag{12}$$

and (with $f = \omega/2\pi$),

$$f_{\tilde{x}} = \frac{f(x_o-x)}{C_o z_o} \ ; \ f_{\tilde{y}} = \frac{f(y_o-y)}{C_o z_o} \tag{13}$$

In expression (11) we see that the image has now been defined in terms of a reconstruction process involving the erroneous phase term. The perturbed image $\hat{B}$ can be obtained by multiplying the data array by the phase term $e^{j\omega R/C_o}$ (a process analogous to applying a spherical wave reference as described by Stroke, 1966; or Leith & Upatnieks, 1967) and spatially Fourier transforming. The spatial frequencies given by relations (13) having been encoded on the data array by the operation in brackets on the righthand side of equation (11), define the lensless Fourier transform hologram (Goodman, 1968, p. 228). The computational advantages of equation (11) over the process defined in (4) can be readily seen.

The image, or reconstructed reflection point is exact (under the contraints of Huygens-Fresnel diffraction theory and the small angle approximation) only for the location $\Delta x = \Delta y = \Delta z = 0$. (It should be noted that at this point $\varepsilon_o = \varepsilon = 0$.) For image points away from $\Delta x = \Delta y = \Delta z = 0$ phase errors as described by equation (12) will contribute (along with the difference $\varepsilon_o - \varepsilon$). The first exponential on the right side of equation (12) is an image phase error that depends on the lateral distances $\Delta x = x_o - x$ and $\Delta y = y_o - y$. The second exponential is a function of the depth difference $\Delta z = z_o - z$. Both phase errors are seen to depend on temporal frequency, and can be calculated from known quantities and applied to the image if desired.

Recovery or reconstruction of the original object information for an entire plane can now be accomplished with a two dimensional Fourier Transform. For the example of a single impulsive point source a monochromatic solution of equation (11) with $\Delta z = 0$, that is $z = z_o$, will yield an image of the object. For objects with vertical extent, however, the monochromatic output of equation (11) for a particular depth plane will also contain image information from object points in other planes. Therefore, the lateral phase factor mentioned above must be applied, followed by frequency averaging of the image data. Frequency summation or averaging will lead to phase cancellation and suppression of out of plane image points.

It is important to recognize that the spatial frequencies of equations (13) are functions of the propagation parameters ($\omega, C_o, z,$ $\Delta x = x_o - x$, and $\Delta y = y_o - y$). They, therefore, establish the spacing and location of the output data when a harmonic analysis, such as an FFT algorithm, is used to reconstruct the holograms. This is discussed in more detail in the following section.

DISCRETE FORMULATION

The discrete form of the temporal Fourier transform of the recorded field of equation (1) expressed in $\tilde{x}, \tilde{y}$ coordinates is given by

$$B(\tilde{x},\tilde{y},o;\omega) = B(\tilde{x},\tilde{y},o;2\pi m\delta f) = B(\tilde{x},\tilde{y},o;2\pi \frac{m}{M\delta t}) =$$

$$\sum_{n=0}^{N-1} \Delta t B(\tilde{x},\tilde{y},o;n\delta t) e^{-j2\pi \left(\frac{mn}{M}\right)}, \quad \text{for } m=0,1,\ldots M-1 \tag{14}$$

where M is the number of output temporal frequency samples spaced $\delta f = \frac{1}{M\delta t}$ apart.

Before writing the discrete version of the spatial Fourier transform, we will define the lensless Fourier transform hologram as follows,

$$H(\tilde{x},\tilde{y},o;\omega) = B(\tilde{x},\tilde{y},o;\omega)e^{j\frac{\omega}{C_o}R} = B(\tilde{x},\tilde{y},o;2\pi \frac{m}{M\delta t})e^{j\frac{\omega}{C_o}R} \tag{15}$$

Discretization of this spatial transform yields,

$$\hat{B}(x_o - x, y_o - y, o; \omega) = \hat{B}[p\delta(\Delta x), q\delta(\Delta y), z_o; \omega] = \sum_{u=0}^{U-1} \sum_{v=0}^{V-1} \delta\tilde{x}\delta\tilde{y}$$

$$H[u\delta\tilde{x}, v\delta\tilde{y}, o; \omega]e^{-j2\pi(p\delta f_{\tilde{x}} u\delta\tilde{x} + q\delta f_{\tilde{y}} v\delta\tilde{y})} \tag{16}$$

$$\text{for } p=0,1,\ldots P-1$$
$$q=0,1,\ldots Q-1$$

If the location of the reference (x,y) corresponds to a recording aperture grid point location $(\tilde{x},\tilde{y})$ the discretized input and output grids will align at this point. Due to the manner in which spatial frequencies are encoded on the hologram, as given by equation (13), the output grid spacing will be a scaled version of the input grid spacing when an FFT algorithm is used,

$$\delta f_{\tilde{x}} = \frac{1}{P\delta\tilde{x}} = \frac{f\delta(\Delta x)}{C_o z_o} \rightarrow \delta(\Delta x) = \frac{C_o z_o}{fP\delta\tilde{x}} \tag{17}$$

$$\delta f_{\tilde{y}} = \frac{1}{Q\delta\tilde{y}} = \frac{f\delta(\Delta y)}{C_o z_o} \rightarrow \delta(\Delta y) = \frac{C_o z_o}{fQ\delta\tilde{y}}$$

where $\delta(\Delta x)$ and $\delta(\Delta y)$ are the output spacing, and P and U are the number of output points in the x_o and y_o directions.

It can be seen from equation (17) that $\delta(\Delta x), \delta(\Delta y)$ will generally not be equal to the input grid spacing $\delta\tilde{x}, \delta\tilde{y}$ and will vary with C_o,

z_0, and f (which will become troublesome when making multifrequency images). Substituting equations (17) into equation (16) gives,

$$\hat{B}\left(p\frac{C_0 z_0}{fP\delta x},\ q\frac{C_0 z_0}{fQ\delta\tilde{y}},\ z_0;\omega\right) = \sum_{u=0}^{U-1}\sum_{v=0}^{V-1}\delta\tilde{x}\delta\tilde{y}\ H(u\delta\tilde{x},v\delta\tilde{y},0;\omega)$$
$$e^{-j2\pi\left(\frac{up}{P}+\frac{vq}{Q}\right)}.$$

(18)

By combining equations (14), (15) and (18) a final expression suitable for discrete calculation is reached,

$$\hat{B}\left(p\frac{C_0 z_0}{fP\ \delta\tilde{x}},\ q\frac{C_0 z_0}{fQ\delta\tilde{y}},\ z_0;\ 2\pi\frac{m}{M\delta t}\right) = \frac{\omega}{j2\pi C_0 z_0}\sum_{u=0}^{U-1}\sum_{v=0}^{V-1}\delta\tilde{x}\delta\tilde{y}$$
$$\left[\sum_{n=0}^{N-1}\delta tB(u\delta\tilde{x},v\delta\tilde{y},0;n\delta t)e^{-j2\pi\frac{mn}{M}}\right]e^{j\frac{\omega}{C_0 z_0}R_0}e^{-j2\pi\left(\frac{up}{P}+\frac{vq}{Q}\right)}.$$

(19)

In order to obtain a particular output spacing $\delta(\Delta x)',\delta(\Delta y)'$ on (x_0,y_0) for given reconstruction parameters C_0, z_0 and f, the FFT output can be resampled directly. An alternative method would be to resample the FFT input to a new grid spacing.

$$\delta\tilde{x}' = \frac{C_0 z_0}{fP\delta(\Delta x)'}\ ;\quad \delta\tilde{y}' = \frac{C_0 z_0}{fQ\delta(\Delta y)'}$$

(20)

All Fourier transforms are done with FFT algorithms on an array processor with a resampling on $\tilde{x}$, $\tilde{y}$ (as given by equation (20) above) via bilinear interpolation done immediately after the temporal FFT to insure the proper output spacing. Calculation of the propagation term should also reflect the resampling to new grid spacing.

DISCUSSION OF RESULTS

Error Analysis

Propagation via an erroneous phase factor leads to the following result,

$$e^{j\frac{\omega}{C_0}R_0} = e^{j\frac{\omega}{C_0}R}e^{j\frac{\omega}{2C_0 z}(\Delta x^2+\Delta y^2)}e^{j\frac{\omega}{C_0}\Delta z(1+\frac{\Delta z}{2z})}$$
$$e^{j\frac{\omega}{C_0}(\varepsilon_0-\varepsilon)}e^{-j\frac{\omega}{C_0 z}(\Delta x\tilde{x}+\Delta y\tilde{y})}$$

(21)

which has been rewritten from equation (10). The correct phase factor is expressed in terms of the erroneous one and four additional phase correction terms. The first two are in terms of Δx, Δy and Δz, and relate the location (x_0,y_0,z_0) to the location (x,y,z). They contribute purely lateral and verticle distortions, respectively. The third additional term represents the difference in all higher order terms between $e^{j\frac{\omega}{C_0}R_0}$ and $e^{j\frac{\omega}{C_0}R}$ and is the true source of

errors in the reconstruction process, since its contribution is
neglected. The fourth term contains the desired spatial frequencies,
which are transformed to produce an image. The individual effects
of these four terms are examined below in a quantitative example.

Synthetic data was calculated for two simple models, shown in
Figures 4a and 4b, in order to study the effects the two distortion
terms of equation (21) had on image quality. The analysis was
restricted to the x-z plane for a spread of 6000 feet; a constant
velocity of 7500 ft/sec was used. Each location on the data array
represents a CDP trace spaced at an interval of 100 feet. The
coordinates are fixed to the center of the spread. The calculations
were carried out for three temporal frequencies (30, 40, 50 Hz) and
an exact Fourier transform algorithm was used for flexibility of
choice in output location.

At the top of Figure 5, the exponent of the last term of (21)

$$e^{-j\frac{\omega}{c_o z_o} [(x_o-x)\tilde{x} + (y_o-y)\tilde{y}]}, \tag{i}$$

the actual spatial phase expected on the array, is plotted for 30
Hz. As expected, this quantity is zero at the array midpoint and
is linear across the array. Calculating for an output spacing of
10 feet, and using a constant amplitude equal to one, the data
were spatially Fourier transformed and the image amplitude and phase
spectra determined. The bottom of Figure 5 is a typical amplitude
spectrum, the phase spectra for all three frequencies being zero.

The next sequence, shown in Figure 6, has the horizontal aberration term;

$$e^{j\frac{\omega}{2cz_o} [(\Delta x)^2 + (\Delta y)^2]} \tag{ii}$$

added to the previous calculation. The top plot is again the spatial
phase across the spread (for 30 Hz). The function is still linear,
but is now non-zero at the array midpoint (x=0), and different in
magnitude for the three temporal frequencies chosen. The bottom
plot is the resulting migration output or image point amplitude,
and is the same as that in Figure 5. Figure 7 shows the output
phase for 30, 40, and 50 Hz. They are flat, as expected, but of
slightly different magnitude. If the three output images are
summed (they are on the same grid spacing, 10 feet), they will not
quite add in phase, thereby resulting in some distortion of the
image. The magnitude of this distortion, as can be seen by inspection of equation (ii) above, increases with increased lateral object
distance from the location of the central reconstruction position
(x,y).

The addition of the vertical distortion term,

$$e^{j\frac{\omega}{c_o} (1 + \frac{\Delta z}{2z})} \tag{iii}$$

with the reflection point now 200 feet above the migration output
depth results in the three spatial phase plots of Figure 8. Due to
the nonlinear nature of these phase curves, severe alteration of the
previous results is now anticipated. The amplitude spectra of
Figure 9 bear this out (note that the actual lateral reflection point
location is at +200 feet). The image phase plots of Figure 10 indicate
that severe mismatch will occur if the three frequencies are summed.

In summary, the results of Figure 5 show that the reflection
point contributions, at the central reconstruction position (x,y),
will add in phase over a range of frequencies. Figures 6 and 7 show
that all other reflection points in the migration output plane will
add out-of-phase, causing some degradation. The amount of this degradation increases with distance from location (x,y). The results,
displayed in Figures 8,9, and 10, demonstrate that reflection points
(in this case, 200 feet) off the migration output plane (the plane
of the reference) will not add in phase and will be significantly
degraded. This last result indicates that summation over a sufficient range of frequencies should result in refinement of depth of
field by suppression of out-of-plane reflection points.

Image Size Scaling

The results, obtained above, were calculated using an exact
Fourier transform algorithm so that the same output spacing could
be easily obtained at all frequencies. Normally, an FFT algorithm
is used, where a typical spatial transform is 64 x 64. In order to
obtain a given output spacing, only a particular input spatial frequency spacing is allowed, as described by equation (20). Since the
actual spatial frequencies, as given by equation (17), will usually
be different, the data must be resampled before FFT.

There are three convenient places to do this resampling. The
first would be to simply resample the lensless Fourier transform
hologram before FFT. This is also an attractive choice because the
phase factors used to produce the LFTH do not need to be completely
recalculated for each frequency. Unfortunately, the bilinear interpolation introduced aliasing, due to rapid changes across the LFTH.
This problem might be cured by increasing the grid density before
interpolation, but quickly becomes a cumbersome task on a minicomputer. It was for these same reasons that the second alternative,
interpolation and summation of the final FFT output (the actual
reconstructed image) was abandoned.

A third method, which was the one finally adopted, was to
immediately resample the plane wave reference hologram of equation
(2) to the necessary spacing. This proved to be attractive from
the standpoint of the bilinear interpolator, since the data at this
stage are slowly varying across the spread. We now, of course, must
completely recalculate the phase factors mentioned above, because
the "R" distances of equation (11) are different for each temporal
frequency since $(\tilde{x},\tilde{y})$ has been changed for $(\tilde{x}',\tilde{y}')$. A demonstration

of the image size change accomplished with this process is founded
in Figures 11 and 12 for data collected over two physical tank models
shown in Figures 13 and 14.

Frequency Summed Images

        The plexiglas anticline data set was used to demonstrate
temporal frequency summation. A band of 30 frequency points from
225 to 262 KHz (28.1 to 32.8 Hz in scaled quantities) was summed for
five different depths. This represents about 20% of the available
bandwidth. All other quantities remained the same, velocity = 7500
ft/sec and the reference location was at x = 4500 feet, y = 1500 feet
(unscaled coordinates - see Figures 14 and 15). Figures 16 through
20 show a monochromatic reconstruction at the center frequency (top)
and the frequency summed output (bottom) for z = 5400, 5200, 5000,
4800, and 4600 feet, respectively. Considerable depth of field
improvement can be seen in each view. This is especially true for
the shallower ones (Figures 19 and 20) where the migration output
plane is above the flat portion of the model, and the velocity
effects of the anticline are not as strong. No attempt was made to
compensate for the higher velocity of the plexiglas on deeper planes,
and water velocity (7500 feet/sec in our scaled system) was used
throughout.

REFERENCES

Gabor, D., 1948, A new microscope principle, Nature, V. 161, p. 777.
Goodman, J.W., 1968, "Introduction to Fourier Optics," McGraw-Hill,
      New York.
Leith, E.N., and Upatnieks, J., 1962, Reconstructed wavefronts and
      communication theory, J. Opt. Soc. Am., V. 52, N. 10, p. 1123.
Leith, E.N., and Upatnieks, J., 1967, Recent advances in holography,
      in "Progress in Optics, Vol. VI," E. Wolf, ed., North-Holland,
      Amsterdam.
Meier, R.W., 1965, Magnification and third-order aberrations in
      holography, J. Opt. Soc. Am., V. 55, N. 8, p. 987.
Stroke, G.W., and Falconer, D.G., 1964, Attainment of high resolu-
      tions in wavefront-reconstruction imaging, Phys. Let., V. 13,
      N. 4, p. 306.
Stroke, G.W., 1965, Lensless Fourier-transform method for optical
      holography, Appl. Phys. Let., V. 6, N. 10, p. 201.
Stroke, G.W., 1966, "An introduction to coherent optics," Academic
      Press, New York.

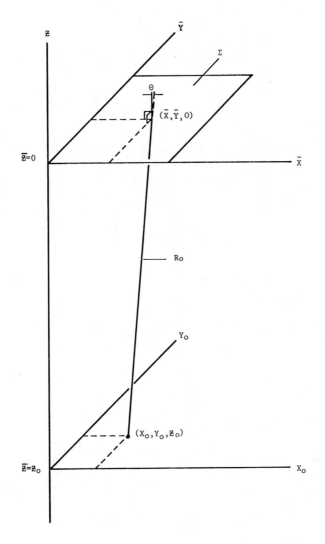

Figure 1. Coordinate geometry for equation 1.

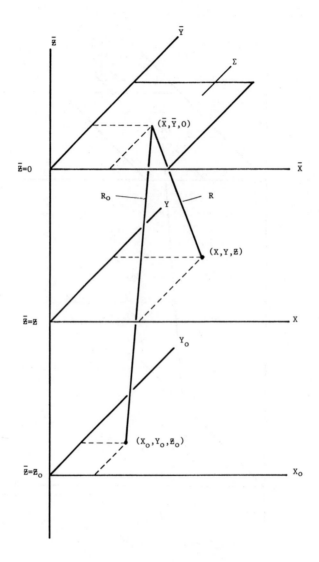

Figure 2.   Relationships among $\bar{X},\bar{Y}$; X,Y and X_0,Y_0 coordinates.

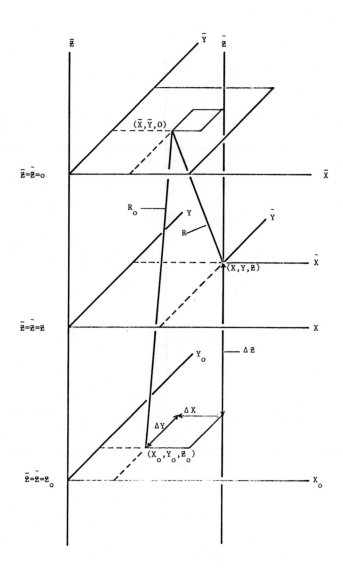

Figure 3. Transform to X, Y, $\tilde{Z}$ coordinates

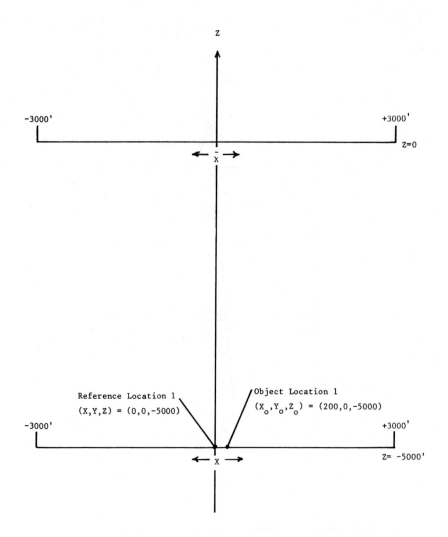

FIGURE 4a. Model for data in Figures 5, 6 and 7.

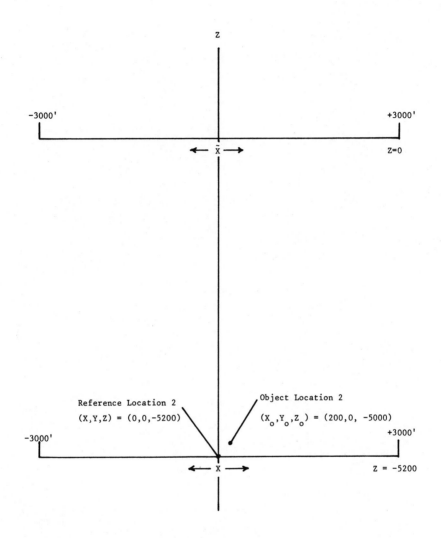

FIGURE 4b. Model data for data in Figures 8,9, and 10.

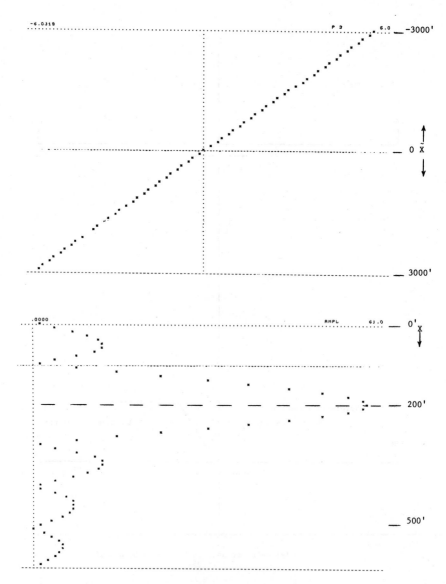

Figure 5.   Top, Recorded phase pattern for Model of 4a.
             Bottom, Resultant image intensity.

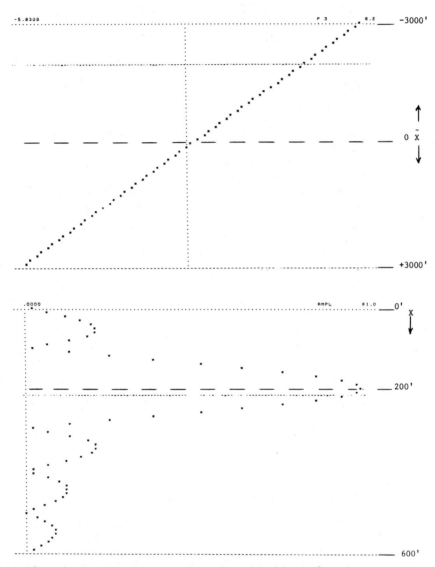

Figure 6.   Top; Recorded phase pattern for Model of 4a, horizontal
distortion term added.   Bottom; Resultant image intensity.

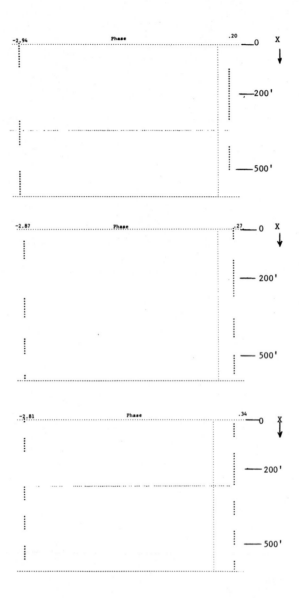

Figure 7. Resultant image phase for Model of 4a with horizontal
           distortion term added. Top, 30 Hz. Middle, 40 Hz.
           Bottom, 50 Hz.

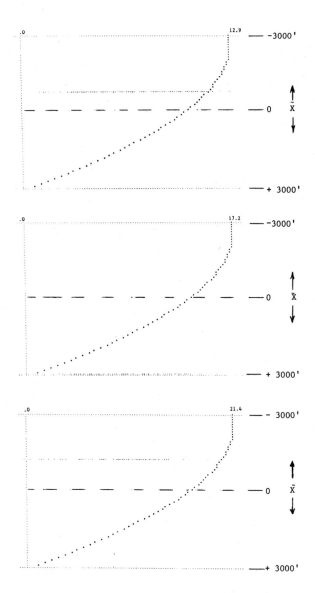

Figure 8. Recorded phase pattern for Model of 4b with horizontal and vertical distortion terms included. Top, 30 Hz. Middle, 40 Hz. Bottom, 50 Hz.

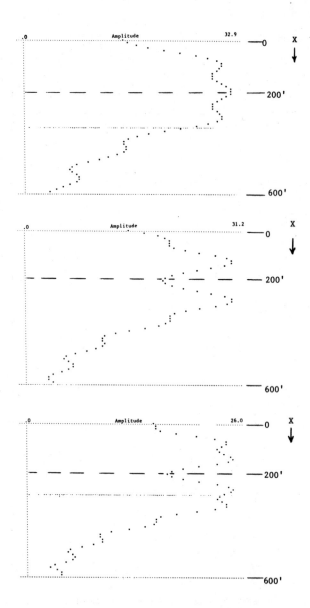

Figure 9. Resultant Image intensities for Model 4b.
Top, 30 Hz. Middle, 40 Hz. Bottom, 50 Hz.

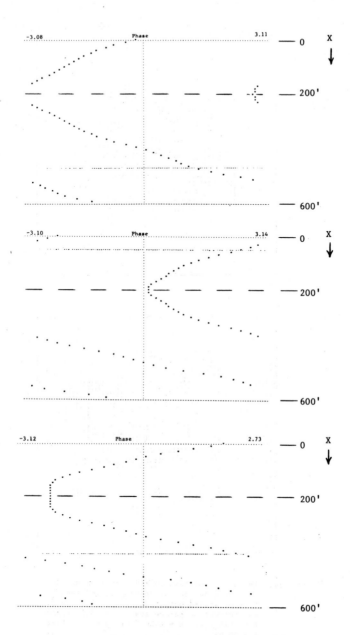

Figure 10.   Resultant Image Phases for Model of 4b.
            Top, 30 Hz. Middle, 40 Hz. Bottom, 50 Hz.

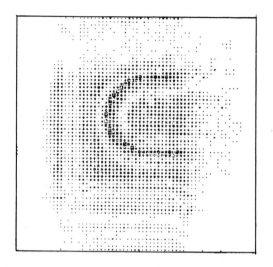

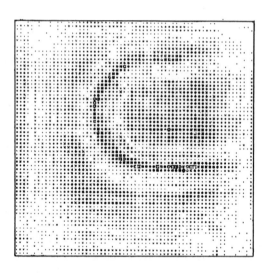

Figure 11.   Image scaling test on RTV channel model SALCOL.
Top, original output spacing of 160 feet.
Bottom, scaled output spacing of 110 feet.
Reconstruction parameters: c = 7058 ft/sec.,
Z = 7015 feet, x = 2400 feet, y = 2400 feet,
f = 30 Hz.

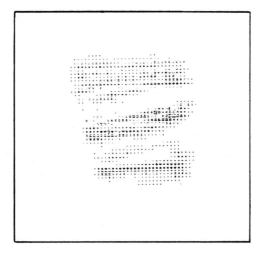

Figure 12.   Image scaling test on plexiglas anticline model.
Top, original output spacing of 100 feet.
Bottom, scaled output spacing of 125 feet.

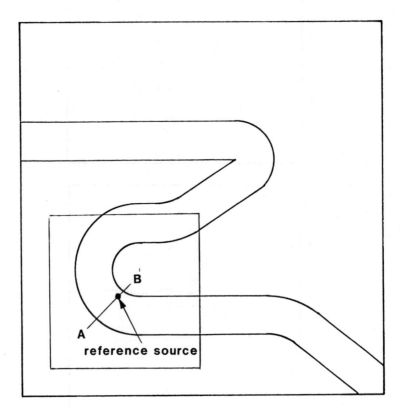

Figure 13.   SAL channel model and recording grid for image scaling test.

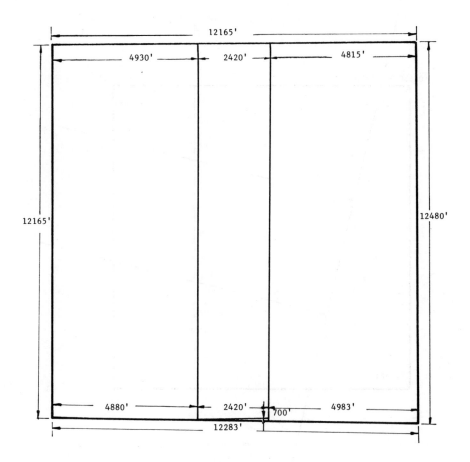

Figure 14a. Plexiglas anticline-top view.

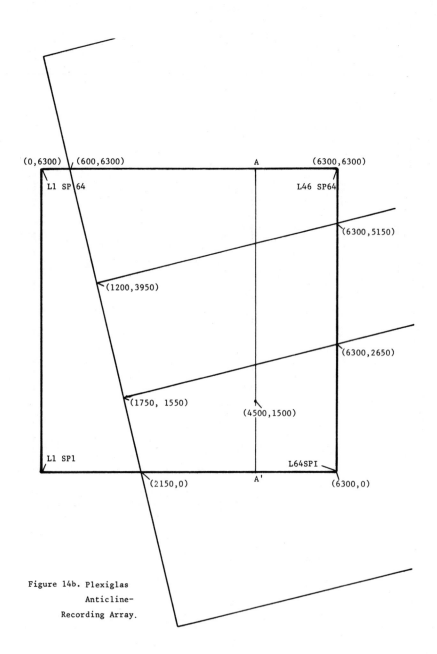

Figure 14b. Plexiglas
Anticline-
Recording Array.

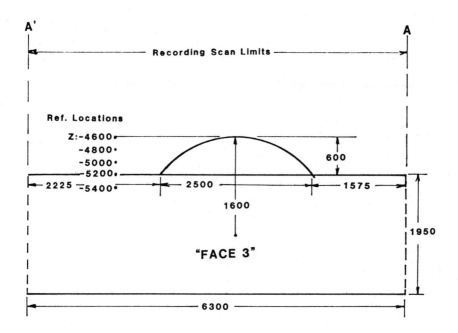

FIGURE 15a.   Section along A'A of Figure 14 showing
Reference positions for Figures 16 to 20.

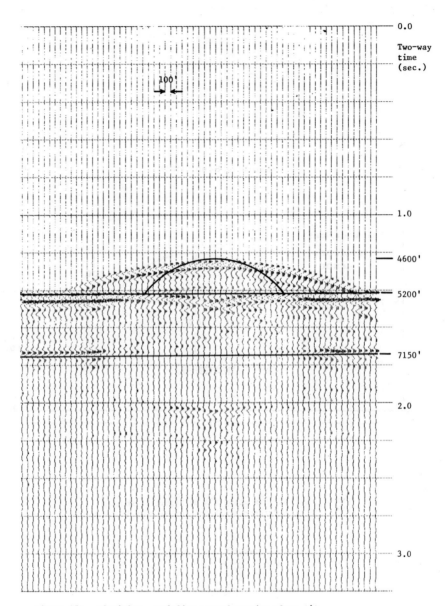

Figure 15b. Plexiglas anticline record section along A'A.

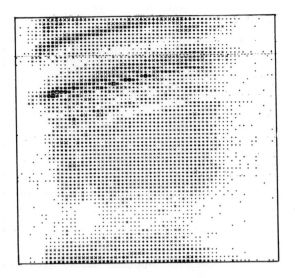

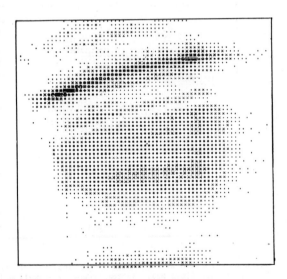

Figure 16. Reconstruction at Z = -5400 feet. Top, center frequency
of 244 KHz (30.5 Hz). Bottom, frequency summation from
225 KHz. (28.1 Hz) to 262 KHz (32.8 Hz).

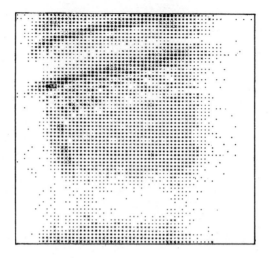

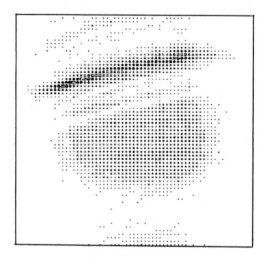

Figure 17. Reconstruction at Z = -5200 feet. Top, center frequency
of 244 KHz (30.5 Hz). Bottom frequency summation from
225 KHz (28.1 Hz) to 262 KHz (32.8 Hz).

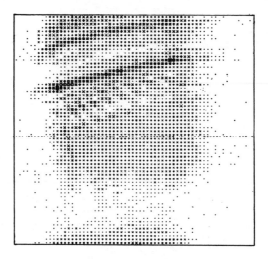

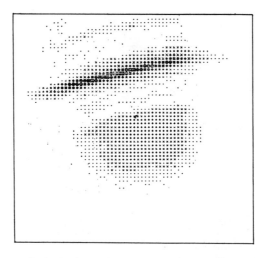

Figure 18. Reconstruction at Z = ⁻5000 feet. Top, center frequency of 244 KHz (30.5 Hz). Bottom, frequency summation from 225 KHz (28.1 Hz) to 262 KHz (32.8 Hz).

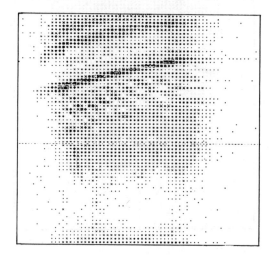

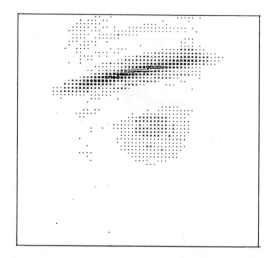

Figure 19. Reconstruction at Z = -4800 feet. Top, center frequency
            of 244 KHz (30.5 Hz). Bottom, frequency summation from
            225 KHz (28.1 Hz) to 262 KHz (32.8 Hz).

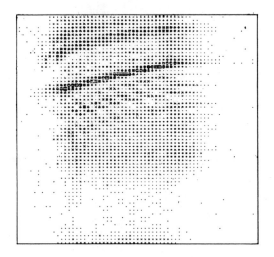

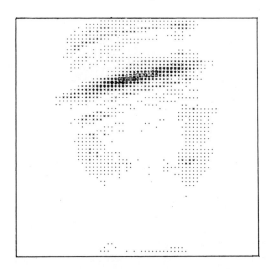

Figure 20. Reconstruction at Z = -4600 feet. Top, center frequency of 244 KHz (30.5 Hz). Bottom, frequency summation from 225 KHz (28.1 Hz) to 262 KHz (32.8 Hz).

HOLOGRAPHIC ENHANCEMENT OF BOUNDARIES IN SEISMIC APPLICATIONS

G. Fitzpatrick, T. Morgan, F. Hilterman, K. Wang, and
A. Haider
Seismic Acoustics Laboratory
University of Houston
Houston, Texas 77004

INTRODUCTION

There are many different ways of forming images from seismic wave data recorded on the earth's surface. One way is to record or produce a hologram at this surface and by an appropriate algorithm or analog procedure reconstruct the hologram to produce an image. This procedure will be employed here when we verify theoretical ideas; however, any method that properly "migrates" seismic waves back into the earth will suffice to take advantage of the techniques we are about to describe.

The purpose of this paper is to develop techniques by which such migrated images or image data may be analyzed in order to understand better the physical conditions (boundaries, discontinuities etc.) in the image space that they represent. To do this, we introduce the idea of an image functional, which in its broadest sense, is a mathematical device for comparing two different sets of image information.

The basic idea is derived from interferometry where one may compare two wavefields ψ_1 and ψ_2 by allowing them to interfere. Thus, if ψ_2 is the data field to be studied and ψ_1 is some "datum" field with which ψ_2 can be compared, the interference of the two produces a "contour" map with fringe intensity given by

$$I = |\psi_1 + \psi_2|^2 \tag{1}$$

The fringes are the contours of constant phase difference between ψ_1 and ψ_2 and along such contours or in such fringe regions I is stationary. Such contouring techniques are particularly useful where ψ_1 and ψ_2 differ significantly, that is when the phase difference between ψ_1 and ψ_2 is of the order 2π or larger. Then several or many contours will appear in the image space. These techniques are clearly useless if $\psi_1 \widetilde{} \psi_2 = \psi$ for then $I \sim |2\psi|^2 \sim 4\psi^*\psi$ is just an ordinary image with no contour information.

Our principal concern in this paper will be with situations where ψ_1 and ψ_2 are very similar and where the small difference between them is to be studied in a way comparable to the large differences encountered in Equation (1).

That is we are interested in situations as depicted in Figure 1 where two component wavefields ψ_1 and ψ_2 at nearly the same temporal frequencies $f_1 \widetilde{} f_2$ are propagated back to their boundaries and focused. Because one deals typically with low resolution systems two different but very similar image volume distributions $I_1 = \psi_1^* \psi_1$ and $I_2 = \psi_2^* \psi_2$ at the boundary of interest are produced. Errors committed in the process of backward propagating ψ_1 and ψ_2 due to a poor "velocity model" say will be similar provided $f_1 \widetilde{} f_2$ ie $\psi_1 \approx \psi_2$. Hence to "eliminate" such errors we must examine only those components for which $\psi_1 \approx \psi_2$ at the boundary in question. Clearly conventional interferometry as expressed by Equation (1) will be of no use.

To make this comparison, we introduce a new type of functional which was also employed in a similar, but much less general context in earlier work (Reference 1). The functional is

$$I = |\psi_1 + z\psi_2|^2$$

where $Z = \alpha + i\beta; \alpha \neq \alpha(x,y), \beta \neq \beta(x,y)$ (2)

is a global complex constant (a phase shifting operator) independent of image coordinates.

Call the spatially varying parts of the waves at the boundary ψ_1 and ψ_2. To analyze this complex boundary condition one searches for a Z value (Equation 2), which minimizes the image functional I. If such a condition can be found then I will be stationary in and around some spatial region or along a "contour " of the object being imaged. Such regions of the object will represent constant boundary conditions and by implication constant physical properties along the contour. In Figure 1, the structure is "two-dimensional" so that the "contours" lie along the axis of such a structure.

Thus, the method attempts to find and display regions where the phase shift $\Delta\phi$ characterizing the difference between ψ_1 and ψ_2 on the boundary is constant. Along such boundaries, the spatial variation of $\Delta\phi$ is zero

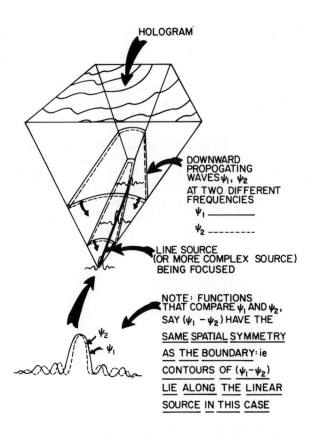

HOLOGRAM

DOWNWARD
PROPOGATING
WAVES ψ_1, ψ_2
AT TWO DIFFERENT
FREQUENCIES

ψ_1 ————

ψ_2 - - - - - - - -

LINE SOURCE
(OR MORE COMPLEX SOURCE)
BEING FOCUSED

NOTE: FUNCTIONS
THAT COMPARE ψ_1 AND ψ_2,
SAY ($\psi_1 - \psi_2$) HAVE THE
SAME SPATIAL SYMMETRY
AS THE BOUNDARY: ie
CONTOURS OF ($\psi_1 - \psi_2$)
LIE ALONG THE LINEAR
SOURCE IN THIS CASE

ψ_2

ψ_1

Figure 1. Backward propagation of two different
frequency components to their line source.

$$\delta(\Delta\phi) = 0 \tag{3}$$

If we can find a Z which allows I to be minimized with respect to $\Delta\phi$, namely

$$\frac{\partial I}{\partial(\Delta\phi)} = 0 \tag{4}$$

we will also have found a region or boundary in the image space along which I does not vary with position

$$\delta(I) = 0 \tag{5}$$

The spatial variation in Equation (5) is identical to that in Equation (3).

It will be shown in the following theoretical development that only special choices of Z and no others, namely, those that characterize the actual boundary conditions, will result in stationary values of I. This allows one, in principle, to assign Z values to each boundary condition present in the image and this Z "decomposition" represents an analysis of the boundaries present. These ideas will be demonstrated by appeal to experimental data taken on simple seismic models.

Theory

Given the image functional

$$I = |\psi_1 + z\psi_2|^2 \tag{6}$$

defining

$$Z_\pm = \alpha + i\beta = \pm e^{i\Delta\phi_o} \tag{7}$$

where the wavefields to be compared are represented by phase-only quantities (Footnote 2)

$$\psi_1 = e^{-i\phi} \tag{8}$$

$$\psi_2 = e^{-i\phi'} \tag{9}$$

$$\Delta\phi = \phi' - \phi$$

Substituting ψ_1 and ψ_2 into Equation (6), one has

$$I = 1 + \alpha^2 + \beta^2 + 2\alpha\cos\Delta\phi + 2\beta\sin\Delta\phi$$

$$I = 2(1 + \alpha\cos\Delta\phi + \beta\sin\Delta\phi) \tag{10}$$

$$\Delta\phi = \phi' - \phi; \quad \Delta\phi = \Delta\phi(x,y,z)$$

Then I has a stationary state when

$$\left.\frac{\partial I}{\partial(\Delta\phi)}\right|_{\Delta\phi_o} = -2\alpha\sin\Delta\phi_o + 2\beta\cos\Delta\phi_o = 0 \tag{11}$$

or

$$\frac{\alpha}{\beta} = \frac{\cos\Delta\phi_o}{\sin\Delta\phi_o} \tag{12}$$

Since

$$\frac{\partial^2 I}{\partial(\Delta\phi)^2} = -2\alpha\cos\Delta\phi - 2\beta\sin\Delta\phi \tag{13}$$

when $\Delta\phi$ is small $(\beta \approx 0)$

$$\frac{\partial^2 I}{\partial(\Delta\phi)^2} \simeq -2\alpha\cos\Delta\phi \approx -2\alpha \tag{14}$$

Therefore, I will be a minimum if α is <u>negative</u>, i.e. when $Z=Z_-$. The choice α positive (Imax) and $Z = Z_+$ will be of no interest as we will now show.

The image functional I for a given α, β and small $\Delta\phi$ is approximately

$$I \approx 1 + \alpha^2 + \beta^2 + 2\alpha(1 - \frac{1}{2}(\Delta\phi)^2) + 2\beta\Delta\phi \tag{15}$$

The significant variations depend on the sine term $(\Delta\phi \gg (\Delta\phi)^2)$ only. Thus, one is led to define a relative contrast measure as (Reference 1). See Appendix I.

$$C(\Delta\phi) = \frac{2\beta \sin \Delta\phi}{1 + \alpha^2 + \beta^2 + 2\alpha\cos\Delta\phi} \tag{16}$$

When $Z = Z+$, $C(\Delta\phi_o) = (\sin^2\Delta\phi_o)/(1 + \cos^2\Delta\phi_o) \approx 0$ which means that the important $\sin (\Delta\phi)$ variation has been almost completely suppressed, thus, we are not interested in the cases where I is a maximum or when $\alpha > 0$. The case of interest is for $Z = Z_-$ then

$$C(\Delta\phi_o) = (-\sin^2\Delta\phi_o)/(1 - \cos^2\Delta\phi_o) \tag{17}$$

$$C(\Delta\phi_o) = -1 \qquad\qquad\qquad\qquad\qquad (18)$$

which brings us to an important and useful result. A relative contrast of $C(\Delta\phi) = -1$ means that the "background" term $1+\alpha^2+\beta^2+2\alpha\cos(\Delta\phi)$ has almost <u>cancelled</u> (not suppressed) the important $2\beta\sin(\Delta\phi)$ variation. This case might at first seem to be uninteresting also (Footnote 3) however, the relative contrast is extremely small and this means it is, or should be, relatively easy to find this region in which (I) is constant at its <u>minimum</u> value. It will be characterized by being "dark" in the reconstruction since

$$I_{min} = 4\sin^2\Delta\phi_o \approx 0 \text{ when } \Delta\phi_o \text{ is small} \qquad (19)$$

Now it is clear that if a certain region in the image space (boundary) is characterized by a $\Delta\phi_o$ constant, then it will be possible, in principle, to find Z values by trial that match this $\Delta\phi_o$. That is, one may vary Z until a minimum in I appears and a very low contrast image is obtained. This will describe the i-th boundary via Z_i. By further variations of Z, one finds still other regions where I is a minimum and contrast is low. Thus, a set of Z's may be found each one of which characterizes a different region or boundary condition on the object or target. (Footnote 4)

Suppose now that we have found such a region by varying Z until I has a distinct minimum in the region or along a boundary and the contrast is very low. We will refer to this as a Case I situation.

Case I: Z_-; $\alpha = -\cos\Delta\phi_o$

$\beta = -\sin\Delta\phi_o$

$\dfrac{\partial I}{\partial(\Delta\phi)} = 0, \ I_{min} = 4\sin^2\Delta\phi_o.$

$C(\Delta\phi) = -1$

Having found such a state by noting the appearance of a minimum in the functional I, one can then use this knowledge to display the image data (for this region) in an entirely different way. In particular, we can use this found Z as an input to generate an image with relative contrast $C(\Delta\phi) = +1$ instead of -1. This represents a case where the "background" term $1+\alpha^2+\beta^2+2\alpha\cos(\Delta\phi)$ adds to the $2\beta\sin(\Delta\phi)$ term rather than subtracting from it.

That is,

Case II: $\alpha = -\cos\Delta\phi_o$

$\beta = +\sin\Delta\phi_o$

$$C(\Delta\phi_o) = \frac{2\sin^2\Delta\phi_o}{2(1-\cos^2\Delta\phi_o)} = +1$$

Note: Case II is <u>not</u>, in general, the condition needed to make I a minimum.

Case II is the way to represent the image data since the term of interest ($2\beta\sin\Delta\phi$) is at least as large as the background in the region or along the boundary of interest.

The above analysis illustrates that in searching an image for a Case I condition, (I_{min}) and low contrast, it is important to realize that the wrong choice of sign for $\Delta\phi_o$ will make a Case I state look like a Case II state and visa versa. For example, suppose that $\Delta\phi_o$ actually needed to make a low contrast minimum appear in I along some boundary in the image space is $\Delta\phi_o$ radians. That is,

$\alpha = -\cos(\Delta\phi_o)$

and

$\beta = -\sin(\Delta\phi_o)$ 　　　　　　　　　　　　　　　　　(20)

are required.

Suppose, instead, that we had erroneously chosen $-\Delta\phi_o$, then

$\alpha = -\cos(-\Delta\phi_o) = -\cos(\Delta\phi_o)$

$\beta = -\sin(-\Delta\phi_o) = +\sin(\Delta\phi_o)$ 　　　　　　　　　(21)

which is a Case II condition, and no minimum in I will appear. Even though an improvement will have occurred in the image because C = +1 for this apparent Case II situation, there is no way to know this is the best that can be done because we have found no low contrast Case I minimum to verify that α,β are the correct choices.

Thus, in searching for the proper Z, both the sign and the magnitude of $\Delta\phi_o$ must be varied until low contrast minima in the functional I appear. When this happens, we know that we have achieved a Case I condition. In order to plot an enhanced Case II image with C = +1, we need only change the sign of the $\Delta\phi_o$ just found.

The basic procedure for analyzing an image to find these boundary regions and characterize each of them with a Z value is now clear. In a later section, experimental verification of these ideas is given.

Applications

The technique used to form our "migrated" or reconstructed images is the Lenseless Fourier Transform Holography Technique. (See the paper by T.R. Morgan in these proceedings)

In that technique, one has a hologram (only the real part is written here)

$$H_1 = \cos\left(\frac{2\pi R}{\lambda} - \phi\right) \tag{22}$$

where amplitude modulation is normalized away and where ϕ is spatial phase information $\phi = \phi(x,y)$ at the earth's surface (extracted from time records as explained elsewhere) and $2\pi R/\lambda$ is the phase of a synthetic point reference located at some point below the earth's surface. The hologram H_1 is called the point source reference hologram. When such a hologram is reconstructed, it yields an amplitude dependent on the image coordinates (u,v).

Reconstruction of H_1

$$\psi_1 \simeq e^{-i\phi(u,v)} = e^{-i\phi} \tag{23}$$

Clearly, if we have another hologram H_2, defined to be

$$H_2 = -\cos\left(\frac{2\pi R}{\lambda} - \phi' + \Delta\phi_o\right) \tag{24}$$

this will reconstruct

$$\psi_2' = -e^{-i\phi'} \cdot e^{i\Delta\phi_o} \tag{25}$$

But $Z_- = -e^{i\Delta\phi_o}$ so we can write this as

$$\psi_2' = Z_- e^{-i\phi'} = Z_- \psi_2 \tag{26}$$

Thus, to study the image functional

$$I = |\psi_1 + Z\psi_2|^2 \tag{27}$$

we should reconstruct the hologram (holographic interferogram)

$$HI = H_1 + H_2 = \cos\left(\frac{2\pi R}{\lambda} - \phi\right) - \cos\left(\frac{2\pi R}{\lambda} - \phi' + \Delta\phi_o\right) \tag{28}$$

As mentioned in the text, there are two cases to consider, depending on the sign of $\Delta\phi_o$.

Reconstructing H_1+H_2 for both $\Delta\phi_o>0$ and $\Delta\phi_o<0$ allows the two complex amplitudes ψ_1 and ψ_2 to interfere in two different ways and their interference can be studied as indicated earlier with the help of the image functional I.

Experiments

Figure 2 illustrates a model from which real data were taken. In Figure 3 is illustrated a series of reconstructed images using $\Delta\phi_o$ values that are both positive and negative ranging from $\Delta\phi_o = \pm2.0$ to $\Delta\phi_o = \pm.800$, that is from slightly more than a quarter wave shift to a phase value near one eighth wave shift.

This Figure illustrates several things. First, as $\Delta\phi_o$ is changed a region near $\Delta\phi_o = \pm1.400$ is encountered where one image shows good contrast ($\Delta\phi_o = +1.400$) and the other ($\Delta\phi_o = -1.400$) shows almost none. By the criteria presented in the text, the high contrast should be identified with a Case II state, while the low contrast images should be identified with a Case I state. Note that since some noise is present and $I_{min} \geq 0$, the Case I plots ($\Delta\phi_o = -1.400$) show a general "low" level background. This sequence indicates that around $\Delta\phi_o = +1.4$, the image should show high contrast and analyses here and elsewhere (Reference 1) indicate that this image should be "enhanced" compared with an ordinary reconstruction. That is, the important $2\beta\sin\Delta\phi_o$ terms in the image will be enhanced.

Figures 4 and 5 illustrate this effect. In Figure 4, the $\Delta\phi_o = +1.4$ reconstruction is plotted for two different choices of threshold values which merely illustrates that the background noise can be eliminated. In Figure 5, the ordinary reconstructed image is displayed for comparison. The same relative threshold scales are used in both Figures 4 and 5.

As can be seen from these two figures, an improvement has occurred in the enhanced version. It is interesting to note that this improvement has occurred all through the image which includes several "different" boundaries.

Another interesting feature of this data is afforded by Figures 6 and 7, where information at a somewhat lower frequency was used. The fact that it is lower frequency is of no significance, however, for the feature we wish to discuss. We have found that if a successful $\Delta\phi_o$ has been found for two frequencies, f_1 and f_2 ($f_2>f_1$) any lower pair of frequencies f_1 and $f_n<f_2$ has a successful $\Delta\phi_o$ given by (see also Appendix III).

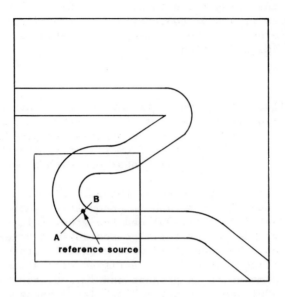

Figure 2b. Meandering Channel Model showing
reference source location and structure. Water
seperates the plexiglass from the RTV rubber.
The Hologram scanning plane was seven inches
above the reference location. The hologram array
was 64x64 and the spacing between samples was .08
inches.

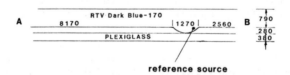

Figure 2a. Meandering Channel Model showing
reference source location and section A-B
through the channel. The small square is
the hologram aperture.

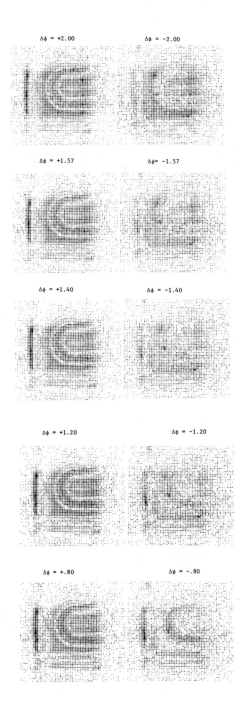

Figure 3. Records 25 (264.89KHz) and 30(271.00KHz) combined for various Δφ values. The positive Δφ are high contrast images and negative Δφ are low contrast images. When Δφ=-1.4, we get what appears to be the image with least contrast. Thus, Δφ=+1.4 produces what should be the "best" image.

Fig 4   Fig 5

Figure 4.   The Δφ=+1.4 image of Figure 3.   In
the lower image we have thrown away the low-
level noise.   Compare these with an ordinary
reconstruction.   (See Figure 5)

Figure 5.   An ordinary reconstruction of record
25.   These images are clearly not as good as the
enhanced images of Figure 4.

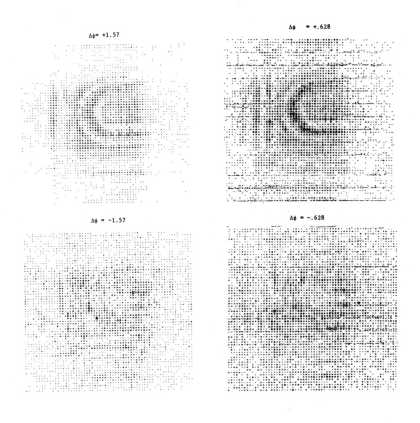

Fig 6   Fig 7

Figure 6. Data at frequencies f_1 = 235.6 KHz and
f_2 = 241.7 KHz compared using $\Delta\phi$ = +1.57. From
this successful $\Delta\phi$ at these frequencies we can
compute another successful $\Delta\phi$ at two different
frequencies (See figure 7).

Figure 7. Data at f_1=235.6KHz and f_n=238.0KHz
are compared using $\Delta\phi$ computed form equation (29)
in the text using f_2=241.7KHz and the $\Delta\phi$ value
in Figure 6 ($\Delta\phi$=$\pm$ 1.57).

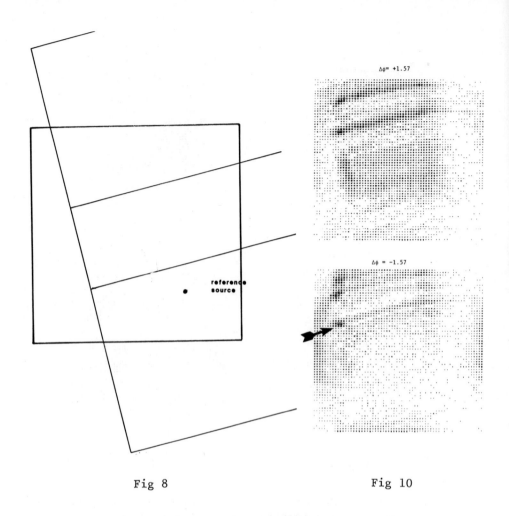

Fig 8   Fig 10

Figure 8. Plexiglas anticline model and reference
source location at the top surface of the model.
The hologram scanning aperture corresponds to
the square. Samples were taken at .1 inch intervals. The hologram sampling plane is located
5.2 inches above the reference location.

Figure 10. In the low contrast plexiglas.
anticline image ($\Delta\phi$=-1.57) for records (2,7)
there appear to be minima along the anticline
boundry.

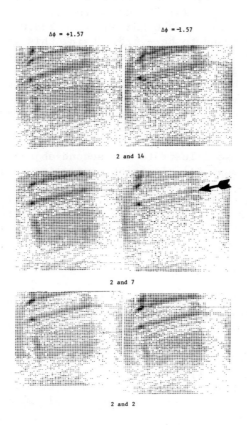

Figure 9. Plexiglas anticline model for various frequency pairs 2(235.60KHz), 7 (241.70KHz) 14(250.24KHz) and the same Δφ. The pairs (2,7) show what appear to be <u>low contrast minima</u> as illustrated here and in Figure 10. Note that when the same records (2,2) or records other than (2,7), that is (2,14), are compared with this Δφ, low contrast minima do not appear for negative Δφ.

$$\Delta\phi_o(1,n) = \left(\frac{f_n-f_1}{f_2-f_1}\right)\Delta\phi_o(1,2) \tag{29}$$

Figure 6 illustrates the successful images for frequencies f_1 = 235.6 KHz and f_2 = 241.70 KHz, with $\Delta\phi_o$ = ±1.57. In Figure 7, the frequencies f_1 = 235.6 KHz and f_n = 238.0 KHz and $\Delta\phi_o$ from Equation (29) is found to be -.628. Using this value, another successful $\Delta\phi_o$ is found for the same physical boundaries, but for two different frequencies. It is clear from this that the meaning of the term "boundary" used here involves both the physical object under study and the particular pair of frequencies we are using to study these physical boundaries (see Appendix II for further discussion).

In Figure 8, we illustrate the second object we studied. This is a two-dimensional plexiglas anticline ("plexicline"). In Figure 9 are a series of runs using different frequency pairs and the same $\Delta\phi_o$, namely $\Delta\phi_o$ = ±1.57. The pairs f_1 = 235.60 and f_2 = 241.70 KHz seem to show a considerable difference for this $\Delta\phi_o$. Note, in particular, the appearance of what appear to be minima in the $\Delta\phi_o$ = -1.57 image (see Figure 10).

Conclusions and Discussion

The foregoing results, though they are of a preliminary nature, indicate that certain boundary conditions in an object can be selectively imaged using information at two different temporal frequencies. Because there are a very large number of frequency pairs or bands that can be compared in any image and because each structural condition represented in the image effects every pair differently, there are a very large number of different ways one may study these images. We have clearly only scratched the surface in this work.

Furthermore, it is important to realize, although we did not discuss it in the text, that one is not limited to temporal frequency pairs. It is possible, as explained in Reference 1, to consider spatial changes that result by changing, say the position of the reference source used in computing the hologram pair. This has the effect of comparing the boundary in question with itself after shifting it in space. Clearly, we can get even more involved by both translating the reference source (in any direction) and comparing different temporal frequencies. The net effect of a set of complex changes of this sort is not easy to analyze or predict, and it may not be clear what value such added complexity can bring. We should keep in mind the fact that such interferometric techniques as described here and in Reference 1, have as their principal goal, a better "map" as an end product. For this reason, techniques of this sort should be judged less on the analytic complexity they introduce, and more on good results they hopefully will yield when the techniques are fully developed.

In spite of the complexities involved, it may eventually be possible to assign a more quantitative significance to these Z values than we have done here. That they are intimately related to the phase shifts actually experienced by the scattered waves, is apparent. But it is not yet clear exactly how one can use such Z values to characterize the physical boundaries.

ACKNOWLEDGEMENTS

The authors thank R. Duffy for the use of data collected over his "buried channel" model. We also thank Michele Maes for typing the final manuscript.

APPENDIX I

Usually one defines the "contrast" in any image as the ratio of all terms in the image that depend on phase; in this case, those that depend on $\Delta\phi$, to the constant background terms. If we did this instead of defining the contrast measure as in Equation (16) in the text, we would have a relative contrast measure given by

$$C'(\Delta\phi) = \frac{2\beta sin\Delta\phi + 2\alpha cos\Delta\phi}{1+\alpha^2+\beta^2}$$

Now it is interesting to note that this contrast function has extreme values for the same $\Delta\phi$ values as the functional I in Equation (6) in the text. That is

$$\frac{\partial C'(\Delta\phi)}{\partial(\Delta\phi)} = 2\beta cos\Delta\phi_o - 2\alpha sin\Delta\phi_o = 0$$

or

$$\frac{\alpha}{\beta} = \frac{cos\Delta\phi_o}{sin\Delta\phi_o}$$

which is Equation (12) in the text. Thus, we find that when we have a Case I situation:

$$\alpha = -cos\Delta\phi_o$$

$$\beta = -sin\Delta\phi_o$$

I has a minimum, and

$$C(\Delta\phi_o) = -1; \quad C'(\Delta\phi_o) = -1 .$$

Both contrast measures $C(\Delta\phi)$ and $C'(\Delta\phi)$ tell us the same thing. In this case, the terms of interest $2\beta sin\Delta\phi$ for $C(\Delta\phi)$ and $2\beta sin\Delta\phi + 2\alpha cos\Delta\phi$ for $C'(\Delta\phi)$ cancel the background terms: $1+\alpha^2+\beta^2+2\alpha cos\Delta\phi$ and $1+\alpha^2+\beta^2$ respectively. Thus, for a Case I situation it makes no difference whether we speak of $C'(\Delta\phi)$ or $C(\Delta\phi)$ since $C'(\Delta\phi) = C(\Delta\phi) = -1$.

A Case II situation, however, is quite different. In this case

$$\alpha = -cos\Delta\phi_o$$

$$\beta = +sin\Delta\phi_o$$

and $C(\Delta\phi) = +1$, i.e., the term $2\beta sin\Delta\phi$ is at least as large as the background $1+\alpha^2+\beta^2+2\alpha cos\Delta\phi$ and adds to it. The contrast measure $C'(\Delta\phi)$ however is

$$C'(\Delta\phi) = \beta^2 - \alpha^2$$

But, for small $\Delta\phi_o$ (which is the case of interest in this discussion), we have

$$\alpha^2 \approx 1 > \beta^2 \approx 0$$

so that

$$C'(\Delta\phi_o) \approx -1$$

for $\Delta\phi_o$ small, which is the same as for Case I above. Clearly in a Case II situation $C(\Delta\phi)$ and $C'(\Delta\phi)$ contrast measures tell us something different and this is why we chose the measure of Equation (16). The measure of Equation (16) gives different results for Cases I and II but $C'(\Delta\phi)$ does not.

We should point out however that in our experiments some of our phase shifts $\Delta\phi_o$ were close to ± 1.57, i.e. one quarter wave. These are not "small" in the sense implied above. In fact in this case $C(\Delta\phi) = -1$ but $C'(\Delta\phi)$ rather than being -1 is $C'(\Delta\phi_o) = \sin^2\Delta\phi_o - \cos^2\Delta\phi_o = 2\sin^2\Delta\phi_o - 1 = +1$. Our technique is designed for smaller phase shifts and when $\Delta\phi_o$ is very small the $C(\Delta\phi)$ measure of contrast is the appropriate one.

APPENDIX II

There are at least three controlling factors that determine the"boundary conditions":

(1)   The frequency pair that we choose
(2)   the $\Delta\phi_0$ we choose, and
(3)   the physical character of the object or boundary.

For a complex object it may happen that there are sufficiently many different boundary conditions that a choice of (1) and (2) can be made for which some improvement occurs throughout the image. What we are more interested in are regions where a significant improvement takes place.

Category (3) above, the"physical boundary", also needs some qualification. Some regions in an object space scatter as "units". That is, waves do not merely scatter from two surfaces in contact but the scattering process can involve the penetration of the wave into a complex region or volume and then back scatter from this region to yield a complex angular spectrum. A good example of this would be the scattering of sound waves from a sphere of one material embedded in another material where the wavelengths involved in both materials are comparable to the radius of the sphere. In this case, the scattering is not merely from the surface of the sphere, but involves penetration into the sphere, scattering from the "front" and "back" surfaces and various refraction and attenuation effects.

The term "boundary" as used in this paper could, in this case, refer to this entire feature (the embedded sphere) as the boundary condition to be studied and we could search for a $\Delta\phi_0$ characterizing this feature as a whole.

This is the sort of thing that was done with the models of Figures 2 and 8 in the text. We essentially treated these two-dimensional objects as single units and looked for a $\Delta\phi_0$ that characterized an entire volume. This is probably why our $\Delta\phi_0$ values were fairly large (in the vicinity of quarter wave phase shifts); however, one is not limited to large phase shifts. It should be stressed again as in Appendix I that our contrast measure $C(\Delta\phi)$ is not particularly relevant for these larger phase shifts. In Reference I the condition of "phase quadrature" which is analogous to what happens here when $\Delta\phi_0$ is near ±1.57 was discussed thoroughly.

If the data contains information on boundaries characterized by smaller phase shifts $\Delta\phi_0$ then these can certainly be analyzed. Our models exhibited mostly large effects,though as in Figure 7 we did analyze one situation with a fairly small phase shift of

$\Delta\phi_o = \pm.628$.   This corresponds to about 1/8 wavelength shift.

APPENDIX III

It is important to realize that we have said nothing about the initial phases $\Delta\phi_i$ of the two frequency components at the source. If, as in our simple models these two frequencies maintain a constant difference everywhere in the space (water) above the model, then any changes we observe upon examining the phase difference after these two frequencies have entered and back-scattered from the model are characteristic of the model and not the source.

In general the $\Delta\phi_o$ we choose depends on the initial phase difference $\Delta\phi_i$

$$\Delta\phi_o = \Delta\phi(physical) \pm \Delta\phi_i$$

where $\Delta\phi(physical)$ is the effect due to the model.

Thus since $\Delta\phi_i$ and $\Delta\phi_o$ are constants, we can always adjust $\Delta\phi_o$ to take into account any effects due to $\Delta\phi_i$.

FOOTNOTES AND REFERENCES

1. G. L. Fitzpatrick, <u>Acoustic Imaging and Holography</u>, Vol. 1, No. 1, Crane-Russak, New York 1978.

2. The functions ψ_1 and ψ_2 are <u>normalized</u> with respect to amplitude variations. That is, in practice $\psi_1 = A_1 e^{-i\phi}$, $\psi_2 = A_2 e^{-i\phi'}$ where A_1 and A_2 are not equal in general. Normalizing ψ_1, $\psi_1/\sqrt{\psi_1^* \psi_1} = e^{-i\phi}$. Thus we use only phase factors in (8 and 9) and the actual holograms which produce ψ_1, ψ_2 are normalized before being combined so as to insure that ψ_1, ψ_2 are "normalized".

3. It was erroneously stated in Ref. 1 that a contrast of -1 was of no interest. We see here that in searching for a high contrast image ($C = +1$) we need first to find the $C = -1$ image. Even though the $C = -1$ is by itself usually not a good image (sometimes adjacent boundaries with different properties can make a $C = -1$ image reasonably "good") it is needed to find the "best" $C = +1$ image.

4. Clearly, the foregoing procedure could have great value in analyzing an image since it would aid in finding the spatial extent of many uniform boundary conditions present. However, such a technique would not work if for any choice of Z there always appeared a minimum in the functional I for a given boundary. That this is not the case can be "proved" as follows: We know that the functional has a minimum <u>only</u> where $\alpha = -\cos\Delta\phi_o$ and $\beta = -\sin\Delta\phi_o$. For any other choice of Z, $Z \neq Z_-$, I does not possess a minimum. When we combine this fact with the reasonable assumption that there are at least a limited number of boundaries, i.e a limited number of Z's characterizing these boundaries, any Z that is not equal to one of these will not produce a minimum in I anywhere in the image space. Hence the appearance of a minimum in I is a unique signal that a "proper" physically relevant Z has been found. These points will be clarified when we examine real data and the degree of uniqueness can be given a more tangible meaning.

LIST OF PARTICIPANTS

Robert Addison
Rockwell International
1049 Camino Dos Rios
Thousand Oaks, CA 91360

Mahfuz Ahmed
University of California
Irvine Medical Center
8401 Middleton
Westminister, CA 92683

Pierre Alais
Paris University
Lab. de Mehanique Physique
Univ. Pierre et Marie Curie
Paris, FRANCE

Leigh Anderson
Exxon
P. O. Box 2189
Houston, TX 77001

Weston A. Anderson
Varian
611 Hansen Way
Palo Alto, CA 94303

Robert Andrews
ONR
Earth Physics Program
Office of Naval Research
Arlington, VA 22217

Khalid Azim
University of Houston
Dept. of Elect. Eng.
Houston, TX 77004

Ralph Barnes
Bowman Gray School of Medicine
Dept. of Neurology
Winston-Salem, N.C. 27103

Kenneth Bates
Hewlett Packard
1301 Page Mill Rd., Bld. IL
Palo Alto, CA 94304

Wayne Begun
Advanced Diagnostic Res. Corp.
P. O. Box 28400
Tempe, Arizona 85282

John E. Beitzel
ARCO
P. O. Box 1346
Houston, TX 77001

P. K. Bhagat
University of Kentucky
Wenner-Gren Research Lab
Lexington, KY 40506

J. Robert Birchak
NL Industries/DST
12950 W. Little York
Houston, TX 77001

David Boyce
General Electric
Electronics Park, Bldg. 7
Syracuse, NY 13221

817

R. A. Broding
Amoco
Amoco Production Company
Tulsa, Oklahoma

Joseph W. Brophy
Babcock & Wilcox
Lynchburg Res. Center
P. O. Box 1260
Lynchburg, VA   24505

C. Bruneel
Univ. de Valenciennes
Le Mont Holly 59326
Valencienne, France

Christoph Burckhardt
Hoffman-Laroche
Aktiengesellschaft
Ch-4002 Basle
Switzerland

Joseph Callerame
Raytheon
28 Seyon Street, Seyon Bldg.
Waltham, Mass.   02154

Richard J. Castle
Chevron Geophysical Company
8435 Westglen
Houston, TX   77036

Thomas Cathey
University of Colorado
359 Pine Brook Hills
Boulder, CO   80302

Ching-Hua Chou
Stanford
3-D Blackwelder, Escondido Vil.
Stanford, CA   94305

R. S. C. Cobbold
University of Toronto
Institute of Biomedical Engrg.
Toronto, Canada   M55   1A4

Dale Collins
Battelle
P. O. Box 999
Richland, WA   99352

Bill D. Cook
University of Houston CC
Department of Mechanical Engrg.
Houston, TX   77004

Douglas Corl
Stanford University
53-D Escondido Village
Stanford, CA   94305

A. J. Cousin
University of Toronto
Dept. of Electrical Engineering
Toronto, Canada

Norman D. Crump
ARCO
P. O. Box 1346
Houston, TX   77001

Steve Curtis
Varian/Ultrasound Div.
2341 South 2300 West
Salt Lake City, Utah   84119

P. Das
Rensselaer Polytechnic Institute
Electrical & Systems Engr.
Room 7012
Troy, New York   12181

B. Delannoy
Universite de Valenciennes
Le Mont Holly 59326
Valenciennes, France

Billie Dopslauf
ARCO
P. O. Box 1346
Houston, TX   77001

Gregory Duckworth
Massachusetts Institute of Tech.
33 Preston Road
Somerville, MA   02143

Andre Duerinckx
IBM T.J. Watson Research Center
P. O. Box 218, RM 81-W08
Yorktown Heights, NY   10598

Wanda English
Duke University
Dept. of Biomedical Engineering
Durham, North Carolina   27706

Kenneth Erikson
Rohe
2722 South Fairview St.
Santa Ana, California   92704

James T. Fearnside
Hewlett Packard Company
1776 Minuteman Drive
Andover, MA   01810

Gerald Fitzpatrick
University of Houston
Seismic Acoustics Lab.
A.P.I. Bldg.
Houston, TX   77004

Francis J. Fry
Indianapolis Center for Advanced
   Research
1219 W. Michigan Ave.
Indianapolis, Indiana   26420

G.H.F. Gardner
Gulf Oil
P. O. Box 2038
Pittsburgh, PA   15230

Bruce Gibson
Western Geophysical
P. O. Box 2469
Houston, TX   77401

D. David Graham
Southwest Research Institute
P. O. Drawer 28510
San Antonio, TX   78284

Dilip K. Guha-Ray
City Hospital, Baltimore
4940 Eastern Ave.
Baltimore, MD   21224

Vaikunth Gupta
University of Kentucky
Dept. of Physiology & Biophysics
Lexington, KY   40536

Tom E. Hall
Battelle
P. O. Box 999
Richland, WA   99352

Amin M. Hanafly
Hewlett-Packard
10 Launching Road
Andover, MA   01810

Michael E. Haran
Bureau of Radiological Health
5600 Fisher Lane
(HFX-210)
Rockville, MD   20857

B. P. Hildebrand
Spectron Development Labs.
3303 Harbor Blvd.
Costa Mesa, CA   92626

C. R. Hill
Institute of Cancer Research
Royal Marsden Hospital
Clifton Avenue
Sutton, Surrey   AM2   5PX

Wayne Hillard
Varian/Ultrasound Div.
2341 South 2300 West
Salt Lake City, Utah   84119

Fred Hilterman
University of Houston
Department of Geology
Houston, TX 77004

E. Aaron Howard
University of Washington
Bioengineering Department
Seattle, Washington 98195

Thomas Hu
University of Houston
Seismic Acoustics Lab.
A.P.I. Bldg.
Houston, TX 77004

Steven Johnson
University of Utah
Department Bioengineering, MEB
Salt Lake City, Utah 84112

Joie P. Jones
University of California at
    Irvine
Dept. of Radiological Sciences
Irvine, California 92716

Ed Karrer
HP Laboratories
1501 Page Mill Road
Palo Alto, CA 94304

Mostafa Kaveh
University of Minnesota
Department of Electrical Engrg.
Minneapolis, Minnesota 55414

L. W. Kessler
Sonoscan, Inc.
530 E. Green St.
Bensenville, IL 60106

Rob Kinney
Varian/Ultrasound Division
2341 South 2300 West
Salt Lake City, Utah 84110

Martin Klein
Klein Associates, Inc.
Klein Drive
Salem, N. H. 03079

James Kosalos
International Submarine Techn.
2733 152nd Ave. NE
Redmond, WA 98052

Richard Krimholtz
Cayson Engineering Ltd.
Caytel Works, Holywell Ind. Est.
Watford, Herts WD1 8RJ, England

Ronald E. Larsen
Southwest Research Institute
P. O. Drawer 28510
San Antonio, TX 78284

John Larson
HP Laboratories
1501 Page Mill Road
Palo Alto, CA 94304

C. Q. Lee
Univ. of Ill. at Chicago Circle
Chicago, Illinois 60680

Paul P. K. Lee
University of Rochester
Department of Radiology
Rochester, NY 14627

S. Leeman
Dept. of Medical Physics
Royal Postgraduate Med. School
and Hammersmith Hospital
London, W12 OHS, England

Sidney Lees
Forsyth Dental Center
140 Fenway
Boston, Massachusetts 02115

Stewart Levin
Western Geophysical
10001 Richmond Ave.
Houston, TX 77001

Luke L. Liu
Mobil R&D
P. O. Box 900
Dallas, TX   75221

Margitta Luetkemeyer
Krupp Atlas Electronics
Sebaldsbruecker Heerstr 235
28 Bremen 44, Postfach 448545
Germany

Hewlett Melton, Jr.
Medical College of Wisconsin
8700 W. Wisconsin Ave.
Milwaukee, WI   53226

Charles R. Meyer
Univ. of Colorado Medical Center
Department of Radiology
4200 E. Avenue
Denver, Colorado   80220

Paul Miller
Bell Labs
555 Union Blvd.
Allentown, PA   18104

Toyokatsu Miyashita
Kyoto Instit. of Tech.
Dept. of Electrical Engineering
Matsugasaki, Sakyo-Ku, Kyoto
Japan

Merle L. Moberly
Radiology Associates of Spokane
North 5901 Lidgerwood
Spokane, Washington

Thomas J. Moran
Air Force Materials Lab
AFML/LLP
Wright-Patterson AFB, OH   45433

Thomas Morgan
University of Houston
Seismic Acoustics Lab
API Bldg.
Houston, TX   77004

Rolf K. Mueller
University of Minnesota
Minneapolis, Minnesota

Tom G. Muir
University of Texas, Austin
Applied Research Lab.
Austin, TX   78712

Jan Ove Nilsson
University of Rochester
470 Kendrick Road
Rochester, NY   14620

Brent Ovard
Varian/Ultrasound Division
2341 South 2300 West
Salt Lake City, Utah   84119

Peder Pedersen
Drexel University
Dept. of Electrical Engineering
Philadelphia, Pennsylvania 19104

E. J. Pisa
Rohe
2722 South Fairview St.
Santa Ana, CA   92704

Jeff Powers
University of Washington
Center for Bio-Engineering
Seattle, Washington   98195

John Powers
Naval Postgraduate School
Dept. of Electrical Engineering
Monterey, CA   93940

Balu Rajagopalan
Mayo Clinic
Dept. Physiology & Biophysics
Rochester, Minnesota   55901

John Reid
Institute of Applied Physiology
  & Medicine
701 16th Ave.
Seattle, Washington   98122

Bruno Richard
Paris University
Lab. de Mechanique Physique
Univ. Pierre et Marie Curie

Americo Rivera
NIH-NIGMS
5333 West Bard Ave.
Bethesda, MD 20205

H.A.F. Rocha
General Electric
P. O. Box 43-37/231
Schenectady, NY 12345

John R. Rogers
Texaco
P. O. Box 425
Bellaire, TX 77401

Gary Ruckgaber
Amoco
Amoco Production Company
Tulsa, Oklahoma

T. Sawatari
Bendix Research Labs
20800 10 1/2 Rd.
Southfield, MI 48075

Carl Schueler
UCSB
Dept. of Electrical & Computer
   Engineering
Santa Barbara, CA 93106

E. J. Seppi
Varian
611 Hansen Way
Palo Alto, CA 94303

P. M. Shah
Exxon
P. O. Box 2189
Houston, TX 77001

Michael Shay
Vector Cable
555 Industrial Road
Sugarland, TX 77478

Khalil Shiraz
MCV Hospital
P. O. Box 615
Richmond, VA 23235

Bikash Sinha
Schlumberger
Doll Research Center
P. O. Box 307
Ridgefield, Conn. 06877

P. Sitaramaswamy
University of Texas/Austin
c/o Professor Elmer Hixson
Dept. of Electrical Engineering
P. O. Box 7728
Austin, TX 78712

Thomas Smith
University of Houston
Seismic Acoustics Lab.
A.P.I. Bldg.
Houston, TX 77004

Richard Soldner
Siemens
860 Hinckley Road
Burlingame, CA 94010

Frank Stenger
University of Utah
Department of Math
Salt Lake City, Utah 84112

Don Straley
Amoco
Amoco Production Company
Tulsa, OK

Steven A. Stubblefield
Texaco
P. O. Box 425
Bellaire, TX 77401

Jerry Sutton
Naval Ocean Systems Center
5213 (B)
San Diego, CA 92152

Kai E. Thomenius
High Stoy Tech. Corp.
P. O. Box 360
Gladstone, N. J.   07934

F. L. Thurstone
Duke University
Dept. of Biomedical Engineering
Durham, N.C.   27706

Bernhard Tittman
Rockwell Institute
1049 Camino dos Rios
Thousand Oaks, CA   91360

Steven M. Tobias
Mobil
7490 Brompton, #347
Houston, TX   77025

Wayne Travis
Cities Service Oil Company
4026 Oberlin
Houston, TX   77005

Thomas Tyler
KUMC
Dept. Diagonstic Radiology
39th & Rainbow Blvd.
Kansas City, KS   66103

David Vilkomerson
Johnson & Johnson
210 Clyde Road
Sommerset, N.J.   08873

Robert Waag
University of Rochester
Dept. of EE & Radiology
University of Rochester
Rochester, NY   14642

Glen Wade
UCSB
Dept. of EE & Computer Engrg.
Santa Barbara, CA   93106

Keith Wang
University of Houston
Dept. of Electrical Engineering
Houston, TX   77004

Arthur B. Weglein
Cities Service Oil Company
1133 North Lewis
Tulsa, OK   74150

John W. Woody
J. W. Woody Company
4519 Hickery Downs
Houston, TX   77084

Wen-Jen Wu
University of Houston
Seismic Acoustics Lab.
A.P.I. Bldg.
Houston, TX   77004

INDEX

A-mode studies, 509
ACSWR (see Acoustic switching ratio)
ACSWR frequency response, 124–128
AIS, 622
ASR, 699, 700
Abbot, J.G., 585
Aberration, 139, 348
Aberration correction, 41
Aberration limited line spread function, 146
Aberration limited transfer function, 143
Abscesses, 574
Absorption effects, 516, 519
Achenbach, J.D., 355
Acoustic impedance log, 655
Acoustic interferograms, 285, 287
Acoustic lens, 58
Acoustic micrographs, 301–313
Acoustic microscopy, 275–299
Acoustic shadow, 307
Acoustic switching ratio (ACSWR) 121–138
Acoustic waves, low frequency imaging, 111–119
Active Model equivalent circuit 123, 124
Adair, R.S., 109
Adaptive array processing, 177–204
Adaptive estimator, 183
Adaptive learning, 40, 323
Addison, R.C., 325, 337, 338
Adler, R., 298

Advanced Research Projects Agency, 354, 377
Advanced Technology Laboratories, Inc., 533
Ahlberg, L., 337
Ahlgren, F., 91
Ahmed, M., 154, 569, 576
Ahuja, A.S., 501
Air Force Office of Scientific Research, 354
Akagi, N., 37
Alais, P., 65, 73, 339
Alumina, 308
Aluminum alloys, 318
Alvarez, R.E., 416
Amplitude detector circuit, 651
Amplitude error function, 251
Amplitude quantization, 249–252
Analog delay calculator, 57–63
Analog image reconstruction device, 219–225
Anchor, image of, 619
Anderson, P.T., 597
Andrews, H.C., 416
Anger, H.O., 416
Annular array, 271
Anticline, 665, 783
Aorta, 560
Aortic root, 555
Aperture weighting functions, 249, 250
Apodization functions, 251
Applied Research Laboratories, 93, 104
Array, 341
  adaptive processing, 177–204
  cylindrical, 57